PRINCIPLES AND PROCEDURES OF STATISTICS

Copyright © 1960 by the McGraw-Hill Book Company, Inc. Printed in the United States of America. All rights reserved. This book, or parts thereof, may not be reproduced in any form without permission of the publishers. Library of Congress Catalog Card Number 59-13216

ISBN 07-060925-x

18 19 20 21 22 KPKP 7 9 8 7 6 5 4 3

Principles and Procedures of
STATISTICS

WITH SPECIAL REFERENCE TO THE BIOLOGICAL SCIENCES

Robert G. D. Steel

Professor of Statistics
North Carolina State College

James H. Torrie

Professor of Agronomy
University of Wisconsin

McGRAW-HILL BOOK COMPANY, INC.

New York   Toronto   London

1960

PREFACE

A tentative outline for this text was prepared by the authors while both were on staff at the University of Wisconsin. Discussions with faculty and students had led to the conclusion that three major considerations are involved in teaching statistics at the graduate level in colleges of agriculture.

There must be comprehensive coverage of statistical principles and procedures in a limited time. Most students can schedule only one or two terms of statistics since statistics is a supporting course except for the few students majoring or minoring in the subject.

The analysis of variance should be presented as quickly as feasible. The beginning graduate student is generally without statistical training yet assistantship duties often involve field, greenhouse, barn, or laboratory experiments which result in measurement data demanding analysis of variance computations. Also, graduate problems may require knowledge of experimental design and the analysis of variance before the student has acquired such knowledge.

Finally, a nonmathematical approach is desirable. The teacher of applied statistics often finds an inadequacy in and fear of mathematics on the part of the student. Hence we decided to avoid algebraic manipulations. Whether or not we have been completely successful will depend largely on how the student feels about our handling of the binomial distribution in Chap. 20.

Statisticians, like mathematicians, use subscripts as an alternative to lengthy descriptions for recurring and important situations. There seems to be little point in avoiding this convenient notation and subscripts are used with two summation notations, namely, Σ and a "dot notation." Greek and English alphabets serve to distinguish between parameters and statistics. The Greek alphabet is given in the Appendix.

In teaching our classes, we have three main aims:

First, to present the student with a considerable number of statistical techniques, applicable and useful in terms of research. A relatively small survey of graduates of a biometry course suggests that the extent to which any statistical method is used depends on how intensively it was treated during the course. Few methods are learned later.

Second, to promote disciplined thinking on the part of the student with respect to the conduct of experiments. For example, the student should be trained to the point where he is able to appreciate that parts of the Salk vaccine trial were not relevant to the intent of the trial; that an experiment is not necessary to determine the relative precision with which a mean is measured for different sample sizes; that charts, averages, ratios, percentages, and nonrandom sampling can be and often are grossly misused.

Third, to promote disciplined thinking on the part of the student with respect to the analysis of experimental data. For example, a student should be able to appraise figures critically, appreciating their fallibility and limitations in terms of natural variation and its effects.

We began teaching from our tentative outline, eliminating inadequacies and inconsistencies as best we could and rearranging and adding material as the result of our teaching experiences. We paid considerable attention to models. We added techniques as they appeared in original sources, for example, those for testing among treatment means. We included some material of a reference nature when we felt it desirable for the student to know of its existence. Finally, realizing that no text can be completely adequate, up to date, and without error, we made a last rewriting and sent the manuscript to the publishers.

Our text generally follows the order of presentation in our courses. Considerable leeway in teaching methods and course content is possible.

In the first five chapters, statistical concepts are presented to provide a basis for understanding the many statistical methods given in the later chapters. These concepts include parent and derived distributions, probability, confidence intervals, tests of hypotheses, type I and II errors, power, sample size, and the analysis of variance. Since these concepts are usually new and difficult to the student, we have stressed them by repetition and elaboration.

In the chapters on statistical methods, we try to show the underlying logic as well. This is done, in part, by discussion of linear models for experimental designs and linear regression.

The placement of Chap. 6 was settled somewhat arbitrarily. The student is certainly not prepared to study it intensively at this point. Some instructors will wish to teach it later in the course while others will touch on it briefly before beginning the analysis of variance and then return to it after Chap. 12.

Discrete variables are discussed after the analysis of variance and covariance have been covered. The application of chi-square to problems involving discrete variables is treated with stress on the fact that chi-square is a continuous distribution and that, consequently, the test criteria here called chi-square are distributed only approximately as chi-square. The discussion of the binomial distribution was intentionally left until after two chapters which deal with discrete data, including binomially distributed data. This is done for two reasons. First, it is possible to overstress the binomial aspects of enumeration data at the expense of the nonparametric potential of the chi-square test criterion. Second, it seems possible to motivate a few more students to an interest in Chap. 20 after they have been exposed to enumeration data and the use of chi-square as a test criterion.

This text is, then, meant for the scientist or scientist-to-be with no special training in mathematics. It attempts to present basic concepts and methods of statistics and experimentation so as to show their general applicability. Most examples are chosen from biological and allied fields since that is where we

PREFACE

are most familiar with applications. The order of presentation is such that the analysis of variance appears as early as is deemed practical.

The authors are indebted to Professor Sir Ronald A. Fisher, Cambridge, to Dr. Frank Yates, Rothamsted, and Messrs. Oliver and Boyd, Ltd., Edinburgh, for permission to reprint Table III from their book "Statistical Tables for Biological, Agricultural, and Medical Research."

The authors are also indebted to Fred Gruenberger and the Numerical Analysis Laboratory of the University of Wisconsin for preparing Table A.1; to E. S. Pearson and H. O. Hartley, editors of *Biometrika Tables for Statisticians*, vol. I, and to the *Biometrika* office for permission to reprint Tables A.2, A.6, A.8, and A.15; to C. M. Thompson and the *Biometrika* office for permission to reprint Table A.5; to D. B. Duncan and the editor of *Biometrics* for permission to reprint Table A.7; to C. W. Dunnett and the editor of the *Journal of the Americal Statistical Association* for permission to reprint Table A.9; to C. I. Bliss for permission to reprint Table A.10; to F. N. David and the *Biometrika* office for permission to reprint Table A.11; to L. M. Milne-Thomson and L. J. Comrie, authors of *Standard Four-figure Mathematical Tables*, and MacMillan and Co. Ltd., London, for permission to reprint Table A.12; to G. W. Snedecor, author of *Statistical Methods*, 4th edition, and the Iowa State College Press for permission to reprint Table A.13; to D. Mainland, L. Herrera, and M. I. Sutcliffe for permission to reprint Table A.14; to F. Mosteller and J. W. Tukey, the editor of the *Journal of the Americal Statistical Association*, and the Codex Book Company, Inc., for permission to reprint Table A.16; to Prasert Na Nagara for permission to reprint Table A.17; to Frank Wilcoxon and the American Cyanamid Company for permission to reprint Table A.18; to Colin White and the editor of *Biometrics* for permission to reprint Table A.19; to P. S. Olmstead, J. W. Tukey, the Bell Telephone Laboratories, and the editor of the *Annals of Mathematical Statistics* for permission to reprint Table A.20.

We are grateful to the many individuals, editors, and publishers who granted permission to use experimental data and other material.

For helpful comments on mimeographed notes, we wish to thank W. J. Drapala, W. T. Federer, and students in our classes.

Special thanks are due to Mrs. Jane Bamberg and Miss Helen Fuller for excellent work in typing the final draft of the manuscript.

Robert G. D. Steel
James H. Torrie

CONTENTS

Preface. v

Selected Symbols xv

Chapter 1. Introduction 1

1.1	Statistics defined	1
1.2	Some history of statistics	2
1.3	Statistics and the scientific method	3
1.4	Studying statistics	4

Chapter 2. Observations 7

2.1	Introduction	7
2.2	Variables	7
2.3	Distributions	8
2.4	Populations and samples	9
2.5	Random samples, the collection of data	9
2.6	Presentation, summarization, and characterization of data . . .	10
2.7	Measures of central tendency	13
2.8	Measures of dispersion	16
2.9	Standard deviation of means	19
2.10	Coefficient of variability or of variation	20
2.11	An example	20
2.12	The linear additive model	20
2.13	The confidence or fiducial inference	22
2.14	An example	23
2.15	The use of coding in the calculation of statistics	25
2.16	The frequency table	26
2.17	An example	27
2.18	Calculation of the mean and standard deviation from a frequency table	28
2.19	Graphical presentation of the frequency table	29
2.20	Significant digits	29

Chapter 3. Probability 31

3.1	Introduction	31
3.2	Some elementary probability	31
3.3	Probability distributions	32
3.4	The normal distribution	35
3.5	Probabilities for a normal distribution; the use of a probability table .	36
3.6	The normal distribution with mean μ and variance σ^2 . . .	38
3.7	The distribution of means	40

CONTENTS

- 3.8 The χ^2 distribution 41
- 3.9 The distribution of Student's t 43
- 3.10 Estimation and inference 44
- 3.11 Prediction of sample results 47

Chapter 4. Sampling from a Normal Distribution 49

- 4.1 Introduction 49
- 4.2 A normally distributed population 49
- 4.3 Random samples from a normal distribution 51
- 4.4 The distribution of sample means 53
- 4.5 The distribution of sample variances and standard deviations . . 54
- 4.6 The unbiasedness of s^2 56
- 4.7 The standard deviation of the mean or the standard error . . 56
- 4.8 The distribution of Student's t 57
- 4.9 The confidence statement 58
- 4.10 The sampling of differences 58
- 4.11 Summary of sampling 63
- 4.12 The testing of hypotheses 64

Chapter 5. Comparisons Involving Two Sample Means . . . 67

- 5.1 Introduction 67
- 5.2 Tests of significance 67
- 5.3 Basis for a test of two or more means 72
- 5.4 Comparison of two sample means, unpaired observations, equal variances . 73
- 5.5 The linear additive model 76
- 5.6 Comparison of sample means; paired observations 78
- 5.7 The linear additive model for the paired comparison . . . 80
- 5.8 Unpaired observations and unequal variances 81
- 5.9 Testing the hypothesis of equality of variances 82
- 5.10 Confidence limits involving the difference between two means . . 84
- 5.11 Sample size and the detection of differences 84
- 5.12 Stein's two-stage sample 86

Chapter 6. Principles of Experimental Design 88

- 6.1 Introduction 88
- 6.2 What is an experiment? 88
- 6.3 Objectives of an experiment 89
- 6.4 Experimental unit and treatment 90
- 6.5 Experimental error 90
- 6.6 Replication and its functions 90
- 6.7 Factors affecting the number of replicates 92
- 6.8 Relative precision of designs involving few treatments . . . 93
- 6.9 Error control 94
- 6.10 Choice of treatments 96
- 6.11 Refinement of technique 96
- 6.12 Randomization 97
- 6.13 Statistical inference 98

Chapter 7. Analysis of Variance I: The One-way Classification . . 99

- 7.1 Introduction 99
- 7.2 The completely random design 99
- 7.3 Data with a single criterion of classification. Equal replication . 101
- 7.4 The least significant difference 106

7.5	Duncan's new multiple-range test	107
7.6	Tukey's w-procedure	109
7.7	Student–Newman–Keuls' test	110
7.8	Comparing all means with a control	111
7.9	Data with a single criterion of classification. Unequal replication	112
7.10	The linear additive model	115
7.11	Analysis of variance with subsamples. Equal subsample numbers	119
7.12	The linear model for subsampling	123
7.13	Analysis of variance with subsamples. Unequal subsample numbers	125
7.14	Variance components in planning experiments involving subsamples	127
7.15	Assumptions underlying the analysis of variance	128

Chapter 8. Analysis of Variance II: Multiway Classifications . . . 132

8.1	Introduction	132
8.2	The randomized complete-block design	132
8.3	Analysis of variance for any number of treatments. Randomized complete-block design	134
8.4	The nature of the error term	137
8.5	Missing data	139
8.6	Estimation of gain in efficiency	142
8.7	The randomized complete-block design: more than one observation per experimental unit	142
8.8	Linear models and the analysis of variance	145
8.9	Double grouping: Latin squares	146
8.10	Analysis of variance of the Latin square	149
8.11	Missing plots in the Latin square	150
8.12	Estimation of gain in efficiency	152
8.13	The linear model for the Latin square	153
8.14	The size of an experiment	154
8.15	Transformations	156

Chapter 9. Linear Regression . . . 161

9.1	Introduction	161
9.2	The linear regression of Y on X	161
9.3	The linear regression model and its interpretation	164
9.4	Assumptions and properties in linear regression	165
9.5	Sources of variation in linear regression	166
9.6	Regressed and adjusted values	167
9.7	Standard deviations, confidence intervals, and tests of hypotheses	169
9.8	Control of variation by concomitant observations	172
9.9	Difference between two regressions	173
9.10	A prediction and its variance	175
9.11	Prediction of X, Model I	177
9.12	Bivariate distributions, Model II	177
9.13	Regression through the origin	179
9.14	Weighted regression analysis	180

Chapter 10. Linear Correlation . . . 183

10.1	Introduction	183
10.2	Correlation and the correlation coefficient	183
10.3	Correlation and regression	187
10.4	Sampling distributions, confidence intervals, tests of hypotheses	188
10.5	Homogeneity of correlation coefficients	190
10.6	Intraclass correlation	191

Chapter 11. Analysis of Variance III: Factorial Experiments . . . 194

11.1	Introduction	194
11.2	Factorial experiments	194
11.3	The 2×2 factorial experiment, an example	199
11.4	The $3 \times 3 \times 2$ or $3^2 \times 2$ factorial, an example	205
11.5	Linear models for factorial experiments	211
11.6	Single degree of freedom comparisons	213
11.7	n-way classifications and factorial experiments; response surfaces . .	220
11.8	Individual degrees of freedom; equally spaced treatments . . .	222
11.9	A single degree of freedom for nonadditivity	229

Chapter 12. Analysis of Variance IV: Split-plot Designs and Analysis . 232

12.1	Introduction	232
12.2	Split-plot designs	232
12.3	An example of a split plot	236
12.4	Missing data in split-plot designs	240
12.5	Split plots in time	242
12.6	The split-plot model	245
12.7	Split plots in time and space	247
12.8	Series of similar experiments	249

Chapter 13. Analysis of Variance V: Unequal Subclass Numbers . . 252

13.1	Introduction	252
13.2	Disproportionate subclass numbers; general	252
13.3	Disproportionate subclass numbers; the method of fitting constants .	257
13.4	Disproportionate subclass numbers; the method of weighted squares of means	265
13.5	Disproportionate subclass numbers; methods for $r \times 2$ tables . . .	269
13.6	Disproportionate subclass numbers; 2×2 tables	272

Chapter 14. Multiple and Partial Regression and Correlation . . . 277

14.1	Introduction	277
14.2	The linear equation and its interpretation in more than two dimensions	278
14.3	Partial, total, and multiple linear regression	279
14.4	The sample multiple linear regression equation	280
14.5	Multiple linear regression equation; three variables	281
14.6	Standard partial regression coefficients	284
14.7	Partial and multiple correlation; three variables	285
14.8	Tests of significance; three variables	287
14.9	Multiple linear regression; computations for more than three variables	289
14.10	The abbreviated Doolittle method	290
14.11	Test of significance of multiple regression	296
14.12	Standard errors and tests of significance for partial regression coefficients	297
14.13	Standard partial regression coefficients	299
14.14	Deletion and addition of an independent variable	299
14.15	Partial correlation	301

Chapter 15. Analysis of Covariance 305

15.1	Introduction	305
15.2	Uses of covariance analysis	305
15.3	The model and assumptions for covariance	308
15.4	Testing adjusted treatment means	310
15.5	Covariance in the randomized complete-block design . . .	311

15.6	Adjustment of treatment means	315
15.7	Increase in precision due to covariance	316
15.8	Partition of covariance	317
15.9	Homogeneity of regression coefficients	319
15.10	Covariance where the treatment sum of squares is partitioned	319
15.11	Estimation of missing observations by covariance	324
15.12	Covariance with two independent variables	325

Chapter 16. Nonlinear Regression 332

16.1	Introduction	332
16.2	Curvilinear regression	332
16.3	Logarithmic or exponential curves	334
16.4	The second-degree polynomial	338
16.5	The second-degree polynomial, an example	338
16.6	Higher degree polynomials	340
16.7	Polynomials and covariance	341
16.8	Orthogonal polynomials	341

Chapter 17. Some Uses of Chi-square 346

17.1	Introduction	346
17.2	Confidence interval for σ^2	346
17.3	Homogeneity of variance	347
17.4	Goodness of fit for continuous distributions	349
17.5	Combining probabilities from tests of significance	350

Chapter 18. Enumeration Data I: One-way Classifications. . . 352

18.1	Introduction	352
18.2	The χ^2 test criterion	352
18.3	Two-cell tables, confidence limits for a proportion or percentage	353
18.4	Two-cell tables, tests of hypotheses	355
18.5	Tests of hypotheses for a limited set of alternatives	358
18.6	Sample size	362
18.7	One-way tables with n cells	364

Chapter 19. Enumeration Data II: Contingency Tables . . 366

19.1	Introduction	366
19.2	Independence in $r \times c$ tables	366
19.3	Independence in $r \times 2$ tables	370
19.4	Independence in 2×2 tables	371
19.5	Homogeneity of two-cell samples	373
19.6	Additivity of χ^2	375
19.7	More on the additivity of χ^2	376
19.8	Exact probabilities in 2×2 tables	379
19.9	Two trials on the same subjects	381
19.10	Linear regression, $r \times 2$ tables	381
19.11	Sample size in 2×2 tables	383
19.12	n-way classification	384

Chapter 20. Some Discrete Distributions 388

20.1	Introduction	388
20.2	The hypergeometric distribution	388
20.3	The binomial distribution	389
20.4	Fitting a binomial distribution	390

20.5	Transformation for the binomial distribution	394
20.6	The Poisson distribution	395
20.7	Other tests with Poisson distributions	397

Chapter 21. Nonparametric Statistics 400

21.1	Introduction	400
21.2	The sign test	401
21.3	Wilcoxon's signed rank test	402
21.4	Two tests for two-way classifications	403
21.5	Tests for the completely random design, two populations	404
21.6	Tests for the completely random design, any number of populations	406
21.7	Chebyshev's inequality	407
21.8	Spearman's coefficient of rank correlation	409
21.9	A corner test of association	410

Chapter 22. Sampling Finite Populations 412

22.1	Introduction	412
22.2	Organizing the survey	413
22.3	Probability sampling	414
22.4	Simple random sampling	415
22.5	Stratified sampling	417
22.6	Optimum allocation	420
22.7	Multistage or cluster sampling	422

Appendix: Tables 427

Index 473

SELECTED SYMBOLS

Symbol	Page first used	Meaning
$\neq$	65	not equal to; e.g., $3 \neq 4$
$>$	22	greater than; e.g., $5 > 2$
$\geq$	23	greater than or equal to
$<$	22	less than; e.g., $3 < 7$
$\leq$	23	less than or equal to
$|\ |$	19	absolute value; e.g., $|-7| = 7$
Σ	14	sum of
$\ldots$	7	indicates a set of obvious missing quantities, e.g., $1, 2, \ldots, 10$
$n!$	380	$n(n-1)\ldots 1$ and called n factorial; e.g., $3! = 3(2)\ 1 = 6$
$^{-}$	13	overbar; used to indicate an arithmetic average or mean
$\hat{\ }$	13	hat; used to indicate an estimate rather than a true value; most often appears over Greek letters
Greek letters		with few exceptions refer to population parameters
μ	13	population mean
σ^2, σ	16, 17	population variance and standard deviation
τ, β, etc.	77	components of population means; commonly used in conjunction with linear models
ε	21	a true experimental error
δ	123	a true sampling error
β	164	population regression coefficient
ρ	183	population correlation coefficient
N, S^2	17, 415, 416	these Latin letters are used as population symbols in Chap. 22

The preceding Greek letters are also used with subscripts where clarity requires. For example:

$\mu_{\bar{x}}$	20	mean of a population of $\bar{x}$'s
$\beta_{yx \cdot z}$	279	regression of Y on X for fixed Z
τ_i	77	a contribution to the mean of the population receiving the ith treatment

Some exceptions to the use of Greek letters for parameters are:

α	70	probability of a Type I error
$1 - \alpha$	71	confidence coefficient
β	70	probability of a Type II error
$1 - \beta$	71	power of a statistical test
χ^2	41	a common test criterion

SELECTED SYMBOLS

Latin letters — used as general symbols, including symbols for sample statistics

Symbol	Page	Description
X	7	a variable (Y is also commonly used)
X_i, X_{ij}	7, 59	individual observations
$X_{i.}, X_{..}$	77, 101	totals of observations
x_i	14	$X_i - \bar{x}$
x_i'	284	$(X_i - \bar{x})/s$
D_j	59	difference between paired observations
$n, n_{..}$	7, 126, 253	total sample size
n_{ij}	126, 253	number of observations in i, jth cell
$\bar{x}, \bar{x}_{..}, \bar{x}_i$	13, 117	sample means, whole or part of sample
$\bar{\bar{x}}$	54	mean of a sample of means
d	59	$\bar{x}_1 - \bar{x}_2$
$s^2, s_{\bar{x}}^2, s_{\bar{d}}^2$	16, 19, 63	sample variances, unbiased estimators of σ^2, $\sigma_{\bar{x}}^2$, and $\sigma_{\bar{d}}^2$
$s, s_{\bar{x}}, s_{\bar{d}}$	17, 19, 59	sample standard deviations
$s_{y \cdot x}^2, s_{y \cdot 1 \ldots k}^2$	169, 297	sample variances adjusted for regression
CL, CI	52	confidence limits or interval
l_1, l_2	23	end points of confidence limits
b	162	sample regression coefficient
$b_{y1 \cdot 2 \ldots k}$	279	sample partial regression coefficient
b'	284	standard regression coefficient
r	183	sample total or simple correlation coefficient
$r_{12 \cdot 3 \ldots k}$	286	sample partial correlation coefficient of X_1 and X_2
$R_{1 \cdot 2 \ldots k}$	280	multiple correlation coefficient between X_1 and the other variables
df, f	18	degrees of freedom
C, CT	18	correction term
SS	17	Σx^2_i, sum of squares
MS	114	mean square
E_a, E_b	234	error mean squares for split-plot design
E_{yy}, E_{xy}, E_{xx}	310	error sums of products in covariance (other letters used for other sources of variation)
*	68	significant, e.g., 2.3*
**	69	highly significant, e.g., 14.37**
ns	69	not significant
lsd	106	least significant difference
RE	93	relative efficiency
CV	20	coefficient of variability $(s/\bar{x})100$
$Q = \Sigma c_i T_i$	215	a comparison where c_i is a constant, T_i is a treatment total, and $\Sigma c_i = 0$
fpc	416	finite population correction
psu	422	primary sampling unit
st	419	stratified, used as subscript
K	215	Σc_i^2
P	22	probability
$p, 1-p$	353	probabilities in a binomial distribution
z, t, F	36, 22, 72	common test criteria
H_0	65	null hypothesis
H_1	65	alternate hypothesis, usually a set of alternatives
∞	49	infinity

Chapter 1

INTRODUCTION

1.1 Statistics defined. Modern statistics provides research workers with knowledge. It is a young and exciting subject, a product of the twentieth century. For the scientist, particularly the biological scientist, statistics began about 1925 when Fisher's *Statistical Methods for Research Workers* appeared.

Statistics is a rapidly growing subject with much original material still not available in texts. It grows as statisticians find answers to more and more of the problems posed by research workers. Men who were among the earliest contributors to statistics are still productive and newcomers find diverse opportunities for their research talents. In the application of statistics, principles are general though techniques may differ, and the need for training in statistics grows as increased application is made in the biological and social sciences, engineering, and industry.

This young, vigorous subject affects every aspect of modern living. For example, statistical planning and evaluation of research have promoted technological advances in growing and processing food; statistical quality control of manufactured products makes automotive and electric equipment reliable; pollsters collect data to determine statistically the entertainment preferences of the public. More and more, the research team has, or has access to, a statistician and approximately 5 billion dollars was spent on research in the United States in 1957.

The extent of statistics makes it difficult to define. It was developed to deal with those problems where, for the individual observations, laws of cause and effect are not apparent to the observer and where an objective approach is needed. In such problems, there must always be some uncertainty about any inference based on a limited number of observations. Hence, for our purposes, a reasonably satisfactory definition would be: *Statistics is the science, pure and applied, of creating, developing, and applying techniques such that the uncertainty of inductive inferences may be evaluated.*

To most scientists, statistics is logic or common sense with a strong admixture of arithmetic procedures. The logic supplies the method by which data are to be collected and determines how extensive they are to be; the arithmetic, together with certain numerical tables, yields the material on which to base the inference and measure its uncertainty. The arithmetic is often routine, carried out on a desk calculator, and need involve no special mathematical training for the user. We shall not be directly concerned with mathematics although there is hardly an area of this subject which has not provided some usable theory to the statistician.

1.2 Some history of statistics. A history of statistics throws considerable light on the nature of twentieth century statistics. The historic perspective is also important in pointing to the needs and pressures which created it.

The term statistics is an old one and statistics must have started as state arithmetic since, to levy a tax or wage a war, a ruler needed to know the wealth and number of his subjects. Presumably all cultures that intentionally recorded history also recorded statistics. We know that Caesar Augustus sent out a decree that all the world should be taxed. Consequently he required that all persons report to the nearest statistician, in that day the tax collector. One result of this was that Jesus was born in Bethlehem rather than Nazareth. William the Conqueror ordered that a survey of the lands of England be made for purposes of taxation and military service. This was called the Domesday Book. Such statistics are history.

Several centuries after the Domesday Book, we find an application of empirical probability in ship insurance, which seems to have been available to Flemish shipping in the fourteenth century. This can have been little more than speculation or gambling but it developed into the very respectable form of statistics called insurance.

Gambling, in the form of games of chance, led to the theory of probability originated by Pascal and Fermat, about the middle of the seventeenth century, because of their interest in the gambling experiences of the Chevalier de Méré. To the statistician and the experimental scientist, the theory contains much of practical use for the processing of data.

The normal curve or normal curve of error has been very important in the development of statistics. The equation of this curve was first published in 1733 by de Moivre. De Moivre had no idea of applying his results to experimental observations and his paper remained unknown till Karl Pearson found it in a library in 1924. However, the same result was later developed by two mathematical astronomers, Laplace, 1749–1827, and Gauss, 1777–1855, independently of each other.

An essentially statistical argument was applied in the nineteenth century by Charles Lyell to a geological problem. In the period 1830–1833, there appeared three volumes of *Principles of Geology* by Lyell, who established the order among the Tertiary rocks and assigned names to them. With M. Deshayes, a French conchologist, he identified and listed fossil species occurring in one or more strata and also ascertained the proportions still living in certain parts of the seas. On the basis of these proportions he assigned the names Pleistocene (most recent), Pliocene (majority recent), Miocene (minority recent), and Eocene (dawn of the recent). Lyell's argument was essentially a statistical one. With the establishment and acceptance of the names, the method was almost immediately forgotten. There were no geological evolutionists to wonder if discrete steps, as implied by the names, were involved or if a continuous process was present and could be used to make predictions.

Other scientific discoveries of the nineteenth century were also made on a

INTRODUCTION

statistical basis with little appreciation of the statistical nature of the technique and with the method unfortunately soon forgotten. This is true in both the biological and physical sciences.

Charles Darwin, 1809–1882, biologist, received the second volume of Lyell's book while on the *Beagle*. Darwin formed his theories later and he may have been stimulated by his reading of this book. Darwin's work was largely biometrical or statistical in nature and he certainly renewed enthusiasm in biology. Mendel, too, with his studies of plant hybrids published in 1866, had a biometrical or statistical problem.

Thus in the nineteenth century, the need of a sounder basis for statistics became apparent. Karl Pearson, 1857–1936, initially a mathematical physicist, applied his mathematics to evolution as a result of the enthusiasm in biology engendered by Darwin. Pearson spent nearly half a century in serious statistical research. In addition, he founded the journal *Biometrika* and a school of statistics. The study of statistics gained impetus.

While Pearson was concerned with large samples, large-sample theory was proving somewhat inadequate for experimenters with necessarily small samples. Among these was W. S. Gosset, 1876–1937, a student of Karl Pearson and a scientist of the Guinness firm of brewers. Gosset's mathematics appears to have been insufficient to the task of finding exact distributions of the sample standard deviation, of the ratio of the sample mean to the sample standard deviation, and of the correlation coefficient, statistics with which he was particularly concerned. Consequently he resorted to drawing shuffled cards, computing, and compiling empirical frequency distributions. Papers on the results appeared in *Biometrika* in 1908 under the name of "Student," Gosset's pseudonym while with Guinness. Today "Student's t" is a basic tool of statisticians and experimenters, and "studentize" is a common expression in statistics. Now that the use of Student's t distribution is so widespread, it is interesting to note that the German astronomer, Helmert, had obtained it mathematically as early as 1875.

R. A. Fisher, 1890–   , was influenced by Karl Pearson and Student and made numerous and important contributions to statistics. He and his students gave considerable impetus to the use of statistical procedures in many fields, particularly in agriculture, biology, and genetics.

J. Neyman, 1894–   , and E. S. Pearson, 1895–   , presented a theory of testing statistical hypotheses in 1936 and 1938. This theory promoted considerable research and many of the results are of practical use.

In this brief history, we shall mention only one other statistician, Abraham Wald, 1902–1950. His two books, *Sequential Analysis* and *Statistical Decision Functions*, concern great statistical achievements not treated in this text although one application is illustrated in Chap. 18.

It is in this century then, that most of the presently used statistical methods have been developed. The statistics of this text is a part of these methods.

1.3 Statistics and the scientific method. It is said that scientists use "scientific method." It would be difficult to define scientific method since the scientist uses any methods or means which he can conceive. However, most of these methods have essential features in common.

Without intending to promote a controversy, let us consider that these are:
1. A review of facts, theories, and proposals with a view to
2. Formulation of a logical hypothesis subject to testing by experimental methods and
3. Objective evaluation of the hypothesis on the basis of the experimental results.

Much could be written about these essential features. How does one arrive at a hypothesis? How does one design an experiment? How does one objectively evaluate a hypothesis? We shall be content with saying very little.

Science is a branch of study which deals with the observation and classification of facts. The scientist must, then, be able to observe an event or set of events as the result of a plan or design. This is the experiment, the substance of the scientific method. Experimental design is a field of statistics.

Objective evaluation of a hypothesis poses problems. Thus it is not possible to observe all conceivable events and, since exact laws of cause and effect are generally unknown, variation will exist among those which are observed. The scientist must then reason from particular cases to wider generalities. This process is one of uncertain inference. It is a process which enables us to disprove hypotheses that are incorrect but does not permit us to prove hypotheses that are correct. The only thing we can muster by way of proof is *proof beyond a reasonable doubt*. Statistical procedures are methods which lead us to this sort of proof.

A part of the possible information necessarily leads only to uncertain inference. Chance is involved in supplying the information and is the cause of the uncertainty. Today's statistician, applying the laws of chance, is able to place a precise and objective measure on the uncertainty of his inference. Actually, this is done for the totality of his inferences rather than for the individual inference. In other words, he follows a procedure that assures him of being right in 9 inferences out of 10, 99 out of 100, or anything short of being right all the time. Why not be right all or very close to all the time? The drawback is one of cost. Cost may be an increase in sample size, the penalty of a wrong decision, or the vagueness of the inference necessary to include the correct answer.

Scientific method is not disjoint hypothesis-experiment-inference sequences which fit into neat compartments. Instead, if the scientist fails to disprove a hypothesis, he may wonder if his theory does not embrace facts beyond the scope of inference of his experiment or if, with modification, it cannot be made to embrace such facts. Thus he is led through the cycle again. On the other hand, he may wonder if all the assumptions involved in his hypothesis are necessary; he formulates his hypothesis with fewer assumptions and starts through the cycle again.

In summary, statistics is a tool applicable in scientific method, for which it was developed. Its particular application lies in the many aspects of the design of an experiment from the initial planning to the collection of the data, and in the analysis of the results from the summarizing of the data to the evaluation of the uncertainty of any statistical inference drawn from them.

1.4 Studying statistics. No attempt will be made to make professional statisticians of those who read and study this book. Our aims are to promote clear, disciplined thinking, especially where the collection and interpretation of numerical data are concerned, and to present a considerable number of statistical techniques of general applicability and utility in research. Computations are required in statistics but this is arithmetic, not mathematics nor statistics.

Statistics implies, for most students, a new way of thinking—thinking in terms of uncertainties or improbabilities. Here as elsewhere, students vary in ability and when confronted with statistics for the first time, some may find a mental hurdle which can be emotionally upsetting. We believe we have done everything possible, consistent with our aims, to minimize the problems of learning statistics.

Many students will find that they learn statistics best by direct application to their own problems; few will find, in a course of one or two terms, use for all the material covered in this text. Consequently, many students will need considerable reflection and discussion in order to benefit most from a course based on this text. Questions and exercises are provided to provoke reflection and to offer some opportunity to apply and gain familiarity with techniques.

Finally, it is necessary to keep in mind that statistics is intended to be a tool for research. The research will be in genetics, marketing, nutrition, agronomy, and so on. It is the field of research, not the tools, that must supply the "whys" of the research problem. This fact is sometimes overlooked and the user is tempted to forget that he has to think, that statistics can't think for him. Statistics can, however, help the research worker to design his experiments and to evaluate objectively the resulting numerical data. It is our intention to supply the research worker with statistical tools that will be useful for this purpose.

References

1.1 Fisher, R. A.: "Biometry," *Biometrics*, **4:** 217–219 (1948).

1.2 Fisher, R. A.: "The expansion of statistics," *J. Roy. Stat. Soc., Series A.*, **116:** 1–6 (1953).

1.3 Fisher, R. A.: "The expansion of statistics," *Am. Scientist*, **42:** 275–282 and 293 (1954).

1.4 Freeman, Linton C., and Douglas M. More: "Teaching introductory statistics in the liberal arts curriculum," *Am. Stat.*, **10:** 20–21 (1956).

1.5 Hotelling, Harold: "The teaching of statistics," *Ann. Math. Stat.*, **11:** 1–14 (1940).

1.6 Hotelling, Harold: "The impact of R. A. Fisher on statistics," *J. Am. Stat. Assoc.*, **46:** 35–46 (1951).

1.7 McMullen, Launce: Foreword to *"Student's" Collected Papers*, edited by E. S. Pearson and John Wishart, *Biometrika* Office, University College, London, 1947.

1.8 Mahalanobis, P. C.: "Professor Ronald Aylmer Fisher," *Sankhya*, **4:** 265–272 (1938).

1.9 Mainland, Donald: "Statistics in clinical research; some general principles," *Ann. New York Acad. Sciences*, **52:** 922–930 (1950).

1.10 Mather, Kenneth: "R. A. Fisher's *Statistical Methods for Research Workers*, an appreciation," *J. Am. Stat. Assoc.*, **46:** 51–54 (1951).

1.11 Menger, Karl: "The formative years of Abraham Wald and his work in geometry," *Ann. Math. Stat.*, **23:** 13–20 (1952).

1.12 Pearson, E. S.: "Karl Pearson, an appreciation of some aspects of his life and work, Part I: 1857–1906," *Biometrika*, **28**: 193–257 (1936).

1.13 Pearson, E. S.: "Karl Pearson, an appreciation of some aspects of his life and work, Part II: 1906–1936," *Biometrika*, **29**: 161–248 (1938).

1.14 Reid, R. D.: "Statistics in clinical research," *Ann. New York Acad. Science*, **52**: 931–934 (1950).

1.15 Tintner, G.: "Abraham Wald's contributions to econometrics," *Ann. Math. Stat.*, **23**: 21–28 (1952).

1.16 Walker, Helen M.: "Bi-centenary of the normal curve," *J. Am. Stat. Assoc.*, **29**: 72–75 (1934).

1.17 Walker, Helen M.: "Statistical literacy in the social sciences," *Am. Stat.*, **5**: 6–12 (1951).

1.18 Walker, Helen M.: "The contributions of Karl Pearson," *J. Am. Stat. Assoc.*, **53**: 11–27 (1958).

1.19 Wolfowitz, J.: "Abraham Wald, 1902–1950," *Ann. Math. Stat.*, **23**: 1–13 (1952).

1.20 Yates, F.: "The influence of *Statistical Methods for Research Workers* on the development of the science of statistics," *J. Am. Stat. Assoc.*, **46**: 19–34 (1951).

1.21 Youden, W. J.: "The Fisherian revolution in methods of experimentation," *J. Am. Stat. Assoc.*, **46**: 47–50 (1951).

Chapter 2

OBSERVATIONS

2.1 Introduction. Observations are the raw materials with which the research worker deals. For statistics to be applicable to these observations, they must be in the form of numbers. In crop improvement, the numbers may be plot yields; in medical research, they may be times to recovery under various treatments; in industry, they may be numbers of defects in various lots of a product produced on an assembly-line basis. Such numbers constitute *data* and their common characteristic is *variability* or *variation*.

This chapter is concerned with the collection, presentation, summarization, and characterization of data. Populations, samples, a linear model, and a statistical inference are discussed.

2.2 Variables. Statements such as "Marcia is blonde," or "He weighs well over 200 pounds" are common and informative. They refer to characteristics which are not constant but which vary from individual to individual and thus serve to distinguish or describe.

Characteristics which show variability or variation are called *variables*, *chance variables*, *random variables*, or *variates*.

Since much of our discussion must be general, we shall employ some symbols. Instead of writing variable each time, let X denote the variable and let X_i (read as X sub i) denote the ith observation. Here we have no particular observation in mind. When the time comes to refer to a specific observation, we replace i by a number. For example, if three children in a family have weights of 52, 29, and 28 lb and X denotes weight, $X_1 = 52$ lb, $X_2 = 29$ lb, and $X_3 = 28$ lb. In more general and abstract terms, denote a set of observations by $X_1, X_2, \ldots, X_n$. Here X_n refers to the last term, the subscript n informs us of the total number, and the three periods between X_2 and X_n refer to the other observations, if any. In our example, $n = 3$. The symbols are seen to be a shorthand.

Observations and variables may be either *qualitative* or *quantitative*.

Qualitative observations are made when each individual belongs to one of several mutually exclusive (it can't be in more than one) categories, generally nonnumerical. Examples are flower color, ability to taste certain chemicals, and nationality.

Quantitative observations result in numerical measurements. Examples are heights, weights, and the number of heads that appear when ten coins are tossed.

Observations and variables may also be classified as *continuous* or *discrete* (discontinuous).

A continuous variable is one for which all values in some range are possible. Height and weight are obvious examples. Height may be measured to the nearest ¼ in. but this is not to say that heights exist only for ¼ in. values. We are limited by the measuring device in obtaining values of a continuous variable.

A discrete or discontinuous variable is one for which the possible values are not observed on a continuous scale. Examples are the number of petals on a flower, the number of householders in a city block, and the number of insects caught in the sweep of a net. The average number of dots appearing on two dice is also a discrete variable. Here the possible values go from 1 to 6 by increases of one-half.

Exercise 2.2.1 Classify the following variables as qualitative, quantitative, continuous, discrete: eye color, insect counts, number of errors per pupil in a spelling test, tire-miles to first puncture, times between refills of a fountain pen permitted to run dry under normal use, possible yields of corn from a given field, number of children born in the nearest hospital on New Year's Day, possible outcomes from tossing 50 coins, number of fish in a pond.

Exercise 2.2.2 In 10 tosses of 7 coins, the numbers of heads were: 2, 6, 2, 2, 5, 3, 5, 3, 3, 4. If the observations are denoted by $X_1, X_2, \ldots, X_n$, what is the value of n? What is the value of X_2? Of X_7? For what values of i does X equal 2? 3? 4? Distinguish between X_{i-1} and $X_i - 1$. What do X_{i-1} and $X_i - 1$ equal for $i = 2$?

2.3 Distributions. Values of a variable serve to describe or classify individuals or to distinguish among them. Most of us do more than simply describe, classify, or distinguish because we have ideas about the *relative frequencies* of the values of a variable. Thus, in Minnesota blondeness is not a rare value for a person's complexion; most people would not put much credence in stories about a 65-lb house cat; a baby that weighed 7½ lb at birth would be considered ordinary to other than his family. One's mind associates a measure with the value of a variable, a measure of its ordinariness or lack of it, or of the probability of the occurrence of such a value. In statistics, we say the variable has a *distribution* or a *probability* or *frequency distribution*. Thus for a fair coin, the probability of it falling heads is the same as for tails, namely ½; this is a statement of the probability distribution of a discrete variable. The statement that a certain percentage of adult weights is less than a stated value becomes a distribution for a continuous variable when we have the percentages for each and every weight. The terms *chance* and *random variable* are used more particularly for variables possessing frequency distributions.

Exercise 2.3.1 From your experience, classify the following events as occurring with high, medium, or low relative frequency: a 2-lb baby; someone who is liked by everyone (Is this the same as someone who is disliked by no one?); a 3,400-lb horse; obtaining 27 heads in 50 tosses of a coin.

Exercise 2.3.2 If a thumbtack is tossed, will it land on its head as often as not? Is a busy signal on the telephone likely to occur as often as one time in ten trials? What percentage of freshman women would you expect to weigh less than 120 lb?

OBSERVATIONS

(After thinking about Exercises 2.3.1 and 2.3.2, you should realize that the idea of distributions, of their means and of their variability, is not entirely new to you.)

2.4 Populations and samples. The individual's first concern with a body of data is whether it is to be considered as all possible data or as part of a larger body. While this sounds easily resolved, it is of great importance and failure to make a clear distinction has resulted in loose thinking and in ambiguous notation in some writings.

A *population* or *universe* consists of all possible values of a variable. These values need not all be different nor finite in number. Examples are the birth weights of pigs in one litter, the number of heads for each of 500 tosses of 10 coins, all possible values of corn yield per acre in Iowa. The variable may be continuous or discrete, observable or unobservable. When all values of a population are known it is possible to describe it without ambiguity since all is known about it.

A *sample* is a part of a population. (In some situations, a sample may include the whole of the population.) Usually, it is desired to use sample information to make an inference about a population. For this reason, it is particularly important to define the population under discussion and to obtain a representative sample from the population defined. This is not a trivial point.

The birth weights of pigs in a litter and the number of heads for each of 500 tosses of 10 coins may both be samples from indefinitely large populations and there is usually more value in considering them as samples than as populations.

A sample must be representative of the population if it is to lead to valid inferences. To obtain a representative sample, we embody in the rules for drawing the sample items, the principle of *randomness*. Thus in conducting a public opinion poll, conclusions meant to be applicable to the population of American adults would rarely be valid if the sample were so nonrandom as to include only women or only New Englanders. Throughout this text, the word sample will imply a random sample. Since randomness is somewhat difficult to define, we shall content ourselves with illustrations as we proceed.

Exercise 2.4.1 Would you object to regarding a sample of herring from the Grand Banks as from a possible infinite population? From an existing infinite population? Explain.

Exercise 2.4.2 Could you regard a day's trout catch from a small river as a sample from an infinite population? Explain.

2.5 Random samples, the collection of data. It has been amply demonstrated that the individual cannot draw a random sample without the use of a mechanical process. The mechanical process used to obtain a random sample or to introduce randomness into an experiment or survey usually involves a table of random numbers such as Table A.1. This table consists of the numbers 0, 1, 2, ..., 9 arranged in a 100 by 100 table; there are 10,000 random numbers. They were turned out by a machine and there was no reason to expect any one number to appear more often than another nor any sequence of numbers more often than another, except by chance. There are 1,015 0's, 1,026 1's, 1,013 2's, 975 3's, 976 4's, 932 5's, 1,067 6's, 1,013 7's, 1,023 8's, 960 9's; 5,094 are even and 4,906 are odd. We shall illustrate the use of this table by drawing a random sample of 10 observations from Table

4.1. The data of Table 4.1 have been ranked according to magnitude and assigned item numbers. Ranking is unnecessary for drawing random samples; item numbers could have been assigned in an arbitrary fashion.

To obtain a random sample of ten weights, draw 20 consecutive digits from Table A.1 and record them as ten pairs. These will be item numbers corresponding to weights. You may start anywhere in the table but a more satisfactory way is to poke at one of the pages, read the four numbers most nearly opposite your finger tip, and use these to locate a starting point. Thus, if the numbers turn out to be 1,188 (intersection of row number 10 and column number 20), proceed to row number 11, column number 88. Record the 20 digits obtained by moving in any direction. To the right, you obtain 06, 17, 22, 84, 44, 55. For convenience, drop down one line and proceed in reverse to get 09, 15, 30, 59. These are item numbers and the sample weights are 20, 30, 32, 51, 39, 41, 25, 29, 35, 42.

This is a random procedure and is equivalent to drawing from a bag of 100 beans marked with the hundred butterfat yields, each bean being replaced in the bag and the beans being thoroughly mixed before each draw. Note that each item may be drawn any number of times from 0 to 10. Sampling is always from the same population and the probability of drawing any item is practically the same. Either procedure gives the same results as if the drawings were made from an infinitely large population.

A sample drawn in this way is a *completely random* sample. In experimental work and in sample surveys there are often valid reasons for restricting randomness to some degree. The use of restrictions will be discussed in later chapters.

Exercise 2.5.1 If you had a round pond stocked with fish, a boat, a net, and a helper, could you devise a workable scheme for sampling this finite population at random?

Exercise 2.5.2 How would you draw a completely random sample of, say 100, phone numbers from a phone book? Can you think of a two-stage scheme that would involve less effort?

Exercise 2.5.3 In drawing a sample of 100 from a large population containing as many men as women, would you recommend a completely random sample or one with 50 men and 50 women? (What are the objectives of your sampling?)

Exercise 2.5.4 In describing the process of drawing a random sample using Table A.1, it was stated that "... the probability of drawing any item is practically the same." Why is the word "practically" used?

2.6 Presentation, summarization, and characterization of data.

There are many ways to present data. They include tables, charts, and pictures.

For qualitative data, enumeration is a common way to summarize and present the results. Thus if a small town makes a survey of mode of transportation, the results might be summarized and presented as percentages. To catch the eye, a pie chart, Fig. 2.1, is useful.

Such devices as pie and bar charts showing similar information about how you spend your time, your dollar, where your taxes go, are easily read. They may or may not be reliable. For example, Fig. 2.1 may be based on a sample or a census, on miles traveled or number of trips. The possibility of deceiving the unwary is obvious.

Charts may give actual numbers or percentages. Figure 2.2 illustrates the use of bars and vertical lines. If no scale is indicated, the reader sees relative frequencies and will be unaware of sample size unless specified. When totals are given, both frequencies and relative frequencies are available and can be presented by use of two scales. Relative frequencies or proportions are essentially probabilities, implied by a glance at either the bars or columns as a height or an area.

Graphic presentations are more eye-catching than frequency tables listing numbers of people, insects, etc., possessing certain characteristics, and can be just as informative.

The histogram and frequency polygon of Fig. 2.3 are common methods of presenting a considerable amount of information collected as quantitative data. The *histogram* pictures the data with *class values*, midpoints of class intervals, along the horizontal axis and with rectangles above the class intervals to represent frequencies. The histogram presents data in a readily understandable form such that one sees at a glance the general nature of the

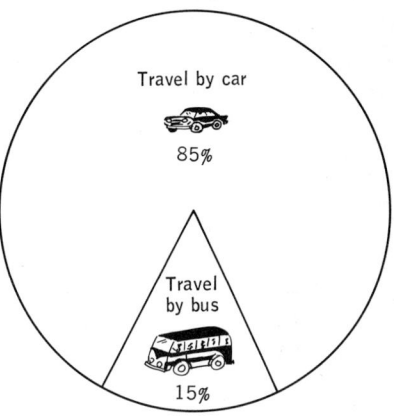

FIG. 2.1 Car versus bus, Ourtown, population 17,000

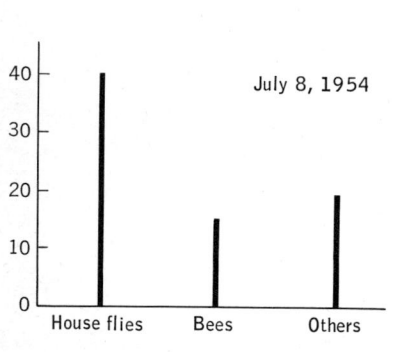

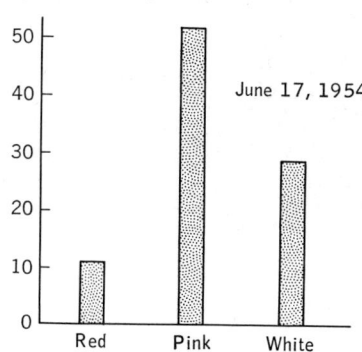

FIG. 2.2 Presentation of discrete data

distribution. If it is desired to compare an observed distribution with a theoretical distribution, the theoretical one can be superimposed on the histogram and discrepancies ascertained.

The *frequency polygon* is prepared by locating the midpoint of each class interval and marking a point above it at a height determined by the frequency for the interval. These points are then connected by straight lines. The frequency polygon tends to imply the smooth curve of the population from which the sample was drawn. The histogram and frequency polygon for the data in Table 2.1 are shown in Fig. 2.3.

It is important in the preparation of both the histogram and frequency polygon that the number of classes be sufficiently large that the general shape of the distribution can be readily ascertained, yet not so small that too much detail is given.

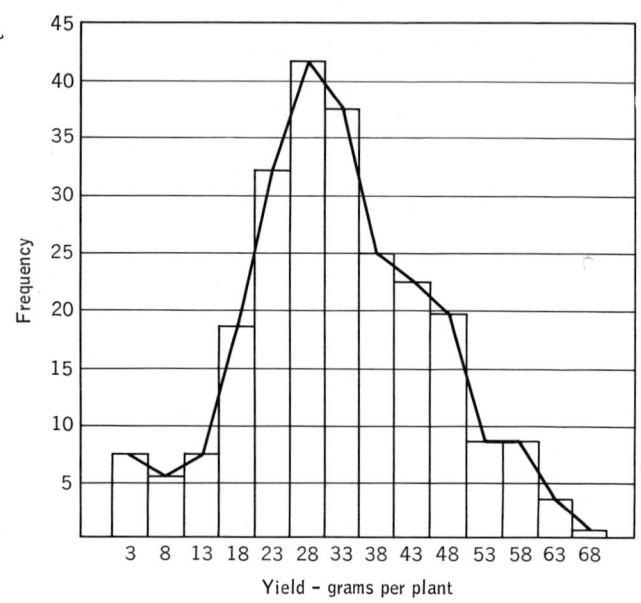

FIG. 2.3 Histogram and frequency polygon for the data of Table 2.1

TABLE 2.1

FREQUENCY TABLE FOR YIELD OF 229 SPACED PLANTS OF RICHLAND SOYBEANS

Yield in grams:	3	8	13	18	23	28	33	38	43	48	53	58	63	68
Frequency:	7	5	7	18	32	41	37	25	22	19	6	6	3	1

TABLE 2.2

NUMBER OF HEADS

Class value	Frequency
5	4
4	15
3	29
2	30
1	17
0	5
Total	100

OBSERVATIONS

Finally, data are presented in *frequency tables*. Here there is generally more information for the serious reader but at the expense of a loss in number of readers. Tables 2.1 and 2.2 are examples for continuous and discrete data. Frequency tables are further discussed beginning with Sec. 2.16.

While charts and graphs often summarize and characterize data, this may also be done by presenting several numbers as a summary and characterization of the data. In particular, we refer to a number locating the center and one measuring the spread of the observations.

More generally, some *measure of location* or of *central tendency* is required, that is, a number that locates a central or ordinary value. In addition, a *measure of spread*, of *dispersion* or of *variation* is needed to tell how far from the central value the more or less extreme values may be found.

Exercise 2.6.1 Look through a copy of *Fortune*, an appropriate government publication, and the current volume of a technical journal for ideas on the presentation of data. Is it clear, without reference to the text, whether a sample or a population is involved? Does the presentation involve frequencies or relative frequencies? Is it completely understandable without reference to the text?

Exercise 2.6.2 Compare a copy of a company's annual report with a copy of a technical journal. In which do you find relatively more charts? More tables? Do charts always have a scale on the vertical axis?

Exercise 2.6.3 Prepare a frequency table, relative frequency table, and bar chart for the numbers of integers in Table A.1.

Exercise 2.6.4 Prepare a frequency table, frequency polygon, and histogram for the data of Table 4.1.

2.7 Measures of central tendency. Expressions such as "of average height" are vague but informative. They relate an individual to a central value. When an experimenter collects data at some expense in time, energy, and money, he cannot afford to be vague. He wants a definite measure of central tendency.

The most common measure of central tendency and the one which is the "best" in many cases is the *arithmetic mean* or the arithmetic *average*. Since there are other types of means, it must be clear in any instance what mean is referred to. The arithmetic mean will be represented by two symbols: μ (the Greek letter mu) for a population, and $\bar{x}$ (read as x bar) for a sample. It is important to make this distinction since a population mean is a fixed quantity, not subject to variation, whereas a sample mean is a variable since different samples from the same population tend to have different means. Other symbols may be found in statistical writings. Thus m may represent the population mean and $\bar{x}$, $\hat{x}$, or $\bar{X}$ the sample mean. The mean is in the same units as the original observations, for example, cm or lb.

Quantities such as the mean are called *parameters* when they characterize populations and *statistics* in the case of samples.

Consider a die. Its six faces have one, two, three, four, five, and six dots. All possible numbers of dots appearing on the upturned face of a die constitute a finite population. By definition,

$$\mu = \frac{1 + 2 + 3 + 4 + 5 + 6}{6} = 3\tfrac{1}{2} \text{ dots}$$

If a sample from this population has four observations, say 3, 5, 2, and 2, then by definition

$$\bar{x} = \frac{3+5+2+2}{4} = \frac{12}{4} = 3 \text{ dots}$$

This computation can be symbolized by

$$\bar{x} = \frac{X_1 + X_2 + X_3 + X_4}{4}$$

where X_1 is the value of the first observation, namely, 3, X_2 is the value of the second, and so on. In the general situation with n observations, X_i is used to represent the ith and $\bar{x}$ is given by

$$\bar{x} = \frac{X_1 + X_2 + X_3 + \cdots + X_i + \cdots + X_n}{n}$$

The notation for $\bar{x}$ is further shortened to

$$\bar{x} = \frac{\sum_{i=1}^{n} X_i}{n} \quad \text{or} \quad \frac{\sum_i X_i}{n} \tag{2.1}$$

This is a sentence with a subject $\bar{x}$, a verb $=$, and an object Σ (the Greek letter sigma, capitalized). The sentence reads "x bar equals the sum of the X's divided by n." This is our definition of the sample mean. The letter i, used to tag the ith individual, is called the index of summation and goes from $i = 1$, written under the summation sign Σ, to n, written above. The range of summation is from 1 to n. Where all n values are to be summed, 1 and n are usually omitted.

Sample *deviates* are defined as the signed differences between the observations and the mean by the equation $x_i = X_i - \bar{x}$. Lower-case x is used to denote a deviate. For our example, $x_1 = 3 - 3 = 0$, $x_2 = 2$, $x_3 = -1 = x_4$.

An interesting property of the arithmetic mean is that the sum of the deviates is zero, that is,

$$\sum_i (X_i - \bar{x}) = \sum_i x_i = 0 \tag{2.2}$$

For the example,

X_i	$\bar{x}$	$X_i - \bar{x}$
3	3	0
5	3	+2
2	3	−1
2	3	−1

$$\sum_i (X_i - \bar{x}) = 0$$

This is a demonstration, not a proof.

An alternative to the mean as a measure of central tendency is the *median*. It may also be used to supplement the mean. The median is that value for which 50% of the observations, when arranged in order of magnitude, lie on each side. If the number of values is even, the median is the average of the two middle values. For example, for the illustration sample 3, 6, 8, and 11, $(6 + 8)/2 = 7$ is the median. The numbers 3, 6, 8, and 30 have the same median. For data distributed more or less symmetrically, the mean and median will differ but slightly. However, where an average of the incomes of a group of individuals is required and most incomes are low, the mean income could be considerably larger than the median income, and quite misleading.

Certain types of data show a tendency to have a pronounced tail to the right or to the left. Such distributions are said to be *skewed* and the arithmetic mean may not be the most informative central value.

Another measure of central tendency is the *mode*, the value of most frequent occurrence.

Other measures of central tendency are averages of certain *quartiles, deciles,* and *percentiles*, points which divide distributions of ranked values into quarters, tenths, and hundredths, respectively. For example, 10% of the observations are less than the first decile. The median is the 2d quartile, 5th decile, and 50th percentile.

For positive numbers, the geometric mean or harmonic mean may be useful. Their main uses are, respectively, in calculation of relative values such as index numbers and in averaging ratios and rates. They are obtained from equations

$$G = \sqrt[n]{X_1 X_2 \cdots X_n}$$
$$= (X_1 X_2 \cdots X_n)^{1/n} \tag{2.3}$$

and

$$\frac{1}{H} = \frac{1}{n} \sum_i \left(\frac{1}{X_i}\right) \tag{2.4}$$

Exercise 2.7.1 Comment on the statement "Fifty per cent of Americans are below average intelligence."

Exercise 2.7.2 For the sample of Exercise 2.2.2, compute $\bar{x}$ and all x_i's. Does $\Sigma x_i = 0$?

Exercise 2.7.3 What is the median of the data in Table 2.1? Table 2.2? Table 2.4?

Exercise 2.7.4 Geometric means are useful in dealing with rates and ratios. If I invest $100 and have $120 at the end of the year, reinvest and have $144 at the end of the second year, then I have been getting 20% on my investment. Clearly the rate of growth is 1.2. The geometric mean is appropriate and is $\sqrt{1.2(1.2)} = 1.2$. Population growth rates, for stable birth and death rates with no migration, give a biological situation where the geometric mean is appropriate. Find an example where the geometric mean is appropriate. Are the values constant or variable? Compute the geometric mean.

Exercise 2.7.5 A company's production of crude oil increased yearly from 1945 to 1955. In 1945, production was 2,350 bbl per day and in 1955, 4,780 bbl per day. What is the geometric mean? For what year would you regard this as an estimate of production?

2.8 Measures of dispersion.

A measure of central tendency provides a partial summary of the information in a set of data. The need for a measure of variation is apparent. Three possible sets of data with a common mean are given here; note how different they are in variability.

$$8, 8, 9, 10, 11, 12, 12$$
$$5, 6, 8, 10, 12, 14, 15$$
$$1, 2, 5, 10, 15, 18, 19$$

The mean, like other measures of central tendency, tells us nothing about variation.

The concept of a measure of spread is a difficult one. How informative is it to say that the three sets of data have spreads 4, 10, and 18, that is, $X_7 - X_1$? The second three sets

$$8, 9, 10, 10, 10, 11, 12$$
$$5, 7, 9, 10, 11, 13, 15$$
$$1, 5, 8, 10, 12, 15, 19$$

of data have spreads 4, 10, and 18 also, and the mean is still 10, but the first sets have observations spread more into the tails while in the second three they are concentrated more toward the mean. It seems desirable to have a definition which uses all observations and gives a small value when they cluster closely about a central value and a large one when they are spaced widely.

Consider the numbers

$$5, 6, 8, 10, 12, 14, 15$$

By our definition, these are no more variable than the numbers

$$105, 106, 108, 110, 112, 114, 115$$

Thus our definition does not depend on the size of the numbers in the sense of relating the measure of dispersion to the mean; it will give the same value for these two sets of numbers.

The resulting numerical measure of spread should admit interpretation in terms of the observations. Hence, it should be in the same unit of measurement. Its function will be to serve as a unit in deciding whether an observation is an ordinary or an unusual value in a specified population. The mean or a hypothesized value will be a starting point for measuring. For example, if a person is many units of spread from the mean of a population, his likeness may be preserved in bronze or he may be committed to an institution, otherwise he's the man in the street; if a male student is 5 ft 4 in., he is not likely to belong to the population of male student basketball players because, in height, he is too many units of spread from the mean of that population.

Finally, the measure should possess mathematical properties with which we need not be concerned at the moment.

The measure which is best on most counts and most common is the *variance* or its square root, the *standard deviation*. The variance will be represented by two symbols: σ^2 (the Greek letter sigma) for a population and s^2 for a sample. These are read as sigma square or sigma squared and s square(d). Standard

OBSERVATIONS

deviations for populations and samples are represented by σ and s, respectively. σ^2 and σ are parameters, constants for a particular population; s^2 and s are statistics, variables which change from sample to sample from the same population.

The *variance* or *mean square* is defined as the sum of the squares of the deviates divided by one less than the total number of deviates. This definition serves for a finite population and for a sample. The variance is a unit which is the square of the original unit, for example, (cm)^2.

For a finite population of N individuals, the variance is defined symbolically by Eq. (2.5).

$$\sigma^2 = \frac{(X_1 - \mu)^2 + (X_2 - \mu)^2 + \cdots + (X_N - \mu)^2}{N - 1}$$

$$= \frac{\sum_i (X_i - \mu)^2}{N - 1} \qquad (2.5)$$

(S^2 is used by those who reserve σ^2 for a definition with N as divisor.)

For the die example of a population,

$$\sigma^2 = \frac{1}{6 - 1} [(1 - 3\tfrac{1}{2})^2 + (2 - 3\tfrac{1}{2})^2 + (3 - 3\tfrac{1}{2})^2$$
$$+ (4 - 3\tfrac{1}{2})^2 + (5 - 3\tfrac{1}{2})^2 + (6 - 3\tfrac{1}{2})^2]$$
$$= 3\tfrac{1}{2}$$

The sample variance or mean square is defined by

$$s^2 = [(X_1 - \bar{x})^2 + (X_2 - \bar{x})^2 + \cdots + (X_n - \bar{x})^2]/(n - 1)$$
$$= \sum_i (X_i - \bar{x})^2/(n - 1) \qquad (2.6)$$

Notice that $(n - 1)s^2 = \sum_i (X_i - x)^2$. The numerator of s^2 is referred to as the *sum of squares* and is denoted by SS. For the numbers 3, 6, 8, and 11, the sum of squares is $(3 - 7)^2 + (6 - 7)^2 + (8 - 7)^2 + (11 - 7)^2 = (-4)^2 + (-1)^2 + (+1)^2 + (+4)^2 = 34$, and the variance is $34/3 = 11.33$.

The square root of the sample variance is called the *standard deviation* and is denoted by s. *Root mean square deviation* is a less-used but descriptive term for s. For our example, $s = \sqrt{34/3} = 3.4$ units of the original observations.

The quantity $SS = \sum_i (X_i - \bar{x})^2 = \sum_i x_i^2$ may be called the *definition formula* for the sum of squares, since it tells us that the sum of squares is the sum of the squared deviations from the arithmetic mean.

The definition formula for the sum of squares reduces to a *working formula* for computation, namely, Eq. (2.7).

$$\sum_i (X_i - \bar{x})^2 = \sum_i X_i^2 - (\sum_i X_i)^2/n \qquad (2.7)$$

The numbers are squared and added for the first term, added and the total squared for the second. On desk computers, $\sum_i X_i^2$ and $\sum_i X_i$ can be obtained simultaneously. The quantity $(\sum_i X_i)^2/n$ is called the *correction term* or *factor*, or *correction* or *adjustment for the mean*, and is represented by C. The term "correction for the mean" points out that since SS is a measure of variation about $\bar{x}$, the correction term must be subtracted from $\sum_i X_i^2$, often called the unadjusted sum of squares. As a correction, the quantity has nothing to do with mistakes. The form $n\bar{x}^2$, shown in Eq. (2.8), is less desirable computationally since it introduces the necessity of rounding at an earlier stage.

$$(\sum_i X_i)^2/n = n\bar{x}^2 \qquad (2.8)$$

The validity of Eq. (2.7) can be demonstrated for an arithmetic example. For our illustration sample,

X_i	X_i^2	$X_i - \bar{x}$	$(X_i - \bar{x})^2$
3	9	-4	16
6	36	-1	1
8	64	$+1$	1
11	121	$+4$	16
$\sum_i$: 28	230	0	34

Thus $SS = \sum_i (X_i - \bar{x})^2 = 34$, by the definition formula, and $SS = \sum_i X_i^2 - (\sum_i X_i)^2/n = 230 - 28^2/4 = 34$, by the working formula. When the rounding of numbers, such as the mean or the $(X_i - \bar{x})$'s, is necessary, there may be minor discrepancies. The computing formula is to be preferred since it is likely to be least affected by rounding errors. Equation (2.8) may also be checked by example.

An important property of the sum of squares is that it is a minimum, that is, if $\bar{x}$ is replaced by any other value, the sum of squares of the new deviations will be a larger value. It is not feasible to demonstrate this for all possible values.

The quantity $(n - 1)$ is termed *degrees of freedom* and is denoted by *df* or *f*.

Sums of squares are divided by degrees of freedom to give unbiased (see Sec. 3.10) estimates of σ^2. The degrees of freedom are $n - 1$ since $\bar{x}$, a sample statistic, is used in computing s^2. It will be seen later that more than one statistic may be needed to compute s^2 in some circumstances. In general, a degree of freedom is deducted from n for each statistic so used.

Another measure of dispersion is the *range*, the difference between the highest and lowest values. It is not ordinarily a satisfactory statistic for a critical evaluation of data since it is influenced by extreme or unusual values. However, for samples of size 2, it is essentially the same as the variance, and for samples of less than 10 it is often satisfactory. Under some conditions, techniques using the range are especially desirable.

A measure of variation with intuitive appeal, because it uses all observations,

OBSERVATIONS

is the mean of the *absolute values*. The absolute mean deviation or, as it is frequently termed, the *average deviation* is calculated as in Eq. (2.9).

$$\frac{\sum_i |X_i - \bar{x}|}{n} \tag{2.9}$$

The vertical bars tell us to take all deviates as positive. For the values 3, 6, 8, and 11, the average deviation is 2.5.

Other measures of dispersion use percentiles. Thus the difference between points cutting off 85% and 15% of the ranked observations has appeal; it does not depend on the extremes as does the range.

Exercise 2.8.1 Obtain a small set of data from the field most familiar to you and compute $\bar{x}$ and s^2. Check Eqs. (2.7) and (2.8) with your data. Compute the average deviation by Eq. (2.9).

2.9 Standard deviation of means. The sampling of populations and characterization of samples have been under discussion. The reader may have been thinking of such characters as height and weight, with plants and animals supplying populations of interest. It must also be remembered that sample means and standard deviations are themselves subject to variation and form populations of sample means and sample standard deviations.

As a matter of observation or intuition, one expects sample means to be less variable than single observations. If one takes two series of means each based upon a different number of observations, say 10 and 20, he expects the variation among the means of the smaller samples to be greater than that among those of the larger ones. Fortunately, there is a known relationship between the variance among individuals and that among means of individuals. This relation and that for standard deviations are

$$\sigma_{\bar{x}}^2 = \frac{\sigma^2}{n} \quad \text{and} \quad \sigma_{\bar{x}} = \sqrt{\frac{\sigma^2}{n}} = \frac{\sigma}{\sqrt{n}} \tag{2.10}$$

where $\sigma_{\bar{x}}^2$ is the variance of the population of $\bar{x}$'s obtained by sampling the *parent population* of individuals with variance σ^2. Similar relations are used for sample values, namely,

$$s_{\bar{x}}^2 = \frac{s^2}{n} \quad \text{and} \quad s_{\bar{x}} = \sqrt{\frac{s^2}{n}} = \frac{s}{\sqrt{n}} \tag{2.11}$$

These relations will be illustrated by sampling in Chap. 4; they are valid for all populations.

The necessity of the subscript $\bar{x}$ is clear. The subscript X is also used at times. Thus μ_X is the mean of the population of individuals and s_X^2 is a variance computed from a sample of individuals.

The value of these relations is apparent. Given σ^2, we can compute the variance of a population of sample means directly and for any sample size. Also, from a single sample yielding a single $\bar{x}$, we can compute a sample variance which estimates the variance in the population of $\bar{x}$'s.

The standard deviation of a mean, often called the *standard error* or *standard error of a mean*, is seen to be inversely proportional to the square root of the number of observations in the mean. It can be calculated if an s^2 or s is available; more than one $\bar{x}$ is not required. For sample observations 3, 6, 8, and 11, $s_{\bar{x}} = s/\sqrt{n} = 3.4/\sqrt{4} = 1.7$. For computational convenience, $s_{\bar{x}}$ is usually calculated as $\sqrt{s^2/n} = \sqrt{11.33/4} = \sqrt{2.8333} = 1.7$.

This is an estimate of σ^2/n, the variance in the population consisting of sample means of four observations from the population of individuals. If we had obtained a number of sample means of four observations and used these means to compute the variance of the means, we would have an estimate of the same quantity; we have a single mean.

Exercise 2.9.1 Compute $s_{\bar{x}}^2$ and $s_{\bar{x}}$ for the data you used in Exercise 2.8.1.

2.10 Coefficient of variability or of variation. A quantity of use to the experimenter in evaluating results from different experiments involving the same character, possibly conducted by different persons, is the coefficient of variability. It is defined as the sample standard deviation expressed as a percentage of the sample mean as shown in Eq. (2.12).

$$\text{CV} = \frac{100s}{\bar{x}} \tag{2.12}$$

To know whether or not a particular CV is unusually large or small requires experience with similar data. The CV is a relative measure of variation, in contrast to the standard deviation which is in the same units as the observations. Since it is the ratio of two averages, it is independent of the unit of measurement used, for example, it is the same whether pounds or grams are used to measure weight.

2.11 An example. To develop a new sanitary engineering technique, amounts of hydrogen sulfide produced from sewage after 42 hr at 37°C for nine runs were collected and are presented by Eliassen (2.2). These are given in Table 2.3. The technique was for sweeping hydrogen sulfide from an anaerobic culture medium by means of an inert gas and the trapping of hydrogen sulfide for quantitative analysis with the complete exclusion of air.

These data constitute a sample of nine observations from the population of all possible values obtainable by this technique. Variability among observations is presumably caused by such things as the samples of sewage, the samples of the anaerobic culture used, the operator's technique, and a host of other factors both known and unknown. It is clear that we cannot hope to obtain μ and σ for this abstract population, but we can estimate them. This is done by calculating $\bar{x}$ and s; these values and others are given in Table 2.3.

An $\bar{x}$ is a sample of one observation from the population of all possible means of samples of the same size, namely, nine. The *derived population* has mean $\mu_{\bar{x}} (= \mu_X)$ and $\sigma_{\bar{x}}^2 (= \sigma_X^2/n)$. Thus from a sample of observations on one population, it has been possible to calculate a mean value $\bar{x}$, an estimate of μ and $\mu_{\bar{x}}$, and two variances, s^2 and $s_{\bar{x}}^2$, estimates of σ^2 and $\sigma_{\bar{x}}^2$.

2.12 The linear additive model. A common model, that is, an explanation of the make-up of an observation, states that an observation consists of a mean plus an error. This is a linear additive model. The mean may involve the single parameter μ or be composed of a sum of parameters. Assumptions about the parameters and the errors depend upon the problem. A minimum assumption is that the errors be random, that is, that the population of X's be sampled at random.

TABLE 2.3

HYDROGEN SULFIDE PRODUCED IN ANAEROBIC FERMENTATION OF SEWAGE AFTER 42 HR AT 37°C

i = run	X_i = H$_2$S, ppm	$x_i = X_i - \bar{x}$	
1	210		−8
2	221	+3	
3	218	0	
4	228	+10	
5	220	+2	
6	227	+9	
7	223	+5	
8	224	+6	
9	192		−26
Totals	1,963	+35	−34
		+1 (due to rounding)	

$$\bar{x} = \frac{\sum_i X_i}{9} = \frac{1,963}{9} = 218 \text{ ppm (after rounding)}$$

$$s^2 = \frac{\sum_i X_i^2 - (\sum_i X_i)^2/9}{9-1} = \frac{429,147 - (1,963)^2/9}{8} = 124.36$$

$$s = \sqrt{124.36} = 11.1 \text{ ppm}$$

$$s_{\bar{x}}^2 = \frac{s^2}{9} = \frac{124.36}{9} = 13.82$$

$$s_{\bar{x}} = \sqrt{13.82} = 3.7 \text{ ppm}$$

$$\text{CV} = \frac{11.1(100)}{218} = 5\% \text{ (approx.)}$$

Such a model is applicable to the problem of estimating or making inferences about population means and variances. The simplest linear additive model is given by Eq. (2.13), namely,

$$X_i = \mu + \epsilon_i \tag{2.13}$$

together with a definition of the symbols appearing in the equation. It states that the ith observation is one on the mean μ but is subject to a sampling error, ϵ_i (epsilon sub i).

The ϵ_i's are assumed to be from a population of ϵ's with mean zero. Independence of the sampling errors is a theoretical requirement for valid inferences about a population and is assured by drawing the sample in a random manner.

The sample mean is

$$\bar{x} = \sum_i X_i/n = \frac{\sum_i (\mu + \epsilon_i)}{n} = \mu + \frac{\sum_i \epsilon_i}{n}$$

For random sampling, $(\sum_i \epsilon_i)/n$ is expected to be smaller as sample size increases, because positive and negative ϵ's should tend to cancel and because the divisor n is increasing. This is about the same as saying that the variance of means can be expected to be smaller than that of individuals or that means of large samples are less variable than means of small ones. In summary, the sample mean $\bar{x}$ is a good estimate of the population mean μ, since $\bar{x}$ should lie close to μ if the sample is large enough and not unusual.

We still require an estimate of the variance of the epsilons, the real individuals. The model shows that we have a finite sample of ϵ's, and these lead to an estimate of σ^2. However, we cannot compute the ϵ's since we do not know μ. On the other hand, we can estimate the ϵ's by corresponding e_i's, calculated as $(X_i - \bar{x})$'s, and combine them to give a sample variance computed by Eq. (2.6). The calculable $(X_i - \bar{x})$, in preference to a deviation from some other number, has the property, previously stated, that $\sum_i (X_i - \bar{x})^2$ is a minimum. It is sometimes stated that $\bar{x}$ is a *minimum sum of squares* estimate of μ.

Exercise 2.12.1 Write out a model for the height of an individual. If males and females are included in the sample, we may use two models or modify the original model. How could you do the latter?

2.13 The confidence or fiducial inference. For sample data, it is possible to infer something about the location of μ, the population mean. This is an inductive process where we reason from a part to the whole. Since $\bar{x}$ is a variable, one hesitates to say that μ is at $\bar{x}$. One would feel happier if he could give an interval, perhaps with $\bar{x}$ as center, and say he was reasonably confident that μ was in the interval. We can do this but, for a more definitely stated degree of confidence, we must involve Chaps. 3 and 4. There we find the sample quantity t defined as

$$t = \frac{\bar{x} - \mu}{s_{\bar{x}}} \quad (2.14)$$

with μ the only unknown. Clearly t may be positive or negative. Table A.3 contains values of t associated with a theoretical t distribution. For example, opposite $df = 17$ and under $P = .05$ is the number 2.110; the table heading says that larger sample values of t, either positive or negative, occur with probability .05, that is, about five times in one hundred for random samples of size $17 + 1 = 18$ observations from a normal distribution. (This distribution will be discussed in Chap. 3 and later chapters.) This may be written symbolically as

$$P(t < -2.110, t > +2.110) = .05$$

or

$$P(-2.110 < t < +2.110) = 1 - .05 = .95 \quad (2.15)$$

OBSERVATIONS

where P stands for probability, $>$ says that the preceding number or quantity exceeds the following, and $<$ says the converse. The parentheses contain a double inequality. (The symbol $\geq$ and $\leq$, "greater than or equal to" and "less than or equal to," may also be used in this case since the equals sign is meaningful only in the case of finite populations.) Substitution of the definition of t and some algebraic manipulation lead to another symbolic statement:

$$P(\bar{x} - t_{.05}s_{\bar{x}} < \mu < \bar{x} + t_{.05}s_{\bar{x}}) = .95 \qquad (2.16)$$

where $t_{.05} = 2.110$ for samples of 18. The parentheses of Eq. (2.16) contain no numbers other than $t_{.05}$. Once the sample is obtained and numerical values of $\bar{x}$ and $s_{\bar{x}}$ are substituted, it becomes untruthful to say "the probability that the population mean lies between the two numbers given is .95." Now μ either is or is not in the numerical interval and Eq. (2.16) must be interpreted as "the procedure which leads me to say that μ lies between the two numbers will result in my making 95 correct statements out of 100, on the average." To distinguish the case where numbers rather than symbols are used, place the subscript F for fiducial after P, for example, $P_F(3 < \mu < 6) = .95$. In somewhat different words, if the sample is not unusual, that is, is the sort to be expected about 95 times out of 100 in random sampling, then μ lies between 3 and 6.

If we fail to make a distinction between statements made before and after the sampling, we are doing the equivalent of tossing a coin, observing the face showing, say a head, and stating that the probability of that face is .5. Probability is not involved since a head is certainly showing. There was a probability of .5 before the coin was tossed, not after the result was observed. Similarly, before the sample was drawn and the computations performed, there was a probability of .95 that the population mean would lie in the interval to be obtained.

For the numbers 3, 6, 8, and 11, $\bar{x} = 7$, $s_{\bar{x}} = 1.7$ and $t_{.05}$ for 3 $df = 3.182$. Hence $l_1 = 7 - (3.182)(1.7) = 1.6$, and $l_2 = 7 + (3.182)(1.7) = 12.4$, where l_1 and l_2 denote the lower and upper limits, respectively, of the confidence interval. If this were a random sample from an appropriate population, then the mean would lie in the interval 1.6 to 12.4 unless a one in twenty mischance in sampling had occurred.

The adjectives confidence and fiducial, in statistics, involve two ideas concerning inference. For cases which we consider, the two procedures lead to the same numerical values.

Exercise 2.13.1 Is "$P(3 \leq 10 \leq 7) = .95$" true or false? What is the correct probability for the double inequality?

Exercise 2.13.2 Construct a confidence interval for the mean of the population sampled in Exercise 2.8.1. Write out a carefully worded confidence statement for your example.

2.14 An example. An example will illustrate the calculation and interpretation of the statistics discussed. The data in Table 2.4 are malt extract values for Kindred barley grown at 14 locations in the Mississippi Valley Barley Nurseries during 1948. The population for which any inference is to be made can be considered as the malt extract values for Kindred barley

TABLE 2.4
MALT EXTRACT VALUES ON MALTS MADE FROM KINDRED BARLEY GROWN AT 14 LOCATIONS IN THE MISSISSIPPI VALLEY BARLEY NURSERIES DURING 1948, IN PER CENT OF DRY BASIS
(One modification)

Malt extract values		Deviation from mean $x = X - \bar{x}$	Deviates squared x^2
Original	X		
77.7	77.7	1.7	2.89
76.0	76.0	0	0
76.9	76.9	0.9	0.81
74.6	74.6	−1.4	1.96
74.7	74.7	−1.3	1.69
76.5	76.5	0.5	0.25
74.2	75.0	−1.0	1.00
75.4	75.4	−0.6	0.36
76.0	76.0	0	0
76.0	76.0	0	0
73.9	73.9	−2.1	4.41
77.4	77.4	1.4	1.96
76.6	76.6	0.6	0.36
77.3	77.3	1.3	1.69
Totals 1,063.2	$\sum_i X = 1{,}064.0$	0	$\sum_i x^2 = 17.38$

SOURCE: Unpublished data obtained from and used with the permission of A. D. Dickson, U.S. Dept. Agr. Barley and Malt Laboratory, Madison, Wis.

grown during 1948 in the region covered by the Mississippi Valley Nurseries. One original value has been modified to facilitate calculations.

The first step in the calculations is to obtain $\sum_i X_i$ and $\sum_i X_i^2$. These two values, 1,064 and 80,881.38, respectively, are obtained simultaneously on the common desk computers. From these, $\bar{x}$ and s^2 are obtained from Eqs. (2.1), (2.6), and (2.7). The sample mean is

$$\bar{x} = \frac{\sum_i X_i}{n} = \frac{1{,}064}{14} = 76\%$$

and the mean square or variance is

$$s^2 = \frac{SS}{df} = \frac{\sum_i X_i^2 - (\sum_i X_i)^2/n}{n-1} = \frac{80{,}881.38 - (1{,}064)^2/14}{13} = \frac{17.38}{13} = 1.337$$

The standard deviation or square root of the variance is

$$s = \sqrt{s^2} = \sqrt{1.337} = 1.16\%$$

Calculation of the sum of squares by means of the definition formula is illustrated in the last column of Table 2.4.

The sample mean, $\bar{x} = 76\%$, and sample standard deviation, $s = 1.16\%$, provide us with the "best" estimates of the corresponding unknown population parameters, namely, μ and σ.

Since our primary interest is in the population from which the sample was drawn, we calculate an interval and state that the population mean lies therein, a statement which we are reasonably confident is correct. We first obtain the standard error of our mean, $s_{\bar{x}}$, by Eq. (2.11). We get

$$s_{\bar{x}} = \sqrt{\frac{s^2}{n}} \quad \text{or} \quad \frac{s}{\sqrt{n}} = \sqrt{\frac{1.337}{14}} \quad \text{or} \quad \frac{1.16}{\sqrt{14}}, \quad \text{respectively}$$

$$= 0.31\%, \text{ in either case}$$

$$\text{CV} = \frac{s(100)}{\bar{x}} = \frac{1.16(100)}{76.0} = 1.5\%$$

The 95% confidence interval is, from Eq. (2.16),

$$\bar{x} \pm t_{.05} s_{\bar{x}} = 76 \pm (2.160)(0.31) = (75.3\%, 76.7\%)$$

The tabular value of t has 13 degrees of freedom for this problem. Now we say that the true malt extract value of our population lies in the interval 75.3 to 76.7%, unless a one in twenty mischance has occurred. In other words, we are using a random sampling procedure to locate a fixed point, the population mean; if we don't get an unusual sample, our resulting interval will contain that parameter.

2.15 The use of coding in the calculation of statistics. Frequently the calculation of statistics can be facilitated by *coding*. Coding is intended to reduce the amount of work and promote accuracy. Coding consists of replacing each observation by a number on a new scale by use of one or more of the operations of addition, subtraction, multiplication, and division.

The arithmetic mean is affected by every coding operation. For example, if the variates are coded by first multiplying by 10 and then subtracting 100, the mean of the coded numbers must be increased by 100 and divided by 10. The rule is to apply the inverse operation, the inverses of multiplication and subtraction being division and addition, in the reverse order.

The standard deviation is affected by multiplication and division only. Addition or subtraction of a value for each observation does not affect the standard deviation since a shift in the origin of the observations does not affect their spread. Since both multiplication and division change the unit of measurement, the standard deviation calculated from coded variates is decoded by applying the inverse operation to the result.

We illustrate the use of coding in calculating the mean and standard deviation of the data in Table 2.4. The coded values will be obtained by subtracting 70 from each observation and multiplying the result by 10. Since all observations are in the 70's, subtraction of 70 is an obvious choice to give small, positive numbers; multiplication by 10 eliminates the decimal. Thus the first variate, $X_1 = 77.7$, is replaced by its coded value, $X_1' = (77.7 - 70)10 = 77$. The mean and the standard deviation of the coded numbers are then obtained from $\sum_i X'_i = 840$ and $\sum_i X'^2_i = 52{,}138$. We obtain

$$\bar{x}' = \frac{\sum_i X'_i}{14} = \frac{840}{14} = 60$$

and

$$s' = \sqrt{\frac{\sum_i X'^2_i - (\sum_i X'_i)^2/n}{n-1}} = \sqrt{\frac{52{,}138 - (840)^2/140}{13}} = 11.6$$

To decode $\bar{x}'$, apply the inverse operation in the reverse order. Thus

$$\bar{x} = \frac{\bar{x}'}{10} + 70 = \frac{60}{10} + 70 = 76\%$$

as in Sec. 2.14. To decode s',

$$s = \frac{s'}{10} = \frac{11.6}{10} = 1.16\%$$

as in Sec. 2.14. The accuracy of the calculation is in no way affected by the coding process; rather, it is simplified by the use of a more convenient set of numbers.

Exercise 2.15.1 The numbers 10.807, 10.812, ..., were coded as 7, 12, ... The mean of 17 coded observations was 14.6. What was the mean of the original observations? What part of the coding affected the standard deviation? How?

2.16† The frequency table. The frequency table was mentioned in Sec. 2.6. When the sample consists of a large number of observations it is desirable to summarize the data in a table showing the frequency with which each numerical value occurs or each class is represented. Thus both continuous and discrete data can be summarized in frequency tables. This reduces the mass of raw data into a more manageable form and provides a basis for its graphical presentation. Statistics such as the mean and standard deviation can be calculated from frequency tables with much less work than from the original values. This is true even if a desk calculator is available.

For discrete variables such as the number of heads observed in the toss of five coins, the class values to be used are generally obvious. Thus a frequency table for the number of heads occurring in the tossing of five coins one hundred times is given in Table 2.2. Where the number of possible classes is large it may be desirable to reduce their number.

For continuous variables, classes other than observed values have to be chosen in some arbitrary manner. The choice depends on factors such as the number of observations, the range of variation, the precision required for statistics calculated from the table, and the degree of summarization necessary to prevent small irregularities from obscuring general trends. The last two points usually work in opposition; the greater the number of classes the greater

† The remainder of the chapter is concerned with large quantities of data, their presentation, and the calculation of $\bar{x}$ and s. The methods of the preceding sections are applicable but lengthier for such data.

is the precision of any calculations made from the table, but if the number of classes is too great the data are not summarized sufficiently.

A rule of use in determining the size of the class interval when high precision is required in calculations made from the resulting frequency table is to make the interval not greater than one-quarter of the standard deviation. If this rule is strictly adhered to, the data are sometimes not sufficiently summarized for graphical presentation. If the size of the class interval is increased to one-third to one-half of a standard deviation, the resulting frequency table will usually be a sufficient summary for graphical presentation and adequate for most data; the lack of precision in any statistics calculated from the table will be small enough to be ignored.

Since the standard deviation is not known at the time a frequency table is being prepared, it is necessary to estimate it. Tippett (2.4) prepared detailed tables showing the relationship between the sample range and the population standard deviation for normal populations. This has been condensed in a short table by Goulden (2.3) and is reproduced as Table A.2. This table is useful in estimating σ.

Exercise 2.16.1 The range of percentage of dry matter in 48 samples of Grimm alfalfa was 7.6%. Estimate the standard deviation.

2.17 An example. To illustrate the use of Table A.2 and the preparation and use of a frequency table, consider the yields, to the nearest gram, of 229 spaced plants of Richland soybeans, reported by Drapala (2.1). Yields range from 1 to 69 grams. From Table A.2 the ratio range/σ for $n = 229$ is about 5.6. Thus σ is estimated as $68/5.6 = 12.2$ grams. One-third of the estimate is 4 grams; one-half is 6 grams. Since 12.2 grams is an approximation, the calculated class interval can be adjusted somewhat. It is convenient to make a class interval an odd number rather than an even number since the midpoint of such an interval falls conveniently between two possible values of the variable. Select 5 grams as class interval. If greater accuracy in the calculation of the statistics from the table is required, use an interval of 3 grams, about one-quarter of the estimated σ. In setting up class values, it is desirable to make the lower limit of the first class slightly less than the smallest value in the sample. Select 3 as the midpoint of the first interval. Table 2.1 is the resulting frequency table.

After the class interval is chosen, set up the necessary classes and sort the observations accordingly. Two methods are commonly used. The first is the tally score method, illustrated below for the first few classes of Table 2.1.

Class range	Class value or midpoint	Tally score	Frequency
1–5	3	𝙽𝙷𝙻 //	7
6–10	8	𝙽𝙷𝙻	5
11–15	13	𝙽𝙷𝙻 //	7

This method consists of making a stroke in the proper class for each observation and summing these for each class to obtain the frequency. It is customary, for convenience in counting, to place each fifth stroke through the preceding four, as shown. The method has the disadvantage that when there is lack of agreement in checking frequencies, the source is difficult to find.

The second and perhaps safer method is to write the value of each observation on a card of size convenient for handling. These entries must be carefully checked. Class ranges are then written on other cards and arranged in order on a desk and the observation cards are sorted into classes. Checking is accomplished readily by examining the cards in each class. The frequency in each class is determined by counting cards. If punching and sorting equipment is available, the value of each observation can be punched on a card and the cards mechanically sorted into classes.

If the yields of Table 2.1 were recorded to the nearest tenth of a gram, the class limits would be .5–5.5, 5.5–10.5, 10.5–15.5, etc. When an even number of observations equals a class limit, for example, 5.5, half of the observations are assigned to each class; if the number of such observations is odd, the odd observation is assigned to one of the two classes at random.

Effects of grouping can be seen by considering the five values, 6, 7, 7, 9, and 9 grams, in the second class of Table 2.1. The arithmetic mean of these values is 7.6 grams but in calculations from Table 2.1, they are given a value of 8 grams. If each class were examined for this type of effect, about half would show a negative error and half a positive error and thus they tend to cancel each other.

2.18 Calculation of the mean and standard deviation from a frequency table. These calculations will be illustrated for the data of Table 2.1. Coding is included. First, prepare a table such as Table 2.5. The first column, X_i, consists of actual class values. The second column, X'_i, is formed by replacing actual values by coded values. For an odd number of classes, zero is assigned to the middle one to facilitate the arithmetic and, in any case, the new interval is one unit. Column 3 is the frequency; the last two columns are needed in the computations.

The three column totals, Σf_i, $\Sigma f_i X'_i$, and $\Sigma f_i X'^2_i$, are needed for the calculation of the mean and standard deviation. They correspond to n, $\sum_i X_i$, and $\sum_i X_i^2$. Note that i may refer to the class of the frequency table, $i = 1, \ldots, 14$, or to the observation for the data, $i = 1, \ldots, 229$.

The arithmetic mean $\bar{x}$ and the standard deviation s in actual rather than coded units are calculated from the coded values by Eqs. (2.17) and (2.18):

$$\bar{x} = a + I \frac{\sum_i (fX')}{\sum_i f_i} \qquad (2.17)$$

$$s = I \sqrt{\frac{\sum_i fX'^2 - (\sum_i fX')^2 / \sum_i f_i}{\sum_i f_i - 1}} \qquad (2.18)$$

where I is the class interval and a is the assumed origin, namely, the X_i value corresponding to the coded class value of 0.

2.19 Graphical presentation of the frequency table. This has been discussed in Sec. 2.6. The histogram and frequency polygon for the data of Table 2.1 are shown in Fig. 2.3.

2.20 Significant digits. With statistical computations comes the question of accuracy. How many figures are justifiable in the end result of a sequence of computations?

TABLE 2.5

CALCULATION OF $\bar{x}$ AND s FROM A FREQUENCY TABLE

Class value or midpoint of class range		Frequency	Frequency multiplied by coded class value	Frequency multiplied by square of coded class value
Actual	Coded			
X_i	X_i'	f_i	$f_i X_i'$	$f_i X_i'^2$
3	−6	7	−42	252
8	−5	5	−25	125
13	−4	7	−28	112
18	−3	18	−54	162
23	−2	32	−64	128
28	−1	41	−41	41
33	0	37	0	0
38	+1	25	25	25
43	+2	22	44	88
48	+3	19	57	171
53	+4	6	24	96
58	+5	6	30	150
63	+6	3	18	108
68	+7	1	7	49
		$\sum_i f_i = 229 = n$	$\sum_i f_i X_i' = -49$	$\sum_i f_i X_i'^2 = 1{,}507$

$$\bar{x} = a + I \frac{\left(\sum_i f_i X_i'\right)}{\sum_i f_i} \qquad s'^2 = \frac{\sum_i f_i X_i'^2 - \left(\sum_i f_i X_i'\right)^2 / \sum_i f_i}{\sum_i f_i - 1}$$

$$= 33 + 5\frac{(-49)}{229} \qquad = \frac{1{,}507 - (-49)^2/229}{228} = 6.56$$

$$= 31.93 \text{ grams} \qquad s = I\sqrt{s'^2} = 5\sqrt{6.56} = 12.80 \text{ grams}$$

From any observation, two items of information are available: the unit of measurement and the number of units in the measurement. Where a continuous variable is involved, the number of units is clearly the closest value obtainable from the available scale. Thus if a chalk board measures 3.7 m long, the unit is $\frac{1}{10}$ m and the board measures between 36.5 and 37.5 of these units. The number is said to have two significant digits.

Computations carried out on an automatic desk computer are so simple that it is very possible that the final result of a sequence of calculations will appear more precise than it really is. Rules concerning numbers of significant digits resulting from the application of the arithmetic operations are available but somewhat impractical. In most statistical work, it is best to carry more figures, say not less than two extra, into the final computations than seem necessary and then to round the result to a meaningful number of digits, relative to the accuracy of the original measurements.

References

2.1 Drapala, W. J.: "Early generation parent-progeny relationships in spaced plantings of soybeans, medium red clover, barley, sudan grass, and sudan grass times sorghum segregates," Ph.D. Thesis, Library, University of Wisconsin, Madison, 1949.

2.2 Eliassen, Rolf: "Statistical analysis in sanitary engineering laboratory studies," *Biometrics*, **6:** 117–126 (1950).

2.3 Goulden, C. H.: *Methods of Statistical Analysis*, 2d ed., John Wiley & Sons, Inc., New York, 1952.

2.4 Tippett, L. H. C.: *Biometrika*, **17:** 386 (1926).

A proposed laboratory exercise in connection with Chap. 2

Purpose. (1) To give the student practice in drawing random samples and computing sample statistics. (2) To obtain, on a class basis, empirical evidence about certain sample statistics and to compare this with theoretical results. (To be used in connection with Chaps. 3 and 4.)

1. Draw 10 random samples of size 10 from Table 4.1 using the table of random numbers, Table A.1. Record these in 10 two-digit columns, leaving space below for a dozen or so entries.
2. For each sample, calculate
    i. The sum of the observations
    ii. The mean
    iii. The correction term
    iv. The (adjusted) sum of squares
    v. The mean square
    vi. The variance and standard deviation
    vii. The coefficient of variation
    viii. The variance and standard deviation of the mean
3. Compute the mean of the 10 sample means.
4. Compute $s_{\bar{x}}^2$ and $s_{\bar{x}}$ from the 10 $\bar{x}$'s.

Chapter 3

PROBABILITY

3.1 Introduction. Events which are common or unlikely are ones whose probabilities of occurrence are large or small, respectively. The whole range of probabilities is used by most people in one way or another. One says, "The fire might have been caused by carelessness" when not at all sure of the cause; or one says, "The fire was almost certainly caused by carelessness," when feeling strongly about it.

The statistician replaces the informative but imprecise words "might" and "almost certainly" by a number lying between zero and one; he indicates precisely how probable or improbable an event is. Statistics is used to reason from the part, the sample, to the whole, the population. Clearly we can't, with incomplete information, expect to be correct in every inference. Chance plays a part and the laws of exact causality do not hold. Statistics supplies us with procedures which permit us to state how often we are right on the average. Such statements are called probability statements. In this chapter, we consider some probability notions and illustrate the use of tables to obtain probabilities associated with the occurrence of statistical events.

3.2 Some elementary probability. The use of a ratio or number to represent a probability is not peculiar to statisticians. A sports writer may predict that a team has, for example, three chances to two of beating the opposing team. The reader may interpret this as meaning a close score is to be expected with the local team in the lead, or as meaning that if this particular game could be played many times, then the local team would win about three-fifths or 60% of the games. Expressions such as three-to-two, often written 3:2, are commonly referred to as *odds* and are converted to probabilities by forming fractions with these numbers as numerators and their sum as the common denominator.

Statisticians assign numbers between zero and one as the probabilities of events. These numbers are, basically, relative frequencies. Thus odds of three-to-two for the local team say that the probability of the local team winning is $3/(3 + 2) = 3/5 = .6$, and of losing, or of the visiting team winning, is $2/5 = .4$. The sum of the probabilities of the complete set of events is one when the occurrence of one event excludes the occurrence of all others. Only one team can win if we do not permit a tie. The probabilities of zero and one are associated with events that are respectively certain not to and certain to occur.

Two facts about probabilities have been stated. They are:

1. The probability P of an event E_i lies between zero and one or may be zero or one. Symbolically,

$$0 \leq P(E_i) \leq 1 \tag{3.1}$$

2. The sum of the probabilities of the events in a mutually exclusive (the occurrence of one event excludes the occurrence of any other) set is one. Symbolically,

$$\sum_i P(E_i) = 1 \tag{3.2}$$

A deck of playing cards contains 26 red and 26 black cards consisting of 13 spades (black), 13 hearts (red), 13 diamonds (red), and 13 clubs (black); P(spade) $= P$(heart) $= P$(diamond) $= P$(club) $= \frac{1}{4}$. Drawing a diamond in a single trial excludes drawing a heart, a spade, and a club. These events are mutually exclusive and P(spade) $+ P$(heart) $+ P$(diamond) $+ P$(club) $= 1$. On the other hand, drawing a red card in a single draw ($P = \frac{1}{2}$) does not exclude drawing a diamond since the former includes the latter; these events are not mutually exclusive and we cannot add their probabilities.

So far, the reader has been presented with nothing strange or new unless it is the use of symbolism or notation. Even the card probabilities must have been obvious. Yet in their calculation, a classical definition of probability has been used, namely:

> If a chance event can occur in n mutually exclusive and equally likely ways, and if m outcomes have a certain property A, then the probability of A is the fraction m/n,

or
$$P = \frac{\text{number of successes}}{\text{total number of events } (= \text{ successes } + \text{ failures})} \tag{3.3}$$

Probabilities of events associated with discrete variables are involved in many sampling problems, for example, opinion polls, studies of genetic characters, and problems where counts are observed. They are clearly not applicable without modification to problems with continuous variables such as weight.

3.3 Probability distributions. In the statement $P(H) = \frac{1}{2} = P(T)$, concerning the outcome of the toss of a fair coin, probabilities are given for all events. This statement is a *probability distribution* for a population, the possible events resulting from tossing a coin. Devices such as bar charts and frequency tables also serve as probability distributions provided they associate a probability with each of the possible events or values of the random variable.

It is often possible to present a mathematical formula which, in a single statement, gives the probability associated with each and every chance event. Thus for a fair coin, if we let $X = 0$ for a tail and $X = 1$ for a head, the equation

$$P(X = X_i) = \tfrac{1}{2}, \qquad X_i = 0, 1 \tag{3.4}$$

(read as: the probability that the random variable X takes the particular value X_i is one-half for X_i equal to zero and for X_i equal to one) constitutes a probability distribution. For tossing a fair die, the probability distribution would be

$$P(X = X_i) = \tfrac{1}{6}, \qquad X_i = 1, 2, \ldots, 6 \tag{3.5}$$

PROBABILITY   33

Table A.1 is a very large sample for a population with probability distribution

$$P(X = X_i) = 1/10, \qquad X_i = 0, 1, 2, \ldots, 9 \tag{3.6}$$

Similarly, if we think of only odd and even numbers, we can relate Table A.1 to Eq. (3.4).

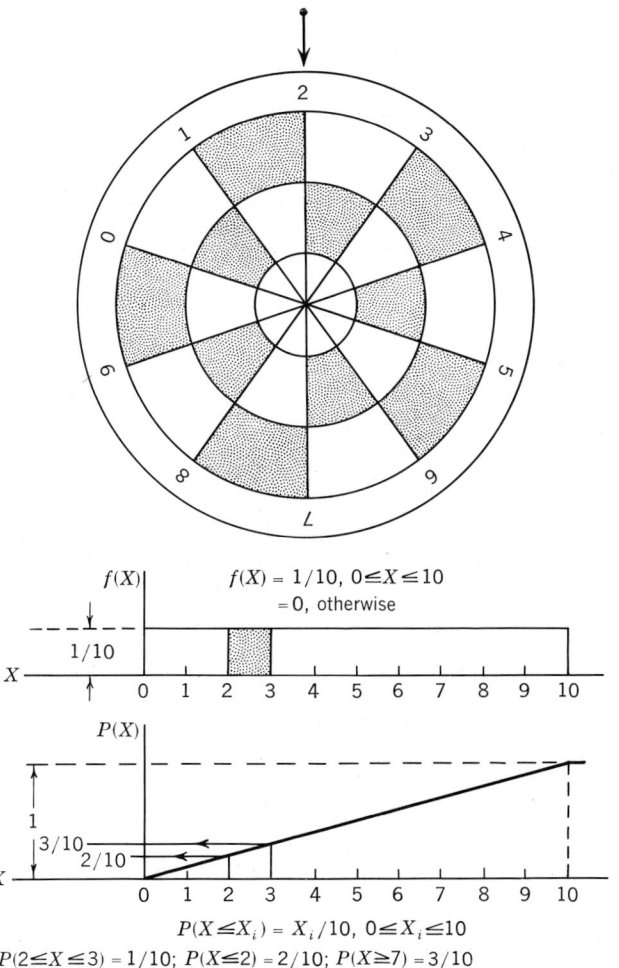

FIG. 3.1 A wheel of fortune and associated probability distribution

All problems do not deal with discrete variables. For continuous variables, charts and histograms are only approximate probability distributions. Since much of our sampling will be of continuous variables, we will spend the rest of the chapter discussing continuous probability distributions.

Consider the wheel of fortune, Fig. 3.1. The stopping point is defined as that point opposite the fixed arrow. How many stopping points are there?

The wheel could be marked into ten sectors and the stopping point be defined as the number nearest the arrow. But each sector could be further divided into ten sectors to give 100 stopping points, and so on. Clearly there is not a finite number of stopping points and, as a result, our classical definition of probability, Eq. (3.3), fails because we have no number for the denominator.

For problems involving continuous variables and an indefinitely large number of events, it is impossible to assign a probability to each event. It becomes necessary, then, to take a different approach to probability and probability statements. For the wheel of fortune divided into ten sectors with the numbers $0, 1, 2, \ldots, 9$ marked on successive dividing lines, we talk of probabilities associated with sectors or intervals but not about probabilities associated with points. The pointer must, of course, stop on one of the indefinitely many points. For probabilities associated with such intervals, a probability distribution is often conveniently associated with an expression in or *a function of X*, written $f(X)$. The symbol $f(X)$ is a generic term like apple; we need something additional to be fully informed. For $f(X)$, we need an equation, for example, Eq. (3.7); for apple, we need a name, for example, MacIntosh.

A probability distribution function is easily interpreted graphically. Referring to Fig. 3.1,

$$f(X) = 1/10, \quad 0 \leq X \leq 10 \tag{3.7}$$

is a function which describes a frequency polygon or histogram. A name may also be used, for example, this is the *uniform* distribution. Here, any value of X between 0 and 10 is possible, whereas there is only one value of X for each interval of a histogram. Areas under this curve (the term is used to include straight lines) are associated with probabilities. For example, the total area is $(1/10)10$ or one; the shaded area between 2 and 3 is one-tenth; there is no area under the curve before zero nor after ten. (The numbers zero and ten are the same for this example.) The associated function $P(X)$ or its graph is used to give *cumulative probabilities*. For example, to find the probability that the pointer will stop between two and three, that is, $P(2 \leq X \leq 3)$, read up from $X = 3$ to the sloping line, then over to the $P(X)$ axis to obtain three-tenths. Repeat for $X = 2$ to obtain two-tenths. Now subtract the probability of obtaining a value less than two from that of obtaining a value less than three to obtain the probability of a value between two and three, that is,

$$P(2 \leq X \leq 3) = P(X \leq 3) - P(X \leq 2)$$

Probability distributions are characterized in the same manner as data in general. The mean and variance, μ and σ^2, are the measures most used. These are parameters. The calculation of μ and σ^2 for continuous variables is beyond the scope of this text. When unknown, they are estimated by sample statistics.

As we proceed, we shall be greatly concerned with cumulative probability distributions, $P(X)$'s. For us, this will mean extracting a number from a table. We will have little use for $f(X)$ distributions except by name.

Exercise 3.3.1 Given $f(X) = 1/10$, $0 \leq X \leq 10$. What is $P(2 \leq X \leq 7)$? $P(1 \leq X \leq 9)$? $P(2 \leq X \leq 4 \text{ or } 6 \leq X \leq 8)$? $P(2 \leq X \leq 4 \text{ and } 3 \leq X \leq 7)$?

3.4 The normal distribution. Important in the theory and practice of statistics is the *normal distribution*. Many biological phenomena result in data which are distributed in a manner sufficiently normal that this distribution is the basis of much of the statistical theory used by the biologist. In fact, the same is true in many other fields of application. Its graph, the normal curve, also called Laplacian or Gaussian, is a bell-shaped curve. The amount of humping, heaping or central tendency depends on the size of σ^2, a small σ^2 giving a higher hump than would a large σ^2 (see Fig. 3.2). The mathematical

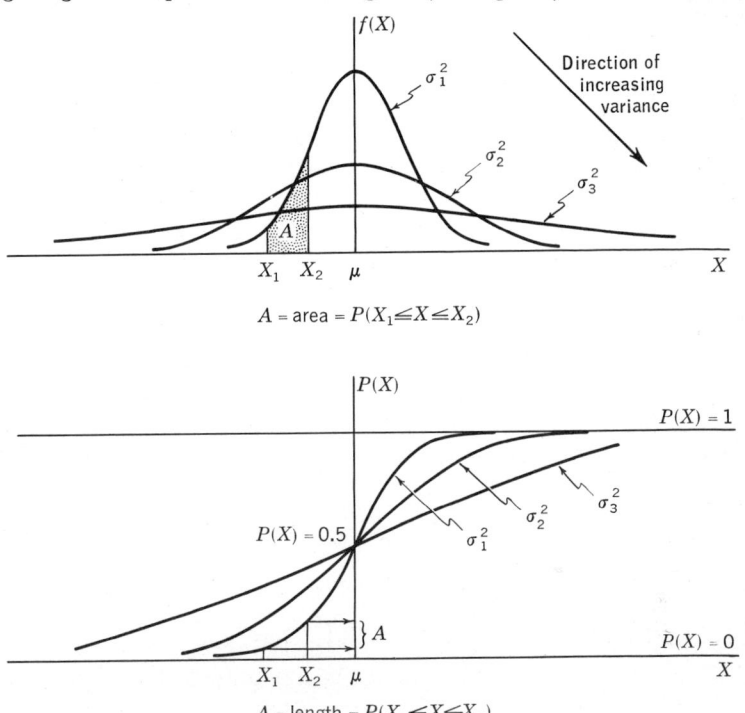

FIG. 3.2 Normal, $f(X)$, and cumulative normal, $P(X)$, distributions

formula for the probability distribution does not give probabilities directly but describes the curves of the upper part of Fig. 3.2. The magnitude of the area A between X_1 and X_2 gives the probability that a randomly drawn individual will lie between X_1 and X_2, that is, $P(X_1 \leq X \leq X_2) = A$. The total area under the curve and above the X axis is one. The curve is symmetric and one-half of the area lies on each side of μ.

Cumulative normal distributions, curves giving the probability that a random X will be less than a specified value, that is, $P(X \leq X_i)$, are shown in the lower portion of the figure for the curves of the upper part. Read up from X_i and over to the $P(X)$ axis to find probabilities. Since cumulative probabilities are given, it is necessary to obtain two values of $P(X)$ to evaluate expressions such as $P(X_1 \leq X \leq X_2)$. There is no simple mathematical expression for $P(X)$ for the normal distribution.

3.5 Probabilities for a normal distribution; the use of a probability table.

Instead of a curve, a table is used to obtain probabilities for a normal distribution. Table A.4 gives probabilities for a normal distribution with $\mu = 0$ and $\sigma^2 = 1$. The variable is denoted by z. Values of z are given to one decimal place in the z column and to a second in the z row. This table can be used to obtain probabilities associated with any normal distribution provided that the mean and variance are known. Its use will now be illustrated for several problems involving a normal distribution with $\mu = 0$ and $\sigma^2 = 1$.

Case 1. To find the probability that a random value of z will exceed z_1, that is, $P(z \geq z_1)$. Let us find $P(z \geq 1.17)$.

Procedure. Find 1.1 in the z column and .07 in the z row, $1.17 = 1.1 + .07$. The probability is found at the intersection of the row and column, thus $P(z \geq 1.17) = .1210$. In other words, about 12 times in a hundred, on the average, we can expect to draw a random value of z greater than 1.17 (see Fig. 3.3.i).

Case 1a. To find the probability that a random value of z will be less than z_1, that is, $P(z \leq z_1)$. Let us find $P(z \leq 1.17)$.

Procedure. Since the area under the curve is 1,

$$P(z \leq 1.17) = 1 - .1210 = .8790$$

(see Fig. 3.3.i).

Case 1b. To find the probability that a random value of z will be less than some negative z_1, that is, $P(z \leq z_1)$ where $z_1 \leq 0$. Let us find $P(z \leq -1.17)$.

Procedure. The normal curve is symmetric. Since we are discussing a normal curve with zero mean,

$$P(z \leq -1.17) = P(z \geq 1.17) = .1210$$

See Fig. 3.3.iv for $z = -1.05$ and $+1.05$.

Case 2. To find the probability that a random value of z lies in an interval (z_1, z_2), that is, $P(z_1 \leq z \leq z_2)$. Let us find $P(.42 \leq z \leq 1.61)$.

Procedure. Find $P(z \geq .42)$ and $P(z \geq 1.61)$. Now

$$P(.42 \leq z \leq 1.61) = P(z \geq .42) - P(z \geq 1.61)$$
$$= .3372 - .0537 = .2835$$

(see Fig. 3.3.ii).

Case 2a. To find the probability that a random value of z will lie in an interval involving two negative numbers, that is, $P(z_1 \leq z \leq z_2)$ where $z_1 < 0$ and $z_2 < 0$. Let us find $P(-1.61 \leq z \leq -.42)$. (Note that it is not necessary to place any sign before the letter z unless some special value is being used as illustration.)

Procedure. Again because of symmetry and because $\mu = 0$,

$$P(-1.61 \leq z \leq -.42) = P(.42 \leq z \leq 1.61) = .2835$$

Case 2b. To find the probability that a random value of z will lie in an interval which includes the mean, that is, $P(z_1 \leq z \leq z_2)$. Let us find $P(-1.61 \leq z \leq .42)$.

Procedure. Since the area under the curve is 1, let us find the areas outside the interval and subtract their sum from one. $P(z \leq -1.61) = P(z \geq 1.61) = .0537$ and $P(z \geq .42) = .3372$. Hence

$$P(-1.61 \leq z \leq .42) = 1 - (.0537 + .3372) = .6091$$

(see Fig. 3.3.iii).

PROBABILITY

A probability that is often required in applications of statistics is that of obtaining a random value numerically greater than some number.

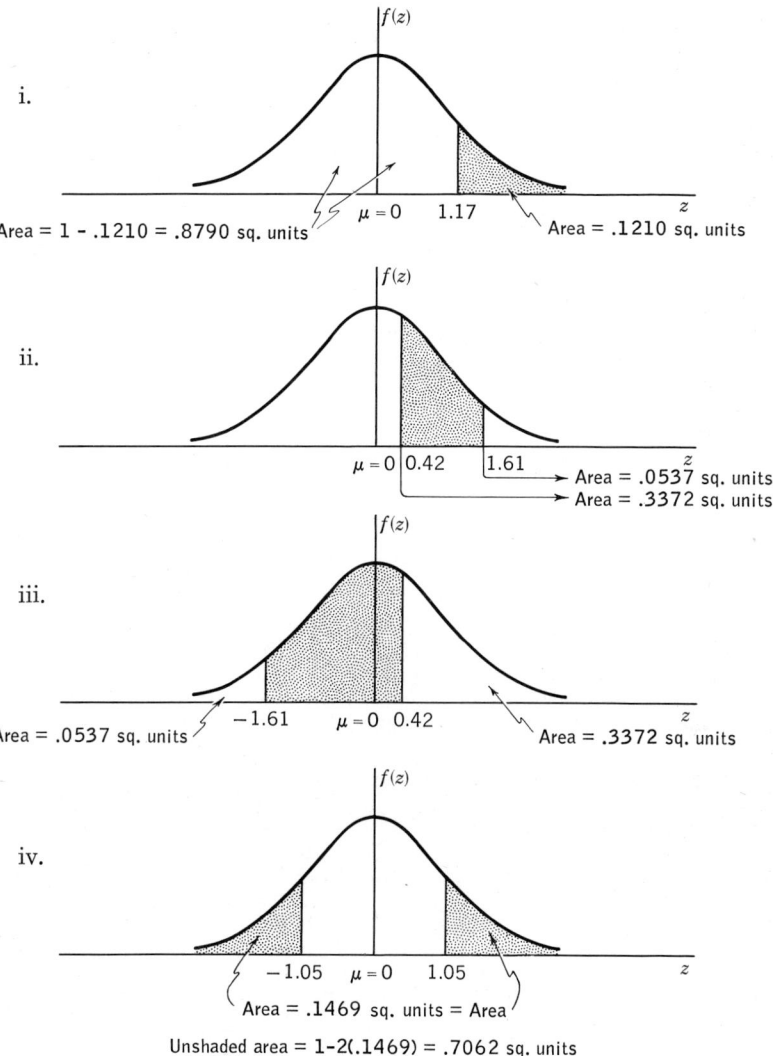

FIG. 3.3   Some probability illustrations

Case 3. To find the probability that a random value of z will numerically exceed z_1, that is, $P(|z| \geq z_1)$. Let us find $P(|z| \geq 1.05)$.
Procedure. Because of symmetry, $P(|z| \geq 1.05) = 2P(z \geq 1.05) = 2(.1469) = .2938$ (see Fig. 3.3.iv). This method is shorter than the one implied in Case 2b.

Case 3a. To find the probability that a random value of z will be numerically less than z_1, that is, $P(|z| \leq z_1)$. Let us find $P(|z| \leq 1.05)$.

Procedure. Using the fact that the area under the curve is 1,

$$P(|z| \le 1.05) = 1 - P(|z| \ge 1.05)$$
$$= 1 - 2(.1469) = .7062$$

(see Fig. 3.3.iv). The procedure of Case 2*b* is also available.

In statistics, one is often required to find values of a statistic such that random values will exceed it in a given proportion of cases, that is, with given probability. This amounts to constructing one's own table with desired P values in margins, say, and values of the variable in the body of the table. For example, in dealing with the normal distribution (or any symmetric one), 50% of random z values will exceed the mean on the average.

Case 4. To illustrate, let us find the value of z that will be exceeded with a given probability; for example, let us find z_1 such that $P(z \ge z_1) = .25$.

Procedure. Look through the body of the table for the probability .2500. It is on line $z = .6$, approximately halfway between columns .07 and .08. The value of z is between .67 and .68. Thus $P(z \ge .67) = .25$ (approx.).

A value of z also often required is that which is exceeded or not exceeded numerically with a stated probability.

Case 5. To find the value of z, say z_1, such that $P(|z| \ge z_1)$ equals a given value. Let us find z_1 such that $P(|z| \ge z_1) = .05$.

Procedure. Since the curve is symmetric, find z_1 such that $P(z \ge z_1) = .05/2 = .025$. The procedure for Case 4 gives $z_1 = 1.9 + .06 = 1.96$. Hence $P(|z| \ge 1.96) = .05$.

Case 5a. To find the value of z, say z_1, such that $P(-z_1 \le z \le z_1)$ equals a stated value. Let us find z_1 such that $P(-z_1 \le z \le z_1) = .99$.

Procedure. Since the area under the curve is one, we refer to Case 5 and note that $P(-z_1 \le z \le z_1) = 1 - P(|z| \ge z_1)$. Hence

$$1 - P(-z_1 \le z \le z_1) = P(|z| \ge z_1)$$
$$= 1 - .99 = .01$$

As in Case 5, find z_1 such that $P(z \ge z_1) = .005$; z_1 lies between 2.57 and 2.58. (To three decimal places $z_1 = 2.576$.) Hence,

$$P(-2.576 \le z \le 2.576) = .99$$

Exercise 3.5.1 Given a normal distribution with zero mean and unit variance, find $P(z \ge 1.70)$; $P(z \ge .96)$; $P(z \le 1.44)$; $P(z \le -1.44)$; $P(-1.01 \le z \le .33)$; $P(-1 \le z \le 1)$; $P(|z| \le 1)$; $P(|z| \ge 1.65)$; $P(.45 \le z \le 2.08)$.

Exercise 3.5.2 Find z_0 such that $P(z \ge z_0) = .3333$; $P(z \le z_0) = .6050$; $P(1.00 \le z \le z_0) = .1000$; $P(|z| \ge z_0) = .0100$; $P(|z| \le z_0) = .9500$.

3.6 The normal distribution with mean μ and variance σ^2.

The normal distribution with $\mu = 0$ and $\sigma^2 = 1$ is a particular one and you may have been wondering about tables for the many possible combinations of values of μ and σ^2. Actually, no further tables are necessary.

Example. Suppose we are sampling from a normal population with $\mu = 12$ and $\sigma^2 = 1$ and require $P(X \ge 13.15)$.

PROBABILITY

The distribution is the same as that of the previous section except that it is displaced so that $\mu = 12$ rather than zero. Consequently,

$$P(X \geq 13.15) = P(X - \mu \geq 13.15 - 12)$$
$$= P(z \geq 1.15) = .1251 \text{ from Table A.4.}$$

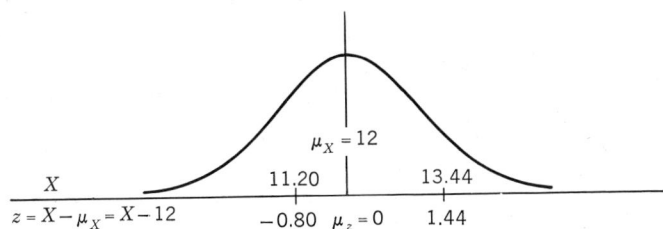

Fig. 3.4 Probabilities for a normal distribution with $\mu = 12$, $\sigma^2 = 1$

In general, we wish to find $P(X_1 \leq X \leq X_2)$. For example,

$$P(11.20 \leq X \leq 13.44) = 1 - P(X \text{ is outside this interval})$$
$$= 1 - [P(X \leq 11.20) + P(X \geq 13.44)]$$
$$= 1 - [P(z \leq -.80) + P(z \geq 1.44)]$$
$$= 1 - [P(z \geq .80) + P(z \geq 1.44)]$$
$$= 1 - (.2119 + .0749) = .7132$$

(see Fig. 3.4). The trick is seen to be the converting of an X variable with a nonzero mean to a z variable with a zero mean by subtracting μ.

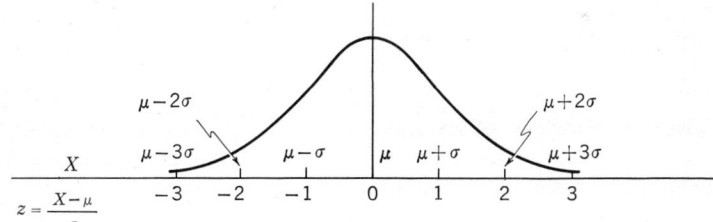

FIG. 3.5 Relation between X and z for computing probabilities for the normal distribution

For the most general case, that is, when $\mu \neq 0$ and $\sigma^2 \neq 1$ (the symbol $\neq$ means "not equal to"), one uses Table A.4 by calculating

$$z = \frac{X - \mu}{\sigma} \qquad (3.8)$$

a deviation from the mean in units of the standard deviation (see Fig. 3.5). Thus we change any normal variable to another with zero mean and unit variance since σ is the new unit of measurement.

Example. If sampling is from a normal distribution with $\mu = 5$ and $\sigma^2 = 4$ or $\sigma = 2$, find the probability of a sample value greater than 7.78. Enter Table A.4 with

$$z = \frac{X - \mu}{\sigma} = \frac{7.78 - 5}{2} = 1.39$$

Then
$$P(X \geq 7.78) = P\left(\frac{X - \mu}{\sigma} \geq \frac{7.78 - 5}{2}\right)$$
$$= P(z \geq 1.39) = .0823$$

In this general case, σ is the unit of measurement. The variable z is a deviation from the mean, namely $X - \mu$, measured in units of standard deviations, that is, $(X - \mu)/\sigma$. The preceding probability statement gives the probability of a random value of z being greater than 1.39, or a random value of X being farther to the right of μ than 1.39σ. The variable z is called a *standard normal deviate*. If a distribution is not known to be normal, such a deviate is simply a *standard deviate*.

The statement

$$P\left(-1 \leq \frac{X - \mu}{\sigma} \leq +1\right) = 1 - 2(.1587) = .6826$$

says that the probability of a random value of $(X - \mu)/\sigma$ lying between -1 and $+1$ is about two-thirds; or that about two-thirds of all values of $(X - \mu)/\sigma$ lie between -1 and $+1$.

Exercise 3.6.1 Given a normal distribution of X with mean 5 and variance 16, find $P(X \leq 10)$; $P(X \leq 0)$; $P(0 \leq X \leq 15)$; $P(X \geq 5)$; $P(X \geq 15)$.

Exercise 3.6.2 Given a normal distribution of X with mean 20 and variance 25, find X_0 such that $P(X \leq X_0) = .025$; $P(X \leq X_0) = .01$; $P(X \leq X_0) = .95$; $P(X \geq X_0) = .90$.

3.7 The distribution of means. When a population is sampled, it is customary to summarize the results by calculating $\bar{x}$ and other statistics. Continued sampling generates a population of $\bar{x}$'s with a mean and variance of its own; and $\bar{x}$ is a sample of one observation from this new population. The population originally sampled is often called the *parent population* or *parent distribution* while a population of sample means, like other populations of statistics, is called a *derived distribution* since it is derived by sampling a parent distribution. It has already been noted that the mean and variance of a population of means of n observations are the mean and $1/n$th of the variance of the parent population, that is, $\mu_{\bar{x}} = \mu$ and $\sigma_{\bar{x}}^2 = \sigma^2/n$. Absence of a subscript indicates a parameter for the parent population.

Consider a type of problem that arises in sampling experiments involving normal populations; namely, what is the probability that the sample mean will exceed a given value?

Example. Given a random sample of $n = 16$ observations from a normal population with $\mu = 10$ and $\sigma^2 = 4$. Find $P(\bar{x} \geq 11)$.

The sample mean is a sample of size one from a normal population with

$$\mu_{\bar{x}} = \mu = 10$$

$$\sigma_{\bar{x}}^2 = \frac{\sigma^2}{n} = \frac{4}{16} = \frac{1}{4} \quad \text{and} \quad \sigma_{\bar{x}} = \sqrt{\frac{1}{4}} = \frac{1}{2}$$

Enter Table A.4 with

$$z = \frac{\bar{x} - \mu_{\bar{x}}}{\sigma_{\bar{x}}} = \frac{11 - 10}{\frac{1}{2}} = 2$$

Now $P(z \geq 2) = .0228$, that is, $P(\bar{x} \geq 11) = .0228$ because 11 is two standard deviations to the right of the mean of the population of $\bar{x}$'s. Also $P(|z| \geq 2) = 2(.0228) = .0456$. Hence $P(-2 \leq z \leq 2) = P(9 \leq \bar{x} \leq 11) = 1 - .0456 = .9544$. A sample $\bar{x}$ less than 9 or greater than 11 would be likely to occur only about five times, 100(.0456), in one hundred or one time in twenty, on the average. This is generally regarded as unusual. About 95 times out of 100 or 19 times out of 20, on the average, random $\bar{x}$'s from this population will lie in the interval (9, 11).

In the previous paragraph, we have defined "unusual" for statistical purposes. If a random event occurs only about one time in twenty, we agree to label the event "unusual." Occasionally we will revise our definition for particular circumstances but this definition is easily the most common.

Exercise 3.7.1 Given that X is normally distributed with mean 10 and variance 36. A sample of 25 observations is drawn. Find $P(\bar{x} \geq 12)$; $P(\bar{x} \leq 9)$; $P(8 \leq \bar{x} \leq 12)$; $P(\bar{x} \geq 9)$; $P(\bar{x} \leq 11.5)$.

Exercise 3.7.2 Given that X is normally distributed with mean 2 and variance 9. For a sample of 16 observations, find $\bar{x}_0$ such that $P(\bar{x} \leq \bar{x}_0) = .75$; $P(\bar{x} \leq \bar{x}_0) = .20$; $P(\bar{x} \geq \bar{x}_0) = .66$; $P(\bar{x} \geq \bar{x}_0) = .05$.

3.8 The χ^2 distribution. We now discuss the distribution of χ^2 (Greek chi; read as chi-square) because of its relation to s^2 and Student's t. Chi-square is defined as the sum of squares of independent, normally distributed variables with zero means and unit variances. Section 3.5 dealt exclusively with a normal variable with zero mean and unit variance while Sec. 3.6 showed how to change any normal variable to one with zero mean and unit variance. Thus we have Eq. (3.9).

$$\chi^2 = \sum_i z_i^2 = \sum_i \left(\frac{X_i - \mu_i}{\sigma_i}\right)^2 \tag{3.9}$$

Equation (3.9) is more general than we presently need since we are discussing sampling a single population with constant σ. Now in sampling from a normal distribution, the quantity $SS = (n-1)s^2$ consists of the sum of squares of $n-1$ independent deviates as stated in Sec. 2.8. It can be shown that such deviates have zero means; division by the common σ assures that they have unit variances. Hence

$$\chi^2 = \frac{(n-1)s^2}{\sigma^2} \tag{3.10}$$

is a particular case of Eq. (3.9) and is the one with which we are to be presently concerned.

The distribution of χ^2 depends on the number of independent deviates, that is, on the degrees of freedom. For each number of degrees of freedom, there is a χ^2 distribution. Some chi-square curves are shown in Fig. 3.6. Obviously χ^2 cannot be negative since it involves a sum of squared numbers. While the peaks are seen to lag behind the degrees of freedom, curves begin to look more and more symmetric as the degrees of freedom increase.

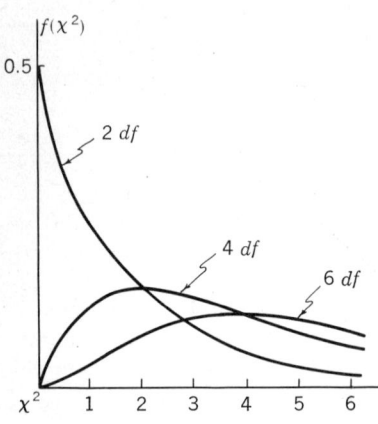

FIG. 3.6 Distribution of χ^2, for 2, 4, and 6 degrees of freedom

It is customary to tabulate only a few values for each of many curves. Thus, we have Table A.5. Probabilities are given at the top of the table, degrees of freedom in the left column, and χ^2 values in the body of the table for the stated combinations of P and df.

Example. To find the random value of χ^2 with 15 degrees of freedom that is exceeded with probability .25, that is, to find χ_1^2 such that

$$P(\chi^2 \geq \chi_1^2) = .25$$

Enter Table A.5 for 15 degrees of freedom and read under the column headed 0.25. Here $\chi^2 = 18.2$ and $P(\chi^2 \geq 18.2) = .25$.

Example. To find the probability with which an observed $\chi^2 = 13.1$ with 10 degrees of freedom will be exceeded.

Enter Table A.5 for 10 degrees of freedom and read across for the number 13.1. It lies between 12.5 and 16.0, χ^2 values exceeded with probabilities .25 and .10. Hence $.25 > P(\chi^2 \geq 13.1) > .10$.

These examples illustrate the problems most often met. They involve the use of only the right-hand tail of the distribution, unlike the z distribution where one may be as interested in both tails as in one.

The distribution of χ^2 with 1 degree of freedom is directly related to the normal distribution. Consider χ^2 with 1 degree of freedom for $P = .10$. In our shorthand, $P(\chi^2 \geq 2.71) = .10$ from Table A.5. Since by definition this χ^2 is the square of a single normal deviate with zero mean and unit variance, $\sqrt{\chi^2}$ must be a normal deviate with zero mean and unit variance. Hence if we enter Table A.4 with $z = \sqrt{2.71} = 1.645$, we should find the probability of obtaining a greater value at random to be $(\frac{1}{2})(.10) = .05$. From Table A.4, $P(z \geq 1.64) = .0505$ and $P(z \geq 1.65) = .0495$. Thus our whole normal table, Table A.4, is condensed in a single line of Table A.5, that for 1 degree of freedom. Note that z values from both tails of the normal distribution go into the upper tail of χ^2 for 1 degree of freedom due to disappearance of the minus sign in squaring, while z values near zero, either positive or negative, go into χ^2 with 1 degree of freedom at the tail near zero. Values near zero are not generally of special interest so that the χ^2 table is generally used with the accent on large values.

Exercise 3.8.1 Find χ_0^2 such that $P(\chi^2 \geq \chi_0^2) = .05$ for 10 degrees of freedom; $P(\chi^2 \geq \chi_0^2) = .01$ for 12 degrees of freedom; $P(\chi^2 \geq \chi_0^2) = .50$ for 25 degrees of freedom; $P(\chi^2 \leq \chi_0^2) = .025$ for 18 degrees of freedom.

Exercise 3.8.2 Find P such that $P(\chi^2 \geq 17.01)$ for 11 degrees of freedom; $P(\chi^2 \geq 6.5)$ for 6 degrees of freedom; $P(\chi^2 \geq 20)$ for 10 degrees of freedom; $P(\chi^2 \leq 3.8)$ for 4 degrees of freedom.

3.9 The distribution of Student's t. William Sealy Gosset, 1876–1937, a brewer or statistician according to your point of view, wrote many statistical papers under the pseudonym of "Student." He recognized that the use of s for σ in calculating z values for use with normal tables was not trustworthy for small samples and that an alternative table was required. He concerned himself with a variable closely related to the variable $t = (\bar{x} - \mu)/s_{\bar{x}}$, an expression involving two statistics, $\bar{x}$ and $s_{\bar{x}}$, rather than with $z = (\bar{x} - \mu)/\sigma_{\bar{x}}$, with one. Now, the statistic

$$t = \frac{\bar{x} - \mu}{s_{\bar{x}}} \tag{3.11}$$

for samples from a normal distribution, is universally known as "Student's t."

Like χ^2, t has a different distribution for each value of the degrees of freedom. Again, we are content with an abbreviated table, Table A.3, with t values rather than probabilities in the body of the table. Table A.3 gives probabilities for larger values of t, sign ignored, at the top. For example, for a random sample of size 16, from the line for $df = 16 - 1 = 15$ and in the column headed 0.05, we find that $P(|t| \geq 2.131) = .05$. Table A.3 gives probabilities for larger values of t, sign not ignored, at the bottom. Thus, for a random sample of size 16, from the line for $df = 15$ and in the column with .025 at the bottom, we find that $P(t \geq 2.131) = .025 = P(t \leq -2.131)$.

The curve for t is symmetric, as implied by the previous examples. It is somewhat flatter than the distribution of $z = (\bar{x} - \mu)/\sigma_{\bar{x}}$, lying under it at the center and above it in the tails. As the degrees of freedom increase, the t distribution approaches the normal. This can be seen from an examination of the entries in Table A.3, since the last line, $df = \infty$, is that of a normal distribution and the entries in any column are obviously approaching the corresponding value.

An important property of t for samples from normal populations is that its components, essentially $\bar{x}$ and s, show no sign of varying together. In other words, if many samples of the same size are collected, $\bar{x}$ and s calculated, and the resulting pairs plotted as points on a graph with axes $\bar{x}$ and s, then these points will be scattered in a manner giving no suggestion of a relation, such as that large means are associated with large standard deviations. For any distribution other than the normal, some sort of relationship exists between sample values of $\bar{x}$ and s in repeated sampling.

Exercise 3.9.1 Find t_0 such that $P(t \geq t_0) = .025$ for 8 degrees of freedom; $P(t \leq t_0) = .01$ for 15 degrees of freedom; $P(|t| \geq t_0) = .10$ for 12 degrees of freedom; $P(-t_0 \leq t \leq t_0) = .80$ for 22 degrees of freedom.

Exercise 3.9.2 Find $P(t \geq 2.6)$ for 8 degrees of freedom; $P(t \leq 1.7)$ for 15 degrees of freedom; $P(t \leq 1.1)$ for 18 degrees of freedom; $P(-1.1 \leq t \leq 2.1)$ for 5 degrees of freedom; $P(|t| \geq 1.8)$ for 6 degrees of freedom.

3.10 Estimation and inference. The discussion so far has been concerned with sampling from known populations. In general, population parameters are not known though there may be hypotheses about their values. The subject of statistics deals largely with drawing inferences about population parameters, uncertain inferences since they are based on sample evidence.

Consider the problem of estimating parameters. For example, one might wish to know the average yield at maturity of a wheat variety, or the average length of time to get over a cold. It is fairly obvious that $\bar{x}$ is an estimate of μ and that s^2 is an estimate of σ^2. Of course, these are not the only estimates of these parameters. How good are these statistics as estimates of parameters? Unless the parameters are known, one cannot know how well a particular sample statistic estimates a parameter; one must be content in knowing how good a sample statistic is on the average, that is, how well it does in repeated sampling, or, how many sample values can be expected to lie in a stated interval about the parameter. For example, consider three statistics or formulas for estimating the parameter μ. (The median, the mean, and the middle of the range, called the mid-range, are three though not necessarily those of this example.) Denote these by $\hat{\mu}_1$, $\hat{\mu}_2$, and $\hat{\mu}_3$, where the ^ (call it "hat") indicates an estimate rather than the parameter itself. The formulas are called *estimators* as well as statistics. All possible values of $\hat{\mu}_1$, $\hat{\mu}_2$, and $\hat{\mu}_3$ might generate distributions as in Fig. 3.7 where $\hat{\mu}_1$ gives fairly consistent values, that is, they have a relatively small variance but they are not centered on μ; $\hat{\mu}_2$ gives consistent values that are centered at μ; $\hat{\mu}_3$ gives values centered at μ but spreading rather widely. The problem is now to choose the "best" formula where "best" must first be defined.

Instead of defining best, consider a number of desirable properties and try to have as many as possible associated with the choice of an estimator. For example, *unbiasedness* requires that the mean of all possible estimates given by an estimator, that is, the mean of the population of estimates, be the parameter being estimated. The mean of a population of $\bar{x}$'s is μ, the parameter being estimated for the parent population so that $\bar{x}$ is an unbiased estimate of μ. The mean of a population of s^2's, namely μ_{s^2}, is σ^2 so that s^2 is an unbiased estimate of σ^2. However, if the divisor of the sum of squares had been n rather than $n-1$, the estimate would not have been unbiased. Bias is not serious if its magnitude is known. This would be the case if n were the divisor in estimating σ^2. Bias may be serious if its magnitude is unknown for then no correction for bias is possible.

Another desirable property is that of having a small variance. *Minimum variance* is most desirable. In Fig. 3.7, $\hat{\mu}_3$ had too large a variance whereas $\hat{\mu}_1$ and $\hat{\mu}_2$ had comparatively small variances and were preferable on this basis. The estimators $\hat{\mu}_1$ and $\hat{\mu}_2$ are said to be more *efficient* than $\hat{\mu}_3$.

In some cases, it may be necessary to compromise in choosing a formula to calculate estimates of a parameter.

Apart from considerations of computing ease, $\bar{x}$ and s^2 are, jointly, the best estimates of the mean and variance of a normal distribution. For some distributions, the median may be a better estimator of the population mean than is the sample mean.

While $\bar{x}$ and s^2 are estimates of μ and σ^2, it would be quite surprising if they actually were μ and σ^2 rather than just in their neighborhoods. This suggests that it might be more appropriate to give an interval about $\bar{x}$ or s^2 and state that we are reasonably confident that μ or σ^2 is in the interval.

For a given μ and σ^2, it is possible to define an interval on the $\bar{x}$ axis and state the probability of obtaining a random $\bar{x}$ in the interval; we wish to reverse the process and, for a given $\bar{x}$ and s^2, define an interval and state the probability that μ is in the interval. Since μ will either be or not be in the

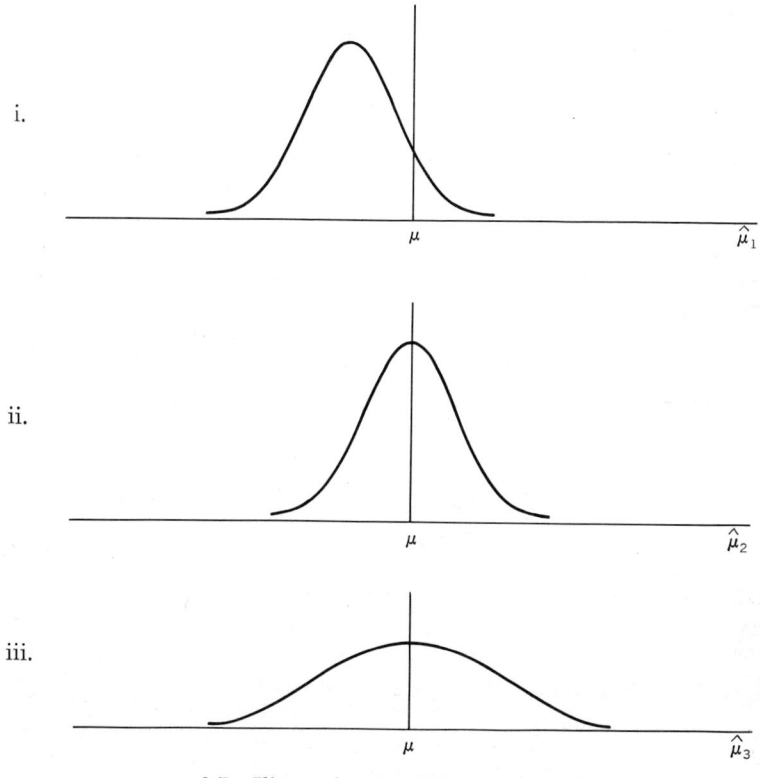

FIG. 3.7 Illustration for choice of estimator

interval, that is, $P = 0$ or 1, the probability will really be a measure of the confidence to be placed in the procedure that led to the statement. This is like throwing a ring at a fixed post; the ring doesn't land in the same position or even catch on the post every time. However, we are able to say that we can circle the post nine times out of ten, or whatever the value should be for the measure of our confidence in our proficiency.

Section 2.13 shows us how to make confidence interval statements. We take a statement such as

$$P\left(-t_{.05} \leq \frac{\bar{x} - \mu}{s_{\bar{x}}} \leq t_{.05}\right) = .95$$

where $\bar{x}$ and $s_{\bar{x}}$ are to be obtained and μ is known, and regard it as a statement

where $\bar{x}$ and $s_{\bar{x}}$ are known and μ is unknown. Some manipulation leads to the statement that
$$P(\bar{x} - t_{.05}s_{\bar{x}} \leq \mu \leq \bar{x} + t_{.05}s_{\bar{x}}) = .95 \qquad (3.12)$$

Equation (3.12) is false if interpreted as though μ were a random variable. It is correct if read as "the probability that the random, to be obtained interval, $\bar{x} \pm t_{.05}s_{\bar{x}}$, will contain the fixed μ, is .95."

An illustration may make this clearer. A sample, number 1, Table 4.3, was drawn from a known population and the statistics $\bar{x}$, s^2, and t calculated. Use of Eq. (3.12) leads to the statement that
$$P(24.77 \leq \mu \leq 48.03) = .95$$

The probability .95 measures the confidence placed in the process that leads to such intervals. The practice of writing P_F instead of P calls attention to the nature of the probability involved.

In a real situation, the experimenter does not know the mean of the population he is sampling; his problem is to estimate the population mean and state the degree of confidence to be associated with the estimate. The confidence interval procedure is a solution to this problem.

Example. In a prolactin assay reported by Finney (3.1), response is measured as the crop-gland weight of pigeons, in 0.1 gram. For a dose of the test preparation of 0.125 mg, four weights obtained are 28, 65, 35, 36. The problem is to estimate the population mean.

Assume that the data are a random sample from a normal population with unknown mean and variance, namely, the population of crop-gland weights for all such possible birds that might be housed and kept for assay purposes under the conditions existing at the laboratory and given a dose of 0.125 mg of the test preparation. Sample statistics are $\bar{x} = 41$ decigrams, $s^2 = 269$, $s_{\bar{x}} = 8.2$ decigrams. Tabulated $t_{.05}$ for 3 degrees of freedom is 3.18. An interval estimate for the population mean is given by $\bar{x} \pm ts_{\bar{x}} = 41 \pm (3.18)(8.2) = (15,67)$ decigrams; we assign .95 as our degree of confidence in the statement that μ lies in the interval (15,67) decigrams. Alternatively, the population mean μ is in the interval (15,67) decigrams unless the particular random sample is an unusual one, that is, in a group from which we will draw a random sample only about five times in a hundred, on the average. Consequently, if μ is not in (15,67) decigrams, it is because our sample is unusual.

In our example, we have an *error rate* of 5%. Obviously, the choice of error rate will depend upon the seriousness of a wrong decision.

If there were a priori reasons to expect the population mean to be 20 decigrams, this sample would not constitute evidence, at the 5% *level of significance*, to throw doubt upon such an expectation. If there were a priori reasons to expect a population mean of 80, this sample would be evidence tending to deny such an expectation at the 5% *level of significance*.

When σ^2 is known, it is possible to choose a sample size such that the confidence interval, when obtained, will be of a predetermined length for the chosen error rate. When σ^2 is unknown, it is possible to determine a sample size such that the experimenter can be reasonably confident (the degree of confidence can be set) that the confidence interval to be calculated for a chosen error rate will not be longer than a predetermined value. In addition, a method of sampling, called sequential, can be used to obtain a confidence interval of fixed length.

Exercise 3.10.1 Finney (3.1) also reported crop-gland weights of pigeons, in 0.1 gram, for doses of the test preparation of 0.250 and 0.500 mg. The resulting weights were 48, 47, 54, 74 and 60, 130, 83, 60, respectively. Define populations for these two sets of data and estimate the population means using 95 and 99% confidence intervals. Notice the difference in lengths for each population.

3.11 Prediction of sample results.

An incorrect statement that one may hear is that a probability statement concerning the population mean tells us something about the distribution of future sample means, for example, that 95% of future sample means will be in a stated interval. Most of these statements are misleading and incorrect. However, it is possible to make a statement about a future sample observation that can be very useful.

Consider a problem in hydrologic forecasting. The use and development of water resources have created a serious demand for advance estimates of the flow or runoff supplied by streams in a watershed. In other words, the prediction of a future event or observation is desired. The population of all possible rates of flow supplied by a particular stream in the watershed is obviously unknown but runoff values for a number of preceding years are a sample from the population. Obviously this is not a random sample but we can begin by assuming that it is. (The experience of checking predictions with results is one method of judging whether the measure of reliability placed on our statements is justified. An alternative method is to use all but the most recent observation in making the prediction, then check the prediction against the unused observation. If this can be done for a sufficient number of similar watersheds, it is possible to pass judgment upon the reliability of the stated error rate before being publicly committed to a probability statement.)

If the population mean is known, we could use this as the predicted value but could do no better; it would be quite pointless to guess what the random value itself was going to be. However, an interval estimate or prediction of the next random value can be prepared with the use of normal or t tables. The use of an interval is an attempt to allow for the random ϵ that is called for in the linear additive model, Eq. (2.13), and is certain to be present.

If μ and σ^2 were known, then the interval and confidence statement could be

$$P(\mu - z_{.01}\sigma \leq X \leq \mu + z_{.01}\sigma) = .99$$

since 99% of random observations lie within $z_{.01}$ standard deviations of the mean (see Sec. 3.6).

In the usual problem, μ and σ^2 are unknown but estimates $\bar{x}$ and s^2 are available. Prediction of the next X, for example, next year's runoff, is necessarily $\bar{x}$. The variance appropriate to this predicted value of X is the sum of the variances, namely, $(s^2/n) + s^2 = [(n + 1)/n]s^2$. Thus the appropriate confidence statement is

$$P\left[\bar{x} - t_{.01}\sqrt{s^2\left(\frac{n+1}{n}\right)} \leq X \leq \bar{x} + t_{.01}\sqrt{s^2\left(\frac{n+1}{n}\right)}\right] = .99 \quad (3.13)$$

where t is the tabulated value of Student's t for the 1% probability level.

Exercise 3.11.1 In Sec. 3.3, the term *function of X* is used. Write out six functions of X which are used in this chapter and which appear to be important in statistics.

Exercise 3.11.2 Distinguish clearly between estimation of a population mean and prediction of a value to be observed. Try to specify situations for which each procedure is more likely to be wanted.

References

3.1 Finney, D. J.: *Statistical Method in Biological Assay*, Table 12.1, Hafner Publishing Company, New York, 1952.

A proposed laboratory exercise in connection with Chap. 3

Purpose. (1) To give the student practice in computing some familiar statistics and in use of associated tables. (2) To build up more empirical evidence about sample statistics for comparison with theoretical results. (See laboratory exercises for Chaps. 2 and 4.)

For each of your 10 random samples (see laboratory exercises for Chap. 2), compute

i. $z = \dfrac{\bar{x} - \mu}{\sigma/\sqrt{n}}$, using $\mu = 40$, $\sigma = 12$

ii. $t = \dfrac{\bar{x} - \mu}{s/\sqrt{n}}$

iii. The 95% confidence interval for μ

iv. The 99% confidence interval for μ

v. $\chi^2 = [(n-1)s^2]/\sigma^2$

Chapter 4

SAMPLING FROM A NORMAL DISTRIBUTION

4.1 Introduction. Calculation of the common statistics $\bar{x}$, s^2, and s as measures of central tendency and dispersion has been discussed. When calculated from random samples, $\bar{x}$ and s^2 are unbiased estimates (Sec. 3.10) of the parent population parameters μ and σ^2 while s is a biased estimate of σ.

The standard deviation of means can be estimated from a sample of observations by the formula $s_{\bar{x}} = s/\sqrt{n}$, where s is the sample standard deviation. The use of $\bar{x}$, $s_{\bar{x}}$, and a tabulated value of Student's t to establish a confidence interval for the population mean was given in Sec. 3.10. The average number as a decimal fraction, of intervals that contain μ, is called the *confidence probability* or *confidence coefficient*.

The results used thus far are based on mathematical theorems and principles. Such results can be demonstrated with a reasonable degree of accuracy by large-scale sampling procedures. This is an *empirical method*. In this chapter, sampling is used as a method of examining the distribution of a number of statistics and the confidence interval procedure. The testing of hypotheses is introduced.

4.2 A normally distributed population. A normal population has a continuous variable with an infinite range; consequently an observation may assume any real value, positive or negative. Table 4.1 consists of yields in pounds of butterfat for 100 Holstein cows, the original data having been modified somewhat to form an approximately normal distribution. The resulting data depart from normality in two major respects: the variable has a finite range and is discrete or discontinuous. The effects due to the finite range and discreteness of the data are small in comparison with sampling variation and accordingly will have little effect upon inferences based on samples.

The salient characteristics of the distribution are depicted in Figs. 4.1 and 4.2. Figure 4.1 is a histogram of the data with pounds on the horizontal axis and frequency on the vertical. The values concentrate at the center and thin out symmetrically on both sides, quickly at first and then less rapidly. Figure 4.2 shows the 100 values cumulatively. For example, to find the number of observations (position in array) less than a certain weight in pounds, draw a line vertically from that weight to the smooth curve implied by the dots representing the observations, then horizontally to the ordinate axis where the

TABLE 4.1

ARRAY OF POUNDS OF BUTTERFAT PRODUCED DURING A MONTH BY 100 HOLSTEIN COWS

The original data were modified to approximate a normal distribution with $\mu = 40$ lb and $\sigma = 12$ lb

Item	Pounds	Item	Pounds	Item	Pounds	Item	Pounds
00	10	25	33	50	40	75	47
01	12	26	33	51	40	76	48
02	14	27	34	52	41	77	48
03	15	28	34	53	41	78	48
04	17	29	34	54	41	79	49
05	18	30	35	55	41	80	49
06	20	31	35	56	42	81	49
07	22	32	35	57	42	82	50
08	23	33	36	58	42	83	50
09	25	34	36	59	42	84	51
10	26	35	36	60	43	85	51
11	27	36	37	61	43	86	52
12	28	37	37	62	43	87	52
13	28	38	37	63	43	88	53
14	29	39	37	64	44	89	54
15	29	40	38	65	44	90	55
16	30	41	38	66	44	91	57
17	30	42	38	67	45	92	58
18	31	43	38	68	45	93	60
19	31	44	39	69	45	94	62
20	31	45	39	70	46	95	63
21	32	46	39	71	46	96	65
22	32	47	39	72	46	97	66
23	32	48	40	73	47	98	68
24	33	49	40	74	47	99	70
						Total	4,000

number is read. The relation of the histogram to the array is that the height of the rectangle in any class of the histogram is proportional to the number of dots lying between the corresponding pair of vertical lines of the array.

Table 4.2 is a frequency distribution of pounds of butterfat for the data of Table 4.1. Each class has a range of 5 lb.

TABLE 4.2

A FREQUENCY DISTRIBUTION OF POUNDS OF BUTTERFAT OF 100 HOLSTEIN COWS

Midpoint or, class mark:	10	15	20	25	30	35	40	45	50	55	60	65	70
Frequency:	2	3	3	4	12	16	20	16	12	4	3	3	2

Exercise 4.2.1 For a normal distribution, is the variable discrete? Continuous? Qualitative? Quantitative? Finite in range? Is there a minimum value? A maximum?

SAMPLING FROM A NORMAL DISTRIBUTION

Exercise 4.2.2 For random samples from normal distributions, do $\bar{x}$ from Eq. (2.1), s^2 from Eq. (2.6), and $s = \sqrt{s^2}$ give biased estimates of μ, σ^2, and σ?

Exercise 4.2.3 What is the distribution of random $\bar{x}$'s where X_i is from a normal distribution? How are the mean and variance of a population of $\bar{x}$'s related to the mean and variance of the parent population?

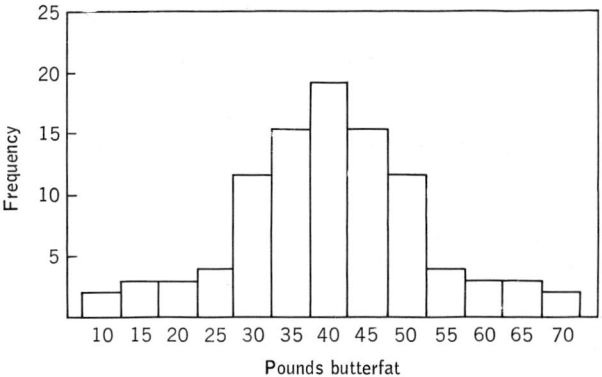

FIG. 4.1 Histogram of the distribution of pounds of butterfat from 100 Holstein cows

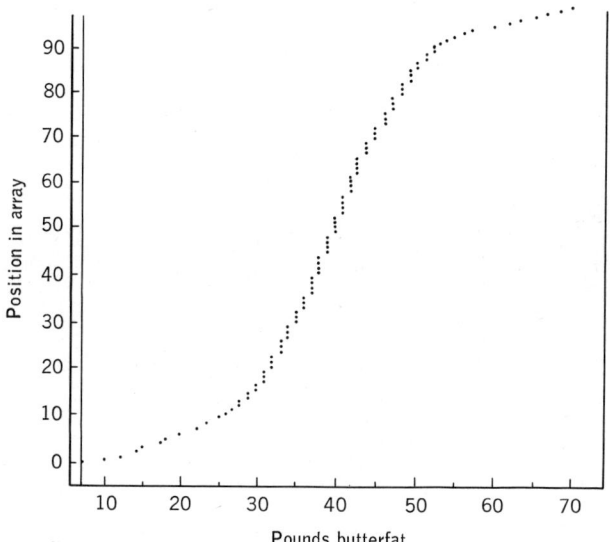

FIG. 4.2 Graphical representation of the array of pounds of butterfat of 100 Holstein cows

4.3 Random samples from a normal distribution.
The drawing of random samples cannot be left to subjective discretion but must be the result of objective and preferably mechanical methods. A table of random numbers

TABLE 4.3
Five Random Samples of 10 Observations from Table 4.1, Together with Sample Statistics

Observation number and formula		Sample number									
		1		2		3		4		5	
		Item	Yield	Item	Yield	Item	Yield	Item	Yield	Item	Yield
1		96	65	39	37	95	63	39	37	94	62
2		37	37	51	40	59	42	63	43	42	38
3		04	17	34	36	54	41	84	51	16	30
4		29	34	81	49	47	39	16	30	08	23
5		84	51	34	36	41	38	17	30	59	42
6		05	18	49	40	81	49	67	45	97	66
7		71	46	47	39	09	25	98	68	98	68
8		35	36	75	47	03	15	97	66	45	39
9		03	15	23	32	15	29	59	42	89	54
10		69	45	96	65	87	52	00	10	50	40
Sum	$= \Sigma X$		364		421		393		422		462
Mean	$= \bar{x}$		36.4		42.1		39.3		42.2		46.2
CT	$= (\Sigma X)^2/10$		13,249.60		18,541.00		17,175.00		20,488.00		23,498.00
SS	$= \Sigma X^2 - (\Sigma X)^2/10$		2,376.40		17,724.10		15,444.90		17,808.40		21,344.40
s^2	$= SS/9$		264.04		816.90		1,730.10		2,679.60		2,153.60
			15,626.00								
s	$= \sqrt{s^2}$		16.2		90.77		192.23		297.73		239.29
$s_{\bar{x}}$	$= \sqrt{s^2/10}$		5.14		9.5		13.9		17.3		15.5
t	$= (\bar{x} - 40)/s_{\bar{x}}$		-0.70		3.01		4.38		5.46		4.89
$t_{.05}s_{\bar{x}}$	$= 2.262 s_{\bar{x}}$		11.6		$+0.70$		-0.16		$+0.40$		$+1.27$
CL (lower)	$= \bar{x} - t_{.05}s_{\bar{x}}$		24.8		6.8		9.9		12.4		11.1
CL (upper)	$= \bar{x} + t_{.05}s_{\bar{x}}$		48.0		35.3		29.4		29.8		35.1
					48.9		49.2		54.6		57.3

For the parent population, $\mu = 40$, $\sigma^2 = 144$.

SAMPLING FROM A NORMAL DISTRIBUTION

such as Table A.1 serves to introduce objectivity. To facilitate drawing random samples by use of a table of random numbers, assign consecutive numbers to the individuals in a population. For example, the 100 yields in Table 4.1 have been assigned the numbers 00 to 99, referred to as items.

Use of a random number table was illustrated in Sec. 2.5. For successive samples, the last two pairs of integers may be used to locate the next starting row and column. Applied to Table 4.1, this sampling procedure assures that each item or yield may be drawn any number of times. Sampling is always from the same population and the probability of drawing any particular item is the same for all items. The procedure is essentially the same as if drawing were from an infinite population.

Table 4.3 gives five random samples, together with certain pertinent calculations, as obtained by the above procedure. These five samples are from 500 random samples of 10 observations from Table 4.1 used for discussion in the remainder of this chapter.

Exercise 4.3.1 The two random sampling procedures given in this section are not precisely equivalent. Why? (*Hint.* See Sec. 2.5 for distribution of digits in Table A.1.)

4.4 The distribution of sample means. Five hundred samples of 10 observations were drawn from Table 4.1. The frequency distribution of the 500 means is given in Table 4.4 for a class interval of 1.5 lb and illustrates several basic features of sampling. First, the distribution of the means is approximately normal. Theory states that the derived distribution of sample means of random observations from a normal population is also normal. Theory also states that even if the parent distribution is considerably anormal, the distribution of means of random samples approaches the normal distribution as the sample size increases. This is very important in practice, for the form of the parent distribution is rarely known. Secondly, the average of the 500 means, 39.79 lb, is very close to $\mu = 40$ lb, the parent population mean. This illustrates unbiasedness. The sample mean is said to be unbiased because the mean of all possible sample means is the parent population mean. Thirdly, the variation of the means is much less than that of the individuals, the sample ranges being 27 lb for means and 60 lb for individuals. Theory states that $\sigma_{\bar{x}}^2 = \sigma^2/n$; here $\sigma_{\bar{x}}^2 = 14.4$ and $\sigma_{\bar{x}} = 3.79$ lb. The corresponding sample relation is $s_{\bar{x}}^2 = s^2/n$. Applying this to the average of the 500 sample variances (see Table 4.5), we obtain $s_{\bar{x}}^2 = 140.4/10 = 14.04$ and $s_{\bar{x}} = 3.75$ lb. Computation with the 500 means gives $s_{\bar{x}} = \sqrt{[\Sigma \bar{x}^2 - (\Sigma \bar{x})^2/500]/499} = 3.71$ lb. Observed and theoretical frequencies of means are compared in Table 4.4. Theoretical frequencies are for a normal distribution with $\mu = 40$ lb and $\sigma = 3.79$ lb.

Exercise 4.4.1 Given a parent normal distribution and a derived distribution of means of 100 observations. What is the relation between the means of the two populations? Between the variances? Are the ranges of the populations the same? If you had a sample for each population, one of 50 observations, the other 50 means, how would you expect the two ranges to compare?

Exercise 4.4.2 To obtain an estimate of $\sigma_{\bar{x}}^2$, one procedure above was to average

TABLE 4.4

FREQUENCY DISTRIBUTION OF 500 MEANS OF RANDOM SAMPLES OF 10 ITEMS FROM TABLE 4.1

Class mark,† lb	Observed frequency	Theoretical frequency	Observed cumulative frequency	Theoretical cumulative frequency
26.5	1	0	1	0.2
28.0	0	0.5	1	0.8
29.5	2	2	3	2.6
31.0	2	5	5	7.5
32.5	14	11	19	18.8
34.0	20	23	39	41.9
35.5	47	39	86	80.6
37.0	65	58	151	138.8
38.5	74	72	225	210.4
40.0	71	79	296	289.6
41.5	78	72	374	361.2
43.0	49	58	423	419.4
44.5	40	39	463	458.1
46.0	24	23	487	481.2
47.5	8	11	495	492.5
49.0	4	5	499	497.4
50.5	0	2	499	499.2
52.0	0	0.5	499	499.8
53.5	1	0	500	500.0
Totals	500	500		

Mean of means: $\bar{\bar{x}} = 39.79$

† The center of a class interval.

the 500 sample variances and divide by 10. Compare this procedure with that of dividing each s^2 by 10 and averaging the 500 results.

4.5 The distribution of sample variances and standard deviations. For each of 500 random samples, the variance and standard deviation were calculated. The procedure and the results are illustrated for five samples in Table 4.3.

The distribution of the 500 sample variances is given in Table 4.5. There

TABLE 4.5

FREQUENCY DISTRIBUTION OF 500 VARIANCES s^2 FOR RANDOM SAMPLES OF SIZE 10 FROM TABLE 4.1

Class mark:	20	40	60	80	100	120	140	160	180	200	220	240	260	280	300	320	340	360	380
Frequency:	11	27	40	46	59	62	55	51	43	27	21	16	18	7	5	7	3	1	1

140.4

$\overline{s^2} = 140.4$ lb², using df, namely 9, as divisor
 (= 126.4 lb², using sample size, namely 10, as divisor)
$\sigma^2 = 144$ lb²

SAMPLING FROM A NORMAL DISTRIBUTION 55

is a heaping of the variances to the left of their mean, denoted by $\bar{s^2}$, and an attenuation to the right. The distribution is skewed. Sample quantities $(n-1)s^2/\sigma^2 = 9s^2/144$ are distributed as χ^2 with $n-1 = 9$ degrees of freedom. The mean variance is $\bar{s^2} = 140.4$ lb², closely approximating the population variance $\sigma^2 = 144$. This illustrates the unbiasedness of s^2 as an estimate of σ^2. The individual s^2's range from 20 to 380 lb².

Table 4.6 gives the distribution of the standard deviations. Notice that the

TABLE 4.6

FREQUENCY DISTRIBUTION OF 500 STANDARD DEVIATIONS s CORRESPONDING TO THE VARIANCES OF TABLE 4.5

Class mark:	4	5	6	7	8	9	10	11	12	13	14	15	16	17	18	19
Frequency:	1	10	14	23	37	42	55	69	66	63	40	26	30	11	11	2

11.47 ↓ (between 11 and 12)

$\bar{s} = 11.47$ lb
$\sqrt{\bar{s^2}} = 11.85$ lb
$\sigma = 12$ lb

process of taking square roots eliminates much of the skewness exhibited by variances. Examination of equally spaced s's and corresponding s^2's shows that variances above the mean increase faster than those below, differences between successive s^2's being consecutive odd numbers.

s	s^2
10	100
11	121
12	144
13	169
14	196

(This is an interesting observation in that it suggests the importance of choosing a scale of measurement for any investigation, the distribution of the observations being highly dependent on the scale. If the distribution is normal or can be made so by a transformation, that is, by the choice of a scale of measurement, statistical techniques based on the normal distribution are applicable; otherwise, they are only approximations.)

The average of the 500 standard deviations, denoted by $\bar{s}$, is 11.47 lb as compared to $\sigma = 12$ lb. The square root of the average of the 500 variances, that is $\sqrt{\bar{s^2}}$, is here $\sqrt{140.4} = 11.85$ lb. It is not surprising that $\bar{s}$ is less than $\sqrt{\bar{s^2}}$ since s underestimates σ. Correction factors to eliminate the bias in s are available; for example, see Dixon and Massey (4.3).

Exercise 4.5.1 Draw histograms for the data of Tables 4.5 and 4.6 and observe the resulting skewness in each.

4.6 The unbiasedness of s^2. It has been stated that $s^2 = \Sigma(X_i - \bar{x})^2/(n-1)$ is an unbiased estimate of σ^2. The average of the 500 s^2's, $\overline{s^2}$, is 140.4 lb².

If the sample variance is defined as $\Sigma(X_i - \bar{x})^2/n$, we have a biased estimate of σ^2, the average of the population of such values being $(n-1)\sigma^2/n$. We could reconstruct each sum of squares by using $(n-1)s^2 = \Sigma(X_i - \bar{x})^2$. However, since the degrees of freedom are the same for all samples, the average of the 500 variances computed with $n = 10$ as divisor is $9\overline{s^2}/10 = 126.4$ lb², a much smaller value than 140.4. The difference between the values obtained using n and $n-1$ here obviously becomes less as n increases.

4.7 The standard deviation of the mean or the standard error. The standard deviation of the mean is one of the most useful statistics. It is calculated as $s_{\bar{x}} = s/\sqrt{n}$ or $s_{\bar{x}} = \sqrt{s^2/n}$ and is a biased estimate of $\sigma_{\bar{x}}$, the standard deviation of means of random samples of size n from a parent population with standard deviation σ. Thus for a sample of size 10 from Table 4.1, $s_{\bar{x}}$ is an estimate of $\sigma_{\bar{x}} = \sigma/\sqrt{10} = 12/\sqrt{10} = 3.79$ lb. For an estimate of $\sigma_{\bar{x}}$ from the 500 samples, extract the square root of the average of the variances divided by $n = 10$. Calling it $(s_{\bar{x}})'$, we have

$$(s_{\bar{x}})' = \sqrt{\overline{s^2}/n} = \sqrt{140.4/10} = 3.75 \text{ lb}$$

This is a better procedure for estimating $\sigma_{\bar{x}}$ from a set of s^2's than that of dividing the average of the 500 standard deviations, an average of biased estimates, by the square root of 10. If the latter is called $(s_{\bar{x}})''$, then we have

$$(s_{\bar{x}})'' = \bar{s}/\sqrt{n} = 11.47/\sqrt{10} = 3.63 \text{ lb}$$

To further justify obtaining $s_{\bar{x}}$ from s, we will use the 500 means to estimate $\sigma_{\bar{x}}$ for comparison. We find

$$s_{\bar{x}} = \sqrt{\frac{\Sigma \bar{x}^2 - (\Sigma \bar{x})^2/500}{499}} = 3.71 \text{ lb}$$

The close agreement between this and $(s_{\bar{x}})' = 3.75$ lb enables us to state with more confidence that the relationship $\sigma_{\bar{x}} = \sigma/\sqrt{n}$ is indeed valid and, accordingly, that each random sample provides an estimate $s_{\bar{x}}$ of the standard error of the mean $\sigma_{\bar{x}}$.

It is important to realize that the variance, population or sample, of a mean decreases inversely as n while the standard deviation of a mean decreases inversely as $\sqrt{n}$. This is clearly shown by example as well as by formula.

n	$\sigma_{\bar{x}}^2$		$\sigma_{\bar{x}}$	
4	$\dfrac{\sigma^2}{4} = \dfrac{144}{4}$	$= 36$	$\dfrac{\sigma}{\sqrt{4}} = \dfrac{12}{\sqrt{4}}$	$= 6$
8	$\dfrac{\sigma^2}{8} = \dfrac{144}{8}$	$= 18$	$\dfrac{\sigma}{\sqrt{8}} = \dfrac{12}{\sqrt{8}}$	$= 4.24$
16	$\dfrac{\sigma^2}{16} = \dfrac{144}{16}$	$= 9$	$\dfrac{\sigma}{\sqrt{16}} = \dfrac{12}{\sqrt{16}}$	$= 3$

SAMPLING FROM A NORMAL DISTRIBUTION

4.8 The distribution of Student's t. Student's t distribution and Student's t were discussed in Sec. 3.9. We are now ready to show that the distribution of our 500 sample t values approximates the theoretical distribution of t for 9 degrees of freedom.

For each of the 500 samples, $t = (\bar{x} - \mu)/s_{\bar{x}} = (\bar{x} - 40)/s_{\bar{x}}$ was calculated. It is seen that t is the deviation of the sample mean from the population mean in units of sample standard deviations of means, a unit of measurement commonly used for making decisions about the usualness or unusualness of a deviation. Since a population of sample means is distributed symmetrically about μ, approximately one-half of the 500 t values should be positive and one-half negative; the mean should be approximately zero. We find 248 are positive, 252 are negative, and the mean is -0.038.

Table 4.7 is a frequency distribution of the observed t values. Unequal class

TABLE 4.7

SAMPLE AND THEORETICAL VALUES OF t FOR 9 DEGREES OF FREEDOM

$$t = \frac{\bar{x} - \mu}{s_{\bar{x}}}$$

Interval of t		Sample				Theoretical		
				Cumulative			Cumulative	
From	To	Frequency	Percentage frequency	One tail†	Both tails‡	Percentage frequency	One tail†	Both tails‡
—	-3.250	2	0.4	100.0		0.5	100.0	
-3.250	-2.821	2	0.4	99.6		0.5	99.5	
-2.821	-2.262	7	1.4	99.2		1.5	99.0	
-2.262	-1.833	12	2.4	97.8		2.5	97.5	
-1.833	-1.383	29	5.8	95.4		5.0	95.0	
-1.383	-1.100	21	4.2	89.6		5.0	90.0	
-1.100	-0.703	63	12.6	85.4		10.0	85.0	
-0.703	0.0	116	23.2	72.8		25.0	75.0	
0.0	0.703	133	26.6	49.6	100.0	25.0	50.0	100.0
0.703	1.100	38	7.6	23.0	50.2	10.0	25.0	50.0
1.100	1.383	30	6.0	15.4	30.0	5.0	15.0	30.0
1.383	1.833	23	4.6	9.4	19.8	5.0	10.0	20.0
1.833	2.262	15	3.0	4.8	9.4	2.5	5.0	10.0
2.262	2.821	6	1.2	1.8	4.0	1.5	2.5	5.0
2.821	3.250	1	0.2	0.6	1.4	0.5	1.0	2.0
3.250	—	2	0.4	0.4	0.8	0.5	0.5	1.0
		500	100.0					

† Percentage of values larger than the entry in the extreme left column.
‡ Percentage of values larger in absolute value than the entry in the extreme left column.

intervals were selected so that the observed frequencies could be compared with the theoretical frequencies tabulated in Table A.3. Thus, the class boundaries are identical with those for tabulated t at the 0.5, 0.3, 0.2, 0.1, 0.05, 0.02, and 0.01 probability levels. Percentage frequencies for sample and theoretical values of t are given to facilitate comparison.

In a population of t values, 2.5% are larger than $+2.262$ and 2.5% are smaller (numerically larger) than -2.262. This is seen from the theoretical percentage frequency. The last column in Table 4.7 combines both tails of the distribution by ignoring the sign of t. This is the column most often referred to for probability levels. Thus 2.262 is referred to as the value of t at the 5% level for 9 degrees of freedom. When only the positive tail of the t distribution is considered, 5% of the t's lie beyond 1.833. Again when both tails are considered, 1% of the t values lie beyond ± 3.250, the t value at the 1% level for 9 degrees of freedom. For the sample values, 20 t's numerically exceed the 5% level and 4 t's numerically exceed the 1% level as compared with an expected 25 and 5, respectively. This shows reasonable agreement between the sample and theoretical values. A comparison of sample and theoretical values at other probability levels also shows reasonable agreement.

Exercise 4.8.1 When is a statistic said to be an unbiased estimator of a parameter? Classify the statistics $\bar{x}$, s^2, s, $s_{\bar{x}}^2$, $s_{\bar{x}}$ as biased or unbiased.

Exercise 4.8.2 Given a single sample of 20 random observations from a normal distribution, how can one estimate the variance of the population of sample means of 20 observations? Of 40 observations?

Exercise 4.8.3 Is the relation $\sigma_{\bar{x}}^2 = \sigma^2/n$ valid if a population is not normal?

Exercise 4.8.4 How does Student's t differ from the corresponding normal z criterion? Is there more than one t distribution? If one had two random samples of the same size from the same normal distribution and computed $(\bar{x}_1 - \mu)/\sqrt{s_2^2/n}$, would it be distributed as t? If the samples were of different size, would the criterion be distributed as t? If the samples were from different populations, would the criterion be distributed as t? (*Hint.* See Sec. 3.9.)

4.9 The confidence statement. We now check the confidence statements based on the samples to see if the stated confidence is justified. For each random sample and any level of probability, a confidence interval is established about the sample mean. The procedure is to solve the two equations $\pm t = (\bar{x} - \mu)/s_{\bar{x}}$ for μ to obtain $\mu = \bar{x} \pm ts_{\bar{x}}$, then a tabulated value of t and the sample $\bar{x}$ and $s_{\bar{x}}$ are substituted to give two values of μ, denoted by l_1 and l_2 and called the *limits* of the confidence interval. Thus $l_1 = \bar{x} - ts_{\bar{x}}$ and $l_2 = \bar{x} + ts_{\bar{x}}$.

For each of the 500 random samples, $l_1 = \bar{x} - 2.262s_{\bar{x}}$ and $l_2 = \bar{x} + 2.262s_{\bar{x}}$ have been calculated, that is, end points for a 95% confidence interval. The 1% t value was also used. Since $\mu = 40$ lb, the number of correct statements regarding μ can be determined. For the 500 samples, the numbers of intervals containing μ were 480 at the 5% level and 496 at the 1% level. These compare favorably with theoretical values of 475 and 495, respectively. The percentage of intervals not including μ is the same as the percentage of sample t values which exceed the 5% and 1% tabular t values.

In actual practice the parameter μ is not known. Accordingly, an experimenter never knows whether μ lies in the confidence interval; he knows only the percentage of correct inferences about μ.

An erroneous idea sometimes held is that a 95% confidence limit about a sample mean gives the range within which 95% of future sample means will fall. This is incorrect since the distribution of sample means is centered on the population mean and not upon a particular sample mean. On the basis of a present sample, a correct statement about a future observation or mean was discussed in Sec. 3.11.

4.10 The sampling of differences. A problem which often confronts an experimenter is that of determining whether there is a real difference in the responses to two treatments or, alternatively, whether the observed difference is small enough to be attributed to chance. An empirical method of approach

TABLE 4.8

THREE SAMPLES OF DIFFERENCES BETWEEN RANDOM OBSERVATIONS FROM TABLE 4.1

Item numbers	Paired observations $X_{1j}\ X_{2j}$	Differences $D_j = X_{1j} - X_{2j}$	Item numbers	Paired observations $X_{3j}\ X_{4j}$	Differences $D_j = X_{3j} - X_{4j}$	Item numbers	Paired observations $X_{5j}\ X_{6j}$	Differences $D_j = X_{5j} - X_{6j}$
97 78	66 48	18	66 72	44 46	-2	21 14	32 29	3
74 69	47 45	2	62 28	43 34	9	63 28	43 34	9
58 81	42 49	-7	15 64	29 44	-15	98 42	68 38	30
48 83	40 50	-10	28 37	34 37	-3	86 05	52 18	34
44 43	39 38	1	00 05	10 18	-8	77 94	48 62	-14
73 15	47 29	18	73 07	47 22	25	79 93	49 60	-11
73 81	47 49	-2	56 57	42 42	0	51 29	40 34	6
93 91	60 57	3	04 25	17 33	-16	99 66	70 44	26
79 46	49 39	10	92 53	58 41	17	39 06	37 20	17
63 21	43 32	11	34 94	36 62	-26	17 62	30 43	-13

Sum = ΣD		44			-19			87
Mean = $\bar{d}$		4.4			-1.9			8.7
ΣD^2		1,036.00			2,229.00			3,633.00
CT = $(\Sigma D)^2 / 10$		193.60			36.10			756.90
SS = $\Sigma D^2 - (\Sigma D)^2/10$		842.40			2,192.90			2,876.10
$s_D^2 = $ SS/9		93.6			243.66			319.57
$s_D = \sqrt{s_D^2}$		9.8			15.6			17.9
$s_{\bar{d}} = \sqrt{s_D^2/10}$		3.06			4.94			5.65
$t = (\bar{d} - 0)/s_{\bar{d}}$		1.44			-0.38			1.54
$t_{.05}s_{\bar{d}} = 2.262 s_{\bar{d}}$		6.92			11.15			12.78
CL = $\begin{cases} l_1 = \bar{d} - t_{.05}s_{\bar{d}} \\ l_2 = \bar{d} + t_{.05}s_{\bar{d}} \end{cases}$		-2.5			-13.1			-4.1
		11.3			9.3			21.5

to this problem is to consider, in the manner of this chapter, the results of a sampling procedure with two dummy treatments, that is, to sample a single population but to consider the resulting data as though from two populations. In this way, we learn what constitutes an ordinary sampling difference and an unusual difference, when no population difference exists.

Signed differences of observations randomly drawn from a normal population will be normally distributed about a mean of zero. The 500 random samples from Table 4.1 were paired at random and signed differences obtained. For each of the resulting 250 samples of 10 differences D, the following statistics were calculated: the mean difference $\bar{d}$, the variance of the differences s_D^2, the standard deviation of the differences s_D, the standard deviation of the mean difference $s_{\bar{d}}$, the t value, and confidence limits for the population mean difference, known to be $\mu_{\bar{d}} = 0$ in this case. This is illustrated in Table 4.8, similar to Table 4.3 for individual values. Table 4.9 is a

TABLE 4.9

FREQUENCY DISTRIBUTION OF 250 MEAN DIFFERENCES $\bar{d}$ FOR SAMPLES OF 10 DIFFERENCES

Class mark	Frequency
−12	4
−10.5	7
−9	7
−7.5	8
−6	12
−4.5	16
−3	30
−1.5	29
0	33
1.5	21
3	28
4.5	17
6	13
7.5	10
9	8
10.5	4
12	2
13.5	0
15	1
	250

frequency distribution of the resulting 250 mean differences $\bar{d}$. The observed distribution is approximately symmetrical with 118 of the mean differences greater than zero and 132 less. These numbers were obtained from Table 4.9, along with the additional information that the class with zero class mark has 14 $\bar{d}$'s positive and 19 negative. The mean of the 250 is −0.533, very close to zero.

The notation X_i, for the ith observation in a sample, is inadequate to distinguish among observations from several samples. Thus, we introduce a second subscript and denote an observation by X_{ij}. Then X_{ij} refers to the jth observation in the ith sample. For example X_{25} (read as X sub 2, 5; commas are used between numerical subscripts only where required for clarity) is the fifth observation of the second sample; and $X_{13} - X_{23}$ is the signed difference resulting from subtracting the third observation in sample 2 from the third in sample 1.

Tables 4.10 and 4.11 are frequency distributions of the 250 sample variances

TABLE 4.10

FREQUENCY DISTRIBUTION OF THE VARIANCES $s_D{}^2$ OF 250 RANDOM SAMPLES OF 10 DIFFERENCES BASED ON TABLE 4.1

Class mark:	60	100	140	180	220	260	300	340	380	420	460	500	540	580	620	660	700	740	
Frequency:	8	14	24	37	40	34	19	16	12	15	13	7	1	4	2	3	0	1	250

$\overline{s_D{}^2} = 272.7 \qquad \overline{2s^2} = \overline{2s^2} = 280.8 \qquad 2\sigma^2 = 288$

TABLE 4.11

FREQUENCY DISTRIBUTION OF THE STANDARD DEVIATIONS s_D OF 250 RANDOM SAMPLES OF 10 DIFFERENCES BASED ON TABLE 4.1

Class mark:	7	8	9	10	11	12	13	14	15	16	17	18	19	20	21	22	23	24	25	26	27	
Frequency:	1	5	4	7	8	17	24	28	29	26	19	13	19	10	16	10	4	4	3	2	1	250

$\overline{s_D} = 16.04 \text{ lb} \qquad \sqrt{\overline{2s^2}} = 16.76 \text{ lb} \qquad \sqrt{\overline{s_D{}^2}} = 16.51 \text{ lb} \qquad \sqrt{2\sigma^2} = 16.97 \text{ lb}$

and standard deviations of 10 differences. The forms of these distributions are similar to those of Tables 4.5 and 4.6 for s^2 and s. Compare appropriate tables and note that the ranges are considerably greater for differences than for individuals. The reason is apparent when the possible range of the differences is considered. The possible range is from $(10 - 70) = -60$ lb to $(70 - 10) = +60$ lb, twice that for individuals. The average of the 250 variances, $\overline{s_D{}^2}$, is 272.7; from Table 4.5, $\overline{2s^2} = 2(140.4) = 280.8$; both are reasonably close to $2\sigma^2 = 2(144) = 288$. The data illustrate an important theorem:

> The variance $\sigma_D{}^2$ of differences of randomly paired observations is twice that of the observation in the parent population.

It then follows that

> The sample variance $s_D{}^2$ of differences of randomly paired observations is an unbiased estimate of $2\sigma^2$.

In practice, where the variance of a difference between two means is often required, random pairing is not done and differences are not used in calculating $\bar{d}$ or $s_D{}^2$. By rearranging the arithmetic, it is apparent that $\bar{d} = \bar{x}_1 - \bar{x}_2$. From s^2, the variance $2\sigma^2$ is estimated by $2s^2 = s_D{}^2$.

Averages of standard deviations of differences are $\overline{s_D} = 16.04$ and $\sqrt{\overline{s_D{}^2}} = \sqrt{272.7} = 16.51$ lb. Again the direct average of standard deviations is less than the square root of the average of the variances but both are reasonably close to $\sigma_D = \sqrt{2\sigma^2} = \sqrt{288} = 16.97$ lb. Standard deviations have a slight bias; variances are unbiased.

It has been stated that $\sigma_{\bar{x}}^2 = \sigma^2/n$ and that $\sigma_D^2 = 2\sigma^2$. Together, these theorems say that the variance of a difference between two means, denoted by $\sigma_{\bar{d}}^2$, is equal to $2\sigma^2/n$ when each mean contains n observations. Thus $\sigma_{\bar{d}} = \sqrt{288/10} = 5.37$ lb. For the 250 samples, $s_{\bar{d}} = \sqrt{s_D^2/n} = \sqrt{272.7/10} = 5.22$ lb or $s_{\bar{d}} = \sqrt{2s^2/n} = \sqrt{280.8/10} = 5.30$ lb.

From a single value of s^2, estimates of the following important parameters, σ^2, $\sigma_{\bar{x}}^2$, σ_D^2, $\sigma_{\bar{d}}^2$, σ, $\sigma_{\bar{x}}$, σ_D, and $\sigma_{\bar{d}}$, are obtainable. The interrelationships in terms of statistics are shown diagrammatically in Fig. 4.3.

For each of the 250 samples of 10 differences, t was calculated as $(\bar{d} - 0)/s_{\bar{d}}$, a deviation from the population mean in units of standard deviations of $\bar{d}$. Notice that $t = \bar{d}/s_{\bar{d}} = (\bar{x}_1 - \bar{x}_2)/s_{\bar{x}_1 - \bar{x}_2}$ since $\bar{d} = \bar{x}_1 - \bar{x}_2$. The distribution of these t values is given in Table 4.12 and is similar to that of Table 4.7 for

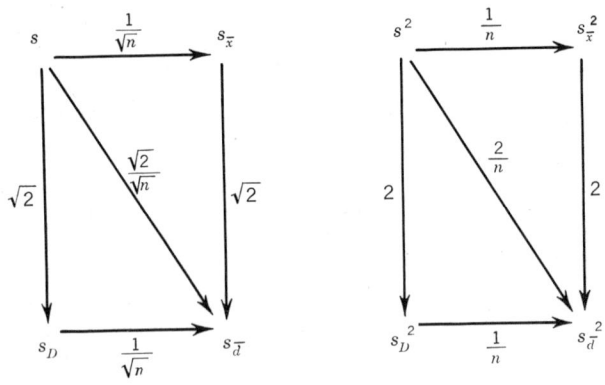

FIG. 4.3 Illustration of relations among standard deviations and variances. (This diagram is not meant to imply the theorem of Pythagoras, that is, that the sum of the squares of the lengths of the sides of a right-angled triangle equals the square of length of the hypotenuse.)

$t = (\bar{x} - \mu)/s_{\bar{x}}$. Of the t values, 118 are positive and 132 are negative; their mean is -0.00013. Fourteen t values exceed the 5% level as compared with an expected number of 12.5; four exceed the 1% as compared with 2.5.

Returning to the problem of determining whether there is a real difference between the responses to two treatments, we see that the sampling procedure discussed has shown what to expect when there is no real difference and the results are attributable to chance alone. The values of $t = \bar{d}/s_{\bar{d}}$ in Table 4.12 are seen to correspond well with Student's t distribution. Thus in the real situation, it is necessary only to calculate the statistic t and find the probability of a value as large as or larger when sampling is random and from a population with zero mean. If the probability of a larger value is small and the experimenter is not sure that sampling is from a population with zero mean, he will probably decide there is a real difference in the responses to the two treatments. The calculation of confidence intervals leads to the same sort of inference, for if the sample t is beyond, say, the 5% probability level, then the 95% confidence interval will not contain zero. If the confidence interval does not contain zero,

SAMPLING FROM A NORMAL DISTRIBUTION

the investigator can have little confidence in a statement that there is no difference between the responses to the treatments.

TABLE 4.12

SAMPLE AND THEORETICAL VALUES OF $t = \dfrac{d}{s_{\bar{d}}}$, 9 df; $\mu_{\bar{d}} = 0$

Interval of t		Sample		Theoretical		
From	To	Frequency	Percentage frequency	Percentage frequency	Cumulative	
					One tail	Both tails
—	−3.250	1	0.4	0.5	100.0	
−3.250	−2.821	0	0.0	0.5	99.5	
−2.821	−2.262	7	2.8	1.5	99.0	
−2.262	−1.833	5	2.0	2.5	97.5	
−1.833	−1.383	14	5.6	5.0	95.0	
−1.383	−1.100	13	5.2	5.0	90.0	
−1.100	−0.703	21	8.4	10.0	85.0	
−0.703	0.0	71	28.4	25.0	75.0	
0.0	0.703	62	24.8	25.0	50.0	100.0
0.703	1.100	24	9.6	10.0	25.0	50.0
1.100	1.383	10	4.0	5.0	15.0	30.0
1.383	1.833	7	2.8	5.0	10.0	20.0
1.833	2.262	9	3.6	2.5	5.0	10.0
2.262	2.821	2	0.8	1.5	2.5	5.0
2.821	3.250	1	0.4	0.5	1.0	2.0
3.250	—	3	1.2	0.5	0.5	1.0
		250	100.0			

Exercise 4.10.1 Given two samples of paired observations:

$$\begin{array}{ll} \text{I} & 10, \ 15, \ 13, \ 12, \ 11 \\ \text{II} & 12, \ 14, \ 14, \ 15, \ 13 \end{array}$$

What is the value of X_{11}? X_{22}? X_{14}? $X_{12} - X_{22}$? $\bar{x}_1 - \bar{x}_2$?

Exercise 4.10.2

Given $s^2 = 36$. What is $s_D{}^2$? $s_{\bar{d}}{}^2$?
Given $s_{\bar{x}}{}^2 = 12$. What is $s_{\bar{d}}{}^2$? s^2?
Given $s_D{}^2 = 50$. What is s^2? $s_{\bar{x}}{}^2$? $s_{\bar{d}}{}^2$?

4.11 Summary of sampling. A summary of the results obtained from the sampling experiment is given in Table 4.13. This summary clearly shows that by sampling it has been possible to demonstrate a number of important characteristics and theorems concerning normally distributed populations. In particular:

1. Means of random samples of n observations are normally distributed with mean μ and standard deviation $\sigma/\sqrt{n}$. (This theorem is approximately true with respect to the normality of means when sampling is from non-normal populations and is always true with respect to standard deviation.)

2. Means of differences of random samples of n observations are normally distributed with mean zero and standard deviation $\sqrt{2\sigma^2/n}$.

3. A random sample provides unbiased estimates of μ, σ^2, $\sigma_{\bar{x}}^2$, σ_D^2, and $\sigma_{\bar{d}}^2$.

4. The statistic $t = (\bar{x} - \mu)/s_{\bar{x}}$ or $t = (\bar{d} - 0)/s_{\bar{d}} = (\bar{x}_1 - \bar{x}_2)/s_{\bar{x}_1 - \bar{x}_2}$ is distributed symmetrically about mean zero and follows the tabulated-distribution of Student's t.

TABLE 4.13

A SUMMARY OF INFORMATION FROM:

1. 500 samples of 10 observations

		Symbol	$\bar{x}$	s^2 divisor		s		$s_{\bar{x}}$	
Sample				$n-1=9$	$n=10$	$\sqrt{s^2}$	$\bar{s}$	$s_{\bar{x}}$	$\sqrt{s^2/10}$	$\bar{s}/\sqrt{10}$
		Value	39.79	140.42	126.38	11.85	11.47	3.71	3.75	3.63
Population		Value	40.00	144		12		3.79		
		Symbol	μ	σ^2		σ		$\sigma_{\bar{x}}$		

2. 250 samples of 10 differences

		Symbol	$\bar{d}$	s_D^2		s_D			$s_{\bar{d}}$		
Sample				$\overline{s_D^2}$	$\overline{2s^2}$	$\bar{s}_D$	$\sqrt{\overline{s_D^2}}$	$\sqrt{\overline{2s^2}}$	$s_{\bar{d}}$	$\sqrt{\overline{s_D^2}/10}$	$\sqrt{\overline{2s^2}/10}$
		Value	-0.53	272.71	280.84	16.04	16.51	16.76	5.16	5.22	5.30
Population		Value	0	288			16.97			5.37	
		Symbol	μ	σ_D^2		σ_D			$\sigma_{\bar{d}}$		

3. t values

Number of samples	Mean	Number		Without regard to sign			
				Number beyond $t_{.05} = 2.262$		Number beyond $t_{.01} = 3.250$	
		Plus	Minus	Observed	Expected	Observed	Expected
500	-0.038	248	252	20	25	4	5
250	-0.00013	118	132	14	12.5	4	2.5

4.12 The testing of hypotheses. In many experiments the investigator has an idea about the results to be expected. He may have a hypothesis, as in the case of responses to two treatments, that the difference between the mean responses is of a certain magnitude, perhaps that for a population of differences, $\mu = 0$. The calculation of a confidence interval supplies a test of the hypothesis $\mu = 0$. If the interval contains $\mu = 0$, there is no sound evidence against the hypothesis, whereas if zero is not in the interval, there is evidence against the hypothesis that $\mu = 0$. The hypothesis may be tested with possibly less effort by calculating an appropriate statistic, such as t, and comparing this value with tabulated values. The table supplies the probability, when the sampling is from the hypothesized population, of obtaining a greater value of the statistic than that obtained.

The procedure to test a proposed hypothesis, then, is this:

1. For a population or distribution, formulate a meaningful hypothesis for which a statistic can be computed.

2. Choose a level of probability based jointly upon the seriousness of failing to reject the hypothesis when it is false and of rejecting it when true.

Steps 1 and 2 are a part of the planning of the experiment. When the experiment is completed or the investigation performed,

3. Compute the statistic and determine whether the value can be attributed to chance, if the hypothesis is true, by finding the probability of a more extreme value. More extreme or unusual values are large values of the statistic in most cases. However, for some statistics, small values are unusual and throw doubt upon the validity of the hypothesis.

4. If the probability at step 3 is such that chance (use the probability level set at step 2) seems inadequate to explain the result, conclude that the hypothesis is incorrect. Otherwise, conclude that the hypothesis is correct.

The hypothesis of step 1 is called the *null hypothesis* and is designated by H_0; for example, $H_0: \mu = 0$. A null hypothesis is one for which it is possible to compute a statistic and the corresponding probability of a more extreme value. In practice, step 3 is often performed by simply comparing the sample statistic with the tabulated value implied by step 2. In the case of step 4, while rejection of the null hypothesis results in the conclusion that an *alternative hypothesis* H_1 is true, failure to reject it does not result so much in the conclusion that the null hypothesis is true as in the conclusion that the true situation is not very different from that stated as the null hypothesis. The alternative hypothesis is usually a set of alternatives; for example, $H_1: \mu \neq 0$.

Exercise 4.12.1 Does a confidence interval for a parameter always contain that parameter? Why does one make 95% confidence statements when 99% confidence statements would lead to more correct inferences? What does a confidence interval for μ tell about future sample observations from the same population? (*Hint.* See the last paragraph of Sec. 4.9.)

Exercise 4.12.2 Seven observers were shown, for a brief period, a grill with 161 flies impaled and asked to estimate the number. The results are given by Cochran (4.2). Based on five estimates, they were 183.2, 149.0, 154.0, 167.2, 187.2, 158.0, 143.0. Define a reasonable population from which these means might have been obtained.

Exercise 4.12.3 Yields of 10 strawberry plants in a uniformity trial are given by Baker and Baker (4.1) as 239, 176, 235, 217, 234, 216, 318, 190, 181, 225 grams. Calculate 95 and 99% confidence intervals for the population mean. Test the hypothesis $\mu = 205$ (chosen arbitrarily) against the alternatives $\mu > 205$ at the 5% level of significance.

Exercise 4.12.4 Suppose we have a normal population with unknown mean but a known variance, $\sigma^2 = 16$. It is desired to draw a random sample of 36 to test the null hypothesis of $\mu = 50$ against the alternatives $\mu > 50$. The test is to be at the 5% level of significance.

The sample is drawn and $\bar{x} = 51$. What is the rejection region? What is the probability of accepting H_0 if it is true? What is the probability of accepting H_0 if H_1 is true and $\mu = 52$? $\mu = 53$? $\mu = 55$? What is the power of the test for the alternatives in the question above? (*Hint.* The sample information is not necessary in answering these questions.)

References

4.1 Baker, G. A., and R. E. Baker: "Strawberry uniformity yield trials," *Biometrics*, **9:** 412–421 (1953).

4.2 Cochran, W. G.: "The combination of estimates from different experiments," *Biometrics*, **10:** 101–129 (1954).

4.3 Dixon, W. J., and F. J. Massey, Jr.: *Introduction to Statistical Analysis*, 2d ed., Table 8*b*, McGraw-Hill Book Company, Inc., New York, 1957.

A proposed laboratory exercise in connection with Chap. 4

Purpose. To obtain evidence relative to the distribution of randomly paired observations, *viz.*, its mean and variance.

1. Pair the 10 random samples to give five samples of 10 pairs. (Since the 10 samples are random, no further mechanical process is required to assure random pairing. It is sufficient to place consecutive samples adjacent to each other.)
2. For each of the five pairs of samples, compute
   i. The signed differences
   ii. The mean of the signed differences. (This is also the difference between the means)
   iii. The variance and standard deviation
   iv. $t = \bar{d}/s_{\bar{d}}$
   v. The 95 and 99% confidence intervals for the mean differences
3. Record class data on summary sheets and summarize as in the text.
4. Plot $(\bar{x}, s)$ pairs as obtained in the laboratory exercise for Chap. 3 to see if any relation is readily apparent (see Sec. 3.9).

Chapter 5

COMPARISONS INVOLVING TWO SAMPLE MEANS

5.1 Introduction. There are few people who do not use statistics in some form. Most people make rough confidence statements on many aspects of their daily lives by the choice of an adjective, an adverb, or a phrase. Experimenters make investigation-based confidence statements to each of which is attached a decimal fraction between zero and one as a measure of the confidence to be placed therein. Confidence statements concerning population means were discussed in Secs. 2.13 and 3.10.

This chapter is devoted to testing hypotheses concerning two population means. Student's t is the principal test criterion although F is introduced in anticipation of a generalization to more than two means and of the analysis of variance. The problem of sample size is introduced.

5.2 Tests of significance. In Chap. 4, 500 random samples of 10 observations were drawn and t values were computed. Equation (5.1) shows that t is the deviation of a normal variable $\bar{x}$ from its mean μ, measured in standard error units, $s_{\bar{x}}$.

$$t = \frac{\bar{x} - \mu}{s_{\bar{x}}} \qquad (5.1)$$

The 500 sample t values were compared with tabulated t values for 9 degrees of freedom. It was found that approximately 5% of the sample t values were numerically equal to or greater than the tabulated value which is in theory exceeded with probability .05. This tabulated value is usually written as $t_{.05}$. In general, cumulative percentages for the 500 samples were in reasonable agreement with corresponding theoretical values.

A sample t can be used to test a hypothesis about μ. If the hypothesis is correct, for example, if the hypothesis $\mu = 40$ is used for the 500 samples, then the long-run percentage of times it is falsely declared to be incorrect is the chosen error rate. This is a penalty exacted because we are reasoning from a sample to a population; we are making an uncertain inference. If the hypothesis is incorrect, for example, if the hypothesis $\mu = 55$ is used for the 500 samples, then the percentage of times it is declared to be incorrect is not the chosen error rate but is presumably larger since rejecting an incorrect hypothesis is desirable. In other words, if the null hypothesis is false then we do not really compute Student's t but a value from another distribution. This value should seem out of place when we compare it with the tabulated values, which apply only when the hypothesis is true.

To make a test of significance with t as the statistic or test criterion, we first formulate a meaningful hypothesis for which Eq. (5.1) can be computed. For example, to compare the yields of a new and a standard variety of corn, an experimenter may set up the null hypothesis that there is no difference between the population means for the two varieties.

The null hypothesis is not necessarily that of no difference. Thus the hypothesis may be that the new variety of corn will outyield the standard by 5 bu per acre. In this case, the null hypothesis could be that the population mean of the new variety exceeds that of the standard by 5 bu per acre.

We choose a specific μ in order to be able to compute the statistic t; the usual alternative is really a set of alternatives, so does not lend itself to a single computation. At the time of choosing the null hypothesis, a probability level or *significance level* is also chosen. The chosen significance level determines a theoretical or tabulated t value such as $t_{.05}$. The sample t is computed and compared with the tabulated value.

If the sample t is greater than the tabulated value with which it is compared, then the null hypothesis is rejected and the alternative accepted. This is to say that the investigator has obtained a sample value which is an unusual one if the null hypothesis is true; since he has gone to some trouble to obtain a representative sample by use of a random procedure, he is not prepared to say that he has an unusual one so concludes that it appears unusual only because he has based his calculations on a false hypothesis; this becomes his conclusion, namely, that the null hypothesis is false and the alternative true. If a significance level of .05 is being used, rejection of the null hypothesis is generally made in a statement such as that there is a significant difference at the 5% level.

If the investigator is always presented with data for which the null hypothesis is true, then he is led to an incorrect inference every time he rejects this hypothesis. The level of significance is thus the average proportion of incorrect statements made when the null hypothesis is true.

In practice, the experimenter may seriously doubt the truth of the null hypothesis from the moment he proposes it. His purpose is simply to give himself a starting point from which he can calculate some meaningful statistic and come to an objective decision. Even when the investigator has considerable faith that the null hypothesis is true, he may simply mean that any real difference is of no economic importance or is otherwise meaningless. Acceptance of the null hypothesis is stated in weaker terms than is the acceptance of the alternative. When no difference has been detected, it may be that one exists which is not large enough to be presently of interest or economic value or that the experiment was not extensive enough to detect it.

In many fields of experimentation, 5 and 1% significance levels are customarily used. If a more discrepant value of the test criterion than that obtained is likely to occur less than 5% of the time but not less than 1% of the time by chance when the null hypothesis is true, then the difference is said to be *significant* and the sample value of the test criterion is marked with a single asterisk. If a more discrepant value of the test criterion than that obtained is likely to occur less than 1% of the time when the null hypothesis is true, the difference is said to be *highly significant* and the sample value of the test criterion is marked with two asterisks. Acceptance of the null hypothesis may be indicated by the letters *ns*.

The levels of 5 and 1% are arbitrary but seem to have been adequate choices in the field of agriculture where they were first used. In the case of small-sized experiments, it is possible that the null hypothesis will not likely be rejected if these levels are required, unless a large real difference exists. This suggests the choice of another level of significance, perhaps 10% for small experiments. If an experimenter uses levels other than the 5 and 1%, this should be stated clearly.

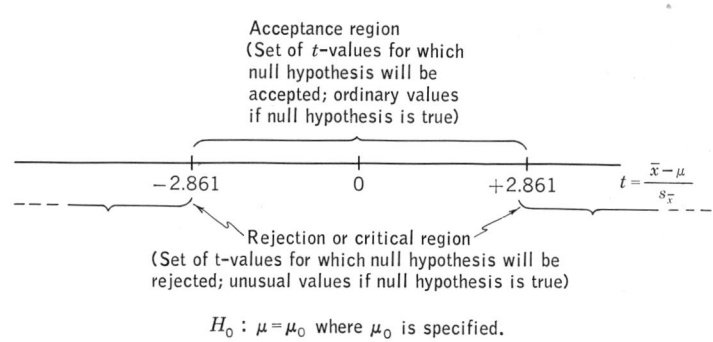

FIG. 5.1   Acceptance and rejection (critical) regions for a particular example

The tabulated value of the test criterion which corresponds to the chosen level of significance thus divides the possible values into two classes, called the *region of acceptance* and *region of rejection* or *critical region*. Figure 5.1 shows this for a test with $t_{.01}$ for 19 degrees of freedom as criterion as in the case of testing the hypothesis, using a 1% level of significance and a sample of 20 observations, that a population has a certain mean.

In using t to test the hypothesis that two population means are the same against the alternative that they are different, $H_1: \mu_1 \neq \mu_2$, there is no more reason to look at $\bar{x}_1 - \bar{x}_2$ than at $\bar{x}_2 - \bar{x}_1$. The region of rejection should then include all values of t numerically greater than $t_{.01}$, for example. The probability of a random t being larger than $t_{.01}$ is .005 and of being smaller than $-t_{.01}$ is .005. Such a test is called a *two-tailed test*.

If the set of alternatives is $\mu_1 > \mu_2$ and the null hypothesis is true, then discrepant values of t will be those for which $\bar{x}_1 - \bar{x}_2$ is large and positive. Large negative values would be unusual but can be attributed only to chance since the alternatives do not admit of such values being due to other than chance. (It may be that $\mu_1 < \mu_2$ but that the experimenter simply has no interest in such a result. In such a case the experimenter, using t as a test criterion and a 5% level of significance, for example, would choose his rejection region to contain all positive values of t greater than $t_{.10}$ since the probability of getting a larger positive value of t than $t_{.10}$ is .05. In brief, if the set of alternatives is $\mu_1 > \mu_2$, use $t_{.10}$ for a test at the 5% level of significance and $t_{.02}$ for a 1% level; similarly if $\mu_1 < \mu_2$. Such a test is called a *one-tailed test*. Table A.3 has been simplified in that probabilities for one-tailed tests may be read directly at the bottom of the table.

Regardless of which hypothesis is accepted, the null or alternative, the conclusion may be in error. An error, if made, may be one of two types depending upon whether the null or an alternative hypothesis is true.

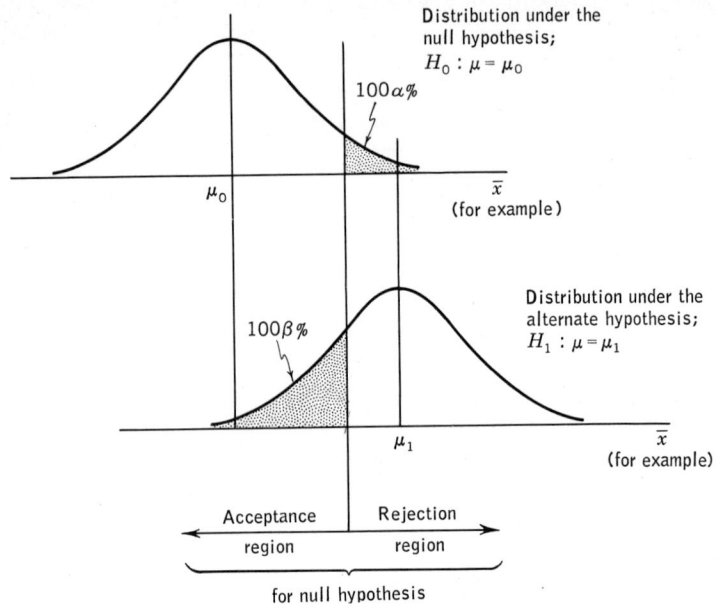

FIG. 5.2 An illustration relative to Type I and Type II errors

An *error of the first kind* (*Type* I) is made when the experimenter rejects the null hypothesis and it is true. The probability of such an error, often designated by α (Greek alpha), is fixed in advance of the conduct of the experiment. If the experimenter is always presented with a sample from the distribution associated with the null hypothesis, he will reject the null hypothesis $100\alpha\%$ of the time. Figure 5.2 shows the situation for a one-tailed test of the null hypothesis that $\mu = \mu_0$ against the single alternative that $\mu = \mu_1$.

An *error of the second kind* (*Type* II) is made when the experimenter accepts the null hypothesis and the alternative is true. The probability of an error of this type, often denoted by β (Greek beta), is determined by the choice of α and the distance between μ_0 and μ_1 (see Fig. 5.2). Thus, if the experimenter i always presented with a sample from the distribution associated with the alternative hypothesis, he rejects this hypothesis when he accepts the null hypothesis; this amounts to $100\beta\%$ of the time. The concept of Type II error is especially important in determining the sample size necessary to detect a difference of a stated magnitude.

COMPARISONS INVOLVING TWO SAMPLE MEANS

The outcomes of possible decisions concerning hypotheses are summarized in Table 5.1.

TABLE 5.1

DECISIONS AND THEIR OUTCOMES, WITH SPECIAL REFERENCE TO TYPE OF ERROR

Data is from a population for which		Decision relative to		Decision is	Probability should be
Null hypothesis H_0 is	Alternative hypothesis H_1 is	H_0 is	H_1 is		
True	False	Accept	Reject	Right	High
True	False	Reject	Accept	Wrong, Type I error	Low
False	True	Accept	Reject	Wrong, Type II error	Low
False	True	Reject	Accept	Right	High

When both types of error are considered in the case of a sample of fixed size, it is seen that a reduction in the probability of a Type I error must be accompanied by an increase in the probability of a Type II error. This is readily apparent in Fig. 5.2. To decrease α, we must move the point separating the acceptance and rejection regions to the right, with a consequent increase in the size of β. This shows the need of considering the seriousness of the different types of error in choosing a level of significance. If it is a serious error to fail to detect a real difference, that is, if accepting the null hypothesis when it is false is serious, then β should be small. Since the population means are fixed, this can be done only by moving the point separating acceptance and rejection regions to the left. This results in an increase in α, that is, an increase in the probability of rejecting the null hypothesis when it is true. In an ideal situation, both α and β are fixed in advance. This determines the required sample size. In practice, too often only α is fixed in advance; this means that β is not generally computed. We can only say that as the sample size increases, β decreases. If one were to fix the point separating the acceptance and rejection regions, then an increase in sample size would decrease both α and β since μ_0 and μ_1 are also fixed but the variance of means, σ^2/n, is decreasing, pulling the distribution closer about its true mean.

Closely related to Type II error is the power of a test. The *power* of a test is its ability to detect the alternative hypothesis when it is true. It is associated with the area under the lower curve of Fig. 5.2 for the rejection region. If the alternative hypothesis is rejected with probability β when it is true, then it is accepted with probability $1 - \beta$. It is obviously desirable to have this probability high.

Exercise 5.2.1 For the five random samples of Table 4.3, compute $t = (\bar{x} - 65)/s_{\bar{x}}$ (called a "noncentral t") as though you were ignorant of the true μ and were hypothesizing that $\mu = 65$ lb. Now find the probability of a larger t value, sign ignored, than that obtained for each of the five cases. What can you say about these five P values and the hypothesis $\mu = 65$ lb?

Exercise 5.2.2 Suppose we have a null hypothesis that $\mu_1 - \mu_2 = 28$ grams and two sample means, $\bar{x}_1$ and $\bar{x}_2$, for testing the hypothesis. What will be the numerator of t to test this hypothesis?

Exercise 5.2.3 Two-tailed tests are at least as common as one-tailed tests. Suppose $H_0: \mu = \mu_0$ is to be tested against $H_1: \mu \neq \mu_0$. Choose a $\mu_1 \neq \mu_0$ for illustrative purposes and prepare a figure similar to Fig. 5.2 showing acceptance and rejection regions and shading areas associated with α and β.

Exercise 5.2.4 Suppose we have a normal distribution with unknown mean but a known variance, $\sigma^2 = 16$. It is desired to draw a random sample of 36 observations to test $H_0: \mu = 50$ against $H_1: \mu > 50$. The test is to be at the 5% level of significance. What is the rejection region? What is the probability of accepting H_0 if true? Of accepting H_0 if H_1 is true and $\mu = 52$? $\mu = 53$? $\mu = 55$? What is the power of the test for each of the alternatives?

5.3 Basis for a test of two or more means.

The variance of sample means is σ^2/n, where σ^2 is the variance of the individuals in the parent population and n is the sample size. This implies that means, where more than one is available, may be used to estimate σ^2. Thus for two means based on n observations each, an estimate of σ^2/n is given by $\Sigma(\bar{x}_i - \bar{\bar{x}})^2/(2 - 1)$, which when multiplied by n gives an estimate of σ^2. A second estimate of σ^2 is available from the individuals. Hence a test of significance of the difference between two sample means, that is, a test of $H_0: \mu_1 - \mu_2 = 0$ versus $H_1: \mu_1 - \mu_2 \neq 0$, could involve the ratio of two such estimates of σ^2. In particular, Eq. (5.2) defines a common test criterion.

$$F = \frac{\text{estimate of } \sigma^2 \text{ from means}}{\text{estimate of } \sigma^2 \text{ from individuals}} \qquad (5.2)$$

The estimate of σ^2 from individuals is not affected by the μ's while that from the $\bar{x}$'s is affected. In other words, if H_1 is true, then the difference between the $\bar{x}_i$'s is increased because of the difference between the μ's; this tends to give a large numerator. Values of F are given in Table A.6 for five probability levels and various pairs of degrees of freedom, one value for the numerator and one for the denominator. For two means, the estimate of σ^2 for the numerator is based on a single degree of freedom. For two samples, the square root of F has Student's t distribution.

For the null hypothesis of no difference, t is defined by Eq. (5.3).

$$t = \frac{\bar{x}_1 - \bar{x}_2}{s_{\bar{x}_1 - \bar{x}_2}} = \frac{d}{s_d} \qquad (5.3)$$

Note that d is used to replace $\bar{x}_1 - \bar{x}_2$ regardless of whether or not the observations are paired. s_d is the standard deviation appropriate to a difference between two random means from a normal population. If t cannot reasonably be attributed to chance and the null hypothesis, we assume that d is too large because $\mu_1 \neq \mu_2$. Calculation of s_d depends on

1. Whether the two populations have a common variance σ^2
2. Whether the values of the σ^2's, or the common σ^2, are known or estimated
3. Whether the two samples are of the same size and
4. Whether the variates are paired

COMPARISONS INVOLVING TWO SAMPLE MEANS

The choice of a rejection region depends upon
1. The level of significance chosen
2. The sample size and
3. Whether a one- or two-tailed test is required

While F may be considered as a ratio of two variances, t may be considered as a ratio of two standard deviations. It is obvious that the denominator of Eq. (5.3) is a standard deviation; it may be shown that the numerator, $\bar{x}_1 - \bar{x}_2$, is related to a standard deviation. In particular, $\bar{x}_1 - \bar{x}_2$ is an estimate of $\sqrt{2}\sigma_{\bar{x}}$ when the two samples are of the same size.

Exercise 5.3.1 Let the numbers 6 and 2 be two observations. Compute s^2 and show that it equals $(6-2)^2/2$.

5.4 Comparison of two sample means, unpaired observations, equal variances. Let μ_1 and μ_2, σ^2, $\bar{x}_1$ and $\bar{x}_2$, s_1^2 and s_2^2, and n_1 and n_2 denote the two population means, the common population variance, the sample means, variances, and sizes, respectively. Using a random sample from each population, we are to test the null hypothesis H_0: $\mu_1 = \mu_2$, assuming that the populations are normally distributed and have a common but unknown variance.

The test criterion is given in Eq. (5.4):

$$t = \frac{(\bar{x}_1 - \mu_1) - (\bar{x}_2 - \mu_2)}{s_{\bar{x}_1 - \bar{x}_2}} = \frac{(\bar{x}_1 - \bar{x}_2) - (\mu_1 - \mu_2)}{s_{\bar{x}_1 - \bar{x}_2}} \quad (5.4)$$

which for the null hypothesis $\mu_1 = \mu_2$ becomes Eq. (5.3). In general the difference $\mu_1 - \mu_2$ may be set equal to any value which the experimenter may hypothesize.

In the test criterion, $s_{\bar{d}}$ is an estimate of $\sigma_{\bar{x}_1 - \bar{x}_2}$ and subject to sampling variation. We first estimate σ^2 by pooling the sums of squares of the two samples and dividing by the pooled degrees of freedom as in Eq. (5.5).

$$s^2 = \frac{(n_1 - 1)s_1^2 + (n_2 - 1)s_2^2}{(n_1 - 1) + (n_2 - 1)} \quad (5.5)$$

This is a *weighted average* of the sample variances and is superior to the arithmetic average which gives equal weight to the sample variances. The weighted and arithmetic averages are the same when the samples are of the same size. The criterion t is distributed as Student's t for random samples from normal populations but considerable departures from normality may not seriously affect the distribution, especially near the commonly used 5 and 1% points.

A criterion involving an estimate of σ^2 was first studied by Student in 1908. He prepared the distribution of a related statistic which he called z. Later, R. A. Fisher worked out the distribution of t, the basis of Table A.3.

Case 1. The test when $n_1 \neq n_2$. Calculate $s_{\bar{d}}$ by the formula

$$s_{\bar{d}} = \sqrt{s^2\left(\frac{1}{n_1} + \frac{1}{n_2}\right)} = \sqrt{s^2\left(\frac{n_1 + n_2}{n_1 n_2}\right)} \quad (5.6)$$

where s^2 is the weighted average of the sample variances, Eq. (5.5).

A numerical example is given in Table 5.2 for data from Watson et al. (5.8).

TABLE 5.2
COEFFICIENTS OF DIGESTIBILITY OF DRY MATTER,
FEED CORN SILAGE, IN PER CENT

X_1 (sheep)	X_2 (steers)
57.8	64.2
56.2	58.7
61.9	63.1
54.4	62.5
53.6	59.8
56.4	59.2
53.2	
ΣX 393.5	367.5
ΣX^2 22,174.41	22,535.87
$\bar{x}$ 56.21%	61.25%

$\Sigma x_1^2 = \Sigma X_1^2 - (\Sigma X_1)^2/n_1 = 22{,}174.41 - 22{,}120.32 = 54.09 = (n_1 - 1)s_1^2$

$\Sigma x_2^2 = \Sigma X_2^2 - (\Sigma X_2)^2/n_2 = 22{,}535.87 - 22{,}509.37 = 26.50 = (n_2 - 1)s_2^2$

$s^2 = \dfrac{\Sigma x_1^2 + \Sigma x_2^2}{(n_1 - 1) + (n_2 - 1)} = \dfrac{54.09 + 26.50}{6 + 5} = 7.33$, an estimate of the common σ^2

$df = (n_1 - 1) + (n_2 - 1) = 11$

$s_{\bar{d}} = \sqrt{s^2 \dfrac{(n_1 + n_2)}{n_1 n_2}} = \sqrt{7.33 \dfrac{(7 + 6)}{42}} = \sqrt{2.27} = 1.51\%$, the standard deviation appropriate to the difference between the sample means

$t = \dfrac{\bar{d}}{s_{\bar{d}}} = \dfrac{56.21 - 61.25}{1.51} = \dfrac{-5.04}{1.51} = -3.33^{**}, df = 11$

For the 95% confidence interval for $\mu_2 - \mu_1$, $\bar{x}_2 - \bar{x}_1 \pm t_{.05} s_{\bar{x}_1 - \bar{x}_2} = 5.04 \pm 2.201(1.51)$ = 5.04 ± 3.32; $l_1 = 1.72\%$ and $l_2 = 8.36\%$.

The confidence interval for $\mu_2 - \mu_1$ rather than $\mu_1 - \mu_2$ was calculated only because $\bar{x}_2 - \bar{x}_1$ is positive.

When the criterion F is used, the results are generally shown in an *analysis of variance* table where every variance or mean square is an estimate of the same σ^2 if the null hypothesis is true (see Table 5.3). To compute the numerator of

TABLE 5.3
ANALYSIS OF VARIANCE OF DATA IN TABLE 5.2

Source of variation	df	Sum of squares	Mean square	F
Sheep vs. steers	1	81.93	81.93	11.18**
Among sheep + among steers	6 + 5	54.09 + 26.50	7.33	
Total	12	162.52		

COMPARISONS INVOLVING TWO SAMPLE MEANS   75

F, recall that $(\bar{x}_1 - \bar{x}_2)^2$ estimates $2\sigma_{\bar{x}}^2 = 2\sigma^2/n$. To estimate σ^2, we must multiply by $n/2$.

When the means are for samples of unequal size, the square of the difference is an estimate of $\sigma^2(n_1 + n_2)/n_1 n_2$; hence for an estimate of σ^2, we multiply by $n_1 n_2/(n_1 + n_2)$. Thus $(56.21 - 61.25)^2(6)(7)/(6 + 7) = 81.93$ is an estimate of σ^2, based on means. (The actual computation was made with totals; the one with means differs slightly due to rounding errors.) Note that $n_1 n_2/(n_1 + n_2)$ is the reciprocal of the multiplier of s^2 when t is the criterion and $s_{\bar{d}}$ is calculated. Now $t^2 = F$ whenever F has 1 degree of freedom associated with the numerator, thus $(3.33)^2$ and 11.19 are equal within rounding errors. Note also that degrees of freedom and sums of squares in the body of the table add to the total. Table A.6 is one of theoretical F values for several probability levels. The first column applies to cases where only two sample means are involved. $F(1, 11) = 4.84$ at the 5% level and 9.65 at the 1% level. Since sample $F = 11.18$ exceeds 9.65, the difference is declared to be highly significant. The same conclusion is reached using t as the test criterion.

Case 2.   The test when $n_1 = n_2 = n$. The procedure given as Case 1 is applicable. The criterion is t as in Eq. (5.4); Eq. (5.6) reduces to Eq. (5.7).

$$s_{\bar{d}} = \sqrt{2s^2/n} \tag{5.7}$$

The degrees of freedom are $2(n - 1)$. A numerical example is worked in Table 5.4 for data from Ross and Knodt (5.4). Since the observed difference between means is significant at only the 5% level, the 95% confidence interval does not contain zero whereas the 99% one does.

When σ^2 is known, the test criterion becomes

$$z = \frac{\bar{x}_1 - \bar{x}_2}{\sqrt{\sigma^2(n_1 + n_2)/n_1 n_2}} \tag{5.8}$$

which is compared with tabulated values in the last line of the table of Student's t. These values are taken from tables of the normal distribution.

Exercise 5.4.1   From an area planted in one variety of guayule, 54 plants were selected at random. Of these, 15 were offtypes and 12 were aberrants. Rubber percentages for these plants were:†

Offtypes: 6.21, 5.70, 6.04, 4.47, 5.22, 4.45, 4.84, 5.88, 5.82, 6.09, 5.59, 6.06, 5.59, 6.74, 5.55;

Aberrants: 4.28, 7.71, 6.48, 7.71, 7.37, 7.20, 7.06, 6.40, 8.93, 5.91, 5.51, 6.36.

Test the hypothesis of no difference between means of populations of rubber percentages. Compute a 95% confidence interval for the difference between population means. Present the results in an analysis of variance table. Compare F and t^2.

Exercise 5.4.2   The weights in grams of 10 male and 10 female juvenile ring-necked pheasants trapped one January in the University of Wisconsin arboretum were:‡

Males: 1,293, 1,380, 1,614, 1,497, 1,340, 1,643, 1,466, 1,627, 1,383, 1,711;

Females: 1,061, 1,065, 1,092, 1,017, 1,021, 1,138, 1,143, 1,094, 1,270, 1,028.

† Data courtesy of W. T. Federer, Cornell University, Ithaca, N.Y.
‡ Data courtesy of Department of Wildlife Management, University of Wisconsin, Madison, Wis.

Test the hypothesis of a difference of 350 grams (chosen for illustrative purposes only) between population means in favor of males against the alternative of a greater difference. This is a one-tailed test.

Exercise 5.4.3 If one had more ($k > 2$) samples, how could one pool the information to estimate a common σ^2? *Hint.* Generalize Eq. (5.5).

5.5 The linear additive model. The linear additive model (see Sec. 2.12) attempts to explain an observation as a mean plus a random element of variation, where the mean may be the sum of a number of components

TABLE 5.4

GAIN IN WEIGHT OF HOLSTEIN HEIFERS

X_1, control	X_2, vitamin A
175	142
132	311
218	337
151	262
200	302
219	195
234	253
149	199
187	236
123	216
248	211
206	176
179	249
206	214
ΣX 2,627	3,303
ΣX^2 511,807	817,583
$\bar{x}$ 187.6 lb	235.9 lb

$n_1 = n_2 = n = 14$

$\Sigma x_1^2 = \Sigma X_1^2 - (\Sigma X_1)^2/n = 511{,}807 - 492{,}938 = 18{,}869 = (n_1 - 1)s_1^2$

$\Sigma x_2^2 = \Sigma X_2^2 - (\Sigma X_2)^2/n = 817{,}583 - 779{,}272 = 38{,}311 = (n_2 - 1)s_2^2$

$s^2 = \dfrac{\Sigma x_1^2 + \Sigma x_2^2}{2(n-1)} = \dfrac{57{,}180}{26} = 2{,}199$, an estimate of the common σ^2

$df = 2(n - 1) = 26$

$s_{\bar{d}} = \sqrt{\dfrac{2s^2}{n}} = \sqrt{\dfrac{2(2{,}199)}{14}} = 17.7$ lb the standard deviation appropriate to the difference between sample means

$t = \dfrac{d}{s_{\bar{d}}} = \dfrac{187.6 - 235.9}{17.7} = \dfrac{-48.3}{17.7} = -2.73^*;\ (t_{.01} = 2.78)$

For the 95% confidence interval, $\bar{x}_2 - \bar{x}_1 \pm t_{.05}s_{\bar{d}} = 48.3 \pm 2.056(17.7) = 48.3 \pm 36.4$; $l_1 = 11.9$ and $l_2 = 84.7$ lb.

For the 99% confidence interval, $\bar{x}_2 - \bar{x}_1 \pm t_{.01}s_{\bar{d}} = 48.3 \pm 2.779(17.7) = 48.3 \pm 49.2$; $l_1 = -0.9$ and $l_2 = 97.5$ lb.

COMPARISONS INVOLVING TWO SAMPLE MEANS

associated with several effects or sources of variation. For samples from two populations with possibly different means but a common variance, the composition of any observation is given by Eq. (5.9).

$$X_{ij} = \mu + \tau_i + \epsilon_{ij} \qquad (5.9)$$

where $i = 1, 2$ and $j = 1, \ldots, n_1$ for $i = 1$ and $j = 1, \ldots, n_2$ for $i = 2$. (τ is Greek tau.) The model explains the jth observation on the ith population as composed of a general mean μ, plus a component τ_i for the population involved, plus a random element of variation. In terms of Sec. 5.4, $\mu + \tau_1 = \mu_1$ and $\mu + \tau_2 = \mu_2$. For convenience, we set $\tau_1 + \tau_2 = 0$ or $\tau_2 = -\tau_1$. In other words, the τ's are measured as deviations. Thus if τ represents an increase, then $\tau_2 = -\tau_1$ represents an equal decrease. While this may not be the truth of the matter, it does not affect the difference between means, that is, 2τ; and the difference is the important aspect of the problem while μ is simply a convenient reference point. The ϵ's are assumed to be from a single population of ϵ's with variance σ^2. Thus σ^2 may be estimated from either or both samples. The difference between this model and that of Sec. 2.12 is that this is more general; it permits us to describe two populations of individuals simultaneously.

Let us look at the sample observations, totals, and means. Notice the notation for sums, namely, $X_{i\cdot}$ and $\epsilon_{i\cdot}$, and for means, namely, $\bar{x}_{i\cdot}$ and $\bar{\epsilon}_{i\cdot}$. The

Sample 1	Sample 2
$X_{11} = \mu + \tau_1 + \epsilon_{11}$	$X_{21} = \mu + \tau_2 + \epsilon_{21}$
$X_{12} = \mu + \tau_1 + \epsilon_{12}$	$X_{22} = \mu + \tau_2 + \epsilon_{22}$
. . . .	. . . .
$X_{1n_1} = \mu + \tau_1 + \epsilon_{1n_1}$	$X_{2n_2} = \mu + \tau_2 + \epsilon_{2n_2}$
$\sum_j X_{1j} = X_{1\cdot} = n_1\mu + n_1\tau_1 + \epsilon_{1\cdot}$	$\sum_j X_{2j} = X_{2\cdot} = n_2\mu + n_2\tau_2 + \epsilon_{2\cdot}$
$\bar{x}_{1\cdot} = \mu + \tau_1 + \bar{\epsilon}_{1\cdot}$	$\bar{x}_{2\cdot} = \mu + \tau_2 + \bar{\epsilon}_{2\cdot}$

dot notation is an alternative to using Σ. Summation is for all values of the subscript for the place occupied by the dot.

To obtain s^2, $\sum_j (X_{1j} - \bar{x}_{1\cdot})^2 = (n_1 - 1)s_1^2$ is calculated. This sum of squares is associated with ϵ's only, since $\mu + \tau_1$ is common to all these observations and so does not affect their variation. We also calculate $\sum_j (X_{2j} - \bar{x}_{2\cdot})^2$, associated only with ϵ's. Since the populations have a common σ^2, that is, since there is a single population of ϵ's, these sums of squares are pooled in estimating σ^2, as seen in Eq. (5.5). No contribution from the τ's enters our estimate of σ^2.

Now consider $(\bar{x}_{1\cdot} - \bar{x}_{2\cdot}) = (\tau_1 - \tau_2) + (\bar{\epsilon}_{1\cdot} - \bar{\epsilon}_{2\cdot})$. This is the numerator of t and is used in computing the numerator of F. Under the null hypothesis, the populations have the same mean and $\tau_1 = \tau_2 = 0$; also, $\bar{x}_{1\cdot} - \bar{x}_{2\cdot}$ is a difference of two means of observations from the same population and has a variance which is a multiple of σ^2. Thus if the ϵ's are normally distributed, both t and F are appropriate criteria for testing the null hypothesis.

The quantity $(\bar{\epsilon}_{1\cdot} - \bar{\epsilon}_{2\cdot})$ should be small since it is the algebraic sum of two quantities each with zero mean and with small variance, namely, σ^2/n_i. When the null hypothesis is false, the quantity $(\tau_1 - \tau_2) = 2\tau_1 \neq 0$ since $\tau_2 = -\tau_1 \neq 0$; an alternative is true and the numerator of t or F is enlarged, leading to a larger value of the criterion than would be expected by chance if the null hypothesis were true. Thus large values of the criterion are unusual if the null hypothesis is true, so the probability of detecting a real difference is seen to increase as the real difference increases.

5.6 Comparison of sample means; paired observations. Frequently observations are paired. For example, two rations may be compared using two animals from each of 10 litters of swine by assigning the animals of each litter at random, one to each ration; or the percentage of oil in two soybean varieties grown on paired plots at each of 12 locations may be compared. If the members of the pair tend to be positively correlated, that is, if members of a pair tend to be large or small together, an increase in the ability of the experiment to detect a small difference is possible. The information on pairing is used to eliminate a source of extraneous variance, that existing from pair to pair. This is done by calculating the variance of the differences rather than that among the individuals within each sample. The number of degrees of freedom on which the estimate of σ_D^2 is based is one less than the number of pairs. If the ability of the experiment to detect a real difference is to be improved, then the variance of differences must be sufficiently less than that of individuals, that is, σ_D^2 must be less than $2\sigma^2$ to compensate for the loss of degrees of freedom due to pairing; for random pairing, $\sigma_D^2 = 2\sigma^2$.

The test criterion is t as in Eq. (5.3) with $s_{\bar{d}}$ computed as in Eq. (5.10).

$$s_{\bar{d}} = \sqrt{\frac{\sum_j (X_{1j} - X_{2j})^2 - [\sum_j (X_{1j} - X_{2j})]^2/n}{n(n-1)}} = \sqrt{\frac{\sum_j D_j^2 - (\sum_j D_j)^2/n}{n(n-1)}} \quad (5.10)$$

In the divisor, $n - 1$ is the degrees of freedom and n is the number of sample differences or pairs.

It can be shown that

$$s_D^2 = s_1^2 + s_2^2 - \frac{2\Sigma x_1 x_2}{n-1} \quad (5.11)$$

Thus s_D^2 is equal to the sum of the mean squares of X_1 and X_2 less twice $\Sigma x_1 x_2/(n-1)$, called the *covariance* (see Sec. 9.2). If high values of X_1 are associated with high values of X_2, then the covariance will be a positive quantity and s_D^2 will be an estimate of $\sigma_D^2 < 2\sigma^2$. The same is true if both are large and negative. Thus, the more similar the members of a pair are as compared with members of different pairs, the larger will be the covariance term and the greater the reduction in $s_{\bar{d}}^2$ as compared with the case of no pairing.

Tests of hypotheses with paired variates are seen to reduce to testing that the mean of differences is a specified number, often zero. An example is worked in Table 5.5 for data from Shuel (5.6).

The null hypothesis tested is that the mean of the population of differences is zero; the alternatives are that the mean is not zero. The test criterion is distributed as t when the assumption that differences are normally distributed is correct and the null hypothesis is true. Tabulated $t_{.01}$ for 9 degrees of freedom and a two-tailed test is 3.3. Here the observed difference is hard to explain on the basis of random sampling from the population associated with the null hypothesis. The null hypothesis is rejected on the basis of the evidence presented.

TABLE 5.5

SUGAR CONCENTRATION OF NECTAR IN HALF HEADS OF RED CLOVER KEPT AT DIFFERENT VAPOR PRESSURES FOR 8 hr

Vapor pressure		Difference $X_1 - X_2$
4.4 mm Hg X_1	9.9 mm Hg X_2	
62.5	51.7	10.8
65.2	54.2	11.0
67.6	53.3	14.3
69.9	57.0	12.9
69.4	56.4	13.0
70.1	61.5	8.6
67.8	57.2	10.6
67.0	56.2	10.8
68.5	58.4	10.1
62.4	55.8	6.6
ΣX 670.4	561.7	108.7
		$1{,}226.07 = \sum_j (X_{1j} - X_{2j})^2$
$\bar{x}$ 67.0	56.2	10.8

$$s_{\bar{d}}^2 = \frac{\Sigma(X_{1j} - X_{2j})^2 - [\Sigma(X_{1j} - X_{2j})]^2/n}{n(n-1)} = \frac{1{,}226.07 - 1{,}181.57}{(9)(10)} = 0.4944$$

$s_{\bar{d}} = .703$

$$t = \frac{\bar{d}}{s_{\bar{d}}} = \frac{10.8}{.703} = 15.4^{**}, \quad \text{for 9 } df$$

The 99% confidence interval for the population mean difference is calculated as $\bar{d} \pm t_{.01}s_{\bar{d}}$ = 10.8 ± 3.3(.703). Hence $l_1 = 8.5$ and $l_2 = 13.1\%$.
Note that $\Sigma(X_{1j} - X_{2j}) = \Sigma X_{1j} - \Sigma X_{2j}$ and that $\overline{x_1 - x_2} = \bar{x}_1 - \bar{x}_2$.

When σ^2 is known, observed t is compared with tabulated t on the last line of the t table. The F criterion for paired observations is discussed in Chap. 8.

Exercise 5.6.1 The cooling constants of freshly killed mice and those of the same mice reheated to body temperature were determined by Hart (5.2) as

Freshly killed: 573, 482, 377, 390, 535, 414, 438, 410, 418, 368, 445, 383, 391, 410, 433, 405, 340, 328, 400

Reheated: 481, 343, 383, 380, 454, 425, 393, 435, 422, 346, 443, 342, 378, 402, 400, 360, 373, 373, 412

Test the hypothesis of no difference between population means.

Exercise 5.6.2 Peterson (5.3) determined the number of seeds set per pod in lucerne. The data follow:

Top flowers	4.0	5.2	5.7	4.2	4.8	3.9	4.1	3.0	4.6	6.8
Bottom flowers	4.4	3.7	4.7	2.8	4.2	4.3	3.5	3.7	3.1	1.9

Test the hypothesis of no difference between population means against the alternatives that top flowers set more seeds.

5.7 The linear additive model for the paired comparison.

The linear additive expression for the composition of any observation is given by Eq. (5.12).

$$X_{ij} = \mu + \tau_i + \beta_j + \epsilon_{ij} \qquad (5.12)$$

X_{ij} is the observation on the ith sample for the jth pair, $i = 1, 2$ and $j = 1, 2, \ldots, n$. Again we have a general mean μ, a component τ_i peculiar to the sample, and a random element ϵ_{ij}; in addition, there is a component β_j peculiar to the pair of observations. The τ_i's and β_j's contribute to the variability of the X's, provided they are not all equal to zero. For convenience, we let $\Sigma \tau_i = 0$ and $\Sigma \beta_j = 0$. This model admits of a different population mean for each observation, but these means are closely related by construction; X_{ij} is the only observation from the population with mean $\mu + \tau_i + \beta_j$ but τ_i is present in $n - 1$ other observations and β_j is present in one other observation. The ϵ's are considered to be from, at most, two populations corresponding to the subscript i but need not be assumed to be from a single population as in the case of the model of Eq. (5.9). Because of the relations among population means, that is, the $(\mu + \tau_i + \beta_j)$'s, it is possible to estimate an appropriate variance for testing the difference of sample means. To see this clearly, set up a table as follows:

	Sample 1 X_{1j}	Sample 2 X_{2j}	Difference $X_{1j} - X_{2j}$
	$X_{11} = \mu + \tau_1 + \beta_1 + \epsilon_{11}$	$X_{21} = \mu + \tau_2 + \beta_1 + \epsilon_{21}$	$(\tau_1 - \tau_2) + (\epsilon_{11} - \epsilon_{21})$
	. . . . . .	. . . . . .	. . . . . .
	$X_{1n} = \mu + \tau_1 + \beta_n + \epsilon_{1n}$	$X_{2n} = \mu + \tau_2 + \beta_n + \epsilon_{2n}$	$(\tau_1 - \tau_2) + (\epsilon_{1n} - \epsilon_{2n})$
Totals:	$X_{1.} = n\mu + n\tau_1 + \Sigma\beta_j + \epsilon_{1.}$	$X_{2.} = n\mu + n\tau_2 + \Sigma\beta_j + \epsilon_{2.}$	$n(\tau_1 - \tau_2) + (\epsilon_{1.} - \epsilon_{2.})$
Means:	$\bar{x}_{1.} = \mu + \tau_1 + \bar{\epsilon}_{1.}$	$\bar{x}_{2.} = \mu + \tau_2 + \bar{\epsilon}_{2.}$	$(\tau_1 - \tau_2) + (\bar{\epsilon}_{1.} - \bar{\epsilon}_{2.})$

It is obvious that the differences in the last column have a variation associated only with differences (algebraic sums) of ϵ's since $(\tau_1 - \tau_2)$ is a constant in all. The variance of the sample differences is an estimate of $\sigma_1^2 + \sigma_2^2$ if the ϵ_1's and ϵ_2's have different variances, or of $2\sigma^2$ if a common variance exists. The numerator of the test criterion, either t or F, involves the mean of the differences and, consequently, has a contribution from the difference between treatment means, namely, $\tau_1 - \tau_2 = 2\tau_1$ if such is present, in addition to a contribution due to the ϵ's. If then the numerator is much larger than the denominator, the largeness is customarily attributed to a real treatment difference rather than to an unusual chance event. Note that if the β_j's are real, they will contribute to any variance computed directly from either sample.

This test has one important property not possessed by the previous tests involving the hypothesis of a difference between population means. Theory tells us that the algebraic sum of normally distributed variables is normally distributed. The application of this to the present case is to the effect that the differences are normally distributed provided the errors are, *regardless of whether or not the ϵ_1's and the ϵ_2's have a common variance*.

A value in pairing not previously mentioned concerns the scope of inference. It is seen that the variation from pair to pair can be large. If we deliberately make this variation large, we widen the scope of our inference. Thus our pairs of swine can come from many litters involving different sires and dams, and possibly, different breeds; our soybeans may have been grown at quite different locations; our inference is broadened as a result.

5.8 Unpaired observations and unequal variances. Given a sample from each of two populations with $\sigma_1^2 \neq \sigma_2^2$, that is, with unequal variances, it is desired to test the hypothesis that $\mu_1 = \mu_2$ using the sample estimates of the variances.

The appropriate $s_{\bar{d}}$ is computed as in Eq. (5.13).

$$s_{\bar{d}} = \sqrt{\frac{s_1^2}{n_1} + \frac{s_2^2}{n_2}} \tag{5.13}$$

Neither the sums of squares nor the degrees of freedom are pooled as when s_1^2 and s_2^2 were estimates of a common σ^2 and the observations were unpaired, as in Eq. (5.5). Now we compute $t' = \bar{d}/s_{\bar{d}}$, the prime indicating that the criterion is not distributed as Student's t.

For determining a significant value of t' for a given significance level, Cochran and Cox (5.1) give a sufficiently accurate approximation. The value of t' such that larger observed values are to be judged significant is calculated as in Eq. (5.14).

$$t' = \frac{w_1 t_1 + w_2 t_2}{w_1 + w_2} \tag{5.14}$$

where $w_1 = s_1^2/n_1$, $w_2 = s_2^2/n_2$, and t_1 and t_2 are the values of Student's t for $n_1 - 1$ and $n_2 - 1$ degrees of freedom, respectively, at the chosen level of significance. This value t' corresponds to a tabulated value. This approximation errs slightly on the conservative side in that the value of t' required for significance may be slightly too large.

It is seen that t' lies between the tabulated t values for $n_1 - 1$ and $n_2 - 1$ degrees of freedom. Hence the calculation is needed only for those cases where the difference is borderline. When $n_1 = n_2 = n$, t' is the tabular t for $n - 1$ degrees of freedom.

Data from Rowles (5.5) are used in Table 5.6 to exemplify the procedure.

5.9 Testing the hypothesis of equality of variances. In Sec. 5.2, the choice of a critical or rejection region was seen to depend, in part, upon the set of alternative hypotheses. The tests just discussed have been treated from the point of view of a two-tailed test, the set of alternatives being that a difference existed. For one-tailed tests of such hypotheses, the test criterion is the same but the rejection regions differ.

TABLE 5.6

FINE GRAVEL IN SURFACE SOILS, PER CENT

Good soil	Poor soil
5.9	7.6
3.8	0.4
6.5	1.1
18.3	3.2
18.2	6.5
16.1	4.1
7.6	4.7

ΣX	76.4	27.6
ΣX^2	1,074.60	150.52
$\bar{x}$	10.91	3.94

$$\Sigma x_1^2 = \Sigma X_1^2 - (\Sigma X_1)^2/n_1 = 1{,}074.60 - 833.85 = 240.75$$

$$s_1^2 = \frac{\Sigma x_1^2}{n_1 - 1} = \frac{240.75}{6} = 40.12$$

$$\Sigma x_2^2 = \Sigma X_2^2 - (\Sigma X_2)^2/n_2 = 150.52 - 108.82 = 41.70$$

$$s_2^2 = \frac{\Sigma x_2^2}{n_2 - 1} = \frac{41.70}{6} = 6.95$$

$$s_{\bar{d}} = \sqrt{\frac{s_1^2}{n_1} + \frac{s_2^2}{n_2}} = \sqrt{\frac{40.12}{7} + \frac{6.95}{7}} = \sqrt{6.72} = 2.59\%$$

$$t' = \frac{\bar{d}}{s_{\bar{d}}} = \frac{10.91 - 3.94}{2.59} = \frac{6.97}{2.59} = 2.69*$$

Compare t' with tabular t for 6 degrees of freedom ($= 2.45$ at 5%). Computationally $t' = [(s_1^2/n_1)t_{.05} + (s_2^2/n_2)t_{.05}]/[s_1^2/n_1 + s_2^2/n_2]$ but $t_{.05}$ is the same value in each case and t' reduces to this $t_{.05}$.

The 95% confidence interval is $\bar{x}_1 - \bar{x}_2 \pm t_{.05} s_{\bar{d}} = 6.97 \pm 2.45(2.59) = 6.97 \pm 6.35$. Hence, $l_1 = 0.62$ and $l_2 = 13.32\%$.

The test criterion t may be viewed as a comparison of standard deviations, its square as a comparison of two variances. Such a test, the ratio of two variances, can be generalized so that any two independent variances, regardless of the number of degrees of freedom in each, may be compared under the null hypothesis that they are sample variances from populations with a common variance.

Such a test is useful for the purpose of deciding whether or not it is legitimate to pool variances, as was done in Sec. 5.4 when computing a variance for testing the hypothesis of the equality of population means using samples with unpaired observations. An appropriate criterion for testing the hypothesis of homogeneity is denoted by F or $F(m,n)$, where m and n are the degrees of freedom for the estimates of variance in numerator and denominator, respectively.

If we examine F in Eq. (5.2), we realize that large values occur when the set of alternatives $\mu_1 \neq \mu_2$ holds and that the criterion does not distinguish between $\mu_1 < \mu_2$ and $\mu_1 > \mu_2$. Thus our F test is one-tailed relative to F and variances but two-tailed relative to means, that is, $H_1: \sigma^2$ (based on means) $> \sigma^2$ (based on individuals) is equivalent to $H_1: \mu_1 > \mu_2$ or $\mu_2 > \mu_1$, that is, to $H_1: \mu_1 \neq \mu_2$. Near-zero values of t when squared become positive, near-zero values of F; clearly these call for acceptance rather than rejection of the null hypothesis. Thus F could be small, significantly so, if the numerator were small relative to the denominator. We must explain such values, for our problem, by attributing them to chance. In some problems of testing the homogeneity of two variances, there may be no reason to assume which variance will be larger, for example, σ_1^2 or σ_2^2 for the populations implied in Table 5.6; thus there is no reason to compute s_1^2/s_2^2 rather than s_2^2/s_1^2. Here is a case where the alternatives are $\sigma_1^2 \neq \sigma_2^2$. This clearly calls for a two-tailed F test.

Consider testing the null hypothesis $\sigma_1^2 = \sigma_2^2$ against the set of alternatives $\sigma_1^2 \neq \sigma_2^2$. Determine s_1^2 and s_2^2 and calculate F by Eq. (5.15).

$$F = \frac{\text{the larger } s^2}{\text{the smaller } s^2} \quad (5.15)$$

This F is compared with tabulated values in Table A.6 where the degrees of freedom for the larger mean square are given across the top of the table and those for the smaller along the side. For the set of alternatives $\sigma_1^2 \neq \sigma_2^2$, the stated significance levels are doubled. If calculated F is larger than $F_{.025}$, we claim significance at the 5% level; if larger than $F_{.005}$, at the 1% level, and so on. This test is a two-tailed test with respect to F since we do not specify which σ^2 is expected to be larger.

For the example of Sec. 5.8, compare the two sample variances by $F = 40.12/6.95 = 5.77$ for 6 degrees of freedom in both numeraor and denominator. The tabulated F value is 5.82 at the 5% level for the desired alternatives ($F_{.025} = 5.82$) and our test just misses significance.

The F tables are tabulated for convenience in making one-tailed tests since the associated alternatives are more common; in the t tests of this chapter, it was seen that the numerator was expected to be larger when the null hypothesis was false, that is, the numerator variance had to be large to deny the null hypothesis. If you square any tabulated value of $t_{.05}$ or $t_{.01}$, you will find this value in the column of the F table headed by 1 degree of freedom and opposite the appropriate number of degrees of freedom.

Exercise 5.9.1 Test the homogeneity of the variances for the two populations sampled in Table 5.2, Table 5.4, Exercise 5.4.1, and Exercise 5.4.2. Does the assumption of a common variance seem justified in all cases?

5.10 Confidence limits involving the difference between two means. It is often of interest to establish a range within which we state that the true difference $\mu_1 - \mu_2$ lies. Equation (5.16) is appropriate for a 95% confidence interval.

$$l_1 = d - t_{.05}s_d \quad \text{and} \quad l_2 = d + t_{.05}s_d \tag{5.16}$$

Examples are given in Tables 5.2, 5.4, 5.5, and 5.6.

If a test of significance at the 5% level is applied and t is less than $t_{.05}$, then the confidence interval will include zero; if t is greater than $t_{.05}$, then the confidence interval will not include zero. The examples illustrate this point.

Exercise 5.10.1 Compute 95 and 99% confidence intervals for the data in Exercises 5.4.1, 5.4.2, 5.6.1, and 5.6.2.

5.11 Sample size and the detection of differences. The ideas of Sec. 5.2, namely, Type I and Type II errors and the power of a test, are essential to a clear understanding of the problem of sample size. We now discuss this problem.

In many experiments, the experimenter is interested in determining whether a treatment is superior to a control. A one-tailed t test is appropriate. Consider Fig. 5.3. Suppose we have a parent distribution with unknown mean μ and known variance σ^2. Then A, B, C, D are illustrations of possible derived distributions of $\bar{x}$'s with different means, $\mu_0, \ldots, \mu_3$ and two possible variances σ^2/n_i, $i = 1$ or 2, depending upon the sample size. The shaded areas refer to the population with the larger variance, that is, the smaller sample size. In particular:

A: shows the distribution of $\bar{x}$ when H_0 is true. The shaded area, say $100\alpha\%$, gives the probability of a *Type* I *error*. The line dividing the two areas separates the $\bar{x}$ axis into an acceptance and rejection region. The test procedure is based on H_0 regardless of what the true μ is. If we are always presented with data for which H_0 is true, then we wrongly conclude that H_0 is false $100\alpha\%$ of the time.

B: shows the distribution of $\bar{x}$ when H_0 is false and μ_1 is the true parameter. If we falsely conclude that H_0 is true, we make a *Type* II *error*. We do this when $\bar{x}$ falls in the acceptance region determined by the test criterion used to test H_0. The shaded area of the curve measures the probability of a Type II error, over 50% in this case. If it were important to detect a real difference as large as $\mu_1 - \mu_0$, then we would do better, on the average, to toss a coin to choose between μ_0 and μ_1 for, using the coin, we would declare in favor of μ_1 50% of the time.

The power of the test with respect to the alternative B is measured by the unshaded area under B. This is associated with the critical region as determined by the test of H_0. When, in fact, $\mu = \mu_1$, this area is a measure of the ability of the test to detect μ_1. In this case, if the experimenter is presented with data for which μ_1 is true, then he will detect μ_1 less than half of the time.

C: Now the true μ is μ_2. It is far enough from μ_0 that it will be detected 50% of the time. If $\mu = \mu_2$, tossing a coin without collecting any data relative to the experiment is still as satisfactory a test procedure as one based on H_0.

D: If $\mu = \mu_3$, we have a good chance of detecting this fact as indicated by the unshaded area under the curve. The power of the test is high and the probability of a Type II error is low.

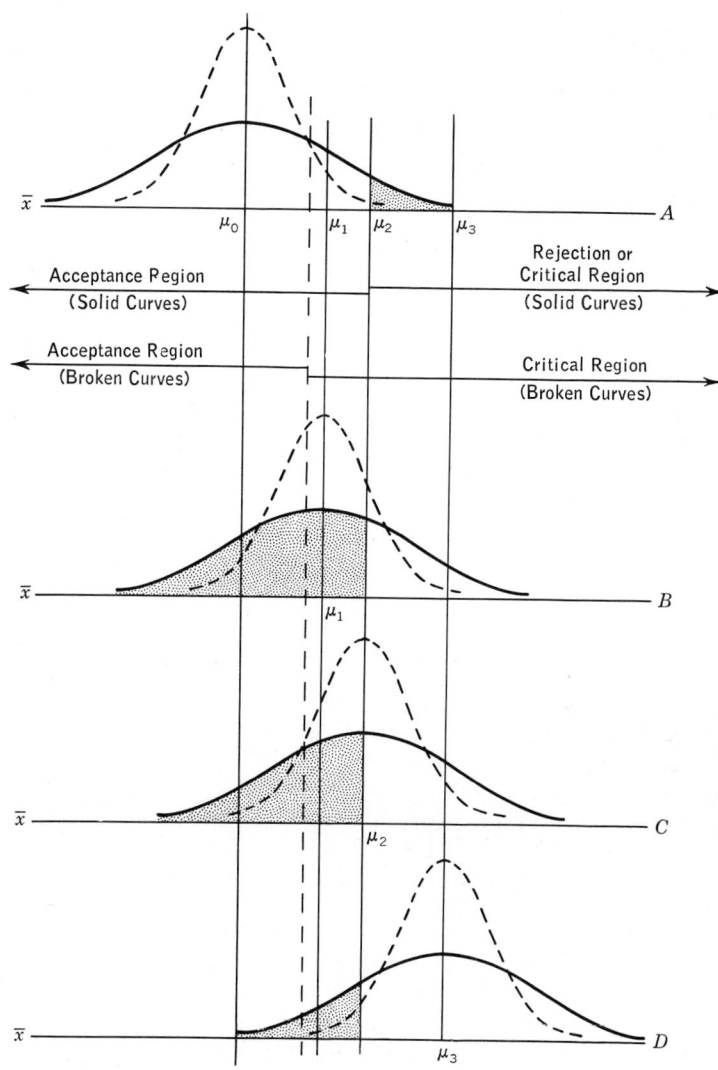

FIG. 5.3 An illustration relative to sample size

If we wish reasonable assurance of detecting μ_1 when $\mu = \mu_1$, it will be necessary to take a larger sample. This leads to a new distribution of $\bar{x}$ with a smaller variance since $\sigma_{\bar{x}}^2 = \sigma^2/n$; for example, one of the distributions shown by the broken curves. The acceptance region no longer extends so far to the right. See the broken curve for A.

Now the rejection region includes a new portion of the $\bar{x}$ axis to the left of the original rejection region. Our chances of detecting an alternative to μ_0 when an alternative is true are improved. This can be seen from the broken curves of B, C, and D. Type II error decreases.

In choosing a sample size to detect a particular difference, one must admit the possibility of either a Type I or Type II error and choose the sample size accordingly. Consider Fig. 5.4. Here are two d distributions, that for which $H_0: \mu = \mu_0$ is true and that for which $H_1: \mu = \mu_1$ is true. The sample size has been adjusted to give a variance $\sigma_{\bar{d}}^2$, such that the point $\bar{d}_0$ divides the $\bar{d}$ axis to give an area of 5% to the right of $\bar{d}_0$ and under the H_0 distribution and an area of 80% to the right of $\bar{d}_0$ and under the H_1 distribution. In other

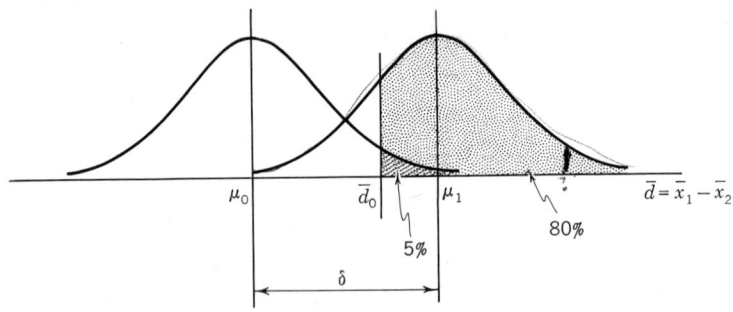

FIG. 5.4 Choosing a sample size for desired protection

words, we have a one-tailed test with an error rate of 5% associated with H_0 and will accept H_1 80% of the time when it is true; the probability of a Type I error is .05, of a Type II error is .20, and the power of the test is .80. The problem is to find a sample size to accomplish our aims. In practice, we rarely know σ^2 so must rely upon an estimate. Completion of the problem is deferred to Sec. 8.14.

5.12 Stein's two-stage sample. Where attention is concentrated on estimation rather than testing, a procedure by Stein (5.7), for determining the required number of observations, is available for continuous data.

The problem is to determine the necessary sample size to estimate a mean by a confidence interval guaranteed to be no longer than a prescribed length; the procedure is to take a sample, estimate the variance, then compute the total number of observations necessary. The additional observations are then obtained and a new mean based on all observations is computed.

A confidence interval is of length $2t_\alpha s_{\bar{x}}$. (Where σ^2 is known, no initial sample is required to estimate the sample size. This can be computed as soon as the required confidence interval length is decided upon, by setting the length equal to $2z_\alpha \sigma/\sqrt{n}$ and solving for n.) When σ^2 is unknown, we take a sample and estimate n by Eq. (5.17).

$$n = \frac{t_1^2 s^2}{d^2} \quad (5.17)$$

t_1 is the tabulated t value for the desired confidence level and the degrees of freedom of the initial sample and d is the half-width of the desired confidence interval, which gives the total number of observations required. The additional observations are obtained and a new $\bar{x}$ is computed. This $\bar{x}$ will be within distance d of μ unless the total procedure has resulted in an unusual sample, unusual enough that more extreme ones are to be found no more than $100\alpha\%$ of the time, due to chance. The procedure is readily applicable to obtaining sample size for a confidence interval for a difference between population means by using the appropriate s^2.

Example. In a field study, R. T. Clausen† wished to obtain a 95% confidence interval of not more than 10 mm for leaf length of mature leaves of *Sedum lucidum* on some individual plants. On a plant near Orizaba, Mexico, a sample of five leaves gave lengths 22, 19, 13, 22, and 23 mm, for which $\bar{x} = 19.8$ and $s = 4.1$ mm. Hence, the total sample size should be

$$n = \frac{(7.71)(16.7)}{25} = 5.2 \text{ observations}$$

where $7.71 = F(1,4) = t^2_{.05}$, $16.7 = s^2$, and $25 = (10/2)^2$. The confidence interval just misses being within the acceptable standard of length; with one more observation and the new mean of all six, we would simply say that the new $\bar{x}$ was within 5 mm of μ with confidence .95.

† Cornell University, Ithaca, N.Y.

References

5.1 Cochran, W. G., and G. M. Cox: *Experimental Designs*, 2d ed., John Wiley & Sons, Inc., New York, 1957.

5.2 Hart, J. S.: "Calorimetric determination of average body temperature of small mammals and its variation with environmental conditions," *Can. J. Zool.*, **29:** 224–233 (1951).

5.3 Peterson, H. L.: "Pollination and seed set in lucerne," *Roy. Vet. and Agr. Coll. Yearbook*, pp. 138–169, Copenhagen, 1954.

5.4 Ross, R. H., and C. B. Knodt: "The effect of supplemental vitamin A upon growth, blood plasma, carotene, vitamin A, inorganic calcium, and phosphorus content of Holstein heifers," *J. Dairy Sci.*, **31:** 1062–1067 (1948).

5.5 Rowles, W.: "Physical properties of mineral soils of Quebec," *Can. J. Research*, **16:** 277–287 (1938).

5.6 Shuel, R. W.: "Some factors affecting nectar secretion in red clover," *Plant Physiol.*, **27:** 95–110 (1952).

5.7 Stein, C.: "A two-sample test for a linear hypothesis whose power is independent of the variance," *Ann. Math. Stat.*, **16:** 243–258 (1945).

5.8 Watson, C. J., et al.: "Digestibility studies with ruminants. XII. The comparative digestive powers of sheep and steers," *Sci. Agr.*, **28:** 357–374 (1948).

Chapter 6

PRINCIPLES OF EXPERIMENTAL DESIGN

6.1 Introduction. This chapter is an introduction to the planning and conduct of experiments in relation to aims, analysis, and efficiency.

If we accept the premise that new knowledge is most often obtained by careful analysis and interpretation of data, then it is paramount that considerable thought and effort be given to planning their collection in order that maximum information be obtained for the least expenditure of resources. Probably the most important function of the consulting statistician is providing assistance in designing efficient experiments that will enable the experimenter to obtain unbiased estimates of treatment means and differences and of experimental error.

That the statistician can make a real contribution at the planning stage of an experiment cannot be overemphasized. Frequently the statistician is presented with data that provide only biased estimates of treatment means and differences and experimental error, that do not provide answers to the questions initially posed, in which certain treatments do not provide pertinent information, for which the conclusions which can be drawn do not apply to the population the experimenter had in mind, and such that the precision of the experiment is not sufficiently great to detect important differences. Often by some slight change in the design and with less effort, the experiment would have provided the information desired. The experimenter who contacts the consulting statistician at the planning stage of the experiment rather than after its completion improves the chances of achieving his aims.

6.2 What is an experiment? Different definitions are available for the word *experiment*. For our purposes, we consider an experiment as a planned inquiry to obtain new facts or to confirm or deny the results of previous experiments, where such inquiry will aid in an administrative decision, such as recommending a variety, procedure, or pesticide. Such experiments fall roughly into three categories, namely, preliminary, critical, and demonstrational, one of which may lead to another. In a preliminary experiment, the investigator tries out a large number of treatments in order to obtain leads for future work; most treatments will appear only once. In a critical experiment, the investigator compares responses to different treatments using sufficient observations of the responses to give reasonable assurance of detecting meaningful differences. Demonstrational experiments are performed when extension workers compare a new treatment or treatments with a standard.

PRINCIPLES OF EXPERIMENTAL DESIGN

In this text, we are concerned almost entirely with the critical type of experiment. In such an experiment, it is essential that we define the population to which inferences are to apply, design the experiment accordingly, and make measurements of the variables under study.

Every experiment is set up to provide answers to one or more questions. With this in mind, the investigator decides what treatment comparisons provide relevant information. He then conducts an experiment to measure or to test hypotheses concerning treatment differences under comparable conditions. He takes measurements and observations on the experimental material. From the information in a successfully completed experiment, he answers the questions initially posed. Sound experimentation consists of asking questions that are of importance in the field of research and in carrying out experimental procedures which answer these questions. In this chapter, we are primarily concerned with the procedures.

To the statistician the experiment is the set of rules used to draw the sample from the population. This makes definition of the population most important. The set of rules is the experimental procedure or experimental design. For example, the use of unpaired observations and the use of paired observations are experimental designs for two-treatment experiments.

6.3 Objectives of an experiment. In designing an experiment, state objectives clearly as questions to be answered, hypotheses to be tested, and effects to be estimated. It is advisable to classify objectives as major and minor since certain experimental designs give greater precision for some treatment comparisons than for others.

Precision, sensitivity, or *amount of information* is measured as the reciprocal of the variance of a mean. As n increases, $\sigma_{\bar{x}}^2 = \sigma^2/n$ decreases. A comparison of two sample means becomes more sensitive, that is, can detect a smaller difference between population means, as the sample sizes increase.

It is of paramount importance to define the population for which inferences are to be drawn and to sample that population randomly. Suppose that the major objective of an experiment is to compare the values of several rations for swine in a certain area. Suppose also that the farmers in the area handle pigs of different breeds, that some use self-feeders and that others hand-feed. If the experimenter used only one breed of swine in his experiment or fed only by self-feeders, his sample could hardly be considered as representative of the population unless he had previous information that breed and method of feeding had little or no effect on differences due to rations. If no information is available on the effect of breed and method of feeding, it would be extremely hazardous to make inferences from an experiment based upon one breed and one method of feeding to other breeds and methods. To make such recommendations, the experimenter would have to include all local breeds and feeding methods as factors in the experiment. By doing so, the scope of his experiment is increased.

Another example is that of an experiment planned to compare the effectiveness of several fungicides in controlling a disease in oats. Suppose that, in the area where these recommendations are to be applied, several varieties are in common use and that the quality of seed varies. To be able to make adequate recommendations, the experimenter must compare the fungicides on several varieties and several qualities of seed for each variety. This is essential if the experimenter is to determine whether or not a single fungicide can be recommended for all varieties and qualities of seed used in the area. In general, worthwhile inferences about an extensive population cannot usually be obtained from a single isolated experiment.

6.4 Experimental unit and treatment. An *experimental unit* or *experimental plot* is the unit of material to which one application of a treatment is applied; the *treatment* is the procedure whose effect is to be measured and compared with other treatments. These terms are then seen to be very general. The experimental unit may be an animal, ten birds in a pen, a half-leaf, and so on; the treatment may be a standard ration, a spraying schedule, a temperature–humidity combination, etc. When the effect of a treatment is measured, it is measured on a *sampling unit*, some fraction of the experimental unit. Thus the sampling unit may be the complete experimental unit, such as an animal on a treatment ration, or a random sample of leaves from a sprayed tree, or the harvest from 6 ft of the middle row of a three-row experimental unit.

In selecting a set of treatments, it is important to define each treatment carefully and to consider it with respect to every other treatment to assure, as far as possible, that the set provides efficient answers related to the objectives of the experiment.

6.5 Experimental error. A characteristic of all experimental material is variation. *Experimental error* is a measure of the variation which exists among observations on experimental units treated alike. Variation comes from two main sources. First, there is the inherent variability which exists in the experimental material to which treatments are applied. Secondly, there is the variation which results from any lack in uniformity in the physical conduct of the experiment. In a nutrition experiment with rats as the experimental material, the individuals will have different genetic constitutions unless highly inbred; this is variability inherent in the experimental material. They will be housed in cages subject to differences in heat, light, and other factors; this constitutes a lack of uniformity in the physical conduct of the experiment. The relative magnitudes of the variation from these two sources will be quite different for various fields of research.

It is important that every possible effort be made to reduce the experimental error in order to improve the power of a test, to decrease the size of confidence intervals, or to achieve some other desirable goal. This can be accomplished by attacking the two main sources of experimental error. Thus, we can:

1. Handle the experimental material so that the effects of inherent variability are reduced, and
2. Refine the experimental technique.

These methods will be discussed in sections succeeding a discussion of replication and the factors affecting the number of replicates.

6.6 Replication and its functions. When a treatment appears more than once in an experiment, it is said to be *replicated*. The functions of replication are:

1. to provide an estimate of experimental error,
2. to improve the precision of an experiment by reducing the standard deviation of a treatment mean,

3. to increase the scope of inference of the experiment by selection and appropriate use of quite variable experimental units, and

4. to effect control of the error variance.

An estimate of experimental error is required for tests of significance and for confidence interval estimation. An experiment in which each treatment appears only once is said to consist of a single replicate or replication. From such an experiment, no estimate of experimental error is available. Here, it is possible to explain an observed difference as a difference between treatments or between experimental units; it is impossible to have objective assurance as to which explanation is correct. In other words, when there is no method of estimating experimental error, there is no way to determine whether observed differences indicate real differences or are due to inherent variation. The experiment is not self-contained and any inference must be based upon prior experience. (*Exception.* A single replication or even a particular fraction of a replicate of an experiment involving a large number of factors or types of treatments may be used and will provide an estimate of error when certain assumptions hold.)

As the number of replicates increases, estimates of population means, namely, the observed treatment means, become more precise. If a difference of five units is detectable using four replicates, an experiment of approximately 16 replicates will detect half this difference or 2.5 units since the standard deviations are in the ratio of $2:1$, being $\sigma/\sqrt{4}$ and $\sigma/\sqrt{16}$, respectively. The word "approximate" is used because precision, especially in small experiments, depends in part upon the number of degrees of freedom available for estimating experimental error. Also, increased replication may require the use of less homogeneous experimental material or a less careful technique, thus giving a new parent population with a larger experimental error. However, increased replication usually improves precision, decreasing the lengths of confidence intervals and increasing the power of statistical tests.

In certain types of experiment, replication is a means of increasing the scope of inference of an experiment; the sampled population is less restricted in definition and the inference is broadened. For example, suppose we wish to determine whether there is any real difference in the performance of two varieties of a crop in a given area and that there are two major soil types in this area. If the object of the experiment is to draw inferences for both soil types, it is obvious that both should be in the experiment. It is also important that the area included within each replicate, that is, in each pair of plots on which the pair of varieties is planted, be of one soil type and as uniform as possible; it is not necessary and may be undesirable to have very uniform conditions between or among replicates, especially where rather broad populations are involved.

In many field experiments, the experiment is repeated over a period of years. The reason is obvious, namely, that conditions vary from year to year and make it important to know the effect of years on the different treatments since recommendations are usually made for future years. Likewise, different locations are used to evaluate treatments under the different environmental conditions present in the population, that is, in the area for which recommendations are to be made. Both repetitions in time (years) and space (location) can be considered as broad types of replication. The purpose is to increase the scope of inference. The same principle is frequently used in laboratory experiments, namely, that the entire experiment is repeated several times, possibly by different people, to determine the repeatability of treatments under possibly different conditions which may exist from time to time in the laboratory.

6.7 Factors affecting the number of replicates. The number of replications for an experiment depends upon several factors, of which the most important is probably the degree of precision required. The smaller the departure from the null hypothesis to be measured or detected, the greater is the number of replicates required.

It is important in any experiment to have the correct amount of precision. There is little point in using ten replicates to detect a difference that four will find in most cases; likewise there is little value in performing an experiment where the number of replications is not sufficient to detect important differences except occasionally.

To measure any departure from the null hypothesis, experimental error, that is, variation among observations on experimental units, must supply the unit of measurement. At times it is not practical to make an observation on the complete experimental unit, for example, in the case of chemical determinations such as that of the protein content of a forage, and the unit is sampled. Usually the variation among experimental units is large in comparison with the variation among samples from a single unit. There is little point in making a large number of chemical determinations on each experimental unit since experimental error must be based on variation among the experimental units, not on variation among samples from within the units.

Certain material is naturally more variable than others. Consider the problem of soil heterogeneity. Certain soils are more uniform than others and, for the same precision, less replication is required on uniform soil than on variable soil. Also, different crops grown at the same location show unequal variabilities.

The number of treatments affects the precision of an experiment, and thus the number of replications required for a stated degree of precision. This is of little importance except in small experiments, less than 20 degrees of freedom for error, since it implies improving the estimate of σ^2 at the expense of n in $s_{\bar{x}}^2 = s^2/n$. If only two treatments are to be compared, then more replicates are required per treatment than if ten treatments are to be compared with the same precision, since σ^2 is estimated with more degrees of freedom in the latter case.

The experimental design also affects the precision of an experiment and the required number of replications. Where the number of treatments is large and requires the use of more heterogeneous experimental units, the experimental error per unit increases. Appropriate designs can control part of this variation.

Unfortunately the number of replicates is very often determined largely by the funds and time available for the experiment. There is little point in an experiment if the required precision is not obtainable with available funds. The solution is to postpone the experiment until sufficient funds are available

PRINCIPLES OF EXPERIMENTAL DESIGN

or to reduce the number of treatments so that sufficient replication and precision are available for the remaining ones. The practical number of replicates for an experiment is reached when the cost of the experiment in material, time, etc., is no longer offset by an increase in information gained.

Worthy of mention is the fact that replication does not reduce error due to faulty technique. Also, mere statistical significance may give no information as to the practical importance of a departure from the null hypothesis. The magnitude of a real departure that is of practical significance can be judged by one with technical knowledge of the subject; this magnitude together with a measure of how desirable it is to detect it serves to determine the required precision and, eventually, the necessary number of replicates.

Since so much has been said about the appropriate amount of replication, the question arises as to methods of ascertaining this amount. The problem of determining sample size was discussed in Secs. 5.11 and 5.12. Further discussion is given in Sec. 8.14. With the right kind of information, the experimenter can usually find an available method for determining the replication needed for his purposes.

6.8 Relative precision of designs involving few treatments. The precision of an experiment is affected by the number of degrees of freedom available for estimating experimental error. The degrees of freedom depend on the number of replicates, number of treatments, and experimental design. The increase in precision may be very worthwhile when fewer than 20 degrees of freedom are involved. Thus observe that $t_{.05}$ for 5 degrees of freedom is 2.57, for 10 degrees of freedom is 2.23, for 20 degrees of freedom is 2.09, and for 60 degrees of freedom is 2.00. The value of t at any probability level decreases noticeably for each additional degree of freedom up to 20; beyond this point the decrease is slow.

Fisher's (6.3) procedure, as given by Cochran and Cox (6.1), compares two designs with respect to precision. The term *relative efficiency* is used. The efficiency of design 1 relative to design 2 is given by Eq. (6.1).

$$\text{RE} = \frac{(n_1 + 1)/(n_1 + 3)s_1^2}{(n_2 + 1)/(n_2 + 3)s_2^2} = \frac{(n_1 + 1)(n_2 + 3)s_2^2}{(n_2 + 1)(n_1 + 3)s_1^2} \tag{6.1}$$

where s_1^2 and s_2^2 are the error mean squares of the first and second designs, respectively, and n_1 and n_2 are their degrees of freedom. If the number of observations in a treatment mean differs for the two experiments being compared, replace s_1^2 and s_2^2 by $s_{\bar{x}_1}^2$ and $s_{\bar{x}_2}^2$.

Suppose we wish to compare a design allowing 5 degrees of freedom to estimate σ^2 with one allowing 10 degrees of freedom; this could be a comparison of a paired experiment with a nonpaired one for two treatments and six replications. The paired design is more precise than the nonpaired only if the efficiency of the former relative to the latter is greater than one, that is, if the relative efficiency

$$\frac{(n_1 + 1)(n_2 + 3)s_2^2}{(n_2 + 1)(n_1 + 3)s_1^2} = \frac{(6)(13)s_2^2}{(11)(8)s_1^2} = \frac{.886\, s_2^2}{s_1^2}$$

is greater than one. In other words, pairing pays if $.886\, s_2^2$ is greater than s_1^2.

Exercise 6.8.1 Analyze the data of Table 5.4 as though the observations had not been paired. What is the efficiency of this "unpaired" design relative to that of the "paired" design which was used?

6.9 Error control. *Error control* can be accomplished by:
1. Experimental design
2. The use of concomitant observations
3. The choice of size and shape of the experimental units

1. *Experimental design.* The use of experimental design as a means of controlling experimental error has been widely investigated since about the end of the first quarter of the present century. This is an extensive subject and only the basic principles will be discussed here. For more extensive treatment of the subject, the reader is referred to Cochran and Cox (6.1), Federer (6.2), and Kempthorne (6.4).

Control of experimental error consists of designing an experiment so that some of the natural variation among the set of experimental units is physically handled so as to contribute nothing to differences among treatment means. For example, consider a two-treatment experiment that uses a pair of pigs from each of 10 litters. If we apply the treatments, one to each member of a pair, then the mathematical description is $X_{ij} = \mu + \tau_i + \rho_j + \epsilon_{ij}$ and the difference between means is $\bar{x}_{1.} - \bar{x}_{2.} = (\tau_1 - \tau_2) + (\bar{\epsilon}_{1.} - \bar{\epsilon}_{2.})$; there is no contribution due to variation among the pairs, that is, due to the set of ρ's. Alternately if there is no pairing and 10 pigs are simply chosen at random from the 20 and assigned to a specific treatment, then the description is $X_{ij} = \mu + \tau_i + \epsilon_{ij}$ where each ϵ_{ij} now includes a contribution from litters and is an element of a population with a presumably larger variance than that associated with the ϵ's of the first model. Thus choice of a design with non-random pairs has controlled the experimental error σ^2, provided that pairs are an additional source of variation; that is, provided the natural variation between observations on individuals of the same pair is less than that between observations on individuals of different pairs. Common sense and acuity in recognizing sources of variation are seen to be basic in choosing a design.

Where the experimental units are grouped into complete blocks, that is, where each contains all treatments, such that the variation among units within a block is less than that among units in different blocks, the precision of the experiment is increased as a result of error control. Such blocks of similar outcome are also called *replicates*. The design is known as a *randomized complete-block design*. Experimental error is based upon the variation among units within a replicate after adjustment for any observed, over-all treatment effect. In other words, variation among replicates and variation among treatments do not enter into experimental error.

As the number of treatments in an experiment increases, the number of experimental units required for a replicate increases. In most instances, this results in an increase in the experimental error, that is, in the variance in the parent population. Designs are available where the complete block is subdivided into a number of incomplete blocks such that each incomplete block contains only a portion of the treatments. The subdivision into incomplete blocks is done according to certain rules, so that the experimental error can be estimated among the units within the incomplete blocks. Precision is increased to the extent that the experimental units within an incomplete block are more uniform than the incomplete blocks within a replicate. Such designs are called *incomplete-block designs*. The reader is referred to Cochran and Cox (6.1) and Federer (6.2).

The *split-plot design* is an incomplete-block design where the precision of certain comparisons is increased at the expense of others. The over-all precision is the same as the basic design used. Some split-plot designs are considered in Chap. 12.

In experiments where comparisons among all treatments are of essentially the same importance, a different type of incomplete-block design is used. The treatments are assigned to the experimental units so that every treatment occurs with every other treatment the same number of times within its set of incomplete blocks. Such designs are known as *balanced incomplete blocks*. Another group, known as *partially balanced lattices*, is available where each treatment occurs only with certain other treatments, not all, in its set of incomplete blocks.

The best design to use in any given situation is the simplest design available which gives the required precision. There is no point in using an involved design if an increase in precision is not obtained.

2. *The use of concomitant observations.* In many experiments, the precision can be increased by the use of accessory observations and an arithmetic technique called *covariance*. Covariance is used when the variation among experimental units is, in part, due to variation in some other measurable character or characters not sufficiently controllable to be useful in assigning the experimental units to complete or incomplete blocks on the basis of similar outcome. Covariance is discussed in Chap. 15.

3. *Size and shape of experimental units.* As a rule, large experimental units show less variation than small units. However, an increase in the size of the experimental unit often results in a decrease in the number of replicates that can be run since, usually, a limited amount of experimental material is available for any given experiment. Adequate replication of small plots is generally easier to obtain than adequate replication of large plots.

In field plot experiments, the size and shape of the experimental unit or plot, as well as that of the complete or incomplete block, are important in relation to precision. *Uniformity trial* studies, that is, studies of data from experiments where there were no treatments, conducted with many crops and in many different countries, have shown that the individual plot should be relatively long and narrow for greatest precision; the block, whether complete or incomplete, should be approximately square. For a given amount of variability among experimental units this tends to maximize the variation among the blocks while minimizing that among plots within blocks. Large variation among blocks indicates that their use has been helpful because this variation is eliminated from experimental error and, also, does not contribute to differences among treatment means. When blocks are square, the differences among blocks tend to be large. It is desirable to have small variation among plots within blocks and, at the same time, have the plot representative of the block. In fields where definite fertility contours appear, the most precision is obtained when the long sides of the plots are perpendicular to the contours or parallel to the direction of the gradient.

For some types of experiment, the experimental units are carefully selected to be as uniform as possible; for example, rats from an inbred line might be the experimental units in a nutrition experiment. As a result, experimental error is reduced, but at the same time, the scope of the inference is also reduced. The response obtained with selected experimental units may differ widely from that with unselected units and inferences must be made with this in mind.

6.10 Choice of treatments. In certain types of experiment, the treatments have a substantial effect on precision. This is especially true of factorial experiments, discussed in Chap. 11.

In certain types of experiment, the amount or rate of some factor is important. Suppose that the experimenter is measuring the effect of increased levels of some nutrient on the response of a plant. It is important to include several levels to determine if the response is linear or curvilinear in nature. Here, choice of the number of levels and their spacing is important for appropriate answers to the questions asked.

In general, the more the investigator knows about his treatments, the better is the statistical test procedure that the statistician can devise. This knowledge often dictates the kind and amount of any particular treatment in a set of treatments. This, in turn, may influence the precision of the experiment.

6.11 Refinement of technique. The importance of careful technique in the conduct of an experiment is self-evident. It is the responsibility of the experimenter to see that everything possible is done to assure the use of careful technique as no analysis, statistical or other, can improve data obtained from poorly performed experiments. In general, variation resulting from careless technique is not random variation and not subject to the laws of chance on which statistical inference is based. This variation may be termed *inaccuracy*, in contrast to lack of precision. Unfortunately, accuracy in technique does not always result in high precision because precision is also concerned with the random variability among experimental units, which may be quite large.

Some points of technique to be considered in conducting an experiment follow. It is important to have uniformity in the application of treatments. This applies equally to the spreading of fertilizer, spraying of fruit trees, cutting of a forage crop to a fixed height for all plots, and the filling of test tubes to a fixed level. Control should be exercised over external influences so that all treatments produce their effects under comparable and desired conditions. For example, to compare the effects of several treatments it may be necessary to create an epidemic by the use of an artificial inoculum. Efforts should be made to have the epidemic as uniform as possible. Again, if it is impossible to have uniform conditions, it may still be possible to have them uniform within each block. For example, in field crop experiments, if a complete experiment cannot be set out or harvested in one day, it is desirable to do complete blocks in any day; in laboratory experiments where it is necessary to use several technicians, it is desirable that each should manage one or more complete sets of treatments.

Suitable and unbiased measures of the effects of the several treatments or of differences among them should be available. Often the required measurements are obvious and easily made; in other instances, considerable research is necessary to secure a reliable measurement to express treatment effects. Thus the sanitary engineering data of Table 2.3 came after repeated trials and revisions of technique because it had been decided that no technique would be used until the coefficient of variation was reduced to the order of 5%. Care should always be exercised to prevent gross errors which can occur in experimentation; adequate supervision of assistants and close scrutiny of the data will go far in preventing them.

Faulty technique may increase the experimental error in two ways. It may introduce additional fluctuations of a more or less random nature and possibly subject to the laws of chance. Such fluctuations, if substantial, should reveal themselves in the estimate of the experimental error, possibly by observation of the coefficient of variation. If an experimenter finds that his estimates of experimental error are consistently higher than those of other workers in the same field, he should carefully scrutinize his technique to determine the origin of the errors. The other way in which faulty technique may increase experimental error is through nonrandom mistakes. These are not subject to the laws of chance and may not always be detected by observation of the individual measurements. Statistical tests are also available for their detection but are not discussed in this text. Faulty technique may also result in measurements that are consistently biased. This does not affect experimental error or differences among treatment means but does affect the values of treatment means. Experimental error cannot detect bias. The experimental error estimates precision or repeatability, not the accuracy, of measurements.

A point sometimes overlooked is that a measurement may be subject to two principal sources of variation, one of which is considerably larger than the other. For illustration, consider the amount of protein per acre produced by a forage crop. The amount is a function of yield per acre and percentage of protein. The variation among measurements of yield of forage is much greater than that among determinations of protein percentage. As a consequence, more effort should be placed in reducing the part of experimental error associated with yield of forage than that associated with protein percentage. In general, where there are several sources of variation, it is most profitable to try to control the greater source.

6.12 Randomization. The function of randomization is to assure that we have a valid or unbiased estimate of experimental error and of treatment means and differences among them. Randomization is one of the few characteristics of modern experimental design that appears to be really new and the idea may be credited to R. A. Fisher. Randomization generally involves the use of some chance device such as the flipping of a coin or the use of random number tables. Randomness and haphazardness are not equivalent and randomization cannot overcome poor experimental technique.

To avoid bias in comparisons among treatment means, it is necessary to have some way of assuring that a particular treatment will not be consistently favored or handicapped in successive replicates by some extraneous sources of variation, known or unknown. In other words, every treatment should have an equal chance of being assigned to any experimental unit, be it unfavorable or favorable. Randomness provides the equal chance procedure. Cochran and Cox (6.1) state "Randomization is somewhat analogous to insurance, in that it is a precaution against disturbances that may or may not occur, and that may or may not be serious if they do occur."

Systematic designs, where the treatments are applied to the experimental units in a nonrandom but selected fashion, often result in either underestimation or overestimation of experimental error. Also, they can result in inequality of precision in the various comparisons among the treatment means. This is especially obvious in many field experiments. Numerous studies have shown that adjacent plots tend to be more alike in their productivity than plots which are some distance apart. Such plots are said to give correlated error components or residuals. As a result of this fact, if the treatments are arranged in the same systematic order in each replicate, then there can be considerable differences in the precision of comparisons involving different treatments. The precision of comparisons among treatments which are physically close is greater than those among treatments which are some distance apart. Randomization tends to destroy the correlation among errors and make valid the usual tests of significance.

6.13 Statistical inference. As we have seen, the object of experiments is to determine if there are real differences among our treatment means and to estimate the magnitude of such differences if they exist. A statistical inference about such differences involves the assignment of a measure of probability to the inference. For this, it is necessary that randomization and replication be introduced into the experiment in appropriate fashion.

Replication assures us the means of computing experimental error.

Randomization assures us a valid measure of experimental error.

A choice between an experiment with appropriate randomization and a systematic one with apparently greater but unmeasurable precision is like a choice between a road of known length and condition and a road of unknown length and condition which is known only to be shorter. It may be more satisfactory to know how far one has to go and what the going will be like rather than to start off on a road of unknown condition and length with only the assurance that it is the shorter road. Until further study of systematic designs has been made, it would seem advisable to consult a statistician prior to using one.

References

6.1 Cochran, William G., and Gertrude M. Cox: *Experimental Designs*, 2d ed., John Wiley & Sons, Inc., New York, 1957.

6.2 Federer, W. T.: *Experimental Design*, The Macmillan Company, New York, 1955.

6.3 Fisher, R. A.: *The Design of Experiments*, 4th ed., Oliver & Boyd, Ltd., Edinburgh, 1947.

6.4 Kempthorne, O.: *The Design and Analysis of Experiments*, John Wiley & Sons, Inc., New York, 1952.

6.5 Brownlee, K. A.: "The principles of experimental design," *Industrial Quality Control*, **13:** 12–20 (1957).

6.6 Greenberg, B. G.: "Why randomize?", *Biometrics*, **7:** 309–322 (1951).

Chapter 7

ANALYSIS OF VARIANCE I: THE ONE-WAY CLASSIFICATION

7.1 Introduction. The analysis of variance was introduced by Sir Ronald A. Fisher and is essentially an arithmetic process for partitioning a total sum of squares into components associated with recognized sources of variation. It has been used to advantage in all fields of research where data are measured quantitatively.

In Chap. 7, we consider the analysis of variance where treatment is the only criterion for classifying the data. We discuss confidence interval estimation, the general topic of comparing treatment means, the linear additive model and variance components, the assumptions underlying the analysis of variance and tests of significance, residual mean square or experimental error, and sampling error.

In Chaps. 8, 11, 12, and 13, additional designs and aspects of analysis of variance are discussed.

7.2 The completely random design. This design is useful when the experimental units are essentially homogeneous, that is, the variation among them is small, and grouping them in blocks would be little more than a random procedure. This is the case in many types of laboratory experiments where a quantity of material is thoroughly mixed and then divided into small lots to form the experimental units to which treatments are randomly assigned, or in plant and animal experiments where environmental effects are much alike.

Randomization, the process which makes the laws of chance applicable, is accomplished by allocating the treatments to the experimental units entirely at random. No restrictions are placed upon randomization as when a block is required to contain all treatments. The choice of numbers of observations to be made on the various treatments is not considered to be a restriction on randomization. Every experimental unit has the same probability of receiving any treatment, that is, if there are n experimental units then any one of the n treatments, clearly not all different, has the same probability of falling on any experimental unit.

Randomization is carried out by use of a random number table. Suppose 15 experimental units are to receive five replications of each of three treatments. Assign the numbers 1 to 15 to the experimental units in a convenient manner, for instance, consecutively. Locate a starting point in a table of random numbers, for example, row number 10 and column number 20 of

Table A.1, and select 15 three-digit numbers. Reading vertically, we obtain:

118	701	789	965	688	638	901	841	396	802	687	938	377	392	848
1	8	9	15	7	5	13	11	4	10	6	14	2	3	12

Ranks are then assigned to the numbers; thus 118 is the smallest, rank number 1, and 965 is the largest, rank number 15. These ranks are considered to be a random permutation of the numbers 1 to 15 and the first five are the numbers of the experimental units to which treatment 1 is assigned. Thus units 1, 8, 9, 15, and 7 receive treatment 1, and so on.

The procedure is also applicable when the treatments are replicated unequal numbers of times, say 6, 6, and 3. Three-digit numbers are used since they are less likely to include ties than are two-digit numbers. In any case, ties may be broken by using extra numbers.

Random number tables may be used in alternative fashion. For example, two-digit numbers less than 90 may be divided by 15 and the remainder recorded. This gives the 15 numbers 00, 01, ..., 14 with equal frequency. Duplicates are discarded and other numbers obtained to replace them. The numbers 90, 91, ..., 99 are not used since they would cause 00, 01, ..., 09 to be present in greater frequency than 10, 11, ..., 14.

If, during the course of the experiment, all experimental units are to be treated similarly, for example, hoeing field plots, this should be done in random order if order is likely to affect the results, as when a technique improves as the result of practice.

The analysis for a completely random design is also applicable to data where "treatment" implies simply a variable of classification and when it may even be necessary to assume randomness. For example, one might have weight measurements on the adults of a certain species of fish caught in several lakes (the treatments), and wish to know if adult weight differs from lake to lake.

Advantages. The completely random design is flexible in that the number of treatments and replicates is limited only by the number of experimental units available. The number of replicates can vary from treatment to treatment though it is generally desirable to have the same number per treatment. The statistical analysis is simple even if the number of replicates varies with the treatment and if the various treatments are subject to unequal variances, usually referred to as nonhomogeneity of experimental error. Simplicity of analysis is not lost if some experimental units or entire treatments are missing or rejected.

The loss of information due to missing data is small relative to losses with other designs. The number of degrees of freedom for estimating experimental error is maximum; this improves the precision of the experiment and is important with small experiments, that is, with those where degrees of freedom for experimental error are less than 20 (see Sec. 6.8).

Disadvantages. The main objection to the completely random design is that it is often inefficient. Since randomization is unrestricted, experimental error includes the entire variation among the experimental units except that due to treatments. In many situations it is possible to group experimental units so that variation among units within groups is less than that among units in different groups. Certain designs take advantage of such grouping, exclude variation among groups from experimental error, and increase the precision of the experiment. Some of these designs are discussed in Chaps. 8, 11, and 12.

7.3 Data with a single criterion of classification. The analysis of variance for any number of groups with equal replication. Table 7.1 gives the nitrogen content in milligrams of red clover plants inoculated with cultures of *Rhizobium trifolii* plus a composite of five *Rhizobium meliloti* strains, as reported by Erdman (7.9). Each of five red clover cultures, *R. trifolii*, was

TABLE 7.1

NITROGEN CONTENT OF RED CLOVER PLANTS INOCULATED WITH COMBINATION CULTURES OF *Rhizobium trifolii* STRAINS AND *Rhizobium meliloti* STRAINS, IN MILLIGRAMS

R. trifolii strain

Computation	3DOk1	3DOk5	3DOk4	3DOk7	3DOk13	Composite	Total
	19.4	17.7	17.0	20.7	14.3	17.3	
	32.6	24.8	19.4	21.0	14.4	19.4	
	27.0	27.9	9.1	20.5	11.8	19.1	
	32.1	25.2	11.9	18.8	11.6	16.9	
	33.0	24.3	15.8	18.6	14.2	20.8	
$\sum_j X_{ij} = X_i.$	144.1	119.9	73.2	99.6	66.3	93.5	$596.6 = X..$
$\sum_j X_{ij}^2$	4,287.53	2,932.27	1,139.42	1,989.14	887.29	1,758.71	12,994.36
$(X_i.)^2/r$	4,152.96	2,875.20	1,071.65	1,984.03	879.14	1,748.45	12,711.43
$\sum_j x_{ij}^2$	134.57	57.07	67.77	5.11	8.15	10.26	282.93
$\bar{x}_i.$	28.8	24.0	14.6	19.9	13.3	18.7	

tested individually with a composite of five alfalfa strains, *R. meliloti*, and a composite of the red clover strains was also tested with the composite of the alfalfa strains, making six treatments in all. The experiment was conducted in the greenhouse using a completely random design with five pots per treatment.

Computations. Arrange the data as in Table 7.1. Let X_{ij} denote the *j*th observation on the *i*th treatment, $i = 1, 2, \ldots, t$ and $j = 1, 2, \ldots, r$. Treatment totals require one subscript and the total for the *i*th treatment may be denoted by $X_i.$, the dot indicating that all observations for the *i*th treatment have been added to give this total. Letters t and r are used for the number of treatments and number of replicates of each treatment; here $t = 6$ and $r = 5$.

For each treatment, obtain $X_i.$ and $\sum_j X_{ij}^2$ simultaneously on a computing machine as in lines 1 and 2 of the computations of Table 7.1. These values are then totaled; thus $\sum_i X_{i.} = X_{..}$ and $\sum_i (\sum_j X_{ij}^2) = \sum_{i,j} X_{ij}^2$. In line 3, each treatment total is squared and divided by $r = 5$, the number of replicates per treatment.

Obtain the correction term C of Eq. (7.1) and the total sum of squares (adjusted for the mean), of Eq. (7.2). The correction term is the squared sum of all the observations divided by their number. For these data

$$C = \frac{X_{..}^2}{rt} = \frac{(\sum_{i,j} X_{ij})^2}{rt} \tag{7.1}$$

$$= \frac{(596.6)^2}{(5)(6)} = 11{,}864.38$$

$$\text{Total SS} = \sum_{i,j} X_{ij}^2 - C \tag{7.2}$$

$$= 12{,}994.36 - 11{,}864.38 = 1{,}129.98$$

The sum of squares attributable to the variable of classification, that is, treatments, usually called the *between* or *among groups sum of squares* or *treatment sum of squares*, is found by Eq. (7.3).

$$\text{Treatment SS} = \frac{X_{1.}^2 + \cdots + X_{t.}^2}{r} - C \tag{7.3}$$

$$= \frac{(144.1)^2 + \cdots + (93.5)^2}{5} - 11{,}864.38 = 847.05$$

The sum of squares among individuals treated alike is also called the *within groups sum of squares, residual sum of squares, error sum of squares,* or *discrepance,* and is generally obtained by subtracting the treatment sum of squares from the total, as in Eq. (7.4). This is possible because of the additive property of sums of squares.

$$\text{Error SS} = \text{total SS} - \text{treatment SS} \tag{7.4}$$
$$= 1{,}129.98 - 847.05 = 282.93$$

Error sum of squares may also be found by pooling the within treatments sums of squares as shown in Eq. (7.5). These sums of squares are shown in the second last line of Table 7.1. Each component has $r - 1 = 4$ degrees of freedom. The sum of the components is 282.93. The additive nature of sums of squares is demonstrated. This is an excellent computational check. It also provides information relative to the homogeneity of error variance.

$$\text{Error SS} = \sum_i \left(\sum_j X_{ij}^2 - \frac{X_{i.}^2}{r} \right) \tag{7.5}$$

$$= \left(4{,}287.53 - \frac{144.1^2}{5}\right) + \cdots + \left(1{,}758.71 - \frac{93.5^2}{5}\right) = 282.93$$

The numerical results of an analysis of variance are usually presented in an analysis of variance table such as Table 7.2, symbolic, or Table 7.3, example, for the data just discussed.

The error mean square is denoted by s^2 and frequently referred to as a generalized error term since it is an average of the components contributed by the several populations or treatments. It is an estimate of a common σ^2, the variation among observations treated alike. That there is a common σ^2 is an assumption and s^2 is a valid estimate of σ^2 only if this assumption is true. The individual components are based on only a few degrees of freedom, consequently can vary widely about σ^2, and thus are not as good estimates as is the

TABLE 7.2

ANALYSIS OF VARIANCE: ONE-WAY CLASSIFICATION WITH EQUAL REPLICATION
(Symbolic)

Source of variation	df	Definition	Working	Mean squares†	F
Treatments	$t-1$	$r\sum_{i}(\bar{x}_{i\cdot} - \bar{x}_{\cdot\cdot})^2$	$\sum_{i}\dfrac{X_{i\cdot}^2}{r} - \dfrac{X_{\cdot\cdot}^2}{rt}$	√	√
Error	$t(r-1)$	$\sum_{i,j}(X_{ij} - \bar{x}_{i\cdot})^2$	by subtraction	√	
Total	$rt-1$	$\sum_{i,j}(X_{ij} - \bar{x}_{\cdot\cdot})^2$	$\sum_{i,j}X_{ij}^2 - \dfrac{X_{\cdot\cdot}^2}{rt}$		

† A mean square is a sum of squares divided by the corresponding degrees of freedom.

TABLE 7.3

ANALYSIS OF VARIANCE FOR DATA OF TABLE 7.1

Source of variation	df	Sum of squares	Mean square	F
Among cultures	5	847.05	169.41	14.37**
Within cultures	24	282.93	11.79	
Total	29	1,129.98		

pooled estimate. The principle is the same as that for means: a mean of 24 observations is a better estimate of μ than one of four observations because the former has a smaller variance. Similarly, a sample variance based on 24 observations is a better estimate of σ^2 than one based on four observations because the former variance has a smaller variance. The validity of the assumption that each of the components of error is an estimate of the same σ^2 can be tested by the χ^2 test of homogeneity, discussed in Sec. 17.3.

Treatment mean square is an independent estimate of σ^2 when the null hypothesis is true. F is defined as the ratio of two independent estimates of the same σ^2. This explains the divisor r used in calculating the sum of squares for treatments; sums have variance $r\sigma^2$ where r is the number of observations in each sum.

The F value is obtained by dividing the treatment mean square by the error mean square, that is, $F = 169.4/11.79 = 14.37**$. Thus the F value gives the treatment mean square as a multiple of the error mean square. These mean squares are on a comparable basis since each estimates the variation among individual observations. The calculated F value is compared with the tabular F value for 5 and 24 degrees of freedom to decide whether to accept the null hypothesis of no difference between population means or the alternative hypothesis of a difference. The tabular F values for 5 and 24 degrees of freedom are 2.62 and 3.90 at the .05 and .01 probability levels, respectively. Since calculated F exceeds 1% tabular F, we conclude that the experiment provides evidence of real differences among treatment means.

The F tables, for the analysis of variance, are entered with the numerator degrees of freedom, that is, treatment degrees of freedom, along the top of the table and the denominator degrees of freedom along the side. This is because the set of alternative hypotheses admits only that treatment differences exist and, consequently, increase the estimate of variance based on treatment means or totals so that only large values of the test criterion are to be judged significant. If the treatment mean square is less than error, the result is declared nonsignificant no matter how small the ratio. A significant F implies that the evidence is sufficiently strong to indicate that all the treatments do not belong to populations with a common μ. However, it does not indicate which differences may be considered statistically significant.

Note that both the degrees of freedom and sums of squares in the body of the table add to the corresponding values in the total line. Mean squares are not additive. The property of additivity of sums of squares is characteristic of well-planned and executed experiments. It leads to certain short cuts in the arithmetic of the analysis of variance, such as finding the error sum of squares by subtracting the treatment sum of squares from the total sum of squares as in Table 7.3 and, in general, as a residual. Experiments not planned and executed so as to possess this property will generally involve far more computations and have a lower precision per observation.

The standard error of a treatment mean and that of a difference between treatment means are given by Eqs. (7.6) and (7.7), respectively. Numerical values are for data of Table 7.1.

$$s_{\bar{x}} = \sqrt{\frac{s^2}{r}} = \sqrt{\frac{11.79}{5}} = 1.54 \text{ mg} \tag{7.6}$$

$$s_{\bar{d}} = \sqrt{\frac{2s^2}{r}} = \sqrt{\frac{2(11.79)}{5}} = 2.17 \text{ mg} \tag{7.7}$$

These statistics are useful for comparing differences among treatment means as discussed in Secs. 7.4 to 7.8 and for computing confidence intervals for treatment means and differences among pairs of treatment means. Further applications are discussed in Secs. 2.13 and 5.10. The coefficient of variability is given by Eq. (7.8).

$$\text{CV} = \frac{\sqrt{s^2}}{\bar{x}} 100 = \frac{\sqrt{11.79}}{19.89} 100 = 17.3\% \tag{7.8}$$

It was shown in Sec. 5.4 that the analysis of variance could be used in place of the t test to compare two treatments where the design was completely random.

The one-tailed F test with 1 and n degrees of freedom corresponds to the two-tailed t test with n degrees of freedom. This t test does not specify the direction of the difference between two treatment means for the alternative hypothesis; thus it is like the one-tailed F test which specifies which mean square is to be the larger as the result of differences of unspecified direction between treatments. These tests can be shown to be algebraically equivalent; in particular $t^2 = F$. The relation is shown graphically in Fig. 7.1. Small numerical values of t, when squared, become small values of F, positive quantities. Large numerical values of t, when squared, become large values of F.

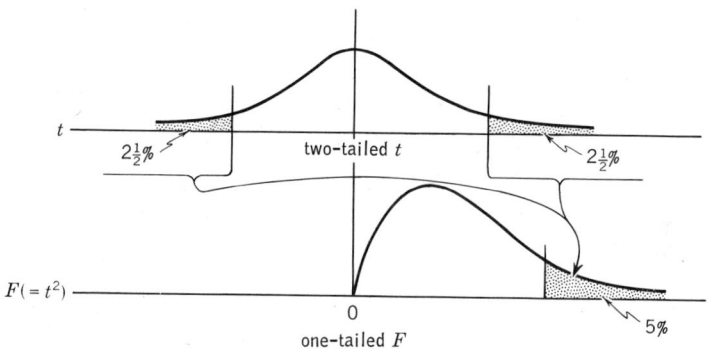

FIG. 7.1 Relation between two-tailed t and one-tailed F (curves are only approximate)

Exercise 7.3.1 F. R. Urey, Department of Zoology, University of Wisconsin, made an estrogen assay of several solutions which had been subjected to an *in vitro* inactivation technique. The uterine weight of the mouse was used as a measure of estrogenic activity. Uterine weights in milligrams of four mice for each of control and six different solutions are shown in the accompanying table.

Control	1	2	3	4	5	6
89.8	84.4	64.4	75.2	88.4	56.4	65.6
93.8	116.0	79.8	62.4	90.2	83.2	79.4
88.4	84.0	88.0	62.4	73.2	90.4	65.6
112.6	68.6	69.4	73.8	87.8	85.6	70.2

Compute the means and the analysis of variance for these data. Compute error sum of squares directly and demonstrate the additivity of sums of squares. Calculate the coefficient of variability.

Exercise 7.3.2 Calculate the sum of squares among treatments 1 through 6, that is, ignoring control. This has 5 degrees of freedom. Calculate the sum of squares for comparing control with the mean of all other observations. This has 1 degree of freedom. Note that the sum of these two sums of squares is the treatment sum of squares in the analysis of variance. Such comparisons are said to be independent and will be discussed further in Chap. 11.

7.4 The least significant difference. If the F value for treatments is not significant, the evidence is against rejecting the null hypothesis and specific treatment comparisons should not usually be made. Exceptions occur when certain comparisons have been planned and the experiment conducted accordingly. A valid test criterion for planned comparisons of paired means, a criterion in considerable vogue in both the past and present, uses the *least significant difference* or *lsd*. The 5% *lsd* is given by Eq. (7.9).

$$lsd(.05) = t_{.05}s_{\bar{d}} \qquad (7.9)$$

$t_{.05}$ is the tabular value of t for error degrees of freedom and $s_{\bar{d}} = \sqrt{2s^2/r}$ with s^2 the error variance and r the number of observations per mean. For *lsd*'s at other probability levels, use tabular t for the level chosen.

The *lsd* is basically a Student's t test using a pooled error variance. Since the *lsd* need be calculated only once and takes advantage of the pooled error variance, its use is seen to be a timesaver as compared with making individual t tests. For the difference between two means to be significant at the 5% level the observed differences must exceed the *lsd*. This is readily apparent from the relation $t = (\bar{x}_1 - \bar{x}_2)/s_{\bar{d}}$ or $ts_{\bar{d}} = \bar{x}_1 - \bar{x}_2$.

Let us suppose that in planning the *Rhizobium* experiment, the investigator had decided to compare treatment means for 3DOk1 and 3DOk5, 3DOk4 and 3DOk7, 3DOk13 and composite at a 5% probability level; these means are given at the bottom of Table 7.1. He computes the 5 and 1% *lsd* values as

$$lsd\,(.05) = t_{.05}s_{\bar{d}} = 2.064\sqrt{\frac{2(11.79)}{5}} = 4.5 \text{ mg}$$

$$lsd\,(.01) = t_{.01}s_{\bar{d}} = 2.797\sqrt{\frac{2(11.79)}{5}} = 6.1 \text{ mg}$$

The observed differences are $\bar{x}_1 - \bar{x}_2 = 28.8 - 24.0 = 4.8$; $\bar{x}_3 - \bar{x}_4 = 14.6 - 19.9 = -5.3$; and $\bar{x}_5 - \bar{x}_6 = 13.3 - 18.7 = -5.4$ mg; all lie between 4.5 and 6.1 mg, so are significant at the 5% but not at the 1% level. Here we have an example of a valid use of the *lsd*.

Notice that each mean appears in only one comparison. Such comparisons are said to be *orthogonal* or *independent* and the stated probability level is considered to be valid for each comparison. In Chap. 11, orthogonal comparisons are discussed more rigorously and it is shown that a single treatment may appear in more than one comparison. Sometimes the investigator plans to make comparisons such as 3DOk1 versus 3DOk5 and 3DOk5 versus 3DOk4 where 3DOk5 appears in both. These are nonindependent comparisons. When nonindependent comparisons are made with a criterion such as the *lsd*, the investigator is apt to be left with the impression that he has more information than is truly available or that each item of information has the reliability implied by the confidence coefficient. This nonindependence should not necessarily discourage the investigator from making such comparisons. The meaningfulness of the comparison should be the prime consideration, and this should be determined on a sound biological or other basis.

Since the *lsd* can be and often is misused, some statisticians hesitate to recommend it. The most common misuse is to make comparisons suggested by the data, comparisons not initially planned. The making of such comparisons is often termed "testing effects suggested by the data" or "arguing after the facts." For the tabulated confidence levels to be valid, the *lsd* should be used only for independent or nonindependent comparisons planned before data have been examined.

Cochran and Cox (7.4) point out that in the extreme situation where the experimenter compares only the difference between the highest and lowest treatment means by use of the t test or the *lsd*, the differences will be substantial even when no effect is present. It can be shown that with three treatments, the observed value of t for the greatest difference will exceed the tabulated 5% level about 13% of the time; with 6 treatments the figure is 40%, with 10 treatments 60%, and with 20 treatments 90%. Thus, when the experimenter thinks he is making a t test at the 5% level, he is actually testing at the 13% level for three treatments, the 40% level for six treatments, and so on.

In summary, the *lsd* is a useful but often misused test criterion. Its advantages lie in the fact that it is a single and easy to compute value. When the comparisons are independent or planned, the tabulated confidence coefficient is on a per-comparison basis. Its biggest disadvantage lies in the fact that it is often used to make many unplanned comparisons. It is not satisfying for all possible paired comparisons as is seen from the preceding paragraph; however, the tabulated confidence level is valid as an average. Some protection against this nonsatisfying situation may be had by making such comparisons only if F is significant, a procedure which again alters the significance level.

For the experimenter who wishes to compare all possible pairs of treatment means, procedures are given in Secs. 7.5, 7.6, and 7.7.

7.5 Duncan's new multiple-range test. The uncritical use of the *lsd* and the need for sound methods of making comparisons among treatment means, especially nonindependent comparisons, have led to considerable recent research. An approach to the problem of comparing each treatment mean with every other treatment mean would be to develop a method calling for a set of significant differences of increasing size, the size to depend upon the closeness of the means after ranking, with the smallest value for adjacent means and the largest for the extremes. Such a method would overcome the obvious disadvantage of the *lsd* in this respect.

In 1951, Duncan (7.5) developed a multiple comparisons test to compare each treatment mean with every other treatment mean. The procedure consists of three stages and, where the number of treatments is 10 or more, can be time-consuming. In 1955, Duncan (7.6) developed the *new multiple-range test* which combines in one step the three stages of his previous multiple comparisons test. The new multiple-range test, while not as powerful as the earlier multiple comparisons test, has strong appeal because of its simplicity.

The 5% multiple-range test is illustrated below for the *Rhizobium* data, Table 7.1. The problem is to determine which of the 15 differences among the six treatment means are significant and which are not. For this procedure, it is not necessary to compute an F value and proceed only if this is significant;

the investigator may use the procedure regardless of the significance of F. The procedure is:
1. Determine
$$s_{\bar{x}} = \sqrt{(\text{error mean square})/r}$$
For our data,
$$s_{\bar{x}} = \sqrt{11.79/5} = 1.54 \text{ mg}$$

with error degrees of freedom $n_2 = 24$. *Significant studentized ranges* for the 5 and 1% levels are given in Table A.7. For the 5% multiple-range test, enter Table A.7 at row $n_2 = 24$ and extract the significant studentized ranges for $p = 2$ to 6, the sizes referring to the number of means involved in a comparison. Values not in the table can be obtained by interpolation. The significant ranges are then multiplied by $s_{\bar{x}}$ to give the *least significant ranges* below.

Value of p	2	3	4	5	6
SSR	2.92	3.07	3.15	3.22	3.28
R_p = LSR	4.5	4.7	4.9	5.0	5.1

2. Rank the means. It is convenient to space these in a manner approximating the differences between means. We obtain

3DOk13	3DOk4	Composite	3DOk7	3DOk5	3DOk1
13.3	14.6	18.7	19.9	24.0	28.8

3. Test the differences in the following order: largest minus smallest, largest minus second smallest, ..., largest minus second largest, then second largest minus smallest, second largest minus second smallest, and so on down to second smallest minus smallest. With one exception, each difference is declared significant if it exceeds the corresponding least significant range, otherwise it is declared not significant. The sole exception is that no difference between two means can be declared significant if the two means concerned are both contained in a larger subset with a nonsignificant range. This exception points up the fact that we do have a range test rather than a test of two means; a range test is a test of the homogeneity of the sample of means for that size of sample and one that uses range as a test criterion.

From the table of step 1, since 3DOk1 to 3DOk13 is the range of six means, it must exceed 5.1 to be significant; because 3DOk5 to 3DOk13 is the range of five means it must exceed 5.0, and so on.

In detail, we have

$$\text{3DOk1-3DOk13} = 15.5 > 5.1; \text{ significant}$$
$$\text{3DOk1-3DOk4} = 14.2 > 5.0; \text{ significant}$$
$$\cdots$$
$$\text{3DOk1-3DOk5} = 4.8 > 4.5; \text{ significant}$$
$$\text{3DOk5-3DOk13} = 10.7 > 5.0; \text{ significant}$$
$$\text{3DOk5-3DOk4} = 9.4 > 4.9; \text{ significant}$$
$$\cdots$$
$$\text{3DOk5-3DOk7} = 4.1 < 4.5; \text{ not significant}$$

and so on.

ANALYSIS OF VARIANCE I: THE ONE-WAY CLASSIFICATION

In practice, a short cut can be used repeatedly, especially when the number of means is large. Instead of first finding the difference between the largest and smallest mean, from the largest mean subtract the least significant range for $p = 6$, thus $28.8 - 5.1 = 23.7$. All means less than 23.7 will then be declared significantly different from the largest since R_p decreases with the number of means in the range being compared. Since R_p decreases with a decrease in subset size, it is necessary to compare the range of the remaining means with the appropriate R_p. For our example, this is $28.8 - 24.0 = 4.8 > 4.5$; this difference is declared significant.

The results of the test can be summarized as follows:

```
   13    4     C     7     5     1
   13.3  14.6  18.7  19.9  24.0  28.8
   ─────────────────────
```

Any two means not underscored by the same line are significantly different. Any two means underscored by the same line are not significantly different.

In summary, the new multiple-range test is easy to apply; it takes into account the number of treatments in the experiment whereas the *lsd* does not; it permits decisions as to which differences are significant and which are not whereas the F test permits no such decisions when F is significant; it uses a set of significant ranges, each range depending upon the number of means in the comparison.

Since the notion of Type I error was not originally intended to apply to multiple comparisons, the idea of significance level is replaced by that of *special protection levels* against finding false significant differences. These levels are based on treatment degrees of freedom; the probability of finding a significant difference between any two means, when the corresponding true means are equal, is less than or equal to the significance level stated in Table A.7. The question of error rate is also discussed by Harter (7.10).

Exercise 7.5.1 Apply Duncan's procedure to the data of Exercise 7.3.1 for the purpose of locating significant differences. Present your results by the underscoring of means as illustrated in Sec. 7.5. Discuss Duncan's procedure relative to the *lsd*; include comments on validity and error rates.

7.6 Tukey's w-procedure. This procedure is also called the honestly significant difference (*hsd*) procedure. In developing multiple comparisons tests, one problem that arises is that, for experiments where many comparisons are to be made, we are almost certain to declare some differences to be significant even when the means are a homogeneous set. This problem has led to the development of procedures where the experiment is the unit used in stating the number of errors of the type where an observed difference is falsely declared to be significant. For such procedures, a 5% level test is one where 5% of the experiments will give one or more false significant differences and 95% will give no significant differences, on the average, when the means are a homogeneous set. This is called an *experimentwise error rate*.

A procedure similar to the *lsd* test in that it requires a single value for judging the significance of all differences, but with the experiment used as the unit for stating the significance level, is given by Tukey (7.15) as the w procedure. The procedure consists of computing

$$w = q_\alpha(p, n_2) s_{\bar{x}} \qquad (7.10)$$

where q_α is obtained from Table A.8 for $\alpha = .05$ or $.01$, p is the number of treatments, and n_2 equals error degrees of freedom. w is used to judge the significance of each of the observed differences.

For the *Rhizobium* data, $p = 6$, $n_2 = 24$, and $w_{.05}(6,24) = 4.37(1.54) = 6.7$ mg. Summarizing the results in the same way as for Duncan's test, we have

$$13.3 \quad 14.6 \quad 18.7 \quad 19.9 \quad 24.0 \quad 28.8$$

Fewer differences have been declared significant by Tukey's procedure than by Duncan's, respectively, 6 and 11. This might have been expected since we are requiring that in 95% of our experiments, we will declare no differences to be significant if the means are truly a homogeneous sample and we are relying on a single value for judging the significance of all observed differences. Because of this, Hartley (7.11) has suggested that an experimentwise error rate might be relaxed to 10% or even a higher value.

Tukey's w may also be used to compute a set of confidence intervals for differences. Thus the true difference of the population means estimated by $\bar{x}_i$ and $\bar{x}_j$ is estimated by Eq. (7.11).

$$(\bar{x}_i - \bar{x}_j) \pm w \qquad (7.11)$$

where w is obtained from Eq. (7.10).

In summary, Tukey's procedure is extremely easy to apply since it requires a single value for judging the significance of all differences, that is, it is a fixed range; it takes into account the number of treatments in the experiment; it permits decisions as to which differences are significant and which are not; the error rate applies on an experimentwise rather than a per-comparison basis; it can be used to make confidence interval statements. (A confidence interval statement implies a significance test, whereas the converse is not true.)

Exercise 7.6.1 Use Tukey's procedure to test all possible paired differences between means for the data of Exercise 7.3.1. Compare the results with those obtained by Duncan's method. In what way do the two tests differ?

Exercise 7.6.2 Compute 95% confidence intervals for differences between all possible pairs of population means for the data of Exercise 7.3.1.

7.7 Student-Newman-Keuls' test. Each of the three persons named in this test contributed to it though it is also referred to as the Newman-Keuls' (7.13) test or simply Keuls' method. It is like Duncan's procedure in that it is a multiple-range test.

The procedure is to compute

$$W_p = q_\alpha(p, n_2) s_{\bar{x}} \qquad (7.12)$$

ANALYSIS OF VARIANCE I: THE ONE-WAY CLASSIFICATION

using q from Table A.8 for $\alpha = .05$ or $.01$, $p = 2, \ldots$, number of means, and $n_2 =$ error degrees of freedom. For the *Rhizobium* data, we obtain:

p	2	3	4	5	6
$q_\alpha(p, n_2)$:	2.92	3.53	3.90	4.17	4.37
W_p :	4.5	5.4	6.0	6.4	6.7

Note that W_6 is the single value required for Tukey's method.

Summarizing the results of applying this procedure, we have

13.3 14.6 18.7 19.9 24.0 28.8

(The dotted line was used because 13.3 and 18.7 mg differ by exactly W_3 when only one decimal place is used.) This test gives nine significant differences more than Tukey's procedure but fewer than Duncan's.

In summary, the Student-Newman-Kéuls' method is relatively easy to apply; it takes into account the number of treatments in the experiment; it is a multiple-range test for judging the significance of a set of differences; it permits decisions as to which differences are significant and which are not; it is a range test procedure and the error rate applies at each stage to tests involving the same number of means. Error rate is seen to apply neither on an experimentwise nor a per-comparison basis since a particular range test may be made for one or more comparisons in a particular experiment.

Exercise 7.7.1 Apply the Student-Newman-Keuls' procedure to the data of Exercise 7.3.1. Compare the results with those obtained by Tukey's procedure and Duncan's procedure.

Exercise 7.7.2 Compare the error rates that apply to Duncan's, Tukey's, and Student-Newman-Keuls' procedures. In particular, discuss how they differ from one another.

7.8 Comparing all means with a control. The objective of an experiment is sometimes to locate treatments which are different or better than some standard but not to compare them, comparison being left for a later experiment. Such a set of comparisons, that is, control against each of the other treatment means, is not an independent set. While the *lsd* is valid, Dunnett (7.7) has given an alternative valid procedure. This procedure requires a single difference for judging the significance of observed differences, and tables are available for one- and two-sided comparisons (see Table A.9). Error rate is on an experimentwise basis. Confidence intervals may also be computed.

Instead of introducing a new example, let us apply Dunnett's procedure to the *Rhizobium* data considering the composite as a standard or control treatment. The procedure is as follows: Consider the means as a set of p treatment means and a control such that p comparisons are to be made. Enter Table A.9 for the appropriate set of alternatives and error rate, with the number of treatments p, error degrees of freedom n_2, and extract a t value (not Student's t). For our illustration, the comparison is to be two-tailed with error rate .05,

$p = 5$ and error degrees of freedom $= 24$. The significant difference is given by Eq. (7.13).

$$d' = t(\text{Dunnett})s_{\bar{d}} \tag{7.13}$$

$$= 2.76\sqrt{2(11.79)/5} = 5.99 \text{ mg, for the example}$$

The only difference to be declared significant is that between 3DOk1 and the composite. The procedures of Duncan, Tukey, and Student-Newman-Keuls also declared this difference to be significant though each procedure required a different value for judging the significance of these two means. Duncan's method also declares the difference between 3DOk13 and the composite and between 3DOk5 and the composite to be significant. These differences are almost at the significance value required by Dunnett's method. The Student-Newman-Keuls' method has the observed difference between 3DOk13 and the composite exactly equal to the significant difference when only one decimal point is used.

This procedure is unlike the ones previously discussed in that it is not intended to be used in judging all possible comparisons. It is intended solely for paired comparisons where one mean is that for the control. The comparisons may be one-sided or two-sided and this determines from which table Dunnett's t is to be obtained. The $\sqrt{2}$ associated with a difference is not a part of the tabulated t as in the case of the other three procedures so that $s_{\bar{d}}$ rather than $s_{\bar{x}}$ must be computed. If one or more differences are falsely declared significant, this is regarded as a single Type I error; thus error rate is on an experimentwise basis. Joint confidence intervals for true differences are given by Eqs. (7.14) and (7.15).

$$\text{CL} = (\bar{x}_i - \bar{x}_0) \pm ts\sqrt{2/r} \tag{7.14}$$

where $\bar{x}_0$ refers to control and t is from Dunnett's two-sided comparison table for a two-sided interval, and

$$\text{CL} = (\bar{x}_i - \bar{x}_0) - ts\sqrt{2/r} \tag{7.15}$$

where t is from the one-sided comparison table for a one-sided interval. For the one-sided interval, one states that the true mean estimated by $\bar{x}_i$ exceeds that estimated by $\bar{x}_0$ by at least the value given by Eq. (7.15).

Exercise 7.8.1 Apply Dunnett's procedure to the data of Exercise 7.3.1; use a one-sided test against the alternative that the mean of the control population is the largest population mean.

Exercise 7.8.2 Compute a set of one-sided confidence intervals for the true differences. How is Eq. (7.15) changed for this computation?

7.9 Data with a single criterion of classification. The analysis of variance for any number of groups with unequal replication. When the number of observations per treatment varies, we have a general case for which Sec. 7.3 provides a special example.

The analysis of variance will now be illustrated for the data of Table 7.4. The original observations were made on a greenhouse experiment in Ithaca, N.Y., and were intended to determine whether or not there are genetic

ANALYSIS OF VARIANCE I: THE ONE-WAY CLASSIFICATION

differences among plants, either among or within locations, location referring to the place of origin of the plant; the data in Table 7.4 are suitable for among-location (more generally, treatment) comparisons only and are for a single character of many that were checked.

TABLE 7.4

LEAF LENGTH AT TIME OF FLOWERING (SUM FOR THREE LEAVES) OF *Sedum oxypetalum* FROM SIX LOCATIONS IN THE TRANS-MEXICAN VOLCANIC BELT, IN MILLIMETERS

Location	r_i	$\sum_j X_{ij}$	$\bar{x}_{i\cdot}$	$\sum_j X_{ij}^2$	$X_{i\cdot}^2/r_i$	$\sum_j x_{ij}^2$
H	1	147	147.00	21,609	21,609.00	
LA	1	70	70.00	4,900	4,900.00	
R	6	634	105.67	71,740	66,992.67	4,747.33
SN	1	75	75.00	5,625	5,625.00	
Tep	3	347	115.67	40,357	40,136.33	220.67
Tis	2	170	85.00	14,500	14,450.00	50.00
Totals	14	1,443	($\bar{x}_{\cdot\cdot}$ = 103.07)	158,731	153,713.00	5,018.00

SOURCE: Unpublished data used with permission of R. T. Clausen, Cornell University, Ithaca, N.Y.

Computations. Compute $\sum_j X_{ij} = X_{i\cdot}$ and $\sum_j X_{ij}^2$, that is, the sum and sum of squares of the observations for each treatment; these are also shown in Table 7.4. Equation (7.16) is the only new computational procedure; Eq. (7.3) arises from this equation when all r_i's are equal.

$$C = \frac{X_{\cdot\cdot}^2}{\Sigma r_i} = \frac{(1,443)^2}{14} = 148,732.07$$

$$\text{Total SS} = \sum_{i,j} X_{ij}^2 - C = 158,731.00 - 148,732.07 = 9,998.93$$

$$\text{Treatment SS} = \sum_i \frac{X_{i\cdot}^2}{r_i} - C = \frac{X_{1\cdot}^2}{r_1} + \cdots + \frac{X_{t\cdot}^2}{r_t} - C \qquad (7.16)$$
$$= 153,713.00 - 148,732.07 = 4,980.93$$

$$\text{Error SS} = \text{total SS} - \text{treatment SS} = 5,018.00$$

Error sum of squares may also be found by pooling the within-treatments sums of squares, given in the last column of Table 7.4. It is of interest to note that three locations do not contribute anything to the error variance since each supplies only one observation. Note that in finding the treatment sum of squares each squared sum is divided by the appropriate r_i before addition.

The analysis of variance is given in Table 7.5. F is not significant; tabulated $F(.05) = 3.69$ for 5 and 8 degrees of freedom. The evidence is not in favor of location differences.

The problem of testing individual differences becomes inconvenient when there is variable replication. If it is desired to make such comparisons, the following procedures are suggested.

For the lsd: compute $s_{\bar{d}}$ as in Eq. (7.17).

$$s_{\bar{d}} = \sqrt{s^2\left(\frac{1}{r_i} + \frac{1}{r_j}\right)} = \sqrt{s^2\frac{(r_i + r_j)}{r_i r_j}} \qquad (7.17)$$

Now $lsd = ts_{\bar{d}}$ where t is Student's t for the chosen significance level and error degrees of freedom, and r_i and r_j are the numbers of observations in the two means being compared. $s_{\bar{d}}$ may also be used for computing confidence intervals. This procedure is statistically sound.

TABLE 7.5

ANALYSIS OF VARIANCE OF THE DATA SUMMARIZED IN TABLE 7.4

Source of variation	df	SS	MS	F
Among locations	5	4,980.93	996.19	1.59
Within locations	8	5,018.00	627.25	
Total	13	9,998.93		

$$s_{\bar{d}} = \sqrt{627.25\left(\frac{1}{r_i} + \frac{1}{r_j}\right)}$$ mm where r_i and r_j are the numbers of observations in the means for the desired comparison.

For Duncan's method: obtain significant studentized ranges and multiply by s rather than $s_{\bar{x}}$ to give a set of intermediate significant ranges. For any desired comparison, multiply the appropriate intermediate value by Eq. (7.18).

$$\sqrt{\frac{1}{2}\left(\frac{1}{r_i} + \frac{1}{r_j}\right)} \qquad (7.18)$$

Kramer (7.12) has proposed this procedure for unequally replicated treatments; its validity has not been verified.

For Tukey's w procedure: obtain a value $w' = q_\alpha(p,n_2)s$ similar to Eq. (7.10) but with s replacing $s_{\bar{x}}$. For any desired comparison, multiply w' by the appropriate values determined by Eq. (7.18). The validity of this procedure has not been verified.

For Student-Newman-Keuls' test: obtain values $W'_n = q_\alpha(p,n_2)s$ similar to Eq. (7.12). For any desired comparison, multiply by the appropriate value given by Eq. (7.18). The validity of this procedure has not been verified.

For Dunnett's procedure: obtain ts where t is from Dunnett's table. For any desired comparison, multiply by

$$\sqrt{\left(\frac{1}{r_i} + \frac{1}{r_j}\right)} \qquad (7.19)$$

The validity of this procedure has not been verified.

ANALYSIS OF VARIANCE I: THE ONE-WAY CLASSIFICATION

Exercise 7.9.1 Wexelsen (7.16) studied the effects of inbreeding on plant weight in red clover. Below is given the average plant weight in grams of non-inbred (F_1) lines and three groups of inbred families arranged in increasing order of inbreeding.

F_1: 254, 263, 266, 249, 337, 277, 289, 244, 265
Slightly inbred: 236, 191, 209, 252, 212, 224
F_2: 253, 192, 141, 160, 229, 221, 150, 215, 232, 234, 193, 188
F_3: 173, 164, 183, 138, 146, 125, 178, 199, 170, 177, 172, 198

Compute the analysis of variance for these data. Obtain error sum of squares directly and demonstrate the additivity of sums of squares. Calculate the coefficient of variability.

Exercise 7.9.2 Apply Dunnett's procedure, modified as suggested in Sec. 7.9, to the data of Exercise 7.9.1. Assume that inbreeding will depress plant weight if it affects it at all and use a one-tailed test.

7.10 The linear additive model.

An observation has been described as the sum of components, a mean and a random element, where the mean, in turn, may be a sum of components. A further basic assumption in the analysis of variance, where tests of significance are to be made, is that the random components are independently and normally distributed about zero mean and with a common variance. For the one-way classification, the linear model begins with Eq. (7.20).

$$X_{ij} = \mu + \tau_i + \epsilon_{ij}, \quad i = 1, \ldots, t \text{ and } j = 1, \ldots, r_i \quad (7.20)$$

The ϵ_{ij}'s are the random components.

Assumptions must also be made about the τ_i's to clarify the model completely. Thus we have:

The fixed model or *Model I*: The τ_i's are fixed and $\sum_i \tau_i = 0$.

The random model or *Model II*: The τ_i's are a random sample from a population of τ's for which the mean is zero.

The distinction is that for the fixed model, a repetition of the experiment would bring the same set of τ's into the new experiment; we concentrate our attention on these τ's. For the random model, a repetition would bring in a new set of τ's but from the same population of τ's; we would be interested in the variability of the τ's since we are not in a position to continue our interest in a specific set. For the fixed model, we draw inferences about the particular treatments; for the random model, we draw an inference about the population of treatments.

Setting $\Sigma \tau_i = 0$ is simply measuring treatment effects as deviations from an over-all mean. The null hypothesis is then easily stated as H_0: $\tau_i = 0$, $i = 1, \ldots, t$ and the alternative as H_1: some $\tau_i \neq 0$.

Once the particular model has been decided upon, it is possible to state what parameters are being estimated when the various mean squares and the F value are computed. For Model II, let σ_τ^2 be the variance in the population of τ's. Then if the experiment were repeated endlessly and the observed mean squares were averaged, we would have the values given in Table 7.6. For an individual experiment, the mean squares are estimates of the corresponding quantities of Table 7.6.

Consider the *Rhizobium* data of Table 7.1 to be data to which Model I applies. Then, from Table 7.3, we have:

σ^2 is estimated by 11.79

$\dfrac{\Sigma \tau_i^2}{t-1}$ is estimated by $\dfrac{169.41 - 11.79}{5} = 31.52$

The value 31.52 is an estimate of the variability of the fixed set of τ's and any variability in this estimate, from experiment to experiment, comes as the result of variability in the estimate of σ^2; there are no sample-to-sample differences in the τ's.

TABLE 7.6

AVERAGE VALUES OF MEAN SQUARES FOR MODELS I AND II, EQUAL REPLICATION

Source of variation	df	Mean square is an estimate of	
		Model I	Model II
Treatments	$t-1$	$\sigma^2 + r\Sigma\tau_i^2/(t-1)$	$\sigma^2 + r\sigma_\tau^2$
Individuals within treatments	$t(r-1) = tr - t$	σ^2	σ^2
Total	$tr-1$		

For an experiment to which Model II applies, variability in the estimate of σ_τ^2 comes as the result of variability in the estimate of σ^2 and sample-to-sample differences in the τ's.

When there is unequal replication of treatments, the coefficient corresponding to r for the average value of the treatment mean square is, for Model II,

$$r_0 = \left(\Sigma r_i - \dfrac{\Sigma r_i^2}{\Sigma r_i}\right)\dfrac{1}{t-1} \qquad (7.21)$$

For Model I, the average value of the treatment mean square depends upon whether the τ's are to be simple deviations from a mean or whether their weighted mean, and consequently weighted sum, is to be zero. The values are given in Table 7.7. The average value of the error mean square is σ^2 regardless of which restriction is placed on the τ's.

TABLE 7.7

AVERAGE VALUE OF TREATMENT MEAN SQUARE FOR MODEL I, UNEQUAL REPLICATION

Impose condition	Treatment mean square is an estimate of
$\Sigma \tau_i = 0$	$\sigma^2 + \dfrac{\Sigma r_i \tau_i^2 - (\Sigma r_i \tau_i)^2/\Sigma r_i}{(t-1)}$
$\Sigma r_i \tau_i = 0$	$\sigma^2 + (\Sigma r_i \tau_i^2)/(t-1)$

Note that $\Sigma r_i \tau_i^2 - \dfrac{(\Sigma r_i \tau_i)^2}{\Sigma r_i} = \Sigma r_i(\tau_i - \bar{\tau})^2$ where $\bar{\tau} = (\Sigma r_i \tau_i)/(\Sigma r_i)$.

ANALYSIS OF VARIANCE I: THE ONE-WAY CLASSIFICATION

That the computed mean squares, for equal replication, are estimates of the quantities given in Table 7.6 is readily shown. The mean square for individuals within treatments is obtained by pooling estimates like that obtained from $X_{i1} = \mu + \tau_i + \epsilon_{i1}, \ldots, X_{ir} = \mu + \tau_i + \epsilon_{ir}$ where only the ϵ's vary. Clearly this will lead us to an estimate of σ^2; the same can be said if r is also a variable. The treatment mean square is based on totals

$$X_{1\cdot} = r\mu + r\tau_1 + \epsilon_{1\cdot}, \ldots, X_{t\cdot} = r\mu + r\tau_t + \epsilon_{t\cdot}$$

or means

$$\bar{x}_{1\cdot} = \mu + \tau_1 + \bar{\epsilon}_{1\cdot}, \ldots, \bar{x}_{t\cdot} = \mu + \tau_t + \bar{\epsilon}_{t\cdot}.$$

If the null hypothesis is true, then the totals are actually of the form $r\mu + \epsilon_{i\cdot}$ and their variance is an estimate of $r\sigma^2$. However, the divisor r is used in computing so that the mean square in the analysis of variance table is an estimate of σ^2. If the null hypothesis is false, then the totals are of the form $r\mu + r\tau_i + \epsilon_{i\cdot}$ and the τ's and ϵ's contribute to the variability of the totals. The contribution attributable to the ϵ's is still σ^2. Squaring the totals gives terms like $r^2\tau_i^2$, which yields $r\tau_i^2$ after division by r. Hence the contribution attributable to the τ's is either $r\Sigma\tau_i^2/(t-1)$ or an estimate of $r\sigma_\tau^2$. Since we assume the τ's and ϵ's to be independent, there is no contribution that involves products of τ's and ϵ's.

When there is unequal replication, totals are of the form $r_i\mu + r_i\tau_i + \epsilon_{i\cdot}$ and means are of the form $\mu + \tau_i + \bar{\epsilon}_{i\cdot}$. The computation of a treatment mean square directly from totals is clearly not valid since $r_i\mu$ is variable and would introduce a contribution due neither to the τ's nor to the ϵ's; computation with means would introduce no such difficulty. It can be shown that

$$\sum_i \frac{X_{i\cdot}^2}{r_i} - \frac{X_{\cdot\cdot}^2}{\sum_i r_i} = \sum_i r_i(\bar{x}_{i\cdot} - \bar{x}_{\cdot\cdot})^2$$

where the left-hand side is the computing formula for the treatment sum of squares. The right-hand side is a weighted sum of squares of the deviations of the treatment means from the over-all mean defined as $X_{\cdot\cdot}/(\Sigma r_i)$ or its equivalent, $(\Sigma r_i \bar{x}_{i\cdot})/(\Sigma r_i)$. The weights are seen to be inversely proportional to the variance, that is, a big variance leads to a small weight and vice versa, as should be the case. It is now clear that if there are no treatment effects, the treatment mean square does lead to an estimate of σ^2; if there are treatment effects, there is also a contribution due to the τ's (see Table 7.7).

When F is the test criterion for testing treatment means in an analysis of variance, the null hypothesis is one of no differences among treatment means and the alternative is that treatment differences, most often unspecified as to size and direction, exist. True differences may be the result of a particular set of fixed effects or of a random set of effects from a population of effects. The basis of the test criterion is the comparison of two independent variances which are estimates of a common variance when the null hypothesis is true. If real treatment differences exist, we have

$$F = \frac{s^2 + r(\Sigma \hat{\tau}_i^2)/(t-1)}{s^2} \quad \text{or} \quad F = \frac{s^2 + rs_\tau^2}{s^2}$$

depending upon the model. On the average, then, the numerator will be

TABLE 7.8

One-week Stem Growths of Mint Plants Grown in Nutrient Solution

	Hours of daylight at																	
	Low night temperatures									High night temperatures								
	8			12			16			8			12			16		
	Pot number			Pot number			Pot number			Pot number			Pot number			Pot number		
Plant number	1	2	3	1	2	3	1	2	3	1	2	3	1	2	3	1	2	3
1	3.5	2.5	3.0	5.0	3.5	4.5	5.0	5.5	5.5	8.5	6.5	7.0	6.0	6.0	6.5	7.0	6.0	11.0
2	4.0	4.5	3.0	5.5	3.5	4.0	4.5	6.0	4.5	6.0	7.0	7.0	5.5	8.5	6.5	9.0	7.0	7.0
3	3.0	5.5	2.5	4.0	3.0	4.0	5.0	5.0	6.5	9.0	8.0	7.0	3.5	4.5	8.5	8.5	7.0	9.0
4	4.5	5.0	3.0	3.5	4.0	5.0	4.5	5.0	5.5	8.5	6.5	7.0	7.0	7.5	7.5	8.5	7.0	8.0
Pot totals = $X_{ij\cdot}$	15.0	17.5	11.5	18.0	14.0	17.5	19.0	21.5	22.0	32.0	28.0	28.0	22.0	26.5	29.0	33.0	27.0	35.0
Treatment totals = $X_{i\cdot\cdot}$	44.0			49.5			62.5			88.0			77.5			95.0		
Treatment means = $\bar{x}_{i\cdot\cdot}$	3.7			4.1			5.2			7.3			6.5			7.9		

source: These data are a part of a larger set and are available through the courtesy of R. Rabson, Cornell University, Ithaca, N.Y.

ANALYSIS OF VARIANCE I: THE ONE-WAY CLASSIFICATION

larger than the denominator if real treatment differences exist. We are trying to detect the population quantities $(\Sigma \tau_i^2)/(t-1)$ or σ_τ^2, depending upon the model. Formally, we have $H_0: \tau_i = 0$, $i = 1, \ldots, t$ versus H_1: some $\tau_i \neq 0$ for the fixed model and $H_0: \sigma_\tau^2 = 0$ versus $H_1: \sigma_\tau^2 \neq 0$ for the random model. This test is called a one-tailed test, the reference to tails being relative to the F distribution.

Exercise 7.10.1 For the data of Exercises 7.3.1 and 7.9.1, estimate the contribution of the treatment effects to the variability measured by the treatment mean square. Decide which model applies and define this contribution in terms of the components of the model.

7.11 Analysis of variance with subsamples. Equal subsample numbers.

In some experimental situations, several observations may be made within the experimental unit, the unit to which the treatment is applied. Such observations are made on *subsamples* or *sampling units*. For example, consider the data of Table 7.8. A large group of plants were assigned at random to pots, four to a pot, the experimental unit; treatments were assigned at random to pots, three pots per treatment. All pots were completely randomized with respect to location during the time spent in daylight and each group of pots was completely randomized within the low or high temperature greenhouse during the time spent in darkness. Observations were made on the individual plants.

Two sources of variation which contribute to the variance applicable to comparisons among treatment means are:

1. The variation among plants treated alike, that is, among plants within pots. Since different treatments are applied to different pots and, consequently, plants, variation among plants will be present in comparisons among treatment means.

2. The variation among plants in different pots treated alike, that is, variation among pots within treatments. This variation may be no greater than that among plants treated alike but usually is. The investigator will generally know whether or not to expect this to be a real source of variation; for example, it may be that the nutrients, light, moisture, etc., vary more from pot to pot than within a pot. A measure of variation computed from pot totals or pot means for the same treatment will, of course, contain both sources of variation.

Mean squares for the two types of variation listed above are generally referred to as *sampling error* and *experimental error*, respectively. If the second source of variation is not real, then the two mean squares will be of the same order of magnitude. Otherwise, the experimental error will be expected to be larger in random sampling since it contains an additional source of variation.

Comparable situations are found in many fields of experimentation. An agronomist may desire to determine the number of plants per unit area in a red clover varietal trial, or an entomologist may wish to determine the number of lice on cattle in an experiment to evaluate insecticides to control lice. In both cases, subsamples can be taken on each experimental unit. In chemical analyses, duplicate or triplicate determinations can be made on each sample. In field and animal experiments the variation among the subsamples or sampling units within an experimental unit is a measure of the homogeneity of the unit, whereas in chemical analyses it is often associated with the repeatability of the technique. This is the variation which leads to the *sampling error*.

In testing a hypothesis about treatment means, the appropriate divisor for F is the experimental error mean square since it includes variation from all sources that contribute to the variability of treatment means, except treatments.

Data that can be classified by a system consisting of a unique order of classification criteria, each criterion being applicable within all categories of the preceding criterion, are said to be in a *hierarchal* or *hierarchical classification*, a *nested classification* or a *within-within-within classification*. Thus, the data of Table 7.8 can be classified according to treatment, the temperature-time combination, then within treatments according to pot, and finally within pots according to plant. There is no other reasonable order. Treatments themselves may be classified in one of two orders, either as hours-within-temperatures or temperatures-within-hours. Treatments form a two-way classification, discussed in Chap. 8.

Computations. Let X_{ijk} be the week's growth for plant k in pot j receiving treatment i, $k = 1, \ldots, 4$, $j = 1, 2, 3$, $i = 1, \ldots, 6$. Thus plant 2 in pot 3 receiving treatment 6 (high night temperature following 16 hr of daylight) has a one-week stem growth of 7.0 cm, that is, $X_{632} = 7.0$ cm. Plant and pot numbers are for convenience in tagging the observations. Thus plants numbered 2 have nothing in common but the number, and pots numbered 3 have nothing in common but the number. On the other hand, plants or pots with the same treatment number received the same treatment so, are expected to respond similarly, while plants or pots with a different treatment number may possibly respond differently.

Denote pot totals by $X_{ij.}$, totals for a particular treatment-pot combination. Thus one-week stem growth of plants in pot 2 receiving treatment 3 (low night temperature following 16 hr of daylight) total 21.5 cm, that is, $X_{32.} = 21.5$ cm. Denote treatment totals by $X_{i..}$ The total one-week stem growth of plants on treatment 4 is 88.0 cm, that is, $X_{4..} = 88.0$ cm. Denote the grand total by $X_{...}$; $X_{...} = 416.5$ cm.

The dot notation ($X_{ij.}$, $X_{i..}$, $X_{...}$) is a common and useful shorthand that can be generalized easily to many experimental situations. The dot replaces a subscript and indicates that all values covered by the subscript have been added; the particular information formerly supplied by the individuals, as indicated by the subscript, has been discarded in favor of a summary in the form of a total; the subscript is no longer required and is replaced by a dot. Thus $X_{11.} = 3.5 + 4.0 + 3.0 + 4.5 = 15.0$ cm, $X_{1..} = 3.5 + 4.0 + \cdots + 3.0 = 15.0 + 17.5 + 11.5 = 44.0$ cm, and so on.

Obtain the 18 pot totals and sums of squares, that is, $\sum_k X_{ijk} = X_{ij.}$ and $\sum_k X_{ijk}^2$ for the 18 combinations of i and j. From these subtotals obtain the correction term C, the total sum of squares, and the sum of squares for pots. These are, respectively,

$$C = \frac{X_{...}^2}{srt} = \frac{(416.5)^2}{4(3)6} = 2{,}409.34$$

where s is the number of subsamples per plot, that is, plants per pot, r is the number of replicates of a treatment, that is, pots per treatment, and t is the number of treatments.

$$\text{Total SS} = \sum_{i,j,k} X_{ijk}^2 - C$$
$$= (3.5)^2 + (4.0)^2 + \cdots + (8.0)^2 - C = 255.91 \text{ with } 71 \text{ df}$$

$$\text{Pots SS} = \frac{\sum_{i,j} X_{ij\cdot}^2}{s} - C$$
$$= \frac{(15.0)^2 + \cdots + (35.0)^2}{4} - C = 205.47 \text{ with } 17 \text{ df}$$

The sum of squares attributable to subsamples (among plants within pots) may be found by subtraction:

$$\text{Within pots SS} = \text{total SS} - \text{pot SS}$$
$$= 255.91 - 205.47$$
$$= 50.44 \text{ with } 71 - 17 = 54 \text{ df}$$

The computations to this point may now be placed in an analysis of variance table (see Table 7.9).

TABLE 7.9

ANALYSIS OF VARIANCE OF MINT PLANT DATA, TABLE 7.8

Source of variation	df	SS	MS	MS is an estimate of
Among pots	17	205.47		
Treatments	5	179.64	35.93	$\sigma^2 + 4\sigma_\epsilon^2 + \left(12\sigma_\tau^2 \text{ or } 12\frac{\Sigma \tau^2}{5}\right)$
Among pots within treatments = experimental error	12	25.83	2.15	$\sigma^2 + 4\sigma_\epsilon^2$
Among plants within pots = sampling error	54	50.44	.93	σ^2
Total	71	255.91		

$$s^2 = .93, \; s_\epsilon^2 = \frac{2.15 - .93}{4} = .30, \; \frac{\Sigma t^2}{5} = \frac{35.93 - 2.15}{12} = 2.82 \text{ (or } = s_\tau^2 \text{ for Model II)}$$

where t_i is an estimate of τ_i.

Pot sum of squares measures variation due to treatments as well as variation among pots treated alike. Partitioning this sum of squares, we obtain:

$$\text{Treatments SS} = \frac{\sum_i X_{i\cdot\cdot}^2}{sr} - C$$
$$= \frac{(44.0)^2 + \cdots + (95.0)^2}{4(3)} - C = 179.64 \text{ with } 5 \text{ df}$$

Note that there are $sr = 12$ observations in each treatment total, $X_{i\cdot\cdot}$, hence the divisor in obtaining the treatment sum of squares. Finally,

Pots within treatment SS = pots SS − treatments SS
$$= 205.47 - 179.64 = 25.83 \text{ with } 17 - 5 = 12 \, df$$

The analysis of variance table is now completed, including mean squares (see Table 7.9). This type of analysis of variance is often presented with the pots line not shown at all or shown as a subtotal below the experimental error line.

The sum of squares obtained by subtraction may also be computed directly. The direct procedure indicates very clearly just what source of variation is involved at each stage and how the degrees of freedom arise.

Sampling error concerns the samples from the experimental unit, that is, the plants within pots. Thus the first pot contributes the following sum of squares:

$$\text{Plants SS for pot 1, treatment 1} = (3.5)^2 + \cdots + (4.5)^2 - \frac{(15.0)^2}{4}$$
$$= 1.25 \text{ with } 3 \, df$$

Similar computations are performed for each of the 18 pots giving 18 sums of squares, each with 3 degrees of freedom. Their total is

$$\text{SS among plants treated alike} = 1.25 + \cdots + 8.75$$
$$= 50.45 \text{ with } 18(3) = 54 \, df$$

The sums of squares and degrees of freedom are pooled to estimate a sampling variance. We assume that there is a common within-pot variance, regardless of the treatment.

Experimental error concerns experimental units treated alike, pots within treatments. Thus, treatment 1 contributes the following sum of squares to experimental error:

Pots SS for treatment 1
$$= \frac{(15.0)^2 + (17.5)^2 + (11.5)^2 - (15.0 + 17.5 + 11.5)^2/3}{4}$$
$$= 4.54 \text{ with } 2 \, df$$

The divisor 4 puts the sum of squares on a per-plant basis so that it may be compared with sampling error. Similar computations are made for the other five treatments and the results added. We obtain

$$\text{SS (pots treated alike)} = 4.54 + \cdots + 8.67$$
$$= 25.83 \text{ with } 6(2) = 12 \, df$$

Here we assume that the variance among pots is the same for all treatments. The sums of squares and degrees of freedom are pooled to give an estimate of this common variance.

Plant-to-plant variation, used to measure sampling error, is also present in pot-to-pot variation since the different pots contain different plants and thus is also present in treatment-to-treatment variation. Thus, both the treatment

ANALYSIS OF VARIANCE I: THE ONE-WAY CLASSIFICATION

variance and the pots-within-treatments variance contain a plant-to-plant variance. Treatment-to-treatment variation has also a contribution attributable to the variation among pots-treated-alike, if such exists, since the effects of different treatments are measured on different pots. It is seen, then, that experimental error is appropriate to comparisons involving different treatments whereas sampling error is not since the treatment mean square has only one possible additional source of variation not possessed by the experimental error, namely, that due to treatments themselves. A valid test of the null hypothesis of no treatment differences is given by

$$F = \frac{\text{treatment mean square}}{\text{error mean square}}$$

$$= \frac{35.93}{2.15} = 16.7^{**} \text{ with 5 and 12 } df$$

Experimental error may or may not contain variation in addition to that among subsamples. This depends upon the environmental differences that exist from unit to unit and whether or not they are greater than those within the unit, the pot in our example. A test is available in

$$F = \frac{\text{experimental error mean square}}{\text{sampling error mean square}}$$

$$= \frac{2.15}{.93} = 2.3^* \text{ with 12 and 54 } df$$

The standard error of a treatment mean is $s_{\bar{x}} = \sqrt{2.15/12} = .42$ cm, and the standard error of a difference between treatment means is $s_{\bar{d}} = \sqrt{2(2.15)/12} = .60$ cm. The coefficient of variability is $(s/\bar{x})100 = (1.47/5.78)100 = 25\%$.

The computations for hierarchal classifications with increasing numbers of classification criteria are the obvious generalization of the computations of this section.

Exercise 7.11.1 Check the computations for the data of Table 7.8. Obtain the sums of squares for sampling error and experimental error directly and thus check the additivity of sums of squares in the analysis of variance.

Exercise 7.11.2 Distinguish between the variability of a pot mean and the variability among pot means (see also Sec. 7.14).

7.12 The linear model for subsampling. The discussion of the mint plant data, Table 7.8, makes it fairly evident that the mathematical expression proposed to explain the data is

$$X_{ijk} = \mu + \tau_i + \epsilon_{ij} + \delta_{ijk} \tag{7.22}$$

where each observation is intended to supply information about the mean of the treatment population sampled, namely, $\mu + \tau_i$. If we treat the τ's as fixed effects, as they obviously are, then we measure them as deviations such that $\Sigma \tau_i = 0$; if we treat them as random, then we assume they are from a population with mean zero and variance σ_τ^2. Two random elements are obtained with each observation. The ϵ_{ij}'s are the experimental unit, that is, pot, contributions and are assumed to be normally and independently distributed with zero mean and variance σ_ϵ^2; the δ_{ijk}'s are the subsample, that is, plant, contributions and are assumed to be normally and independently distributed with zero mean and variance σ^2. The ϵ's and δ's are assumed to be unrelated, that is, the drawing of a particular value of δ does not affect the probability of drawing any particular ϵ.

Now consider the pot totals used in computing experimental error. For our particular set of data, they are given by

$$4\mu + 4\tau_i + 4\epsilon_{ij} + \sum_k \delta_{ijk}$$

for the 18 combinations of i and j. A variance computed with these sums would contain a $4\sigma^2$ and a $(4)^2\sigma_\epsilon^2$, the first quantity involving the variance of sums of δ's and the second involving the variance of a multiple of each ϵ, essentially a coding operation. Thus the coefficients are 4 and 4^2, respectively. The computational procedure used requires us to divide any sum of squares by the number of observations in each squared quantity. Hence the computed experimental error mean square estimates $\sigma^2 + 4\sigma_\epsilon^2$. A similar argument leads to the coefficients of the components of the treatment mean square. The results are presented in Table 7.9. What any F value is intended to detect now becomes clear. From the F value, it appears that both σ_ϵ^2 and a treatment contribution are present in our data. Since the treatment component is present unless we have a very unusual sample, we conclude that the null hypothesis $H_0: \tau_1 = \tau_2 = \cdots = \tau_6 = 0$, appropriate to the fixed model, is false and that at least one τ differs from the others.

Note that the unit of measurement used was ½ cm and the sampling error was .93, giving a standard deviation of .96 cm. Recall that in Sec. 2.16 a class interval not greater than one-quarter of the standard deviation was recommended to keep the loss in information low. Thus, for an estimate of the sampling error with low loss of information due to the choice of size of unit, we are led to propose an impractical unit of measurement. In this case, it is fortunate that we have relatively little use for sampling error. Another point that may be brought out relative to the choice of unit of measurement is this: a treatment such as high night temperature with 8 hrs of daylight, pot 3, contributes nothing to sampling error sum of squares but does contribute 3 degrees of freedom. It is sometimes recommended, in cases like this, that these 3 degrees of freedom be left out of the total.

Exercise 7.12.1 By sampling, construct a mixed model experiment that includes both experimental and sampling error. For simplicity, have three treatments, two experimental units per treatment, and three observations per unit. First decide upon a general mean, say $\mu = 80$. Second, choose a set of fixed effects for treatments, say $\tau_1 = -4$, $\tau_2 = -2$, and $\tau_3 = 6$. Third, obtain a set of six random elements for the six experimental units. For these, sample Table 4.1 and compute $\epsilon_{ij} = (X_{ij} - 40)/4$ so that the population sampled has $\mu_\epsilon = 0$ and $\sigma_\epsilon = 3$. Note that the sample of ϵ's will not normally have $\bar{x}_\epsilon = 0$ or $s_\epsilon = 3$. Finally, obtain 18 random elements for the 18 sampling units. For these, sample Table 4.1 and compute $\delta_{ijk} = (X_{ijk} - 40)/12$ so that the population sampled has $\mu_\delta = 0$ and $\sigma_\delta = 1$.

Write out the analysis of variance showing sources of variation, degrees of freedom, and parameters being estimated. Since all parameters have known values, it is possible to show numerical answers. Compute the analysis of variance and compare the values obtained as estimates with the values being estimated. (If convenient, obtain results on a class basis where each individual does his own sampling. Average the values of the various estimates obtained and compare with parameters.)

Exercise 7.12.2 Repeat Exercise 7.12.1 for the random model. To replace fixed effects, obtain $\tau_i = (X_i - 40)$ by sampling Table 4.1.

7.13 Analysis of variance with subsamples. Unequal subsample numbers.
When samples have unequal subsample numbers, the basic analysis is that of Sec. 7.9; in the computations, the square of any total is divided by the number of observations in the total. For example, consider the numbers in Table 7.10. These might be observations on the product of three manufacturing plants in each of two areas, A and B, and of two plants in area C. For the 14 observations, $n = 14$, $\Sigma X = 95$, $\Sigma X^2 = 659$, $\Sigma x^2 = 14.36$ with 13 degrees of freedom.

TABLE 7.10

OBSERVATIONS ON THE QUALITY OF A PRODUCT MADE IN EIGHT PLANTS IN THREE AREAS

Area	A			B			C	
Plants	I	II	III	I	II	III	I	II
Observations	6	6,8	6,7,8	5,7	6,7	6	7	7,9

Proceeding as in Sec. 7.9, we have

$$\text{Plants SS (ignoring areas)} = 6^2 + \frac{(6+8)^2}{2} + \cdots + \frac{(7+9)^2}{2}$$
$$- \frac{(6+6+8+\cdots+7+9)^2}{14}$$
$$= 5.86 \text{ with } 7 \, df$$

$$\text{Residual SS} = \text{total SS} - \text{plant SS}$$
$$= 14.36 - 5.86 = 8.50 \text{ with } 13 - 7 = 6 \, df$$

Plant sum of squares is further partitioned into a component associated with areas and one associated with plants within areas.

$$\text{Areas SS} = \frac{(6+6+\cdots+8)^2}{6} + \frac{(5+\cdots+6)^2}{5} + \frac{(7+7+9)^2}{3} - C$$
$$= 4.07 \text{ with } 2 \, df$$

Plants within areas SS = plants SS − areas SS
$$= 5.86 - 4.07 = 1.79 \text{ with } 7 - 2 = 5 \, df$$

The resulting analysis of variance is given in Table 7.11.

For these "data," there is no evidence that variation among plants is of a different order of magnitude than that among observations since $F = 0.36/1.42 < 1$. In such a case, one says $s_\epsilon^2 = 0$ rather than that it is a negative value and, especially if the degrees of freedom for experimental error are small, may pool the two errors to obtain a new estimate of a variance suitable for testing areas. Here, the new estimate would be $(1.79 + 8.50)/(5 + 6) = 0.94$ with 11 degrees of freedom.

TABLE 7.11

ANALYSIS OF VARIANCE FOR "DATA" OF TABLE 7.10

Source	df	SS	MS	MS is an estimate of
Areas	2	4.07	2.03	$\sigma^2 + 1.90\sigma_\epsilon^2 + 4.50\sigma_\tau^2$
Plants within areas = experimental error	5	1.79	0.36	$\sigma^2 + 1.64\sigma_\epsilon^2$
Observations within plants = sampling error	6	8.50	1.42	σ^2
Total	13	14.36		

$$s^2 = 1.42 \qquad s_\epsilon^2 = \frac{0.36 - 1.42}{1.64} < 0$$

When unequal numbers of subsamples are involved, computation of the coefficients for the components of variance is far less obvious than when equal numbers prevail. First, let us define r_{ij} as the number of observations at the jth plant in the ith area; for example, $r_{13} = 3$. Now $r_{i\cdot}$ is the total number of observations made in the ith area; for example, $r_{1\cdot} = 1 + 2 + 3 = 6$. Finally, $r_{\cdot\cdot}$ is the total number of observations; here $r_{\cdot\cdot} = 14$. Let k equal the number of areas; here $k = 3$.

The coefficient of σ_ϵ^2 depends upon the line in the analysis of variance. Thus for plants within areas, the coefficient of σ_ϵ^2 is

$$\frac{\left(r_{\cdot\cdot} - \sum_i \left(\sum_j r_{ij}^2/r_{i\cdot}\right)\right)}{df \text{ (experimental error)}} \tag{7.23}$$

$$= \frac{14 - [(1^2 + 2^2 + 3^2)/6 + (2^2 + 2^2 + 1^2)/5 + (1^2 + 2^2)/3]}{5} = 1.64$$

For areas, the coefficient of σ_ϵ^2 is

$$\frac{\sum_i \left(\sum_j r_{ij}^2/r_{i\cdot}\right) - \left(\sum_{i,j} r_{ij}^2\right)/r_{\cdot\cdot}}{df \text{ (areas)}} \tag{7.24}$$

$$= \frac{(1^2 + 2^2 + 3^2)/6 + (2^2 + 2^2 + 1^2)/5 + (1^2 + 2^2)/3 - (1^2 + \cdots + 2^2)/14}{2}$$

$$= \frac{5.8 - 2}{2} = 1.90$$

The coefficient of σ_τ^2 is

$$\frac{r_{\cdot\cdot} - \sum_i r_{i\cdot}^2/r_{\cdot\cdot}}{df \text{ (areas)}} = \frac{14 - (6^2 + 5^2 + 3^2)/14}{2} = 4.50$$

When sampling error is larger than experimental error and chance appears to offer the only explanation of this fact, it is customary to estimate σ_ϵ^2 as zero, that is, $s_\epsilon^2 = 0$. In the more general situation, that is, when $s_\epsilon^2 > 0$, a real problem arises in testing areas for a $\Sigma \tau_i^2$ or σ_τ^2 component since no line in the analysis of variance differs from the area mean square line by some multiple of $\Sigma \tau_i^2$ or σ_τ^2 only. This problem arises because of the unequal subsample sizes. The reader who faces this problem is referred to Anderson and Bancroft (7.1) and Snedecor (7.14) for approximate methods of testing.

7.14 Variance components in planning experiments involving subsamples. In planning experiments which involve subsampling, the question arises as to how to distribute the available time and money, in particular whether to concentrate on many samples with few subsamples or whether to take few samples with more subsamples. The answer depends upon the relative magnitude of experimental and sampling errors and upon costs. Thus, subsamples may involve costly chemical analyses, time-consuming procedures, or destructive tests of expensive items whereas obtaining samples may be of trivial difficulty. On the other hand, it may be that obtaining samples may involve additional equipment, plots, animals, or expensive travel, while subsampling involves none of these. Probably the true situation will be intermediate although, most often, additional subsamples are likely to involve less cost than additional samples.

A thorough consideration of the problem of sampling versus subsampling should consider relative cost as well as the relative magnitude of the variance. Here we examine the problem from the point of view of the variances only.

Where data are available from an experiment involving subsampling, they can be used in planning future experiments. The data from the mint plant experiment reported in Tables 7.8 and 7.9 will be used to illustrate the procedure. Experimental error, the criterion for judging the significance of comparisons among treatment means, consists of two sources of variation, variation among plants treated alike, σ^2, and variation resulting from differences in the environment of pots treated alike, σ_ϵ^2. Both sources of variation contribute to the variance among treatment means. The estimates of σ^2 and σ_ϵ^2 are $s^2 = .93$ and $s_\epsilon^2 = (2.15 - .93)/4 = .30$. Notice that s_ϵ^2, the variation among means of similarly treated pots over and above that due to plants within pots, is greater than the variance of pot means based on the sampling error, that is, on $s^2/4 = .93/4 = .23$.

Suppose that in place of four plants per pot and three pots per treatment, there were six plants per pot and two pots per treatment, the same number of plants per treatment. Experimental error would be an estimate of $\sigma^2 + 6\sigma_\epsilon^2$. Such an experiment should lead to an error variance of the order of

$$s^2 + 6s_\epsilon^2 = .93 + 6(.30) = 2.73$$

with 6 degrees of freedom as compared with 2.15 with 12 degrees of freedom for the experiment conducted. To compare the relative precision of the two designs, consider the variance of a treatment mean. Here we have equal numbers of plants, $4(3) = 12 = 6(2)$, so can compare experimental errors directly. Ignoring degrees of freedom in experimental error, we find the precision of the experiment conducted, relative to that proposed, to be $(2.73/2.15)100\% = 127\%$.

This means that the proposed experiment would require about 27% more plants to detect differences of the same order of magnitude as observed. The variances of a treatment mean are:

1. For the experiment conducted

$$s_{\bar{x}}^2 = \frac{2.15}{12} = .179$$

2. For the experiment proposed with an additional 27% of $12 = 3+$ plants per treatment

$$s_{\bar{x}}^2 = \frac{2.73}{(12+3)} = .182$$

The small difference is due to the fact that 27% of 12 is not exactly 3.

The new F value for comparing treatments should be of the order of

$$F = \frac{s^2 + 6s_\epsilon^2 + 12(\Sigma\hat{\tau}^2/5)}{s^2 + 6s_\epsilon^2}$$

$$= \frac{36.57}{2.73} = 13.40 \text{ with 5 and 6 } df$$

The value for the experiment was 16.71 with 5 and 12 degrees of freedom. Because of the changes in degrees of freedom, 127% is actually an underestimate of the relative precision of the two experiments as shown by Eq. (6.1).

If the experiment were conducted with two plants per pot and six pots per treatment, experimental error would be an estimate of $\sigma^2 + 2\sigma_\epsilon^2$. Error variance should be of the order of

$$s^2 + 2s_\epsilon^2 = .93 + 2(.30) = 1.53$$

with 30 degrees of freedom. The precision of the experiment conducted relative to that proposed is, ignoring degrees of freedom, $(1.53/2.15)100 = 71\%$. In this case 71% of 12 plants should give about the same efficiency; 8 plants would be somewhat less efficient; 9 plants would be somewhat more efficient.

The investigator will need to compare costs, labor, and aims as well as efficiency in choosing among possible designs.

7.15 Assumptions underlying the analysis of variance. In the analysis of variance where tests of significance are made, the basic assumptions are:

1. Treatment and environmental effects are additive.
2. Experimental errors are random, independently and normally distributed about zero mean and with a common variance.

The assumption of normality is not required for estimating components of variance. In practice, we are never certain that all these assumptions hold; often there is good reason to believe some are false. Excellent discussions of these assumptions, the consequences when they are false, and remedial steps

ANALYSIS OF VARIANCE I: THE ONE-WAY CLASSIFICATION

are given by Eisenhart (7.8), Cochran (7.3), and Bartlett (7.2). A brief discussion follows:

Departure from one or more of the assumptions can affect both the level of significance and the sensitivity of F or t. In the case of non-normality, the true level of significance is usually, but not always, greater than the apparent level. This results in rejection of the null hypothesis when it is true, more often than the probability level calls for; that is, too many nonexistent significant differences are claimed. The experimenter may think he is using the 5% level when actually the level may be 7 or 8%. Loss of sensitivity for tests of significance and estimation of effects occurs since a more powerful test could be constructed if the exact mathematical model were known. In other words, if the true distribution of errors and the nature of the effects, their additivity or nonadditivity, were known, we could construct a test better able to detect or estimate real effects.

For most types of biologic data, experience indicates that the usual disturbances resulting from failure of the data to fulfill the above requirements are unimportant. Exceptional cases do occur and procedures to analyze such data will be discussed. In any case, most data do not exactly fulfill the requirements of the mathematical model, and procedures for testing hypotheses and estimating confidence intervals should be considered approximate rather than exact.

Consider the assumption that *treatment and environmental effects are additive*. A common form of nonadditivity occurs when such effects are multiplicative. Consider a simple case where two environments, called blocks in the illustration that follows, and two treatments have effects that are multiplicative. A comparison of additive and multiplicative models is given in Table 7.12, which ignores experimental errors.

TABLE 7.12

ADDITIVE AND MULTIPLICATIVE MODELS

Model	Additive		Multiplicative		Log_{10} (multiplicative becomes additive)	
Block	1	2	1	2	1	2
Treatment 1	10	20	10	20	1.00	1.30
Treatment 2	30	40	30	60	1.48	1.78

For the additive model, the increase from block 1 to 2 is a fixed amount regardless of the treatment; the same is true for treatments. For the multiplicative model, the increase from block 1 to 2 is a fixed percentage regardless of the treatment; the same is true for treatments. [When effects are multiplicative, the logarithms of the data exhibit the effects in additive fashion and an analysis of variance of the logarithms is appropriate.] The new data, the logarithms, are called transformed data; the process of changing the data is called a *transformation*. For other types of nonadditivity, other transformations are available. Transforming data implies that experimental errors are independently and normally distributed on the transformed scale if tests of significance are planned. A test of additivity is given in Sec. 11.9.

The presence of nonadditivity in data results in heterogeneity of error when no transformation is made before analysis. The components of error variance contributed by the various observations do not supply estimates of a common variance. The resulting pooled error variance may be somewhat inefficient for confidence interval estimates of treatment effects and may give false significance levels for certain specific comparisons of treatment means while the significance level for the F test involving all treatment means may be affected very little.

Our second assumption is not independent of the first, as has been seen, and is really a set of assumptions. Consider the *independence of errors*. For field experiments, crop responses on adjacent plots tend to be more alike than responses on nonadjacent plots; the same is also true for laboratory experiment observations made by the same person at about the same time. The result is that tests of significance may be misleading if no attempt is made to overcome the difficulty. In practice, treatments are assigned to the experimental units at random or the order of the observations is determined at random; the effect of the randomization process is to make the errors independent of each other.

In field experiments, convenient systematic designs place the same treatments adjacent to each other in all blocks. Since adjacent plots tend to be more alike, the precision of a comparison is greater for treatments falling on plots close together than for those on plots farther apart. An analysis of variance of such data gives a generalized error term too large for certain comparisons and too small for others. Appropriate error terms for individual comparisons are not available. Cochran and Cox (7.4) summarize the need for randomization very well in the following statement: "Randomization is somewhat analogous to insurance, in that it is a precaution against disturbances that may or may not occur and that may or may not be serious if they do occur. It is generally advisable to take the trouble to randomize even when it is not expected that there will be any serious bias from failure to randomize. The experimenter is thus protected against unusual events that upset his expectations."

Experimental error must be normally distributed. This assumption applies particularly to tests of significance and not to the estimation of components of variance. When the distribution of experimental errors is decidedly skewed, the error component of a treatment tends to be a function of the treatment mean. This again results in heterogeneity of the error term. If the functional relationship is known, a transformation can be found that will give errors more nearly normally distributed. Thus an analysis of variance can be made on the transformed data such that the error term will be essentially homogeneous. Common and useful transformations are the logarithmic, square root, and inverse sine; their use is discussed in Sec. 8.15.

Experimental errors must have a common variance. For example, in a completely random design, the components of error contributed by the several treatments must all be estimates of a common population variance. Here, heterogeneity of error may result from the erratic behavior of the response to certain treatments.

In experiments such as those meant to determine the effectiveness of different insecticides, fungicides, or herbicides, an untreated check may be included to measure the level of infestation and to provide a basis for determining the effectiveness of the treatments. The variation of the individual observations on the check may be considerably greater than that for the other treatments, primarily because the check may have a higher mean and, thus, a greater base for variation. In these situations the error term may not be homogeneous. A remedy is to divide the error term into homogeneous components to test specific treatment comparisons. Sometimes if the means of one or two treatments are much larger than the others and have significantly greater variation, these treatments can be excluded from the analysis.

Violation of any of the other assumptions may result in heterogeneity of experimental error. Suggestions for remedying the situation have been made and depend on the nature of the violation.

References

7.1 Anderson, R. L., and T. A. Bancroft: *Statistical Theory in Research*, McGraw-Hill Book Company, Inc., New York, 1952.

7.2 Bartlett, M. S.: "The use of transformations," *Biometrics*, **3:** 39–52 (1947).

7.3 Cochran, William G.: "Some consequences when the assumptions for the analysis of variance are not satisfied," *Biometrics*, **3:** 22–38 (1947).

7.4 Cochran, William G., and Gertrude M. Cox: *Experimental Designs*, 2d ed., John Wiley & Sons, Inc., New York, 1957.

7.5 Duncan, D. B.: "A significance test for differences between ranked treatments in an analysis of variance," *Virginia J. Sci.*, **2:** 171–189 (1951).

7.6 Duncan, D. B.: "Multiple range and multiple F tests," *Biometrics*, **11:** 1–42 (1955).

7.7 Dunnett, C. W.: "A multiple comparisons procedure for comparing several treatments with a control," *J. Am. Stat. Assoc.*, **50:** 1096–1121 (1955).

7.8 Eisenhart, C.: "The assumptions underlying the analysis of variance," *Biometrics*, **3:** 1–21 (1947).

7.9 Erdman, Lewis W.: "Studies to determine if antibiosis occurs among Rhizobia: 1. Between *Rhizobium meliloti* and *Rhizobium trifolii*," *J. Am. Soc. Agron.*, **38:** 251–258 (1946).

7.10 Harter, H. L.: "Error rates and sample sizes for range tests in multiple comparisons," *Biometrics*, **13:** 511–536 (1957).

7.11 Hartley, H. O.: "Some recent developments in Analysis of Variance," *Comm. on Pure and App. Math.*, **8:** 47–72 (1955).

7.12 Kramer, C. Y.: "Extension of multiple range tests to group means with unequal numbers of replication," *Biometrics*, **12:** 307–310 (1956).

7.13 Newman, D.: "The distribution of range in samples from a normal population, expressed in terms of an independent estimate of standard deviation," *Biometrika*, **31:** 20–30 (1939).

7.14 Snedecor, G. W.: *Statistical Methods*, 5th ed., Iowa State College Press, Ames, Iowa, 1956.

7.15 Tukey, J. W.: "The problem of multiple comparisons," Ditto, Princeton University, Princeton, N.J., 1953.

7.16 Wexelsen, H.: "Studies in fertility, inbreeding, and heterosis in red clover (*Trifolium pratense* L.)," *Norske videnskaps-akad. i Oslo, Mat.-Natur. klasse*, 1945.

Chapter 8

ANALYSIS OF VARIANCE II: MULTIWAY CLASSIFICATIONS

8.1 Introduction. The completely random design is appropriate when no sources of variation, other than treatment effects, are known or can be anticipated. In many situations it is known beforehand that certain experimental units, if treated alike, will behave differently. For example, in field experiments, adjacent plots usually are more alike in response than those some distance apart; likewise, heavier animals in a group of the same age may exhibit a different rate of gain than lighter animals; also, observations made on a particular day or using a certain piece of equipment may resemble each other more than those made on different days or using different equipment. In such situations, where the behavior of individual units may be anticipated in part and the units classified accordingly, designs or layouts can be constructed such that the portion of the variability attributable to the recognized source can be measured and thus excluded from the experimental error; at the same time, differences among treatment means will contain no contribution attributable to the recognized source. In Chap. 5, this principle was used in the comparison of two treatments with paired observations.

This chapter deals with the analysis of variance where two or more criteria of classification are used. Analyses for the randomized complete-block and Latin-square designs are given. In addition, interaction is defined and the use of transformations discussed.

8.2 The randomized complete-block design. This design may be used when the experimental units can be meaningfully grouped, the number of units in a group being equal to the number of treatments or some multiple of it. Such a group is called a *block* or *replicate*. (The term replicate is somewhat unfortunate since we have already defined it simply as a repetition.) The object of grouping is to have the units in a block as uniform as possible so that observed differences will be largely due to treatments. Variability among units in different blocks will be greater, on the average, than variability among units in the same block if no treatments were to be applied. Clearly variability among blocks does not affect differences among treatment means since each treatment appears in every block.

In field experiments, each block usually consists of a compact, nearly square, group of plots. Likewise, in many animal experiments the individual animals are placed in outcome groups or blocks on the basis of such characteristics as initial weight, condition of the animal, breed, sex, or age, as stage of lactation and milk production in dairy cattle, and as litter in hogs.

During the course of the experiment all units in a block must be treated as uniformly as possible in every respect other than treatment. Any changes in technique or other condition that might affect results should be made on the complete block. For example, if harvesting of field plots is spread over a period of several days, all plots of a block should be harvested the same day. Also, if different individuals make observations on the experimental material and if there is any likelihood that observations made on the same plot will differ from individual to individual, and if only one observation is to be made on each experimental unit, then one individual should make all the observations in a block. Again, if the number of observations per unit equals the number of observers, then each observer should make one observation per unit. These practices help control variation within blocks, and thus experimental error; at the same time, they contribute nothing to differences among treatment means. Variation among blocks is arithmetically removed from experimental error.

Notice the balance that exists in this design. Each treatment appears an equal number of times, usually once, in each block and each block contains all treatments. It is sometimes stated that blocks and treatments are orthogonal to one another. It is this property that leads to the very simple arithmetic involved in the analysis of the resulting data. This design is used more frequently than any other design and if it gives satisfactory precision, there is no point in using an alternative.

Randomization. When the experimental units have been assigned to blocks, they are numbered in some convenient manner. Treatments are also numbered and then randomly assigned to the units within any block. A new randomization is carried out for each block.

The actual procedure may be one of those given in Sec. 7.2. For example, if we have eight treatments, we obtain 8 three-digit numbers and observe their ranks. The ranks are considered to be a random permutation of the numbers $1, \ldots, 8$. For example, we might obtain 1, 8, 7, 5, 4, 6, 2, 3 as the random permutation. Now treatment 1 is applied to unit 1, treatment 8 to unit 2, ..., treatment 3 to unit 8. Alternatively, we could apply treatment 1 to unit 1, treatment 2 to unit 8, ..., treatment 8 to unit 3. The other randomization procedures of Sec. 7.2 are as easily generalized.

The randomized complete-block design has many advantages over other designs. It is usually possible to group experimental units into blocks so that more precision is obtained than with the completely random design. There is no restriction on the number of treatments or blocks. If extra replication is desired for certain treatments, these may be applied to two or more units per block. The statistical analysis of the data is simple. If, as a result of mishap, the data from a complete block or for certain treatments are unusable, these data may be omitted without complicating the analysis. If data from individual units are missing, they can be estimated easily so that arithmetic convenience is not lost. If the experimental error is heterogeneous, unbiased components applicable to testing specific comparisons can be obtained.

The chief disadvantage of randomized complete blocks is that when the variation among experimental units within a block is large, a large error term results. This frequently occurs when the number of treatments is large; thus it may not be possible to secure sufficiently uniform groups of units for blocks. In such situations, other designs to control a greater proportion of the variation are available.

8.3 Analysis of variance for any number of treatments; randomized complete-block design. A symbolic summary of the working and definition formulas for the sums of squares and degrees of freedom in the analysis of variance of data from the randomized complete-block design is given in Table 8.1.

TABLE 8.1

FORMULAS FOR THE ANALYSIS OF VARIANCE FOR t TREATMENTS ARRANGED IN A RANDOMIZED COMPLETE-BLOCK DESIGN OF r BLOCKS

Source of variation	df	Sums of squares	
		Definition	Working
Blocks	$r-1$	$t \sum_j (\bar{x}_{.j} - \bar{x}_{..})^2 =$	$\dfrac{\sum_j X_{.j}^2}{t} - C$
Treatments	$t-1$	$r \sum_i (\bar{x}_{i.} - \bar{x}_{..})^2 =$	$\dfrac{\sum_i X_{i.}^2}{r} - C$
Error	$(r-1)(t-1)$	$\sum_{i,j} (X_{ij} - \bar{x}_{.j} - \bar{x}_{i.} + \bar{x}_{..})^2 =$	(total $-$ block $-$ trt)SS
Total	$rt-1$	$\sum_{i,j} (X_{ij} - \bar{x}_{..})^2 =$	$\sum_{i,j} X_{ij}^2 - C$

Let X_{ij} be the observation from the jth block on the ith treatment, $i = 1, \ldots, t$ treatments and $j = 1, \ldots, r$ blocks. Dot notation is used wherever possible. Thus $\sum_j X_{.j}^2$ means obtain sums $X_{.j} = \sum_i X_{ij}$ for each value of j, square them, and add them for all values of j. Represent the grand mean by $\bar{x}_{..}$. Since the variance of means of n observations is σ^2/n, multipliers of t and r shown in the definition SS column result in all mean squares being estimates of the same σ^2 when there are no block or treatment effects. The same reasoning based on totals accounts for the divisors t and r in the working SS column.

Table 8.2 gives the oil content of Redwing flaxseed in per cent for plots located at Winnipeg and inoculated, using several techniques, with spore suspensions of *Septoria linicola*, the organism causing pasmo in flax. The data were reported by Sackston and Carson (8.16). The original data have been coded by subtracting 30 from each observation. One might further code by multiplying each number by 10 to eliminate decimal points; however, coding without multiplication requires no decoding of variances or standard deviations. Note that sums of squares computed in two directions provide a check on the computation of the total sum of squares.

ANALYSIS OF VARIANCE II: MULTIWAY CLASSIFICATIONS

In detail, the computations proceed as follows:

Step 1. Arrange the raw data as in Table 8.2. Obtain treatment totals $X_{i\cdot}$, block totals $X_{\cdot j}$, and the grand total $X_{\cdot\cdot}$. Simultaneously obtain ΣX^2 for each treatment and block, that is, $\sum_j X_{ij}^2$ for each i and $\sum_i X_{ij}^2$ for each j. Obtain the grand total by summing treatment totals and block totals separately. Simultaneously obtain the sum of squares of these totals. These are not shown in Table 8.2, with 788.23 being $\sum_{i,j} X_{ij}^2$.

Step 2. Obtain (adjusted) sums of squares as follows.

$$\text{Correction term} = C = \frac{X_{\cdot\cdot}^2}{rt} \tag{8.1}$$

$$= \frac{(132.7)^2}{24} = 733.72$$

$$\text{Total SS} = \sum_{i,j} X_{ij}^2 - C \tag{8.2}$$

$$= 106.98 + \cdots + 198.43 - 733.72 = 54.51$$

or

$$= 176.50 + \cdots + 213.09 - 733.72 = 54.51$$

$$\text{Block SS} = \frac{\sum_j X_{\cdot j}^2}{t} - C \tag{8.3}$$

$$= \frac{31.6^2 + \cdots + 34.5^2}{6} - 733.72 = 3.14$$

$$\text{Treatment SS} = \frac{\sum_i X_{i\cdot}^2}{r} - C \tag{8.4}$$

$$= \frac{20.4^2 + \cdots + 28.1^2}{4} - 733.72 = 31.65$$

$$\text{Error SS} = \text{total SS} - \text{replicate SS} - \text{treatment SS} \tag{8.5}$$
$$= 54.51 - 3.14 - 31.65 = 19.72$$

The F value for testing the null hypothesis of no treatment differences is $6.33/1.31 = 4.83**$ with 5 and 15 degrees of freedom. It is significant at the 1% level. This is evidence that there are real differences among the treatment means. To determine where the differences lie, general procedures such as discussed in Secs. 7.4 to 7.8 can be used. The appropriate procedure will be determined by the questions initially posed by the experimenter. Other procedures are discussed in Chap. 11.

The sample standard error of the difference between two equally replicated treatment means is given by $s_{\bar{d}} = \sqrt{2s^2/r}$. In this formula, s^2 is error mean square and r is the number of blocks. For the data of Table 8.2, $s_{\bar{d}} = \sqrt{2(1.31)/4} = 0.81\%$ oil, where per cent of oil is the unit of measurement. The coefficient of variability is $CV = s(100)/\bar{x} = 1.14(100)/35.5 = 3.2\%$, where per cent no longer refers to the unit of measurement. If certain treatments receive extra replication, the formula is given by $s_{\bar{d}} = \sqrt{s^2(1/r_1 + 1/r_2)}$.

Table 8.2

Oil Content of Redwing Flaxseed Inoculated at Different Stages of Growth with *S. linicola*, Winnipeg, 1947, in Per Cent

Original observation = 30 + tabled observation

Treatment (stage when inoculated)	Block				Treatment totals		Decoded treatment means
	1	2	3	4	$X_{i.}$	$\sum_j X_{ij}^2$	
Seedling	4.4	5.9	6.0	4.1	20.4	106.98	35.1
Early bloom	3.3	1.9	4.9	7.1	17.2	88.92	34.3
Full bloom	4.4	4.0	4.5	3.1	16.0	65.22	34.0
Full bloom (1/100)	6.8	6.6	7.0	6.4	26.8	179.76	36.7
Ripening	6.3	4.9	5.9	7.1	24.2	148.92	36.0
Uninoculated	6.4	7.3	7.7	6.7	28.1	198.43	37.0
Block totals $\begin{cases} X_{.j} \\ \sum_i X_{ij}^2 \end{cases}$	31.6	30.6	36.0	34.5	132.7		35.5
	176.50	175.28	223.36	213.09		788.23	

Analysis of variance

Source of variation	df	SS	MS	F
Blocks	$r - 1 = 3$	3.14	1.05	
Treatments	$t - 1 = 5$	31.65	6.33	4.83**
Error	$(r-1)(t-1) = 15$	19.72	1.31	
Total	$rt - 1 = 23$	54.51		

Extra replication in a randomized complete-block experiment implies that any treatment appears an equal number of times in all replicates, the number varying from treatment to treatment; r_1 and r_2 will be multiples of r.

The variation among blocks may also be tested. It is not significant in our example. The F test of blocks is valid but requires care in its interpretation. In most experiments, the null hypothesis of no differences among blocks is of no particular concern since blocks are an acknowledged source of variation, often, on the basis of past experience, expected to be large. In some experiments, blocks may measure differences in the order of performing a set of operations, in pieces of equipment, in individuals, etc. In such cases, the F test of blocks may have special meaning.

Certain misconceptions have arisen concerning the significance of blocks. You may hear that if blocks are significant, the experiment is of little value or that blocks must be significant before much credence can be placed upon observed treatment effects. Both points of view are erroneous. If block effects are significant, it indicates that the precision of the experiment has been increased by use of this design relative to the use of the completely random design. Also, the scope of an experiment may be increased when blocks are significantly different since the treatments have been tested over a wider range of experimental conditions. A word of caution should be injected here in that if block differences are very large the problem of heterogeneity of error may exist. This problem is discussed in Secs. 7.15, 8.15, and 11.6. If block effects are small, it indicates either that the experimenter was not successful in reducing error variance by his grouping of the individual units or that the units were essentially homogeneous to begin with.

Exercise 8.3.1 Tucker et al. (8.18) determined the effect of washing and removing excess moisture by wiping or by air current on the ascorbic acid content of turnip greens. The data in milligrams per 100 grams dry weight are shown in the accompanying table.

Treatment	Block				
	1	2	3	4	5
Control	950	887	897	850	975
Washed and blotted dry	857	1,189	918	968	909
Washed and dried in air current	917	1,072	975	930	954

Carry out an analysis of variance on these data. Use Dunnett's procedure to test differences among treatment means.

Exercise 8.3.2 Bing (8.4) compared the effect of several herbicides on the spike weight of gladiolus. The average weight per spike in ounces is given below for four treatments.

Treatment	Block			
	1	2	3	4
2.4–D TCA	2.05	1.56	1.68	1.69
Check	1.25	1.73	1.82	1.31
DN/Cr	1.95	2.00	1.83	1.81
Sesin	1.75	1.93	1.70	1.59

Analyze the data. Use Dunnett's procedure to test differences among treatment means.

8.4 The nature of the error term. In the analysis of variance for a randomized complete-block design, the error sum of squares is found by subtracting the block and treatment sums of squares from the total sum of squares. This is possible since the sums of squares are additive. The error sum of squares can be obtained directly from Eq. (8.6).

$$\text{Error SS} = \sum_{i,j} (X_{ij} - \bar{x}_{i\cdot} - \bar{x}_{\cdot j} + \bar{x}_{\cdot\cdot})^2 \qquad (8.6)$$

This definition formula arises from the model that defines the means of the various populations sampled. There are rt means in the case of the randomized complete-block design, one per cell, with only one observation necessarily made on each population. (Population means are also called *expected values*.) A mean is defined in terms of a general mean μ, a treatment contribution τ_i, and a block contribution β_j; that is, the mean of the i,jth cell is $\mu + \tau_i + \beta_j$. An observation is subject to a random error, errors being from a single population with zero mean and fixed but unknown variance. Thus

$$X_{ij} = \mu + \tau_i + \beta_j + \epsilon_{ij}$$

Using t and b to indicate estimates, we impose the restrictions $\Sigma t_i = 0 = \Sigma b_j$, that is, that treatment and block effects be measured as deviations, and obtain

$$t_i = \bar{x}_{i\cdot} - \bar{x}_{\cdot\cdot} \quad \text{and} \quad b_j = \bar{x}_{\cdot j} - \bar{x}_{\cdot\cdot}$$

Actually, these are the least-squares estimates with μ being estimated by $\bar{x}_{\cdot\cdot}$. In other words, our estimate of $\mu + \tau_i + \beta_j$ is given by

$$\bar{x}_{\cdot\cdot} + (\bar{x}_{i\cdot} - \bar{x}_{\cdot\cdot}) + (\bar{x}_{\cdot j} - \bar{x}_{\cdot\cdot}) = \bar{x}_{i\cdot} + \bar{x}_{\cdot j} - \bar{x}_{\cdot\cdot}$$

and

$$\Sigma[X_{ij} - (\bar{x}_{i\cdot} + \bar{x}_{\cdot j} - \bar{x}_{\cdot\cdot})]^2 = \Sigma(X_{ij} - \bar{x}_{i\cdot} - \bar{x}_{\cdot j} + \bar{x}_{\cdot\cdot})^2$$

is the smallest possible sum of squares, given complete freedom to choose estimates of μ, the τ_i's and β_j's.

Table 8.3 shows the computation of estimates of the rt means and residuals

TABLE 8.3
NATURE OF ERROR TERM IN ANALYSIS OF VARIANCE

Observed values, X_{ij}					Estimates of the means, $\hat{\mu}_{ij}$				Residuals = differences = $X_{ij} - \hat{\mu}_{ij}$				
Treatment	Replicate 1 2 3	Totals	Means		Replicate 1 2 3	Means		Treatment	Replicate 1 2 3		Total SS
1	5 4 3	12	4		4 3 5	4		1	1 1 −2		0 6
2	4 5 6	15	5		5 4 6	5		2	−1 1 0		0 2
3	6 3 9	18	6		6 5 7	6		3	0 −2 2		0 8
4	7 6 8	21	7		7 6 8	7		4	0 0 0		0
5	3 2 4	9	3		3 2 4	3		5	0 0 0		0
Totals Means	25 20 30 5 4 6	75	5		5 4 6	5		Total SS	0 0 0 2 6 8		16

Analysis of variance

Source of variation	df	SS
Replicates	2	10
Treatment	4	30
Error	8	16
Total	14	56

for synthetic data. The estimated mean of the i,jth cell is denoted by $\hat{\mu}_{ij}$ where

$$\hat{\mu}_{ij} = \bar{x}_{i\cdot} + \bar{x}_{\cdot j} - \bar{x}_{\cdot\cdot} \tag{8.7}$$

The sum of squares for residuals, $\sum_{i,j} (X_{ij} - \hat{\mu}_{ij})^2$, is seen to be identical with that obtained from the analysis of variance. Finally, $X_{ij} - \hat{\mu}_{ij} = e_{ij}$ is an estimate of ϵ_{ij}. The resulting error term measures the failure of the treatment differences to be the same in all blocks.

Notice that the sums of residuals are zero for every row and column. This indicates why we have 8 degrees of freedom in error. If one fixes the row and column totals of the residuals, then the table of residuals cannot be filled completely at will. It is necessary to reserve all spaces in the last row and column, for example, for whatever values are necessary to give row and column totals of zero. Essentially this means we are free to choose only $(r-1)(t-1)$ residuals; we have only $(r-1)(t-1)$ degrees of freedom.

8.5 Missing data. Sometimes data for certain units are missing or unusable, as when an animal becomes sick or dies but not as a result of the treatment, when rodents destroy a plot in a field trial, when a flask breaks in the laboratory, or when there has been an obvious recording error. A method developed by Yates (8.20) is available for estimating such missing data. An estimate of a missing value does not supply additional information to the experimenter; it only facilitates the analysis of the remaining data.

Where a *single value* is missing in a randomized complete-block experiment, calculate an estimate of the missing value by the equation

$$X = \frac{rB + tT - G}{(r-1)(t-1)} \tag{8.8}$$

where r and t are the number of blocks and treatments, B and T are the totals of the observed observations in the block and treatment containing the missing unit, and G is the grand total of the observed observations.

The estimated value is entered in the table with the observed values and the analysis of variance is performed as usual with one degree of freedom being subtracted from both total and error degrees of freedom. The estimated value is such that the error sum of squares in the analysis of variance is a minimum. The treatment sum of squares is biased upward by an amount equal to

$$\frac{[B - (t-1)X]^2}{t(t-1)} \tag{8.9}$$

where X is determined from Eq. (8.8).

The standard error of a difference between the mean of the treatment with a missing value and that of any other treatment is

$$s_{\bar{d}} = \sqrt{s^2\left[\frac{2}{r} + \frac{t}{r(r-1)(t-1)}\right]} \tag{8.10}$$

When there are *several missing values*, values are first approximated for all units save one. Reasonable approximations of these may be obtained by computing $(\bar{x}_{i\cdot} + \bar{x}_{\cdot j})/2$ where $\bar{x}_{i\cdot}$ and $\bar{x}_{\cdot j}$ are the means of the known values for the treatment and replicate containing any one of the missing values. Values may also be obtained by inspection. Equation (8.8) is then used to obtain an approximation for the remaining value. With this approximation and the values previously assigned to all but one of the remaining missing plots, again use Eq. (8.8) to approximate this one.

After making a complete cycle, a second approximation is found for all values in the order previously used. This is continued until the new approximations are not materially different from those found in the previous cycle. Usually two cycles are sufficient. The estimated values are entered in the table with the observed values and the analysis of variance is completed. For each missing value, one degree of freedom is subtracted from total and error degrees of freedom. This is because the estimated values make no contribution to the error sum of squares.

To illustrate the procedure, suppose that two values, a and b, are missing from the data of Table 8.2 (see Table 8.4).

TABLE 8.4

MISSING PLOT TECHNIQUE

Treatment	Blocks				Treatment totals	
	1	2	3	4	Observed values	All values
1	4.4	5.9	6.0	4.1	20.4	
2	($a = 4.5$)	1.9	4.9	7.1	13.9	18.4
3	4.4	4.0	4.5	3.1	16.0	
4	6.8	6.6	($b = 7.2$)	6.4	19.8	27.0
5	6.3	4.9	5.9	7.1	24.2	
6	6.4	6.3	7.7	6.7	28.1	
Block totals { Observed values	28.3	29.6	29.0	34.5	122.4	
All values	32.8		36.2			134.1

Computations proceed as follows.

1. Estimate b as

$$b = \frac{\bar{x}_{i\cdot} + \bar{x}_{\cdot j}}{2} = \frac{19.8/3 + 29.0/5}{2} = 6.2$$

2. Estimate a, first cycle, by Eq. (8.8)

$$a_1 = \frac{rB + tT - G}{(r-1)(t-1)} = \frac{4(28.3) + 6(13.9) - 128.6}{(4-1)(6-1)} = 4.5$$

Note that $G = 122.4 + 6.2 = 128.6$.

3. Estimate b, first cycle, as

$$b_1 = \frac{4(29.0) + 6(19.8) - 126.9}{(4-1)(6-1)} = 7.2$$

Again, note that $G = 122.4 + 4.5 = 126.9$.

4. Estimate a, second cycle, as

$$a_2 = \frac{4(28.3) + 6(13.9) - 129.6}{(4-1)(6-1)} = 4.5$$

Again $G = 122.4 + 7.2 = 129.6$.

5. Estimate b, second cycle, as

$$b_2 = \frac{4(29.0) + 6(19.8) - 126.9}{(4-1)(6-1)} = 7.2$$

Here $G = 122.4 + 4.5 = 126.9$ as in the first cycle, since a_1 and a_2 are the same when rounded to one decimal place.

The estimated values of the population means for the missing cells are given by Eq. (8.7). For missing value cells a and b, these estimates are $32.8/6 + 18.4/4 - 134.1/24 = 4.5$ and $36.2/6 + 27.0/4 - 134.1/24 = 7.2$, the same as the missing values to one decimal place. Missing value formulas supply estimates of population means and, thus, the estimates do not contribute to error. In turn, there is no reason for the estimates to contribute to error degrees of freedom.

The analysis of variance is computed as usual after the estimated values have been entered. Total and error degrees of freedom are 21 and 13. Block, treatment, and error mean squares are 0.96, 5.89, and 1.45. Error mean square is an unbiased estimate of σ^2; treatment mean square is biased upward. The F test of the null hypothesis of no differences among treatment means is unlikely to be much in error unless a substantial number of values are missing. This approximate procedure and a little common sense are satisfactory for most problems; an exact method is given by Yates (8.20). An easy and exact method is also given by covariance in Sec. 15.11.

To obtain a standard error for the comparison of two treatment means, each with a missing plot, we use an approximation due to Taylor (8.8). Let a and b be missing plots in treatments 2 and 4. It is necessary to assign an "effective number of replicates" to treatments 2 and 4. A replicate of treatment 2 is counted as 1 when both treatments are observed in the same block, as in blocks 2 and 4; as $(t-2)/(t-1)$ when treatment 2 is present but 4 is not, as in block 3; and as 0 when treatment 2 is missing, as in block 1. The same rule applies to treatment 4. Thus the standard deviation of the difference between the means of treatments 2 and 4 is

$$s_{\bar{d}} = \sqrt{s^2\left(\frac{1}{2.8} + \frac{1}{2.8}\right)}$$

For treatments 2 and 3 where there are no missing values in 3, the same rule is valid and gives

$$s_{\bar{d}} = \sqrt{s^2\left(\frac{1}{3} + \frac{1}{3.8}\right)}$$

Exercise 8.5.1 From the data of Table 8.2, discard one value and regard the resulting data as having one missing plot. Compute a value for the missing plot and complete the analysis of variance. Compute the bias in and adjust the treatment sum of squares.

Exercise 8.5.2 From the data of Exercise 8.3.1 or 8.3.2, discard two or three values to be regarded as missing plots. Compute missing plot values and complete the analysis of variance. Compute the effective number of replicates for comparing two treatment means, each with a missing plot; for two treatment means, only one of which has a missing plot.

8.6 Estimation of gain in efficiency. Whenever a randomized complete-block design is used, it is possible to estimate the relative efficiency compared with that to be expected from a completely random design by the equation

$$E_e(\text{CR}) = \frac{n_b E_b + (n_t + n_e) E_e}{n_b + n_t + n_e} \tag{8.11}$$

where E_b and E_e are the block and error mean squares, and n_b, n_t, and n_e are the block, treatment, and error degrees of freedom. If the degrees of freedom for the randomized block error are under 20, it is important to consider the loss in precision resulting from fewer degrees of freedom with which to estimate the error mean square of the randomized blocks experiment as compared to the completely random design. This is accomplished by multiplying the precision factor, when obtained, by $(n_1 + 1)(n_2 + 3)/(n_2 + 1)(n_1 + 3)$ (discussed in Sec. 6.8) where n_1 and n_2 are the degrees of freedom associated with error for the randomized complete block and the completely random design, respectively.

Applying Eq. (8.11) to the data of Table 8.2, we have

$$E_e(\text{CR}) = \frac{3(1.05) + (5 + 15)1.31}{3 + 5 + 15} = 1.28$$

Using the precision factor, we find

$$E(\text{RB relative to CR}) = \frac{E_e(\text{CR})}{E_e(\text{RB})} \frac{(n_1 + 1)(n_2 + 3)}{(n_2 + 1)(n_1 + 3)} 100$$

$$= \frac{1.28}{1.31} \frac{(15 + 1)(18 + 3)}{(18 + 1)(15 + 3)} 100 = 96\%$$

In this case, information is sacrificed, in theory, by using the randomized complete-blocks design since 96 replicates in a completely random design give as much information as 100 blocks or replicates for a randomized complete-block design. A proof of Eq. (8.11) is given by Cochran and Cox (8.8).

Exercise 8.6.1 Compute the efficiency of the randomized complete-block design relative to the completely random design for the data used in Exercises 8.3.1 and 8.3.2.

8.7 The randomized complete-block design: more than one observation per experimental unit. In many experiments, the complete experimental unit is not always the unit of observation. When more than one observation is made on at least one experimental unit, it becomes necessary to distinguish between sampling error and experimental error (see Sec. 7.11). For the most information per observation and for greatest computational convenience, an equal number of observations per experimental unit is required where possible. This is the case we consider.

Let X_{ijk} represent the kth observation made in the jth block on the ith treatment, $i = 1, \ldots, t$, $j = 1, \ldots, r$, and $k = 1, \ldots, s$ observations. Here, i and j refer to criteria of classification whereas k is a necessary label but does not serve as a criterion of classification. In other words, the kth observation in the i,jth unit is no more like the kth in another unit than it is like the

ANALYSIS OF VARIANCE II: MULTIWAY CLASSIFICATIONS

$(k+1)$st or any other observation. Hence the only meaningful totals are the grand total $X_{...}$, the experimental unit or cell totals $X_{ij.}$, the block totals $X_{.j.}$, and the treatment totals $X_{i..}$.

The computations required for the sums of squares in the analysis of variance are shown symbolically in Table 8.5. Note that the first three lines of computations, those for block, treatment, and experimental error sums of squares, differ from the analysis of variance of a randomized complete-block design based on cell totals, only in that there is a divisor s. Hence tests of hypotheses concerning blocks or treatments and based on experimental error are not affected by the fact that a record of the individual observations is available; experimental unit totals are sufficient for such tests.

TABLE 8.5

THE COMPUTATION OF SUMS OF SQUARES FOR A RANDOMIZED COMPLETE-BLOCK DESIGN WITH SEVERAL OBSERVATIONS PER EXPERIMENTAL UNIT

Source	df	SS
Blocks	$r-1$	$\dfrac{\sum_j X_{.j.}^2}{ts} - \dfrac{X_{...}^2}{rts}$
Treatments	$t-1$	$\dfrac{\sum_i X_{i..}^2}{rs} - \dfrac{X_{...}^2}{rts}$
Experimental error	$(r-1)(t-1)$	$\dfrac{\sum_{i,j} X_{ij.}^2}{s} - \dfrac{X_{...}^2}{rts}$ $-$block SS $-$ treatment SS
Sampling error	$rt(s-1)$	$\sum_{i,j}\left(\sum_k X_{ijk}^2 - \dfrac{X_{ij.}^2}{s}\right)$
Total	$rts-1$	$\sum_{i,j,k} X_{ijk}^2 - \dfrac{X_{...}^2}{rts}$

Sampling error sum of squares may be found by subtracting the sums of squares for blocks, treatments, and experimental error from the total sum of squares or as shown in Table 8.5. This method is to obtain the sum of squares among observations within each cell and sum; for the i,jth cell, the sum of squares among observations is $\sum_k X_{ijk}^2 - X_{ij.}^2/s$. Sampling error measures the failure of the observations made in any experimental unit to be precisely alike.

Experimental error is often expected to be larger than sampling error; in other words, variation among experimental units is often expected to be larger than variation among subsamples of the same unit. When both sources of variation are assumed to be random, experimental error is the appropriate error for testing hypotheses concerning treatments and blocks.

TABLE 8.6
Average Values of Mean Squares for a Randomized Complete-block Analysis

Source	df	Random model	
		No sampling	Sampling
		$X_{ij} = \mu + \tau_i + \beta_j + \epsilon_{ij}$	$X_{ijk} = \mu + \tau_i + \beta_j + \epsilon_{ij} + \delta_{ijk}$
Blocks	$r-1$	$\sigma_\epsilon^2 + t\sigma_\beta^2$	$\sigma^2 + s\sigma_\epsilon^2 + ts\sigma_\beta^2$
Treatments	$t-1$	$\sigma_\epsilon^2 + r\sigma_\tau^2$	$\sigma^2 + s\sigma_\epsilon^2 + rs\sigma_\tau^2$
Residual	$(r-1)(t-1)$	σ_ϵ^2	$\sigma^2 + s\sigma_\epsilon^2$
Sampling error	$rt(s-1)$		σ^2

Fixed model, no interaction

	No sampling	Sampling
	$X_{ij} = \mu + \tau_i + \beta_j + \epsilon_{ij}$	$X_{ijk} = \mu + \tau_i + \beta_j + \epsilon_{ij} + \delta_{ijk}$
Blocks	$\sigma_\epsilon^2 + t\Sigma\beta_j^2/(r-1)$	$\sigma_\epsilon^2 + s\sigma_\epsilon^2 + st\Sigma\beta_j^2/(r-1)$
Treatments	$\sigma_\epsilon^2 + r\Sigma\tau_i^2/(t-1)$	$\sigma_\epsilon^2 + s\sigma_\epsilon^2 + sr\Sigma\tau_i^2/(t-1)$
Residual	σ_ϵ^2	$\sigma^2 + s\sigma_\epsilon^2$
Sampling error		σ^2

Fixed model, interaction

	No sampling	Sampling
	$X_{ijk} = \mu + \tau_i + \beta_j + (\tau\beta)_{ij} + \epsilon_{ij}$	$X_{ijk} = \mu + \tau_i + \beta_j + (\tau\beta)_{ij} + \delta_{ijk}$
Blocks	$\sigma_\epsilon^2 + t\Sigma\beta_j^2/(r-1)$	$\sigma^2 + st\Sigma\beta_j^2/(r-1)$
Treatments	$\sigma_\epsilon^2 + r\Sigma\tau_i^2/(t-1)$	$\sigma^2 + sr\Sigma\tau_i^2/(t-1)$
Residual	$\sigma_\epsilon^2 + \sum_{i,j}(\tau\beta)_{ij}^2/(r-1)(t-1)$	$\sigma^2 + s\sum_{i,j}(\tau\beta)_{ij}^2/(r-1)(t-1)$
Sampling error		σ^2

Mixed model

	No sampling	Sampling
	$X_{ij} = \mu + \tau_i + \beta_j + \epsilon_{ij}$	$X_{ijk} = \mu + \tau_i + \beta_j + \epsilon_{ij} + \delta_{ijk}$
Blocks	$\sigma_\epsilon^2 + t\sigma_\beta^2$	$\sigma^2 + s\sigma_\epsilon^2 + ts\sigma_\beta^2$
Treatments	$\sigma_\epsilon^2 + r\Sigma\tau_i^2/(t-1)$	$\sigma^2 + s\sigma_\epsilon^2 + rs\Sigma\tau_i^2/(t-1)$
Residual	σ_ϵ^2	$\sigma^2 + s\sigma_\epsilon^2$
Sampling error		σ^2

ANALYSIS OF VARIANCE II: MULTIWAY CLASSIFICATIONS 145

When block and treatment effects are assumed to be fixed, one does not necessarily assume that experimental error is random. When nonrandomness is assumed, we say essentially that there are fixed effects for each block-treatment combination over and above the treatment and block contributions. This can also be stated by saying that the differences in responses to the various treatments are not of the same order of magnitude from block to block. Such situations are not uncommon and are discussed as *interactions* in Chap. 11. For example, suppose a field experiment is being conducted, by choice, on a hillside with the top quite dry and the bottom reasonably moist. Blocks are chosen accordingly with block effects being considered as fixed. The experiment is a varietal trial of several locally grown varieties of a forage crop, one of which is considered to be drought-resistant. Treatment effects are also considered to be fixed. Now the drought-resistant variety should be outstanding in the dry block relative to its performance in the moist block. In other words, the size of the difference in response between this and any other variety depends upon the block. This cannot be called a random effect. Hence the contribution which we have been calling experimental error is not random; we relabel it as interaction. Sampling error is an appropriate error for testing hypotheses concerning interaction and hypotheses concerning treatment and block effects.

8.8 Linear models and the analysis of variance. For a proper evaluation of experimental data, the model must be specifically stated. Two common models are the *fixed effects* model or *Model* I and the *random effects* model or *Model* II (see Sec. 7.10).

Another common model is the *mixed model*, which calls for at least one criterion of classification to involve fixed effects and another to involve random effects. Clearly other models are also possible.

From Table 8.6, it is evident what sort of conclusions can be drawn from tests of significance, in the form of ratios of mean squares.

In randomized complete-block design problems where the *fixed effects model* is appropriate, two or more treatments are selected for testing. These are not randomly drawn from a population of possible treatments but are selected, perhaps as those which show the most promise or those most readily available. All treatments about which inferences are to be drawn are included in the experiment. Block effects are also fixed and inferences about treatments or blocks are not intended to apply to blocks not included in the experiment.

For the *fixed model* with no interaction, both block and treatment effects can be tested by residual or error mean square. Even when we have only one sample, $s = 1$, per experimental unit, valid tests are possible. In this case, it is not possible to estimate σ_δ^2. However, when there are fixed interaction effects, it is not possible to make any valid F test when there is no sampling. When s observations are made in each cell, $s > 1$, sampling error is a valid error for testing interaction, treatment, and block mean squares.

For the *random model*, both treatments and blocks are drawn at random from populations of treatment and block effects. Inferences are drawn about the populations of treatments and blocks rather than the particular treatments. Residual mean square is the appropriate error mean square for testing block and treatment effects, whether there is sampling or not. Tests can be made even though only one sample observation is made on the experimental unit; however, σ_δ^2 cannot be estimated.

In most situations, blocks are considered to be representative of a population of blocks since inferences are desired for a range of conditions wider than the particular blocks used. If blocks are assumed to be random and treatments fixed, we have a *mixed model*. Where rather broad inferences are desired to be drawn from an experiment, care should be taken that the blocks used are really representative of the population about which inferences are to be made.

When there is no sampling, block and treatment effects are readily tested. When there is sampling, the residual mean square is appropriate for testing blocks and for testing treatments. If $s = 1$, we cannot estimate σ_δ^2, so cannot test blocks.

8.9 Double grouping: Latin squares. In the Latin-square design, we arrange the treatments in blocks in two different ways, namely, by rows and columns. For four treatments, the arrangement might be

A	D	C	B
B	C	A	D
D	A	B	C
C	B	D	A

Each treatment occurs once and only once in each row and column; each row, like each column, is a complete block. By appropriate analysis, it is possible to remove from error, variability due to differences in both rows and columns.

This design has been used to advantage in many fields of research where two major sources of variation are present in the conduct of an experiment. In field experiments, the layout is usually a square, thus allowing the removal of variation resulting from soil differences in two directions. In the greenhouse or field, if there is a gradient all in one direction, the experiment can be laid out as follows:

A D C B	B C A D	D A B C	C B D A

Here, rows are blocks and columns are positions in the blocks. Marketing experiments lend themselves to this arrangement, with days being rows and stores being columns.

Since rows and columns are general terms for referring to criteria of classification, they may be a kind of treatment. When there is interaction between any two or among all of the criteria, rows, columns, and treatments, computed F is not distributed as tabulated F and no valid tests of significance are possible. In cases where the experimenter is not prepared to assume the absence of interaction, the Latin square should not be used.

Babcock (8.2) used a Latin-square design in an experiment to determine if there were any differences among the amounts of milk produced by the four

ANALYSIS OF VARIANCE II: MULTIWAY CLASSIFICATIONS

quarters of cow udders, quarters being the treatments or letters of the square. In this experiment four times of milking were the rows of the Latin square, and the orders of milking were the columns of the square, order meaning position in time. The Latin square has also been used to advantage in the laboratory, in industry, and in the social sciences.

The chief disadvantage of the Latin square is that the number of rows, columns, and treatments must be equal. Thus, if there are many treatments, the number of plots required soon becomes impractical. The most common square is in the range 5×5 to 8×8, and squares larger than 12×12 are rarely used. Latin squares, like randomized blocks, suffer in that as the block size increases, the experimental error per unit is likely to increase. Small squares provide few degrees of freedom for estimation of experimental error and thus must give a substantial decrease in experimental error to compensate for the small number of degrees of freedom. However, more than one square may be used in the same experiment; for example, two 4×4 squares will give 15 degrees of freedom for error if the treatments respond similarly in both squares. The degrees of freedom for the analysis of s, $r \times r$ squares are

Source	df
Squares	$s - 1 = 1$
Rows within squares	$s(r - 1) = 6$
Columns within squares	$s(r - 1) = 6$
Treatments	$r - 1 = 3$
Error	$s(r - 1)(r - 2) + (s - 1)(r - 1) = 15$
Total	$sr^2 - 1 = 31$

TABLE 8.7

ANALYSIS OF VARIANCE FOR AN $r \times r$ LATIN SQUARE

Source of variation	df	Sums of squares	
		Definition formulas	Computing formulas
Rows	$r - 1$	$r \sum_i (\bar{x}_{i.} - \bar{x}_{..})^2$	$\dfrac{\sum_i X_{i.}^2}{r} - C$
Columns	$r - 1$	$r \sum_j (\bar{x}_{.j} - \bar{x}_{..})^2$	$\dfrac{\sum_j X_{.j}^2}{r} - C$
Treatments	$r - 1$	$r \sum_t (\bar{x}_t - \bar{x}_{..})^2$	$\dfrac{\sum_t X_t^2}{r} - C$
Error	$(r - 1)(r - 2)$	$\Sigma(X_{ij} - \bar{x}_{i.} - \bar{x}_{.j} - \bar{x}_t + 2\bar{x}_{..})^2$	By subtraction
Total	$r^2 - 1$	$\sum_{i,j} (X_{ij} - \bar{x}_{..})^2$	$\sum_{i,j} X_{ij}^2 - C$

Table 8.8

Field Layout Showing Yields, in Kilograms per Plot, of Wheat Arranged in a 4 × 4 Latin Square

Row	Column 1	Column 2	Column 3	Column 4	Row totals $X_{i.}$	$\sum_j X_{ij}^2$
1	C = 10.5	D = 7.7	B = 12.0	A = 13.2	43.4	487.78
2	B = 11.1	A = 12.0	C = 10.3	D = 7.5	40.9	429.55
3	D = 5.8	C = 12.2	A = 11.2	B = 13.7	42.9	495.61
4	A = 11.6	B = 12.3	D = 5.9	C = 10.2	40.0	424.70
Column totals $X_{.j}$	39.0	44.2	39.4	44.6	$\sum_{i,j} X_{ij} = 167.2$	$\sum_{i,j} X_{ij}^2 = 1{,}837.64$
$\sum_i X_{ij}^2$	401.66	503.42	410.34	522.22		

Variety totals and means

	A	B	C	D
Totals = X_t	48.0	49.1	43.2	26.9
Means = $\bar{x}_t$	12.0	12.3	10.8	6.8

Analysis of variance

Source of variation	df	SS	MS	F
Rows	$(r-1) = 3$	1.95	0.65	1.44
Columns	$(r-1) = 3$	6.80	2.27	5.04
Varieties	$(r-1) = 3$	78.93	26.31	58.47**
Error	$(r-1)(r-2) = 6$	2.72	0.45	
Total	$(r^2-1) = 15$	90.40		

Residual

$s = 0.67$ kg $s_{\bar{x}} = 0.34$ kg $s_{\bar{d}} = 0.47$ kg   CV = 6.4%

Randomization in the Latin square consists of choosing a square at random from among all possible Latin squares. Fisher and Yates (8.11) give the complete set of Latin squares for 4 × 4 through 6 × 6 squares, and sample squares up to size 12 × 12. Cochran and Cox (8.8) give sample Latin squares from 3 × 3 through 12 × 12. A method of randomization suggested by Cochran and Cox follows.

The 3 × 3 square: Assign letters to the treatments; this need not be random. Write out a 3 × 3 square, randomize the arrangement of the three columns, and then that of the last two rows.

The 4 × 4 square: Here there are four squares such that one cannot be obtained from another simply by rearranging rows and columns. Hence we randomly select one of the four possible squares and arrange at random all columns and the last three rows.

The 5 × 5 *and higher squares:* By now there are many squares such that one cannot be obtained from another by rearranging rows and columns. Assign letters to the treatments at random. Randomize all columns and all rows.

8.10 Analysis of variance of the Latin square. The degrees of freedom and formulas for sums of squares for an $r \times r$ Latin square are given in Table 8.7. Here X_{ij} represents the observation at the intersection of the ith row and jth column. Row sums and means are given by $X_{i.}$ and $\bar{x}_{i.}$ for $i = 1, \ldots, r$ and column sums and means by $X_{.j}$ and $\bar{x}_{.j}$, $j = 1, \ldots, r$. While this notation is adequate to locate an observation, it tells nothing about the treatment received. We have used X_t and $\bar{x}_t$ to denote treatment totals and means, $t = 1, \ldots, r$.

The statistical analysis of a 4×4 Latin square is illustrated with yield data from a wheat varietal evaluation trial conducted by Ali A. El Khishen, College of Agriculture, Alexandria University, Alexandria, Egypt. The data, analysis, and actual field layout are shown in Table 8.8. The varieties are represented by letters: A = Baladi 16, B = Mokhtar, C = Giza 139, and D = Thatcher. Yields are in kilograms per plot of size 42 sq m, which is 1/100 feddan.

The computational procedure follows.

Step 1. Obtain row totals $X_{i.}$, column totals $X_{.j}$, treatment totals X_t, and the grand total $X_{..}$. Simultaneously find $\sum_j X_{ij}^2$ and $\sum_i X_{ij}^2$ for each value of i and j, respectively. The sum of the four resulting quantities for the i's will equal that for the j's and provides a computational check. This is the total unadjusted sum of squares.

Step 2. Find the correction term and (adjusted) sums of squares.

$$\text{Correction term} = C = \frac{X_{..}^2}{r^2} = \frac{167.2^2}{4^2} = 1{,}747.24$$

$$\text{Total SS} = \sum_{i,j} X_{ij}^2 - C = 1{,}837.64 - 1{,}747.24 = 90.40$$

$$\text{Row SS} = \frac{\sum_i X_{i.}^2}{r} - C = \frac{43.4^2 + \cdots + 40.0^2}{4} - 1{,}747.24$$
$$= 1.95$$

$$\text{Column SS} = \frac{\sum_j X_{.j}^2}{r} - C = \frac{39.0^2 + \cdots + 44.6^2}{4} - 1{,}747.24$$
$$= 6.80$$

$$\text{Treatment SS} = \frac{\sum_t X_t^2}{r} - C = \frac{48.0^2 + \cdots + 26.9^2}{4} - 1{,}747.24$$
$$= 78.93$$

$$\text{Error SS} = (\text{total} - \text{row} - \text{column} - \text{treatment})\text{SS}$$
$$= 90.40 - (1.95 + 6.80 + 78.93) = 2.72$$

Sums of squares are entered in an analysis of variance table and mean squares found. The F value for varieties (treatments) is $26.31/0.45 = 58.47**$, with 3 and 6 degrees of freedom; it greatly exceeds the tabulated 1% value of 9.78. Highly significant differences are said to exist among yields for varieties.

The sample standard error of a treatment mean is $s_{\bar{x}} = \sqrt{s^2/r} = 0.34$ kg, where s^2 is the error mean square and r is the number of experimental units per treatment. The sample standard error of a difference between two treatment means is $s_{\bar{x}_1 - \bar{x}_2} = \sqrt{2s^2/r} = 0.47$ kg. If heterogeneity of error is suspected, the error cannot be subdivided as readily as in the case of the randomized complete-block design. The procedure is illustrated by Cochran and Cox (8.8).

The method for determining which varieties differ depends upon the aims of the experiment and knowledge of the varieties, for example, genetic background. Valid methods have been discussed in Chap. 7.

Exercise 8.10.1 Peterson et al. (8.15) present moisture content of turnip greens and other data for an experiment conducted as a Latin square. The data are shown in the accompanying table for (moisture content $- 80$)%. Treatments are times of weighing since moisture losses might be anticipated in a 70°F laboratory as the experiment progressed.

Plant	Leaf size (A = smallest, E = largest)				
	A	B	C	D	E
1	6.67(V)	7.15(IV)	8.29(I)	8.95(III)	9.62(II)
2	5.40(II)	4.77(V)	5.40(IV)	7.54(I)	6.93(III)
3	7.32(III)	8.53(II)	8.50(V)	9.99(IV)	9.68(I)
4	4.92(I)	5.00(III)	7.29(II)	7.85(V)	7.08(IV)
5	4.88(IV)	6.16(I)	7.83(III)	5.83(II)	8.51(V)

Compute the analysis of variance. What is the standard deviation applicable to a difference between treatment means? Between leaf-size means?

8.11 Missing plots in the Latin square. The principle involved in the estimation of *missing values* is illustrated in Sec. 8.5 for the randomized complete-block design. The formula for a *single missing observation* is

$$X = \frac{r(R + C + T) - 2G}{(r-1)(r-2)} \qquad (8.12)$$

where R, C, and T are the totals of the observed values for the row, column, and treatment containing the missing value, and G is the grand total of the observed values. If *several units*, not comprising a whole row, column, or treatment, are missing, the procedure is to make repeated applications of

Eq. (8.12) as is done with Eq. (8.8) for randomized complete blocks. When all missing units have been estimated, the analysis of variance is performed in the usual manner with one degree of freedom being subtracted from total and error degrees of freedom for each missing value.

As in the case of the randomized complete-block design, the treatment sum of squares is biased upward, this time by the amount

$$\frac{[G - R - C - (r - 1)T]^2}{[(r - 1)(r - 2)]^2} \tag{8.13}$$

In the case of a single missing observation, the sample standard error of the difference between the treatment mean with the missing unit and a treatment mean with all units present is

$$\sqrt{s^2 \left[\frac{2}{r} + \frac{1}{(r-1)(r-2)} \right]} \tag{8.14}$$

If more than one unit is missing, the exact procedure for obtaining the standard error between two means is involved. A useful approximation by Yates (8.20) can be used to determine the number of "effective replicates" for the two treatment means being compared. The effective number of replicates for one treatment being compared with another is determined by summing values assigned as follows: one if the other treatment is present in the corresponding row and column; two-thirds if the other is missing in either the row or column but not both; one-third if the other is missing in both the row and column; and zero when the treatment in question is missing. This is illustrated for the 5×5 Latin square given below with three units missing, one for each of treatments A, B, and C. The missing units are indicated by parentheses.

B	(A)	C	E	D
D	B	E	A	(C)
E	D	A	C	B
C	E	B	D	A
A	C	D	(B)	E

The effective number of replicates for treatments A and B in the comparison of their means is as follows. Starting with column 1, find A. B is in the same column but not the same row with A; assign value $\frac{2}{3}$. For column 2, A is missing; assign value 0. For column 3, B is present in both row and column with A; assign value 1. For column 4, B is absent in the column but present in the row with A; assign value $\frac{2}{3}$. For column 5, B is present in both column and row with A; assign value 1. Similarly for B. The assigned values and effective numbers of replicates are

$$\text{for } A: \tfrac{2}{3} + 0 + 1 + \tfrac{2}{3} + 1 = 2\tfrac{4}{3} = \tfrac{10}{3}$$
$$\text{for } B: \tfrac{2}{3} + \tfrac{2}{3} + 1 + 0 + 1 = 2\tfrac{4}{3} = \tfrac{10}{3}$$

The standard error of the difference between the means for treatments A and B is then $\sqrt{s^2(1/r_A + 1/r_B)} = \sqrt{s^2(3/10 + 3/10)}$ where r_A and r_B refer to effective replication. The effective number of replications for a treatment can differ with the comparison being made.

When all values of one or more rows, columns, or treatments are missing, the analysis is usually more involved. The procedure when one row, column, or treatment is missing is given by Yates (8.21); if two or more are missing, see Yates and Hale (8.22). DeLury (8.9) also gives these procedures. Youden (8.23) discusses the construction of incomplete Latin squares as experimental designs.

Exercise 8.11.1 Discard one of the observations given in Exercise 8.10.1. Compute a missing value for the discarded observation and complete the analysis of variance. Compute the bias in treatment sum of squares. What is the standard deviation applicable to the difference between two treatment means, one being that with a missing observation?

8.12 Estimation of gain in efficiency. The precision of a Latin square relative to a randomized complete-block experiment can be estimated. Two estimates of the relative efficiency can be made, one when rows are considered as blocks and the other when columns are considered as blocks. We estimate the randomized blocks error mean square, if rows are the only blocks, as

$$E_e(\text{RB}) = \frac{n_c E_c + (n_t + n_e)E_e}{n_c + n_t + n_e} \qquad (8.15)$$

where E_c and E_e are the mean squares for columns and error in the Latin square, and n_c, n_t, and n_e are the degrees of freedom for columns, treatments, and error in the Latin square. If columns are the only blocks, replace n_c and E_c in Eq. (8.15) by n_r and E_r, the degrees of freedom and mean square for rows in the Latin square.

Since more degrees of freedom are available for estimating the error mean square of randomized blocks than of Latin squares, this must be considered when computing the relative efficiency of the two designs if less than 20 degrees of freedom are involved in the Latin-square error. This is accomplished by incorporating $(n_1 + 1)(n_2 + 3)/(n_2 + 1)(n_1 + 3)$ into the formula for the efficiency factor, where n_1 and n_2 are degrees of freedom for the Latin square and randomized block designs.

The data in Table 8.8 will be used to illustrate the procedure. The estimated error mean square with rows as blocks is $[3(2.27) + (3 + 6)(0.45)]/(3 + 3 + 6) = 0.91$; with columns as blocks it is $[3(0.65) + (3 + 6)(0.45)]/(3 + 3 + 6) = 0.50$. The adjustment for differences in degrees of freedom for the two designs is $[(6 + 1)(9 + 3)]/(9 + 1)(6 + 3) = 0.933$. Thus, the estimated relative precision using rows as blocks is

$$E(\text{LS to RB}) = \frac{E_e(\text{RB})}{E_e(\text{LS})} \frac{(n_1 + 1)(n_2 + 3)}{(n_2 + 1)(n_1 + 3)} 100$$

$$= \frac{0.91}{0.45}(.933)100 = 189\%$$

If columns were blocks,

$$E(\text{LS to RB}) = \frac{0.50}{0.45}(.933)100 = 104\%$$

The row grouping increased the precision by an estimated 89% and columns by an estimated 4%. Thus, if the randomized complete-block design had been used with the rows of the Latin square as blocks, an estimated 89% more replicates would have been required to detect differences of the same magnitude as detected by the Latin square, whereas 4% more would have been required if columns had been blocks. Rows appear to be inefficient as blocks whereas columns seem reasonably efficient. In any field of research, a number of such comparisons are generally needed before it can be concluded that Latin squares are likely to be more precise, on the average, than randomized blocks.

Exercise 8.12.1 Compute the efficiency of the Latin square relative to the randomized complete-block design for the data in Exercise 8.10.1, using leaf size as blocks. Repeat, using plants as blocks.

8.13 The linear model for the Latin square. Let X_{ij} represent the observation at the intersection of the ith row and jth column. This locates any observation but tells nothing of the treatment applied. A third subscript could be misleading, implying r^3 rather than r^2 observations. For example, treatment 1 appears once in each of the r rows, once in each of the r columns, but only r times in all; hence, $t = 1$ implies a set of i,j values r in number. Similarly for the other values of t.

We express any observation by Eq. (8.16).

$$X_{ij(t)} = \mu + \beta_i + \kappa_j + \tau_{(t)} + \epsilon_{ij} \qquad (8.16)$$

This implies, by the use of (t), that an ordinary three-way classification is not involved.

For valid conclusions, the ϵ's must be random and, if tests of significance or confidence statements are to be made using procedures already discussed, they must also be normally and independently distributed. A model with interactions does not lead to a valid error for testing hypotheses and the Latin square is not a suitable design if interactions are present.

Various assumptions may be made about the components of the means, that is, the β's, κ's, and τ's. Average values of the various mean squares are given in Table 8.9 according to the model assumed.

TABLE 8.9
AVERAGE VALUES OF MEAN SQUARES FOR A LATIN-SQUARE ANALYSIS

Source	df	Average values of mean squares	
		Model I (fixed)	Model II (random)
Rows	$r-1$	$\sigma^2 + r(\Sigma\beta_i^2)/(r-1)$	$\sigma^2 + r\sigma_\beta^2$
Columns	$r-1$	$\sigma^2 + r(\Sigma\kappa_j^2)/(r-1)$	$\sigma^2 + r\sigma_\kappa^2$
Treatments	$r-1$	$\sigma^2 + r(\Sigma\tau_t^2)/(r-1)$	$\sigma^2 + r\sigma_\tau^2$
Residual	$(r-1)(r-2)$	σ^2	σ^2

The mixed model is not shown, but average values of mean squares are obtained by replacing the appropriate contribution, by the corresponding contribution as one changes from fixed to random effects, for example, replacing $(\Sigma \beta_i^2)/(r-1)$ by σ_β^2.

8.14 The size of an experiment. Sample size is discussed in Secs. 5.11, 5.12, and 6.7. These sections deal with obtaining confidence intervals not larger than some stated length and with detecting differences of stated size; problems involving more than two treatments are not considered though most experiments include more than two treatments.

Several approaches have been made to the general problem of size of an experiment. Among these are those of Cochran and Cox (8.8), Harris et al. (8.12), Harter (8.13), Tang (8.17), and Tukey (8.19). Two of these will be discussed.

Calculation of the number of replicates required depends on
1. An estimate of σ^2
•2. The size of the difference to be detected
3. The assurance with which it is desired to detect the difference (Type II error)
4. The level of significance to be used in the actual experiment (Type I error)
5. Whether a one- or two-tailed test is required

Section 5.11 discusses the problem of detecting a difference of a stated size using σ^2 and the z distribution. In practice, an estimate of σ^2 obtained from previous experiments is used. Substitution of s^2 for σ^2 in Eq. (5.17) yields the inequality (8.17) for determining r.

$$r \geq \frac{2(t_0 + t_1)^2 s^2}{\delta^2} \qquad (8.17)$$

where r is the number of replicates per treatment, s^2 is an estimate of σ^2, δ is the true difference to be detected, t_0 is the t value associated with Type I error and t_1 is the t value associated with Type II error; t_1 equals tabulated t for probability $2(1-P)$ where P is the required probability of detecting δ if such a difference exists.

This procedure is an approximation, since it is necessary to assume $s^2 = \sigma^2$. Thus, it is important to have a reliable estimate of σ^2. The probability of detecting a significant difference will be larger than or smaller than the stated probability depending upon how much σ^2 is under- or overestimated. Successive approximations are involved since the number of degrees of freedom associated with t_0 and t_1 depends upon r.

To illustrate the use of Eq. (8.17), suppose we wish to conduct an experiment like that which gave rise to the data of Table 8.2, using again six treatments. We desire to detect differences, regardless of direction, of not more than 2.5% of oil at the 95% level with 90% assurance of detecting a true difference of this size.

From Table 8.2, $s^2 = 1.31$. If we guess four replicates and plan to use a randomized complete-block design, then error degrees of freedom are $(6-1)(4-1) = 15$ degrees of freedom. From the t table, Table A.3,

$t_0 = 2.131$ is the value for probability .05, and $t_1 = 1.341$ is the value for probability $2(1 - .90)$, or .20. Hence

$$r \geq \frac{2(2.131 + 1.341)^2 1.31}{2.5^2} = 5.05 \text{ or } 6$$

Repeat, using $r = 6$, for which degrees of freedom $= 25$

$$r \geq \frac{2(2.060 + 1.316)^2 1.31}{2.5^2} = 4.78 \text{ or } 5$$

It appears that five blocks will suffice. If a one-tailed test is desired, enter Table A.3 for t_0 using the probability of a larger value of t_1 sign considered.

The reader is referred to Cochran and Cox (8.8) for a more detailed discussion. They present tables, based upon this method, which give the number of replicates required to detect a difference, expressed in per cent of the general mean for several levels of Type I and Type II error.

Tukey's (8.19) procedure gives the sample size necessary to give a set of confidence intervals not larger than a specified size for all possible differences between true treatment means. The level of significance is chosen by the investigator and the error rate applies to a Type I error on an experimentwise basis.

Since an experiment is a sample, it is impossible to give complete assurance that, for example, 95% confidence intervals will be invariably less than the specified size. Hence it becomes necessary to state how frequently, or with what assurance, it is desired to have the confidence intervals less than the specified length.

Tukey gives the following formula for computing experiment size.

$$r = \frac{s_1^2 q_\alpha^2(p, n_2) F_\gamma(n_2, n_1)}{d^2} \tag{8.18}$$

where s_1^2 is an available estimate of σ^2 based on n_1 degrees of freedom, q is obtained from Table A.8 for the desired confidence coefficient and the degrees of freedom for the error mean square in the experiment being planned, F_γ is obtained from Table A.6 (one-tailed) for the indicated pair of degrees of freedom, and γ is defined such that $1 - \gamma$ is the assurance we wish to have that the confidence intervals for differences will be less than $2d$. In other words, d is defined as the half-length semiconfidence interval desired. Notice that q and F depend upon the value being sought. This implies that we may have to apply the formula several times.

To illustrate the use of Eq. (8.18) suppose we again use the data of Table 8.2. It is decided to use a set of confidence intervals at the 95% level and to include the same six treatments; the number of treatments is necessary when obtaining q. We wish the 95% semiconfidence interval to be of not more than 2.5% of oil for 90% of our experiment.

From Table 8.2, $s_1^2 = 1.31$ and $n_1 = 15$ degrees of freedom. To obtain q, we must guess n_2. If we guess r will be five replicates and are planning a randomized complete-block design, then $n_2 = (5 - 1)(6 - 1) = 20$ degrees of freedom and $q_\alpha(p,n_2) = 4.45$. We find $F_{.10}(20,15) = 1.92$ and have set $d = 2.5\%$ oil. Hence

$$r = 1.31(4.45)^2(1.92)/(2.5)^2 = 8.0 \text{ blocks}$$

Since 8 blocks are considerably more than expected, we underestimated n_2. It is, then, worthwhile to estimate r again using 7 or 8 blocks to determine n_2. For 7 blocks, $n_2 = (7-1)(6-1) = 30$ degrees of freedom. Now

$$r = 1.31(4.30)^2(1.87)/(2.5)^2 = 7.2 \text{ blocks}$$

It now appears that 7 blocks are not quite sufficient but that 8 blocks are somewhat more than adequate for the desired purpose.

Stein's two-sample procedure (Sec. 5.12) has been generalized by Healy (8.14) for obtaining joint confidence intervals of fixed length and confidence coefficient for all possible differences between population means. The procedure is applicable when the investigator is able to continue the same experiment and does not need to worry about heterogeneity of variance for the two stages of the experiment.

8.15 Transformations. The valid application of tests of significance in the analysis of variance requires that the experimental errors be independently and normally distributed with a common variance. In addition, the scale of measurement should be one for which the linear additive model holds. It is customary to rely upon randomization to break up any dependence of experimental errors whenever it is possible to incorporate a randomization procedure into the investigation. Additivity may be tested by a method given in Sec. 11.9. The other condition which has been given much attention by statisticians is that of stabilizing the variance.

Heterogeneity of error, or heteroscedasticity, may be classified as irregular or regular. The *irregular* type is characterized by certain treatments possessing considerably more variability than others, with no necessarily apparent relation between means and variances. Differences in variability may or may not be expected in advance. For example, in comparing insecticides, untreated or check experimental units are often included. The numbers of insects in the check units are likely to be considerably larger and more variable than those in units where an insecticide offers considerable control. Thus, the check units contribute to the error mean square to a greater degree than do the treated units. Consequently, the standard deviations, based on the pooled error mean square, will be too large for comparison among insecticides and may fail to detect real differences. In other cases, certain treatments may exhibit considerably more variation than others for no apparent reason. This part of the experiment is not under statistical control.

When heterogeneity of error is of the irregular type, the best procedure is to omit certain portions of the data from the analysis or to subdivide the error mean square into components applicable to the various comparisons of interest. The latter procedure is discussed to some extent in Sec. 11.6.

The *regular* type of heterogeneity usually arises from some type of nonnormality in the data, the variability within the several treatments being related to the treatment means in some reasonable fashion. If the parent distribution is known, then the relation between the treatment means and the

ANALYSIS OF VARIANCE II: MULTIWAY CLASSIFICATIONS 157

treatment variances, computed on an individual treatment basis, is known. The data can be transformed or measured on a new scale of measurement so that the transformed data are approximately normally distributed. Such transformations are also intended to make the means and variances independent, with the resulting variances homogeneous. This result is not always attained. When it is impossible to find a transformation that will make the means and variances independent and the variance stable, other methods of analysis, such as weighted analyses, must be used.

The more common transformations are the square root, logarithm, and angular or arcsin transformation. These are discussed briefly below.

Square root transformation, $\sqrt{X}$. When data consist of small whole numbers, for example, number of bacterial colonies in a plate count, number of plants or insects of a stated species in a given area, etc., they often follow the Poisson distribution for which the mean and variance are equal. The analysis of such enumeration data is often best accomplished by first taking the square root of each observation before proceeding with the analysis of variance.

Percentage data based on counts and a common denominator, where the range of percentages is 0 to 20% or 80 to 100% but not both, may also be analyzed by using the square root transformation. Percentages between 80 and 100 should be subtracted from 100 before the transformation is made. The same transformation is useful for percentages in the same ranges where the observations are clearly on a continuous scale, since means and variances may be approximately equal.

When very small values are involved, $\sqrt{X}$ tends to overcorrect so that the range of transformed values giving a small mean may be larger than the range of transformed values giving a larger mean. For this reason, $\sqrt{X + \frac{1}{2}}$ is recommended as an appropriate transformation where some of the values are under 10 and especially when zeros are present.

The appropriate means for a table of treatment means are found by squaring the treatment means computed from the $\sqrt{X}$ values. It must also be remembered that the additive model should hold on the transformed rather than the original scale. A test of additivity is given in Sec. 11.9.

Logarithmic transformation, log X. When variances are proportional to the squares of the treatment means, the logarithmic transformation equalizes the variances. The base 10 is used for convenience, though any base is satisfactory. Effects which are multiplicative on the original scale of measurement become additive on the logarithmic scale.

The logarithmic transformation is used with positive integers which cover a wide range. It cannot be used directly for zero values and when some of the values are less than 10, it is desirable to have a transformation which acts like the square root for small values and like the logarithmic for large values. Addition of 1 to each number prior to taking logarithms has the desired effect. That is, log $(X + 1)$ behaves like the square root transformation for numbers up to 10 and differs little from log X thereafter.

In some experimental work, it is desired to conduct an analysis of variance where the variable is the variance. The logarithmic transformation of the variances prior to analysis is appropriate, that is, we analyze log s^2. Also,

Bartlett's test of the homogeneity of a set of variances requires the use of log s^2. Bartlett's test is given in Sec. 17.3.

Angular or inverse sine transformation, arcsin $\sqrt{X}$ or $\sin^{-1}\sqrt{X}$. This transformation is applicable to binomial data expressed as decimal fractions or percentages, and is especially recommended when the percentages cover a wide range of values. The mechanics of the transformation require decimal fractions but tables of the arcsin transformation are usually entered with percentages (see Table A.10). The tabled values or arcsin values are expressed in either degrees or radians. The variance of the resulting observations is approximately constant, being $821/n$ when the transformed data are expressed in degrees and $0.25/n$ when in radians, n being the common denominator of all fractions. This variance makes it clear that all percentages are intended to be based on an equal number of observations. However, the transformation is often used when the denominators are unequal, especially if they are approximately equal. Bartlett (8.3) suggests that $\frac{1}{4}n$ be substituted for 0%, and $100 - \frac{1}{4}n$ for 100% (n being the divisor).

The square root transformation has already been recommended for percentages between 0 and 20 or 80 and 100, the latter being subtracted from 100 before transformation. If the range of percentages is 30 to 70, it is doubtful if any transformation is needed.

Some general considerations. It is always helpful to examine discrete data to ascertain whether or not there is a correlation between the treatment means and their within-treatment variances. Unfortunately, this is not possible for all designs. If the variation shows little change, the value of any transformation is doubtful. Where there is change in the variation, it is not always clear which transformation is best.

When the appropriate transformation is in doubt, it is sometimes helpful to transform the data for several treatments, including some with small, intermediate, and large means on the original scale, and again examine means and variances for the possibility of a relationship on the transformed scale. The transformation for which the relation is least is likely the most appropriate.

When a transformation is made, all comparisons or confidence interval estimates are made on the transformed scale. If it is not desired to present findings on the transformed scale, then means should be transformed back to the original scale. When the results of an analysis of transformed data are presented using the original scale of measurement, this should be made clear to the reader. It is not appropriate to transform standard deviations or variances, arising from transformed data, back to the original scale.

The experimenter who is interested in further information on the assumptions underlying the analysis of variance and on transformations is referred to papers by Eisenhart (8.10), Cochran (8.5, 8.6, 8.7), and Bartlett (8.3). These papers give many additional references.

Exercise 8.15.1 In evaluating insecticides, the numbers of living adult plum curculios emerging from separate caged areas of treated soil were observed. The results shown in the accompanying table are available through the courtesy of C. B. McIntyre, Entomology Department, University of Wisconsin. Notice that this is a randomized complete-block design and we cannot directly measure the within-treatment variances.

	Treatment					
Block	Lindane	Dieldrin	Aldrin	EPN	Chlordane	Check
1	14	7	6	95	37	212
2	6	1	1	133	31	172
3	8	0	1	86	13	202
4	36	15	4	115	69	217

What transformation would you recommend for these data? Analyze the data using $\log(X + 1)$ as the transformation. Use Duncan's procedure to test all possible pairs of treatment means exclusive of that for check.

Exercise 8.15.2 Aughtry (8.1) presents the following data on the symbiosis of *Medicago sativa* (53)–*M. falcata* (50) cross with strain B. Data are percentages of plants with nodules out of 20 plants per cell. The experiment was conducted as a randomized complete-block design.

Block	P_1	F_1	F_2 lots from each F_1			
	53 50	53 × 50	114–1	114–2	114–3	114–4
1	11 65	47	31	22	16	70
2	16 67	32	40	16	19	63
3	6 76	40	27	20	20	52

What transformation would you recommend for the analysis of these data? Use the angular transformation and conduct the analysis of variance. How does the observed variance compare with the theoretical variance?

References

8.1 Aughtry, J. D.: "The effects of genetic factors in Medicago on symbiosis with Rhizobium," *Cornell Univ. Agr. Expt. Sta. Mem.* 280, 1948.

8.2 Babcock, S. M.: "Variations in yield and quality of milk," *Sixth Ann. Rept. Wis. Agr. Expt. Sta.*, **6**: 42–67 (1889).

8.3 Bartlett, M. S.: "The use of transformations," *Biometrics*, **3**: 39–52 (1947).

8.4 Bing, A.: "Gladiolus control experiments, 1953," *The Gladiolus* 1954, New England Gladiolus Society, Inc.

8.5 Cochran, W. G.: "Some difficulties in the statistical analysis of replicated experiments," *Empire J. Exp. Agr.*, **6**: 157–175 (1938).

8.6 Cochran, W. G.: "Analysis of variance for percentages based on unequal numbers," *J. Am. Stat. Assoc.*, **38**: 287–301 (1943).

8.7 Cochran, W. G.: "Some consequences when the assumptions for the analysis of variance are not satisfied," *Biometrics*, **3**: 22–38 (1947).

8.8 Cochran, W. G., and G. M. Cox: *Experimental Designs*, 2d ed., John Wiley & Sons, Inc., New York, 1957.

8.9 DeLury, D. B.: "The analysis of latin squares when some observations are missing," *J. Am. Stat. Assoc.*, **41**: 370–389 (1946).

8.10 Eisenhart, C.: "The assumptions underlying the analysis of variance," *Biometrics*, **3**: 1–21 (1947).

8.11 Fisher, R. A., and F. Yates: *Statistical Tables for Biological, Agricultural and Medical Research*, 5th ed., Hafner Publishing Company, New York, 1957.

8.12 Harris, M., D. G. Horvitz, and A. M. Mood: "On the determination of sample sizes in designing experiments," *J. Amer. Stat. Assoc.*, **43:** 391–402 (1948).

8.13 Harter, H. L.: "Error rates and sample sizes for range tests in multiple comparisons," *Biometrics*, **13:** 511–536 (1957).

8.14 Healy, W. C., Jr.: "Two-sample procedures in simultaneous estimation," *Ann. Math. Stat.*, **27:** 687–702 (1956).

8.15 Peterson, W. J., H. P. Tucker, J. T. Wakeley, R. E. Comstock, and F. D. Cochran: "Variation in moisture and ascorbic acid content from leaf to leaf and plant to plant in turnip greens," No. 2 in *Southern Coop. Ser. Bull.* 10, pp. 13–17 (1951).

8.16 Sackston, W. E., and R. B. Carson: "Effect of pasmo disease of flax on the yield and quality of linseed oil," *Can. J. Botany*, **29:** 339–351 (1951).

8.17 Tang, P. C.: "The power function of the analysis of variance tests with tables and illustrations of their use." *Stat. Research Memoirs*, **2:** 126–157 (1938).

8.18 Tucker, H. P., J. T. Wakeley, and F. D. Cochran: "Effect of washing and removing excess moisture by wiping or by air current on the ascorbic acid content of turnip greens," No. 10 in *Southern Coop. Ser. Bull.* 10, pp. 54–56 (1951).

8.19 Tukey, J. W.: "The problem of multiple comparisons," Ditto, Princeton University, Princeton, N.J., 1953.

8.20 Yates, F.: "The analysis of replicated experiments when the field results are incomplete," *Empire J. Exp. Agr.*, **1:** 129–142 (1933).

8.21 Yates, F.: "Incomplete latin squares," *J. Agr. Sci.*, **26:** 301–315 (1936).

8.22 Yates, F., and R. W. Hale: "The analysis of latin squares when two or more rows, columns, or treatments are missing," *J. Roy. Stat. Soc. Suppl.*, **6:** 67–69 (1939).

8.23 Youden, W. J.: *Statistical Methods for Chemists*, John Wiley & Sons, Inc., New York, 1951.

Chapter 9

LINEAR REGRESSION

9.1 Introduction. In previous chapters, we developed the idea that an observation is the sum of a population mean and a random component. A set of components was available for means previously discussed and each component was either present in or absent from any particular mean. For example, in a completely random design every population mean contained μ while the ith contained τ_i but no other τ. We now consider population means with one component a fixed multiple of some measurable and variable quantity, called a *concomitant variable*.

In this chapter, we discuss the uses of a concomitant observation, with the assumptions involved in the various uses, and with the necessary computations. Chapters 10, 14, 15, and 16 also deal with these general problems.

9.2 The linear regression of Y on X. In linear regression, Y values are obtained from several populations, each population being determined by a corresponding X value. Randomness of Y is essential for probability theory to

TABLE 9.1

AVERAGE BODY WEIGHT X AND FOOD CONSUMPTION Y FOR 50 HENS FROM EACH OF 10 WHITE LEGHORN STRAINS
(350-DAY period)

Body weight		Food consumption	
X	$X' = X - 4.0$	Y	$Y' = Y - 80$
4.6	0.6	87.1	7.1
5.1	1.1	93.1	13.1
4.8	0.8	89.8	9.8
4.4	0.4	91.4	11.4
5.9	1.9	99.5	19.5
4.7	0.7	92.1	12.1
5.1	1.1	95.5	15.5
5.2	1.2	99.3	19.3
4.9	0.9	93.4	13.4
5.1	1.1	94.4	14.4

$\Sigma x'^2 = \Sigma x^2 = 1.536 \quad \Sigma y'^2 = 135.604 \quad 9\ df$ each

SOURCE: Data courtesy of S. C. King, now at Purdue University, Lafayette, Ind.

apply. Also, it is assumed the Y populations are normal and have a common variance.

The Y variable is termed the *dependent* variable since any Y value depends upon the population sampled. The X variable is called the *independent* variable or *argument*.

The data in Table 9.1, pen records from 1953–1954 New York random sample test, will be used for illustration. The dependent variable is feed

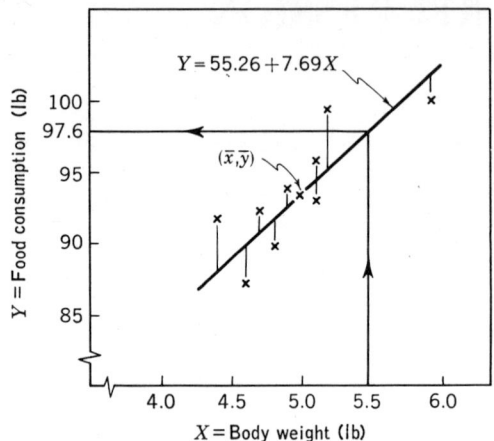

FIG. 9.1 Regression of food consumption Y on body weight X for 10 White Leghorn strains (averages for 50 birds)

consumed Y; it depends upon the variable body weight X. The ten pairs of values from Table 9.1 are plotted in Fig. 9.1. A fairly definite relation exists between the two variables. In particular, a straight line such as has been drawn among these points could serve as a moving average of the Y values. Let us now consider a straight-line graph and its equation.

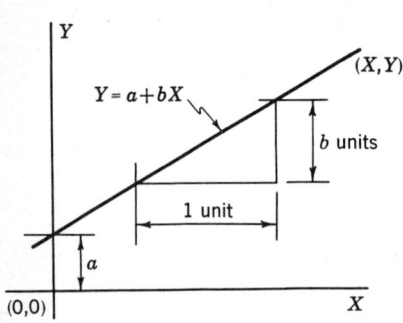

FIG. 9.2 Straight-line graph

The equation of any straight line may be written in the form $Y = a + bX$ (see Fig. 9.2). Any point (X, Y) on this line has an X coordinate or *abscissa* and a Y coordinate or *ordinate* whose values satisfy the equation. Coordinates of points not on the line do not satisfy the equation. When $X = 0$, $Y = a$ so that a is the point where the line crosses the Y axis, that is, a is the Y *intercept*. When a is zero, the line goes through the origin. A unit change in X results in a change of b units in Y so that b is a measure of the *slope* of the line. When b is positive, both variables increase or decrease together; when b is negative, one variable increases as the other decreases. For a straight line, any two points or the slope and the Y intercept uniquely determine the position of the line. Mathematicians call relations such as $Y = a + bX$ *functional relations*. For a value of X, a functional relation assigns a value to Y. This sort of relation has been used in Chap. 3 to assign probabilities, as in Eqs. (3.4), (3.5), and (3.6), and ordinates of a curve as in Eq. (3.7).

When a straight line has to be fitted to data consisting of more than two pairs of values, one must choose that line which "best" fits the data, the one which is the "best" moving average. Our criterion of "best" is the least-squares criterion that requires the sum of the squares of the deviations of the observed points from the straight-line moving average be a minimum. For such a "fitted" line, b is called the *regression coefficient*; the line is called a *regression line*; its equation is called a *regression equation*.

To determine the regression coefficient b, we need the sum of *cross products* of the deviations of the observations from their corresponding means, and the sum of squares of X. Definition and computing formulas for a sum of cross products are given in Eqs. (9.1) and (9.2), respectively.

$$\sum_i x_i y_i = \sum_i (X_i - \bar{x})(Y_i - \bar{y}) \tag{9.1}$$

$$\sum_i x_i y_i = \sum_i X_i Y_i - \frac{\sum_i X_i \sum_i Y_i}{n} \tag{9.2}$$

For coded X and Y, that is, X' and Y', we obtain

$$\Sigma xy = 144.70 - (9.8)(135.6)/10 = 11.812$$

Coding by addition or subtraction of a constant for each variable does not affect sums of cross products. The following relations are also useful:

$$\Sigma xy = \Sigma X_i(Y_i - \bar{y}) \quad \text{and} \quad \Sigma xy = \Sigma(X_i - \bar{x})Y_i$$

The similarity between a sum of cross products and a sum of squares, for both definition and working formulas, is obvious when Y_i and $\bar{y}$ are replaced by X and $\bar{x}$ in the cross-product formula. Sums of squares and cross products are often referred to simply as *sums of products*.

When Σxy is divided by the degrees of freedom, it is called a *covariance*. We have $\Sigma xy/(n - 1) = 11.812/9 = 1.312$. A covariance is a measure of the joint variation of two variables and may be either positive or negative. It is symmetrical in x and y.

The regression coefficient is determined from Eq. (9.3).

$$b = \frac{\Sigma xy}{\Sigma x^2} \tag{9.3}$$

$$= \frac{11.812}{1.536} = 7.69 \text{ lb of feed per pound of bird}$$

For an increase of 1 lb in body weight, food consumption is 7.69 lb.

The regression equation may be written as Eq. (9.4) since the regression line passes through the sample mean.

$$Y - \bar{y} = b(X - \bar{x}) \quad \text{or} \quad y = bx \tag{9.4}$$

We obtain
$$Y - 93.56 = 7.69(X - 4.98)$$
or $\quad Y = 93.56 + 7.69(X - 4.98) \quad$ or $\quad Y = 55.26 + 7.69X$

(see Fig. 9.1). The Y intercept a is $\bar{y} - b\bar{x} = 55.26$.

Lines computed in regression problems are lines about which the pairs of values cluster and are not lines upon which the points fall. The regression line, passing among pairs of values, deals with a *statistical law*, a law which holds *on the average*. In fact, a point on a regression line is an estimate of a mean of a population of Y's, those Y's having the corresponding X value. This statement could serve as a definition of linear regression. Since a statistical law holds on the average, it estimates a mean and, consequently, determines a frequency distribution. Thus, it differs from a mathematical law or functional relation which assigns unique values.

Exercise 9.2.1 The pen records shown below for White Leghorns in the California Random Sample Test† are similar to the data of Table 9.1.

Y (350-day food consumption, lb): 87.8, 93.2, 98.0, 89.8, 94.0, 83.0, 88.3, 82.4, 84.8, 80.2

X (Av. body weight of 50 birds, lb): 4.15, 4.76, 5.23, 4.75, 5.13, 4.24, 4.66, 4.41, 4.50, 4.23

Compute the sample regression equation of consumption Y on body weight X. What is the regression coefficient and what unit applies to it? What is the interpretation of this regression coefficient?

9.3 The linear regression model and its interpretation. By definition, the true regression of Y on X consists of the means of populations of Y values, where a population is determined by the X value. In sampling, it is necessary to assume the form of the line of the means; otherwise it would not be possible to develop a computation procedure. We have assumed a straight line or *linear regression*. Such assumptions are usually made on the basis of theory or experience. Because of computational ease, the straight line is often chosen as an approximation when it fits reasonably well over the range of X involved, even when the true form is known to be nonlinear.

The mathematical description of an observation is given by Eq. (9.5).

$$Y_i = \mu_{Y \cdot x} + \epsilon_i = \mu + \beta x_i + \epsilon_i \tag{9.5}$$

μ and β are parameters to be estimated and x is an observable parameter. The ϵ's are assumed to be from a single population with zero mean and variance σ^2. This variance is another parameter to be estimated.

Once μ and β have been estimated, it is possible to estimate the mean of a population of Y's without ever having observed a single one of the individuals. For example, we observed no Y's for $X = 5.5$ lb for the White Leghorn data. However, we estimate the mean of the population of Y's for $X = 5.5$ lb body weight, using Eq. (9.4), as

$$\hat{Y}_{(5.5 - \bar{x})} = 55.26 + 7.69(5.5) = 97.6 \text{ lb of feed}$$

† Data courtesy of S. C. King, Purdue University, Lafayette, Ind.

Since the notation $\bar{y}_{(5.5-\bar{x})}$ might easily lead one to believe that a sample of Y's with $X = 5.5$ lb had been observed, it is customary to use an alternative to denote an estimate of a population mean. We use $\hat{Y}_x$ or simply $\hat{Y}$, or $\hat{\mu}_{Y \cdot x}$.

Estimates of μ and β are often written as $\hat{\mu}$ and $\hat{\beta}$, thus $\hat{\mu} = 93.56$ lb feed and $\hat{\beta} = 7.69$ lb feed per pound of body weight. Note that $\hat{\mu} = \bar{y}$; $\hat{\beta}$ is often replaced by b.

The regression model may be Model I with fixed X's, or Model II with random X's. For *Model* I, the X's are selected by the investigator; there is no random sampling variation associated with them. The Y values must be random. For *Model* II, the X's as well as the Y's are random. In this case, we draw a random sample of pairs from a bivariate distribution. Equation (9.5) describes any observation for either model.

For linear regression, we assume that the ϵ's are normally and independently distributed with a common variance and that regression is linear. Since the values of X determine which populations of Y's are sampled, the X's must be measured without error. Solutions to the problems which concern us in this chapter are the same for both models, except in the case of prediction of X. This problem is discussed in Sec. 9.11.

Exercise 9.3.1 What model applies to the data of Exercise 9.2.1?

9.4 Assumptions and properties in linear regression. A requirement of regression is that the X's be measured without error. Consider Fig. 9.3, which shows a true regression line. If one makes an observation on Y for $x = x_i$, then Y_i is obtained from the population indicated in the figure and the observer may proceed to use this information meaningfully. If, however, one records the value of x_i as x'_i, stating incorrectly that Y_i is from the population of Y's for which $x = x'_i$, the subsequent computations may result in misleading conclusions. Winsor (9.5) discusses the problem of regression when errors of measurement are present in one or both variates; both models are considered.

Independent variables may be equally or otherwise conveniently spaced for efficiency in the conduct of an experiment, for example, hours of artificial

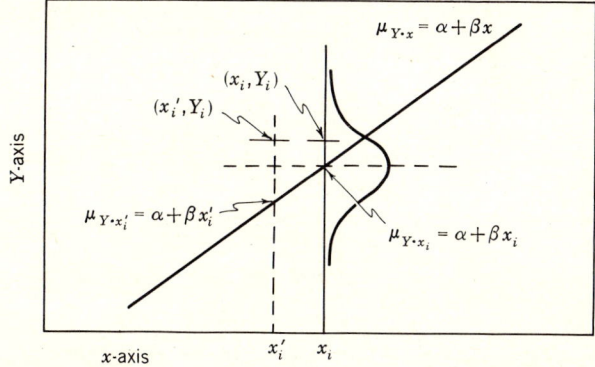

FIG. 9.3 Regression of Y on X; errors of measurement in X

sunlight, temperatures, amounts of treatments, and distances between seedlings.

Measurement of Y without error is not a theoretical requirement provided the error of measurement has a distribution with known mean, generally zero. The observed variance of Y is, then, the sum of the biological or other variance in Y and the variance of the error of measurement. It is, of course, important to keep errors in measurements to a minimum.

When the assumption of a common variance is valid, residual error mean square is applicable to making valid probability inferences about a population mean regardless of the value of X. If variances are not homogeneous, a weighted regression or a transformation of the data so that the variances are homogeneous is necessary. For instance, when entomologists compute probit analyses based on percentage of kill, where the variance is binomial in nature, they use both transformation and weighting.

The linear regression problem possesses the following properties:

1. The point $(\bar{x}, \bar{y})$ is on the sample regression line.
2. The sum of the deviations from regression is zero. (A deviation is the signed difference between the observed value and the corresponding estimate of the population mean. The estimate is the point on the sample regression line with the same X value as the observation.)
3. The sum of the squares of the deviations is a minimum. That is, if we replace the sample regression line as computed in Sec. 9.2 with any other straight line, the sum of the squares of the new set of deviations will be a larger value.

Exercise 9.4.1 For the data in Exercise 9.2.1, which is the dependent variable? The independent variable? Discuss the requirement that "X be measured without error" as it applies to these data.

Exercise 9.4.2 Compute the 10 deviations from the sample regression line, Exercise 9.2.1. Show that the sum of these 10 deviations is zero (within rounding errors). Find the sum of squares of the 10 deviations from regression. Show that the point $\bar{x}, \bar{y}$ is on the sample regression line.

9.5 Sources of variation in linear regression.

The linear regression model, Eq. (9.5), considers an observation as the sum of a mean $\mu_{Y \cdot x} = \alpha + \beta x$, and a random component ϵ. Since, by chance or by intent, different X values are observed, different population means are involved and contribute to the total variance. Thus, the two sources of variation in observations are means and random components. Variation attributable to the means may be considered as attributable to X since X determines the mean.

In terms of the sample regression, an observation Y is composed of a sample mean $\hat{Y}$ determined from the regression line, and a deviation $Y - \hat{Y}$ from this mean (see Fig. 9.4). In itself, $\hat{Y}$ is seen to consist of the sample mean $\bar{y}$, and the deviation $\hat{y} = \hat{Y} - \bar{y} = bx$ attributable to regression. The deviation $d = Y - \hat{Y}$ is an estimate of the chance deviation ϵ and is usually written $d_{y \cdot x}$. Thus Eq. (9.4) may be written as Eq. (9.6).

$$Y - \bar{y} = (\hat{Y} - \bar{y}) + (Y - \hat{Y}) = bx + d_{y \cdot x} \qquad (9.6)$$

The total unadjusted sum of squares of the Y's can be partitioned according

LINEAR REGRESSION   167

to these sources. The sum of squares attributable to the mean is $n\bar{y}^2 = (\Sigma Y)^2/n$; that attributable to regression is $b^2\Sigma x^2 = (\Sigma xy)^2/\Sigma x^2$; and that attributable to chance is $\Sigma d_{y \cdot x}^2$ which is found as a residual sum of squares.

Equation (9.7) is derived from Eq. (9.6) and is comparable to its population counterpart Eq. (9.5).

$$Y = \bar{y} + bx + d_{y \cdot x} \tag{9.7}$$

From Eq. (9.7) it can be shown that

$$\Sigma Y^2 = n\bar{y}^2 + b^2\Sigma x^2 + \Sigma d_{y \cdot x}^2$$

This gives

$$\Sigma Y^2 - \frac{(\Sigma Y)^2}{n} = \frac{(\Sigma xy)^2}{\Sigma x^2} + \text{residual SS} \tag{9.8}$$

The residual SS is found by subtraction. For the example, 135.604 = 90.836 + 44.768. The total sum of squares of Y, $\Sigma y^2 = 135.604$, has been

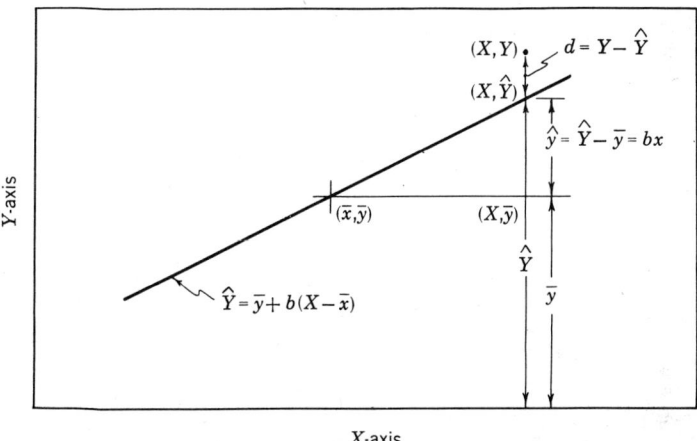

FIG. 9.4   Regression of Y on X; sources of variation in Y

partitioned into a sum of squares attributable to regression and an unexplained portion, the residual sum of squares.

The sum of squares attributable to regression may be written in any one of the following forms

$$\frac{(\Sigma xy)^2}{\Sigma x^2} = b\Sigma xy = b^2\Sigma x^2 = \Sigma \hat{y}^2 = \Sigma(\hat{Y} - \bar{y})^2$$

of which the first is the most common for calculations and the last not at all practical. This quantity has a single degree of freedom.

The residual sum of squares is found by subtraction and has $n - 2$ degrees of freedom; the residual mean square is an estimate of experimental error, the common σ^2.

9.6 Regressed and adjusted values. Values determined by the regression equation, *regression* or *regressed values*, $\hat{Y}$'s, are estimates of population parameters, that is, $\mu_{Y \cdot x} = \mu + \beta x_i$'s. Differences between these and the observed values are estimates of the variation in Y that is not accounted for by variation in X. Observed deviations are shown in Table 9.2 for the White Leghorn data of Table 9.1. The sum of their squares is 44.22, which differs by rounding errors from the value 44.768, found in the preceding section.

TABLE 9.2

REGRESSED CONSUMPTIONS, DEVIATIONS, AND ADJUSTED CONSUMPTIONS, WHITE LEGHORN DATA†

X Body weight, lb	Y Food consumption, lb	$\hat{Y}$ Regressed consumption, lb	$d_{y \cdot x} = Y - \hat{Y}$ Deviation from regression, lb	$d_{y \cdot x}^2$ Squared deviations	$93.56 + d_{y \cdot x}$ Adjusted consumption, lb
4.6	87.1	90.6	−3.5	12.25	90.0
5.1	93.1	94.5	−1.4	1.96	92.2
4.8	89.8	92.2	−2.4	5.76	91.2
4.4	91.4	89.1	+2.3	5.29	95.9
5.9	99.5	100.6	−1.1	1.21	92.4
4.7	92.1	91.4	+0.7	0.49	94.3
5.1	95.5	94.5	+1.0	1.00	94.6
5.2	99.3	95.3	+4.0	16.00	97.6
4.9	93.4	92.9	+0.5	0.25	94.0
5.1	94.4	94.5	−0.1	.01	93.5
	935.6	935.6	0.0	44.22	935.7

† Computations for this table were made with more decimals than indicated and the results rounded

The careful investigator always looks at the individual observations. In a nonregression problem, he will be particularly conscious of those observations which show greatest deviation from the mean. In a regression problem, one must look to deviations from regression for comparable information.

Adjusted values, Eq. (9.9), are those to be expected if all Y values had the mean X value or some other fixed value. The last column of Table 9.2 contains adjusted consumptions, those to be expected if all birds had a body weight of $\bar{x} = 4.98$ lb. These are obtained by adding the deviations to $\bar{y} = 93.56$ lb.

$$\text{Adjusted } Y = \bar{y} + d_{y \cdot x} = Y - bx \qquad (9.9)$$

The latter form is convenient if one does not care to compute the deviations; subtract that part of the observation attributable to regression, namely, $bx = b(X - \bar{x})$. The mean of the Y's and the chance deviation are left (see Fig. 9.4). Differences among adjusted values are identical with those among the deviations; we have simply changed their location. The use of adjusted values replaces the moving standard, the regression value, with a fixed standard, the mean value.

Comparisons among adjusted means are very useful. Suppose two groups

LINEAR REGRESSION

of observations are available, for example, data for each of two years, two locations, or two types of housing. It may be pertinent to compare the group means if they are adjusted to a common X value. It will be necessary to assume a common regression coefficient, an assumption which can be tested as the hypothesis of homogeneous regression coefficients. Procedures for comparing adjusted means are given in Chap. 15 on covariance.

Exercise 9.6.1 D. Kuesel, University of Wisconsin, determined the average percentage of alcohol-insoluble solids Y and the tenderometer reading X for 26 samples of sieved-size Alaska peas. The observations follow.

Y, X: 7.64, 72; 8.08, 78; 7.39, 81; 7.45, 81; 9.56, 81; 7.96, 82; 10.81, 83; 10.70, 83; 10.56, 89; 11.75, 93; 11.56, 96; 11.74, 97; 13.72, 99; 15.08, 103; 16.26, 112; 16.79, 115; 15.40, 118; 15.90, 122; 16.30, 122; 17.56, 133; 17.38, 135; 17.90, 139; 18.80, 143; 19.90, 145; 20.10, 161; 22.01, 165

$$\Sigma x^2 = 18{,}774.62 \qquad \Sigma xy = 2{,}924.50 \qquad \Sigma y^2 = 489.58$$

Plot the 26 pairs of points. Compute the sample regression of Y on X. What unit of measurement applies to b? Draw this line on your plot. Obtain information on how a tenderometer reading is made and comment on the assumption concerning X being measured without error. How do you think you would test the null hypothesis that no variation in Y is attributable to variation in X?

Exercise 9.6.2 Compute two regressed values of alcohol-insoluble solids (for $X = 78$ and $X = 112$) and the deviations from regression. Compute adjusted values of Y for these two X's and compare them.

9.7 Standard deviations, confidence intervals, and tests of hypotheses.

An unbiased estimate of the true variance about regression is given by the residual mean square with $n - 2$ degrees of freedom. It is denoted by $s_{Y \cdot x}^2$ or $s_{y \cdot x}^2$ and defined as

$$s_{y \cdot x}^2 = \frac{\Sigma(Y - \hat{Y})^2}{n - 2} = \frac{\Sigma y^2 - (\Sigma xy)^2/\Sigma x^2}{n - 2} \qquad (9.10)$$

$$= \frac{44.77}{8} = 5.60 \text{ lb}^2$$

for the White Leghorn data. Its square root is called the *standard error of estimate* or the *standard deviation of Y for fixed X* or the *standard deviation of Y holding X constant*.

Figure 9.5 shows that a single standard deviation does not apply to all $\hat{Y}$'s but must depend upon the X value that determines the Y population. If we consider samples of Y values for a fixed set of X values, then $\bar{x}$ is a constant while $\bar{y}$ and b are variables. Variation in $\bar{y}$ raises or lowers the regression line parallel to itself; the effect is to increase or decrease all estimates of means by a fixed value. Variation in b rotates the regression line about the point $\bar{x}, \bar{y}$ and the effect on an estimate depends upon the magnitude of the x that determines the Y population. Variation in b has no effect upon the estimate of the mean if $X = \bar{x}$ but, otherwise, increases it in proportion to the size of x. This is readily seen from the equation that estimates the population mean, that is, $\hat{Y} = \bar{y} + bx$.

A standard deviation applicable to an estimate of a mean must allow for variation in both $\bar{y}$ and b and for the distance x. The variance of $\bar{y}$ is simply an estimate of $\sigma_{y \cdot x}^2/n$, namely $s_{y \cdot x}^2/n$. The regression coefficient $b = (\Sigma xy)/\Sigma x^2$ is a linear function of $(Y - \bar{y})$'s where the coefficient of $(Y_j - \bar{y})$ is $(X_i - \bar{x})/\Sigma x^2$, a constant. The variance of b is the variance $\sigma_{y \cdot x}^2$ multiplied by the sum of the

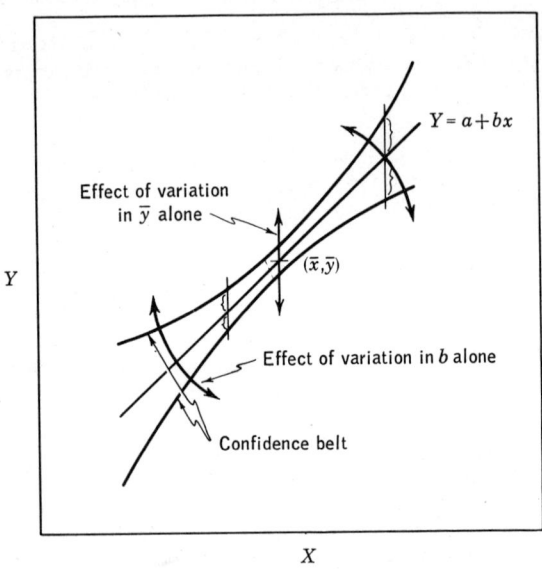

FIG. 9.5 Effect of sampling variation on regression estimates of population means; fixed set of X's

squares of the coefficients of the $(Y_i - \bar{y})$'s. Thus we estimate the variance of b by Eq. (9.11).

$$s_b^2 = \frac{s_{y \cdot x}^2}{\Sigma x^2} \tag{9.11}$$

The required variance of an estimate, $\hat{Y} = \bar{y} + bx$, of a population mean is given by the sum of the variances of $\bar{y}$ and bx. This variance is given by Eq. (9.12).

$$s_{y \cdot x}^2 \left[\frac{1}{n} + \frac{(X - \bar{x})^2}{\Sigma x^2} \right] \tag{9.12}$$

This variance increases as x increases. If t times the standard deviation were plotted in conjunction with the regression line, it would form a confidence belt as in Fig. 9.5.

To set a confidence interval on the estimate of a mean, the variance given in Eq. (9.12) is applicable and the 95% confidence interval is

$$\text{CL}(\hat{Y}) = \bar{y} + bx \pm t_{.05} s_{y \cdot x} \sqrt{\frac{1}{n} + \frac{x^2}{\Sigma x^2}} \tag{9.13}$$

where t is Student's t for $n - 2$ degrees of freedom.

LINEAR REGRESSION

For the White Leghorn data, a confidence interval of the population mean for $X = 5.5$ lb is

$$93.56 + 7.69(5.5 - 4.98) \pm 2.306(2.37)\sqrt{\frac{1}{10} + \frac{(0.52)^2}{1.536}} = 97.6 \pm 2.9$$
$$= (94.7, 100.5) \text{ lb}$$

Sometimes it is desirable to construct a confidence interval for β, the population regression parameter being estimated by b. This interval depends on s_b^2 as given in Eq. (9.11) and is given in Eq. (9.14).

$$b \pm \frac{t_{.05} s_{y \cdot x}}{\sqrt{\Sigma x^2}} \tag{9.14}$$

For the White Leghorn data, we have

$$7.69 \pm \frac{2.306(2.37)}{\sqrt{1.536}} = 7.69 \pm 4.41$$
$$= 3.28, 12.10$$

We conclude that β lies between 3.28 and 12.10 lb feed per pound of bird. The procedure which leads us to this conclusion is such that, on the average, 95% of such conclusions will be correct.

Any relevant null hypothesis about a mean may be tested. To test the null hypothesis that the mean of the population of Y's, for which $x = x_0$, is $\mu_{Y \cdot x_0}$, compute t as in Eq. (9.15).

$$t = \frac{\hat{Y}_x - \mu_{Y \cdot x}}{\sqrt{s_{y \cdot x}^2 (1/n + x_0^2/\Sigma x^2)}} \tag{9.15}$$

This is distributed as Student's t with $n - 2$ degrees of freedom.

To test the null hypothesis that $\beta = \beta_0$, compute t as in Eq. (9.16).

$$t = \frac{b - \beta_0}{\sqrt{s_{y \cdot x}^2 / \Sigma x^2}} \tag{9.16}$$

This is distributed as Student's t with $n - 2$ degrees of freedom. The F test of Table 9.3 is a test of the null hypothesis $\beta = 0$ or of the null hypothesis that variation in X does not contribute to variation in Y.

Exercise 9.7.1 What is $s_{y \cdot x}^2$ for the data of Exercises 9.2.1 and 9.6.1? What is the standard error of estimate?

Exercise 9.7.2 What is the variance of $\bar{y}$ for the data of Exercises 9.2.1 and 9.6.1? What is the variance of b in each case?

Exercise 9.7.3 Estimate the mean of the population of Y's for which $X = 5.00$ lb (data of Exercise 9.2.1) and $X = 90$ (data of Exercise 9.6.1). Compute the 95% confidence interval in one case, the 99% confidence interval in the other, for the estimated population means.

Exercise 9.7.4 Test the null hypothesis that $\beta = 0$ for one of the sets of data. Use Student's t. Compute t^2 and compare with the F value computed to test the null hypothesis that variation in X did not contribute to variability in Y.

9.8 Control of variation by concomitant observations. The sources of variation which affect a variable are not always controllable by an experimental layout. When layout cannot effect control, it may be possible to measure some characteristic of the source of variation. For example, the amount of feed consumed by hens is a variable of economic importance. One would expect it to be affected by other measurable variables such as body weight, and number and weight of eggs laid. For the data of Table 9.1, body weight easily accounts for the greatest amount of variability in feed consumed. The economic importance is obvious.

We now use the data of Table 9.1 to illustrate statistical control of a source of variation by use of a concomitant observation. The standard deviation of Y prior to adjustment for variation in X is $\sqrt{\Sigma y^2/(n-1)} = \sqrt{135.604/9} = 3.88$ lb. We have seen that after adjustment it is $s_{y \cdot x} = 2.37$ lb.

The amount of the Y sum of squares attributable to variation in X is given by Eq. (9.17); also see Eq. (9.8).

$$\text{Reduction in SS} = \frac{(\Sigma xy)^2}{\Sigma x^2} \tag{9.17}$$

$$= \frac{(11.812)^2}{1.536} = 90.836, \text{ for our example}$$

It has 1 degree of freedom. We may also note that the proportion of the Y sum of squares attributable to variation in X is

$$\frac{(\Sigma xy)^2/\Sigma x^2}{\Sigma y^2} = \frac{90.836}{135.604} = .67 \text{ (or 67\%)}$$

The reduced or residual Y sum of squares is found by subtraction and has $(n-2)$ degrees of freedom.

Residual SS for $Y = 135.604 - 90.836 = 44.768$, with 8 *df*

An analysis of variance table may be used to present the results. Table 9.3

TABLE 9.3
ANALYSIS OF VARIANCE FOR DATA OF TABLE 9.1

Source	df	Symbolic SS	Example df	SS	MS	F
X	1	$(\Sigma xy)^2/\Sigma x^2$	1	90.836	90.836	16.22**
Residual	$n-2$	by subtraction	8	44.768	5.60	
Total	$n-1$	Σy^2	9	135.604		

Alternative presentation (example only)

Source	df	Σx^2	Σxy	Σy^2	Reduction	Residual SS	MS	F
Total	$n-1=9$	1.536	11.812	135.604	90.836	44.768	5.60	16.22**

LINEAR REGRESSION 173

shows two possibilities. The first of these is comparable to the tables of Chap. 8; the second is more complete and serves as a basis for generalization to tables for partial and multiple regression (Chap. 14) and covariance (Chap. 15).

Exercise 9.8.1 For the data of Exercise 9.2.1, partition the sum of squares for Y into the component attributable to X and the residual. Was there a significant reduction in variability in Y attributable to variability in X? Present your results in the form of an analysis of variance.

9.9 Difference between two regressions. It may be desired to test the homogeneity of two b's, that is, to determine whether or not they can be considered to be estimates of a common β. For this, t as in Eq. (9.18) is distributed as Student's t with $n_1 + n_2 - 4$ degrees of freedom.

$$t = \frac{b_1 - b_2}{\sqrt{s_p^2(1/\Sigma x_{1j}^2 + 1/\Sigma x_{2j}^2)}} \tag{9.18}$$

Quantities b_1 and $\sum_j x_{1j}^2$ are the regression coefficient and sum of squares for x from the first sample, and similarly for the second sample, and

$$s_p^2 = \frac{\{\Sigma y_{1j}^2 - [(\Sigma x_{1j} y_{1j})^2/\Sigma x_{1j}^2]\} + \{\Sigma y_{2j}^2 - [(\Sigma x_{2j} y_{2j})^2/\Sigma x_{2j}^2]\}}{n_1 - 2 + n_2 - 2}$$

is the best estimate of the variation about regression.

Homogeneity of regression says that the two lines have the same slope but not that they are the same line. To decide about this, it is necessary to test adjusted means. This is dealt with in Chap. 15.

Homogeneity of regression may also be tested by F. Briefly, find the reduction in sum of squares attributable to X when a single regression coefficient is assumed, and the reduction when two coefficients are assumed. The latter cannot be smaller than the former. The difference in the two sums of squares is an additional reduction that will ordinarily be attributable to chance if there is a single β, but will be larger, on the average, if there are two β's.

The procedure is indicated in Table 9.4. Checked spaces are filled by the usual computational procedures. In line 6, column 8, the difference (row 4 minus row 5) will give the same answer as in column 6. The extension of this procedure to test differences among more than two β's is presented in Chap. 15.

Exercise 9.9.1 H. L. Self (Ph.D. thesis, University of Wisconsin, 1954) reported the back-fat thickness Y and slaughter weight X for four lots of Poland China pigs fed different rations. Data for treatment 3 are:

Y (mm): 42, 38, 53, 34, 35, 31, 45, 43
X (lb): 206, 261, 279, 221, 216, 198, 277, 250

and, for treatment 4:

Y: 33, 34, 38, 33, 26, 28, 37, 31
X: 167, 192, 204, 197, 181, 178, 236, 204

Compute the two regression coefficients and test the null hypothesis that they are estimates of a common β.

TABLE 9.4
ANALYSIS OF VARIANCE TABLE FOR TESTING DIFFERENCE BETWEEN TWO β's

Column Line	1	2	3	4	5	6	7	8	9
1	Source of variation	df	Σx^2	Σxy	Σy^2	$\dfrac{(\Sigma xy)^2}{\Sigma x^2}$	df	Residual SS = col 5 − col 6	df
2	Within sample 1	$n_1 - 1$	✓	✓	✓	✓	1	✓	$n_1 - 2$
3	Within sample 2	$n_2 - 1$	✓	✓	✓	✓	1	✓	$n_2 - 2$
4	(Two regressions)					Subtotal	2	Subtotal	$n_1 + n_2 - 4$
5	Within 1 + within 2 (one regression)	$n_1 + n_2 - 2$	✓	✓	✓	✓	1	✓	$n_1 + n_2 - 3$
6	Regression coefficients (two regressions vs. one)					Row 4 − row 5	1		

$$F = \dfrac{\text{line 6, col 6 entry}}{(\text{line 4, col 8 entry})/(n_1 + n_2 - 4)}, \ 1 \text{ and } n_1 + n_2 - 4 \ df$$

LINEAR REGRESSION

Exercise 9.9.2   In assuming a common variance, it is the variance about regression that is referred to. Test the null hypothesis of a common variance about regression.

9.10   A prediction and its variance.

Among the uses of regression, there is prediction. There are times when one wishes to say something about a particular future value such as a time of first frost, maximum rainfall, etc. Or, it may be possible to observe an X value and impossible or impractical to observe the corresponding Y. One gives, as the prediction, the regression estimate of the mean. The important thing is to assign a variance appropriate to an individual rather than to a mean. This is obtained by increasing the variance of the estimate of the mean by the variation for individuals. Thus the variance of a predicted Y is

$$\sigma_{y \cdot x}^2 + \frac{\sigma_{y \cdot x}^2}{n} + x^2 \frac{\sigma_{y \cdot x}^2}{\Sigma x^2} = \sigma_{y \cdot x}^2 \left(1 + \frac{1}{n} + \frac{x^2}{\Sigma x^2}\right)$$

It is estimated by Eq. (9.19).

$$s_{y \cdot x}^2 \left(1 + \frac{1}{n} + \frac{x^2}{\Sigma x^2}\right) \tag{9.19}$$

To set a 95% confidence interval on a predicted value, compute

$$\bar{y} + bx \pm t_{.05} s_{y \cdot x} \sqrt{1 + \frac{1}{n} + \frac{x^2}{\Sigma x^2}}$$

where $t_{.05}$ is tabulated t for $n - 2$ degrees of freedom. For any other confidence coefficient, replace $t_{.05}$ by t for the appropriate probability level.

Consider the data in Table 9.5. It is desired to predict the number of

TABLE 9.5
HORSES ON CANADIAN FARMS, JUNE, 1944–1948

X = year	x	Y = number of horses on Canadian farms (thousands)
1944	−2	2,735
1945	−1	2,585
1946	0	2,200
1947	+1	2,032
1948	+2	1,904

$\Sigma x^2 = 10$ $\Sigma xy = -2,215$ $\Sigma y^2 = 508,702.8$

$\bar{y} = 2,291.2$ horses $b = -221.5$ horses per year

$\dfrac{(\Sigma xy)^2}{\Sigma x^2} = 490,622.5$ reduced MS $= \dfrac{18,080.3}{3} = 6,026.8$

SOURCE: Data from *Canada*, 1950, page 127.

horses for 1949 on the assumption that the decrease in numbers is linear over years.

The regression equation is

$$Y = 2{,}291.2 - 221.5(X - 1946) \qquad (9.20)$$

and $s_{y \cdot x} = 77.6$ horses. The data and regression equation are shown in Fig. 9.6.

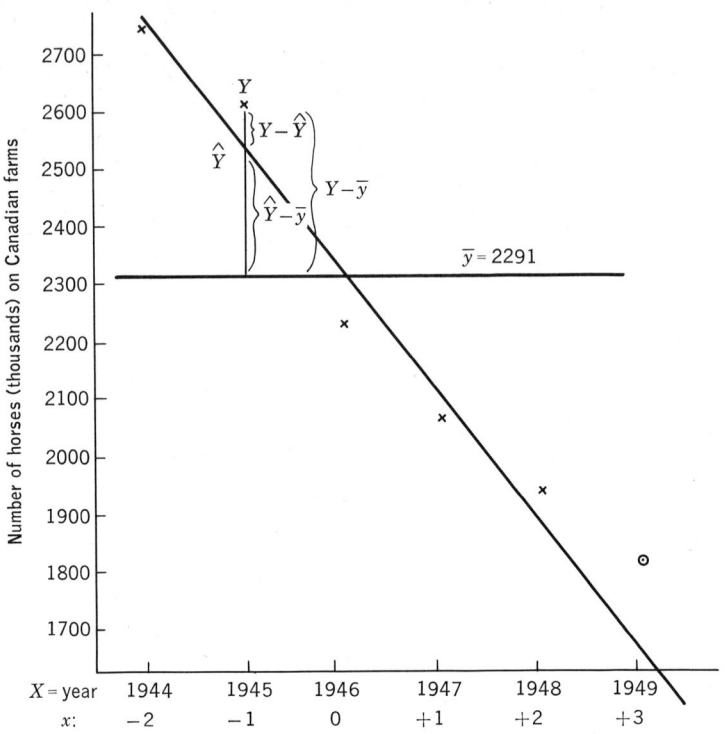

FIG. 9.6 Data and regression line for data of Table 9.5

To predict the number of thousands of horses for 1949, substitute $X = 1949$ in Eq. (9.20). We find $\hat{Y}_3 = 1{,}627$ horses. The standard deviation applicable to this result is

$$s_{y \cdot x}\sqrt{1 + \frac{1}{n} + \frac{x^2}{\Sigma x^2}} = 77.6\sqrt{1 + \frac{1}{5} + \frac{3^2}{10}} = 112.5 \text{ horses}$$

The 95% confidence interval is

$$1{,}627 \pm 3.182(112.5) = (1{,}269,\ 1{,}985)$$

In 1949, the number of thousands of horses was 1,796, well within the confidence interval but $1{,}796 - 1{,}627 = 169$ thousand horses more than the point estimate.

Probability statements, such as discussed in this section, refer to the complete experience of obtaining the sample and observing the future observation. Thus if our sample is an unusual one and we predict many observations, it may be that none of the actual future observations will lie in the predicted interval. This points out the fallacy of using one regression line to make predictions time after time. If random sampling is not repeated, the least one can do is revise the regression line as experience accumulates.

To predict a future mean for n_1 observations, the regression value is used for the predicted mean. The variance appropriate to the prediction is given by replacing the 1 in the parentheses of Eq. (9.19) by $1/n_1$.

In predicting a value or estimating a mean for an X outside the observed range, that is, in extrapolating as opposed to interpolating, one assumes that the relationship will continue to be linear. This assumption may not be sound, especially if a straight line is being used as an approximation and if one extrapolates very far outside the range. In the case at hand, interpolation would not be very informative.

Exercise 9.10.1 Suppose that a lot of 50 hens on the New York random sample test (Table 9.1) has an average weight of 5.3 lb. What is the 95% confidence interval for the prediction of the 350-day food consumption for this particular lot of birds? What is the 95% confidence interval for the mean food consumption for lots with an average weight of 5.3 lb?

9.11 Prediction of X, Model I. At times it is required to predict an X or estimate the mean of a population of X's even though X is a fixed variable in the available data. This is a common situation in biological assay and dosage-mortality problems. In such circumstances one predicts or estimates by solving the regression equation of Y on X, for X. Thus

$$x = \frac{Y - \bar{y}}{b} \quad \text{or} \quad X = \bar{x} + \frac{Y - \bar{y}}{b}$$

This gives a point estimate of X where one usually prefers a confidence interval estimate. Computation of an interval is possible, straightforward, but more lengthy than for a value of the dependent variable. The reader interested in bioassay is referred to Bliss (9.1), Eisenhart (9.2), Finney (9.3), and Irwin (9.4).

9.12 Bivariate distributions, Model II. In many instances, (X, Y) pairs are observed at random; that is, no control over X is exercised. For example, consider height and weight of adult American males, the individuals being drawn at random and a pair of measurements made; or consider an experiment to determine potato yields under a variety of field conditions, where one also observes the rate of nematode infestation. In such cases, sampling is from a bivariate distribution.

Consider the pairs of observations of Table 9.6 on a single strain of guayule, a plant from which rubber is obtained. The variables are oven dry shrub weight and crown circumference; the data are a portion of the original. As dependent variable, choose that one for which means are to be estimated,

TABLE 9.6

Oven Dry Weight and Circumference of Crown for a Random Sample of Guayule Plants

Oven dry weight, grams	Circumference of crown, cm
65	6.5
100	6.3
82	5.9
133	6.3
133	7.3
165	8.0
116	6.9
120	8.1
150	8.7
117	6.6
$\bar{x}$: 118.1	7.06
Σx^2: 8,120.9	7.764
$\Sigma x_1 x_2 = 176.84$	

SOURCE: Data courtesy of W. T. Federer, Cornell University, Ithaca, N.Y.

values are to be predicted, or statistical control by use of the other variable is desired. Thus when circumference is the dependent variable, we have

$$\text{Reduction in SS} = \frac{(176.84)^2}{8,120.9} = 3.851$$

$$\text{Reduced SS} = 7.764 - 3.851 = 3.913$$

$$\text{Per cent reduction} = \frac{3.851}{7.764}\,100 = 49.6\%$$

$$b = \frac{176.84}{8,120.9} = .022 \text{ cm per gram}$$

When weight is the dependent variable,

$$\text{Reduction in SS} = \frac{(176.84)^2}{7.764} = 4,027.9$$

$$\text{Reduced SS} = 8,120.9 - 4,027.9 = 4,093.0$$

$$\text{Per cent reduction} = \frac{4,027.9}{8,120.9}\,100 = 49.6\%$$

$$b = \frac{176.84}{7.764} = 22.78 \text{ grams per cm}$$

It is clear, then, that in random sampling of pairs from a bivariate distribution, there are two regression equations. The two residual sums of squares are measured in terms of perpendiculars to different axes. The lines do not coincide, as can be seen from the fact that the two b's are not reciprocals of each other. The product of the b's is the reduction in sum of squares as a decimal fraction. Thus, $(22.78)(.022)100 = 50.1\%$, differing from 49.6% by rounding errors. Both lines pass through the sample mean $(\bar{x}, \bar{y})$.

The proportion of the sum of squares of the dependent variable that can be attributed to the independent variable is always the same regardless of which variable is independent. This is seen by comparing formulas; we have

$$\frac{(\Sigma xy)^2/\Sigma x^2}{\Sigma y^2} = \frac{(\Sigma xy)^2/\Sigma y^2}{\Sigma x^2}$$

This quantity, often denoted by r^2 and called the *coefficient of determination*, is considered in more detail in the next chapter.

When random sampling is not from a bivariate distribution, it is still possible to estimate means and predict values for either the dependent or the independent variable. This is done with a single regression line, the regression of the dependent variable upon the independent variable. References at the end of Sec. 9.11 deal with the problem of a confidence interval for estimation or prediction of values of the independent variable.

Exercise 9.12.1 Compute the coefficient of determination for the data of Exercise 9.9.1. Interpret this coefficient.

9.13 Regression through the origin. In some situations, theory calls for a straight line that passes through the origin. In such cases, we are given a point on the line, a point for which there is no sampling variation. Clearly such a point must be treated differently from an observed point.

As an example of regression through the origin, consider the data of Table 9.7. Since the regression line must pass through the origin, the required equation may be written $Y = bX$. The regression coefficient is given by

$$b = \frac{\Sigma XY}{\Sigma X^2} \tag{9.21}$$

$$= 3.67 \text{ induced reversions per dose}$$

The regression line is given by

$$Y = 3.67X$$

The sum of the deviations from this line is not zero. The reduction in sum of squares attributable to regression is $(\Sigma XY)^2/\Sigma X^2 = 351{,}819$. The reduced sum of squares is 4,440 with 12 degrees of freedom and the residual mean square for Y is 370. Since no adjustment has been made for the mean, ΣY^2 has 13 degrees of freedom = number of observations, and the residual mean square has 12 degrees of freedom.

Any relevant hypothesis about the value estimated by b may be tested by Student's t test. A test of the hypothesis $\beta = 0$ is also given by $F =$ (reduction in SS)/residual MS.

When there are reservations about the assumption that regression is through the origin, it may be desirable to test this as a hypothesis. To do this, compute the regression line $Y = \bar{y} + bx$. This is given by

$$Y - 148.4 = 3.80(X - 40.68)$$

The reduction attributable to the mean is $(\Sigma Y)^2/n = 286{,}234$ and to b is $(\Sigma xy)^2/\Sigma x^2 = 65{,}670$. The total reduction is $(286{,}234 + 65{,}670) = 351{,}904$; the reduced sum of squares is 4,355 with 11 degrees of freedom and the residual mean square for Y is 396. This is actually higher than the mean square for regression through the origin, though the sum of squares is necessarily lower.

TABLE 9.7

INDUCED REVERSIONS TO INDEPENDENCE PER 10^7 SURVIVING CELLS Y PER DOSE (ERGS/BACTERIUM) $10^{-5}X$ OF STREPTOMYCIN DEPENDENT *Escherichia coli* SUBJECTED TO MONOCHROMATIC ULTRAVIOLET RADIATION OF 2,967 ANGSTROMS WAVELENGTH

X	Y
13.6	52
13.9	48
21.1	72
25.6	89
26.4	80
39.8	130
40.1	139
43.9	173
51.9	208
53.2	225
65.2	259
66.4	199
67.7	255

$\Sigma X = 528.8$ $\Sigma Y = 1{,}929$
$\Sigma X^2 = 26{,}062.10$ $\Sigma Y^2 = 356{,}259$
$\Sigma x^2 = 4{,}552.14$ $\Sigma y^2 = 70{,}025$
$\Sigma XY = 95{,}755.7$ $\Sigma xy = 17{,}289.9$

SOURCE: Data courtesy of M. R. Zelle, Cornell University, Ithaca, N.Y.

To test the hypothesis that regression is through the origin, compute the additional reduction due to fitting the mean. This is $(286{,}234 + 65{,}670) - (351{,}819) = 85$ or $4{,}440 - 4{,}355 = 85$ with a single degree of freedom and is clearly not significant. (The latter computation is the difference between the two residual sums of squares.) The appropriate error mean square is $4{,}355/11 = 396$.

As in the general regression case, it is assumed that deviations from the regression line are normally distributed with a common variance. More extensive data than those presented here indicate that the variance is probably a function of the dose and the wavelength. In this case, a weighted regression analysis is appropriate.

9.14 Weighted regression analysis. At times data are available where the observations have unequal variances. For example, we might require the regression of treatment means on a concomitant variable, when treatment means are based on differing sample sizes, the variances being $\sigma^2/n_1, \ldots, \sigma^2/n_k$.

LINEAR REGRESSION

The assumption of homogeneous variance is not justified and a weighted regression analysis must be performed.

In most weighted regression analyses, the weights depend upon the amounts of information in or the precision of the observations. They are the reciprocals of the variances, that is, $w_i = 1/\sigma_i^2$ where w_i represents the weight for the ith observation. If the observations are means, then $w_i = n_i/\sigma^2$.

It is the relative rather than the actual weights that are important. Hence, when observations have a common variance and we are computing the regression of means on another variable, the weights are the numbers of observations.

For a weighted regression, the total sum of squares of the Y's is a weighted one. In particular,

$$SS(Y) = \Sigma w_i(Y_i - \bar{y})^2, \text{ definition,}$$

$$= \Sigma w_i Y_i^2 - \frac{(\Sigma w_i Y_i)^2}{\Sigma w_i}, \text{ computing,}$$

where $\bar{y} = (\Sigma w_i Y_i)/\Sigma w_i$, a weighted mean. The regression coefficient is

$$b = \frac{\Sigma w_i(X_i - \bar{x})(Y_i - \bar{y})}{\Sigma w_i(X_i - \bar{x})^2}, \text{ definition,}$$

$$= \frac{\Sigma w_i X_i Y_i - [(\Sigma w_i X_i)(\Sigma w_i Y_i)/\Sigma w_i]}{\Sigma w_i X_i^2 - [(\Sigma w_i X_i)^2/\Sigma w_i]}, \text{ computing,}$$

where $\bar{x} = (\Sigma w_i X_i)/\Sigma w_i$. The regression equation is

$$Y = \bar{y} + bx$$

The reduction in sum of squares attributable to regression is

$$\text{Reduction} = \frac{\{\Sigma w_i X_i Y_i - [(\Sigma w_i X_i)(\Sigma w_i Y_i)/\Sigma w_i]\}^2}{\Sigma w_i X_i^2 - [(\Sigma w_i X_i)^2/\Sigma w_i]}, 1 \text{ df}$$

and the reduced sum of squares is found by subtracting the reduction from the total. The weighted sum of deviations from regression is zero. That is,

$$\Sigma w_i(Y_i - \bar{y} - bx_i) = 0$$

Also, the weighted sum of squares,

$$\Sigma w_i(Y_i - \bar{y} - bx_i)^2$$

is a minimum; there is no other regression line giving a smaller weighted sum of squares.

Computations are seen to be lengthier than for the usual regression. However, if columns of $w_i X_i$ and $w_i Y_i$ are used, the weighted sums of products, $\Sigma w_i X_i^2$, $\Sigma w_i Y_i^2$, and $\Sigma w_i X_i Y_i$, are easily obtained from pairs of columns $w_i X_i$ and X_i, $w_i Y_i$ and Y_i, and $w_i X_i$ and Y_i or $w_i Y_i$ and X_i, respectively.

In the special case where the regression is one of means of differing numbers of observations on an independent variable, we have $w_i Y_i = n_i \bar{y}_i$, the total of the observations in the ith mean, a quantity already available. Consequently, entries for the column $w_i Y_i$ have already been computed.

References

9.1 Bliss, C. I.: *The Statistics of Bioassay*, Academic Press, New York, 1952.
9.2 Eisenhart, C.: "The interpretation of certain regression methods and their use in biological and industrial research," *Ann. Math. Stat.*, **10:** 162–186 (1939).
9.3 Finney, D. J.: *Probit Analysis*, 2d ed., Cambridge University Press, 1951.
9.4 Irwin, J. O.: "Statistical methods applied to biological assays," *J. Roy. Stat. Soc. Suppl.*, **4:** 1–48 (1937).
9.5 Winsor, C. P.: "Which regression?" *Biometrics Bull.*, **2:** 101–109 (1946).

Chapter 10

LINEAR CORRELATION

10.1 Introduction. Bivariate distributions were discussed briefly in Sec. 9.12. In sampling from a bivariate population, an observation consists of a random pair of measurements. Here, two regressions are possible and valid although, ordinarily, only one regression is desired. A summary of the data in a sample from a bivariate distribution consists of two means, two variances, and the covariance. The covariance may be replaced without loss of information by the coefficient of determination or its square root, the coefficient of linear correlation. This chapter deals with linear correlation.

10.2 Correlation and the correlation coefficient. Correlation, like covariance, is a measure of the degree to which variables vary together or a measure of the intensity of association. As such, it must be symmetric in the two variables. The sample linear correlation coefficient, also called the simple correlation, the total correlation and the product-moment correlation, is used for descriptive purposes and is defined by Eq. (10.1).

$$r = \frac{\Sigma(X - \bar{x})(Y - \bar{y})/(n - 1)}{\sqrt{\Sigma(X - \bar{x})^2/(n - 1)}\sqrt{\Sigma(Y - \bar{y})^2/(n - 1)}}$$

$$= \frac{\Sigma(X - \bar{x})(Y - \bar{y})}{\sqrt{\Sigma(X - \bar{x})^2 \Sigma(Y - \bar{y})^2}} = \frac{\Sigma xy}{\sqrt{\Sigma x^2 \Sigma y^2}} \qquad (10.1)$$

It is assumed that, in the population, a linear relation exists between the variables. This is a valid assumption when sampling is from a bivariate normal distribution. The correlation coefficient r is an unbiased estimate of the corresponding population correlation coefficient ρ (Greek rho) only when the population parameter ρ is zero. Unlike a variance or a regression coefficient, the correlation coefficient is independent of the units of measurement; it is an absolute or dimensionless quantity. The use of X and Y is no longer intended to imply an independent and a dependent variable.

Some insight for the interpretation of linear correlation may be gained from the *scatter diagrams* of Fig. 10.1 for which the data have been especially manufactured. In part i of the figure, the points cluster about a line through $(\bar{x},\bar{y})$ and parallel to the X axis simply because the variance of X is larger than that of Y. The lack of a tendency to cluster about any line other than one through $(\bar{x},\bar{y})$ and parallel to an axis is typical of data where there is little or no linear correlation. In such cases, each regression line is close to a line through $(\bar{x},\bar{y})$

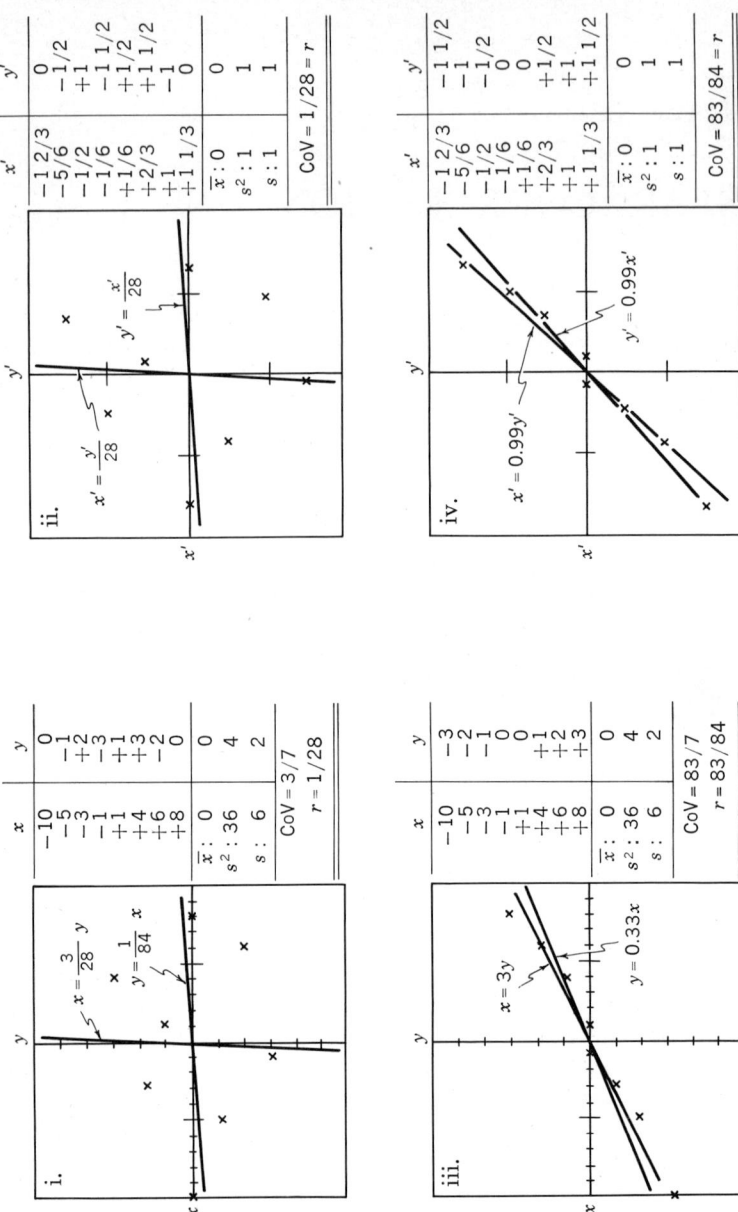

FIG. 10.1 Scatter diagrams to illustrate correlation

and parallel to the axis of the independent variable. Lack of linear correlation is more noticeable where the scatter diagram uses standard deviations as units of measurement, as in part ii. Each datum of part i has been divided by its standard deviation, that is, $x' = x/6$ and $y' = y/2$, so that each variable has unit variance. Variables with zero means and unit variances are called *standard variables*. The points no longer show a tendency to cluster about any axis. The covariance equals the correlation.

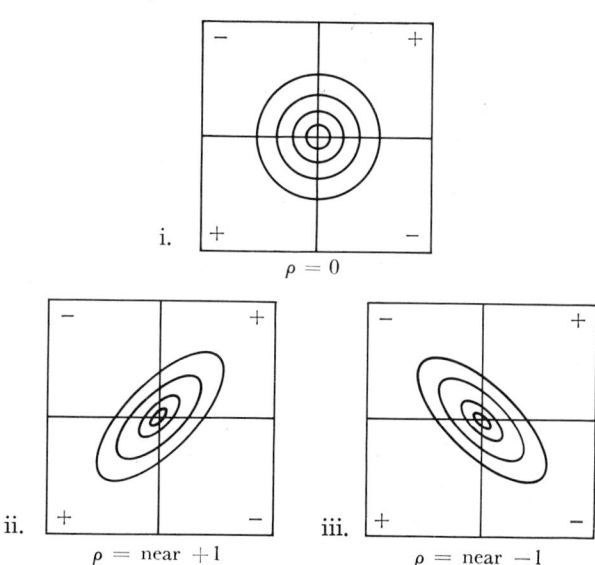

FIG. 10.2 Diagrams to illustrate correlation. Corner signs ($\pm$) refer to the sign of the cross products for that quadrant. (The four quadrants or four parts of a plane are numbered counterclockwise beginning with the upper right quadrant as the first quadrant.) The probability of a point lying in an area between successive circles or ellipses is assumed to be the same for each area. Thus, the three-dimensional frequency distribution would show a heaping up over the intersection of the axes and a gradual lowering as one moved away from this point. This presentation is in lieu of attempting to draw a three-dimensional figure on a two-dimensional page, the third dimension being related to probability.

Where linear correlation is small, r is near zero. Reference to Fig. 10.2, part i, shows this. The sample points would be approximately equally distributed in each of the four quadrants so that xy cross products of the same magnitude occur about equally often in each. Hence Σxy, the numerator of r, tends to be near zero.

Parts iii and iv of Fig. 10.1 make use of the same numbers as do i and ii but they are differently paired. For iii, $\Sigma xy = 83\frac{2}{7}$ while the means and variances are unchanged. Now the points tend to cluster about a line other than an axis. This is typical of data where a high correlation is present. The regression lines are nearly coincident and are tipped toward the x axis because of the large variance of x. Near-coincidence of regression lines is typical of data which exhibit high correlation.

High linear correlation, r near $+1$ or -1, is more readily recognized on scatter diagrams where the variables are standardized, as in Fig. 10.1, part iv. Here the points cluster about a line approximately halfway between the axes. Both regression lines are near this line and the regression coefficients equal the correlation coefficient and are near $+1$ or -1. A unit change in one variable implies approximately a unit change in the other for regression lines in this near-midway position. Reference to Fig. 10.2, part ii, shows why r will be numerically large. Here, large cross products appear more frequently in the upper right and lower left or upper left and lower right corners. This tends to give a numerically large Σxy. In the first instance, Σxy will tend to be positive and lead to a positive correlation; in the second case, Σxy will tend to be negative and give a negative correlation.

Detecting linear correlation visually from plotted points can be difficult. An unfortunate choice of scale may tend to hide a real correlation or indicate a real one when none is present. A change of scale will change the slope of the regression line. If a scatter diagram does not actually use standard deviations as units, at least the standard deviations should be available with the diagram to help prevent false conclusions.

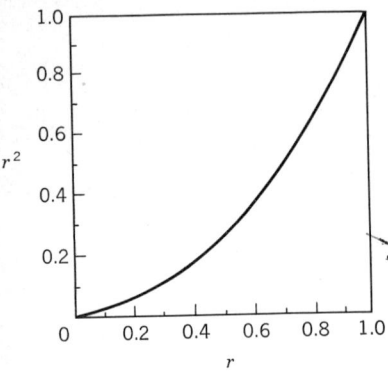

FIG. 10.3 Relation between r and r^2.

In addition to difficulties due to an unfortunate choice of scale, visual detection is further hindered by the fact that the relation between r and the proportion of the total sum of squares explained by regression is not a linear one. Figure 10.3 shows this graphically. For $r = .1$, only 1% of the variation in a dependent variable is explained by the independent variable; for $r = .2$, the percentage is only 4; for $r = .5$, it is only 25. For $r > .5$, the explained percentage increases more rapidly. Apart from actual computation of r, detection is most likely by selecting large or small values of one variable and seeing if the corresponding values of the other variable tend to be at one end of the scale for that variable. Counts of the numbers of points in each of the quadrants are also informative.

Finally, the sampling variation in r is quite large for small-size samples.

It can be shown that r lies between -1 and $+1$, that is, $-1 \leq r \leq 1$. The values $+1$ and -1 indicate perfect linear correlation or a functional relation between the two variables. This places the observations outside the province of statistics. Barring rounding errors, one finds perfect correlation only in moments of carelessness as one might if he correlated height with shoulder height plus height from shoulder to top of head. Extremely high correlations exist but should always be closely scrutinized for the possibility that this sort of thing has been done and that perfect correlation has been missed only because of rounding errors in the computations.

Two quantities closely related to r are the coefficients of determination and alienation. The *coefficient of determination* is r^2, the square of the correlation coefficient. This term may also be used in connection with regression analysis, where the correlation coefficient is not applicable. In such cases, r^2 is the proportion of a total sum of squares that is attributable to another source of variation, the independent variable.

The *coefficient of nondetermination* is given by $1 - r^2 = k^2$ and is the unexplained proportion of a total sum of squares. Usually, this makes it the basis of an error term. Its square root k is called the *coefficient of alienation* while $k' = 1 - k$ has been called the *improvement factor*. These are not very common terms.

Exercise 10.2.1 H. H. Smith collected the following data on flowers of a Nicotiana cross. T = tube length, L = limb length, and N = tube base length.†

T: 49, 44, 32, 42, 32, 53, 36, 39, 37, 45, 41, 48, 45, 39, 40, 34, 37, 35
L: 27, 24, 12, 22, 13, 29, 14, 20, 16, 21, 22, 25, 23, 18, 20, 15, 20, 13
N: 19, 16, 12, 17, 10, 19, 15, 14, 15, 21, 14, 22, 22, 15, 14, 15, 15, 16

Compute the correlation coefficients between T and L, T and N, L and N. What are the coefficients of determination in each case?

Exercise 10.2.2 In studying the use of ovulated follicles in determining eggs laid by ring-necked pheasant, C. Kabat et al. (10.3) present the following data on 14 captive hens.

Eggs laid: 39, 29, 46, 28, 31, 25, 49, 57, 51, 21, 42, 38, 34, 47
Ovulated follicles: 37, 34, 52, 26, 32, 25, 55, 65, 44, 25, 45, 26, 29, 30

Compute the correlation coefficient and coefficient of determination.

10.3 Correlation and regression.

We now have three methods for treating random pairs of observations. These are:

1. Ignore any relation between variables and analyze them separately.
2. Use a regression analysis.
3. Examine the correlation.

Here we are concerned only with the second and third methods.

Correlation measures a co-relation, a joint property of two variables. Where variables are jointly affected because of external influences, correlation may offer the most logical approach to an analysis of the data. Regression deals primarily with the means of one variable and how their location is influenced by another variable. For correlation, random pairs of observations are obtained from a bivariate normal distribution, not too common a distribution; for regression, only the dependent member of each pair need be random and normally distributed. Correlation is associated with descriptive techniques; regression comes close to implying cause and effect relations. Thus, whereas a correlation coefficient tells us something about a joint relationship between variables, a regression coefficient tells us that if we alter the value of the independent variable then we can expect the dependent variable to alter by a certain amount on the average, sampling variation making it unlikely that precisely the stated amount of change will be observed.

† Published with permission of H. H. Smith, Brookhaven National Laboratories.

It has already been noted that

$$r^2 = \frac{(\Sigma xy)^2}{(\Sigma x^2)(\Sigma y^2)}, \text{ square of definition } r$$

$$= \frac{(\Sigma xy)^2/\Sigma x^2}{\Sigma y^2} = \frac{\text{reduction in SS}(Y) \text{ attributable to } X}{\text{total SS}(Y)}$$

$$= \frac{(\Sigma xy)^2/\Sigma y^2}{\Sigma x^2} = \frac{\text{reduction in SS}(X) \text{ attributable to } Y}{\text{total SS}(X)}$$

In addition,

$$r^2 = \left(\frac{\Sigma xy}{\Sigma x^2}\right)\left(\frac{\Sigma xy}{\Sigma y^2}\right) = b_{yx}b_{xy} \tag{10.2}$$

where b_{yx} and b_{xy} are the regression coefficients for the regression of Y on X and of X on Y. Thus, the product of the regression coefficients is the square of the correlation coefficient or the correlation coefficient is the square root of the product of the regression coefficients or their geometric mean. These results are always algebraically correct; they have meaning only when the sample consists of random pairs.

If we standardize our variables, then the regression equation for Y on X becomes

$$\frac{Y - \bar{y}}{s_y} = r \frac{X - \bar{x}}{s_x} \quad \text{or} \quad y' = rx'$$

as stated in Sec. 10.2. Similarly,

$$\frac{X - \bar{x}}{s_x} = r \frac{Y - \bar{y}}{s_y} \quad \text{or} \quad x' = ry'$$

Here r is a regression coefficient. Note that neither of these equations can be obtained by solving the other; a regression coefficient is not a symmetric statement about the relation between two variables.

10.4 Sampling distributions, confidence intervals, tests of hypotheses. Since $-1 \leq r \leq 1$, one cannot expect the sampling distribution of r to be symmetric when the population parameter ρ is different from zero. Symmetry occurs only for $\rho = 0$ and lack of symmetry or asymmetry increases as ρ approaches $+1$ or -1. This means that, for small samples, a normal approximation is not generally suitable for computing a confidence interval for ρ and the confidence interval should not be centered on r. The failure to center on r will increase as observed values approach $+1$ or -1 since these values are the limits of possible values.

Sample r's are quite variable for small samples, particularly for ρ at or near zero. A single pair of values can make a lot of difference to the value of r. This makes it difficult to detect small but real values of ρ using small samples. This is not generally serious since small values of ρ are of little practical use. This was pointed out in Sec. 10.2 when discussing the relation between r and r^2.

Possibly the simplest method for setting confidence limits on ρ is to use the charts prepared by David (10.1). These charts are given as Table A.11. To

LINEAR CORRELATION

set a confidence interval on ρ, draw a vertical line through the observed value of r on the "scale of r." At the points of intersection of this line with the two curved lines corresponding to the sample size, draw lines that cut the "scale of ρ" at right angles. The two ρ values on this scale are the limits of the confidence interval. These easy-to-use charts are sufficiently accurate for most purposes.

A convenient alternative is given by Fisher (10.2). For this, compute the transformation

$$z = .5 \ln \frac{1+r}{1-r} \qquad (10.3)$$

which is approximately normally distributed with approximate mean and standard deviation of $.5 \ln (1 + \rho)/(1 - \rho)$ and $1/\sqrt{n-3}$ regardless of the value of ρ. (ln is the natural logarithm, base e.) Values of z for r from .00 to .99, by steps of .01, are given in Table A.12. Compare z with values from a normal table or the bottom line of a t-table. Confidence limits are set for z and converted to ρ. The conversion formula is $r = (e^{2z} - 1)/(e^{2z} + 1)$ but

TABLE 10.1

HOMOGENEITY AND POOLING OF r's FOR THE CORRELATION OF PER CENT OF RESIN CONTENT AND PER CENT OF RUBBER CONTENT IN GUAYULE

Strain	n_i	r_i	z_i	$z_i - \bar{z}_w$	$(n_i - 3)(z_i - \bar{z}_w)^2$
405	50	0.362	0.379	−0.0913	0.392
407	50	0.419	0.446	−0.0243	0.028
416	50	0.527	0.586	+0.1157	0.629
	150				$\chi^2 = 1.049$, 2 df

$$\bar{z}_w = \frac{\Sigma(n_i - 3)z_i}{\Sigma(n_i - 3)} = 0.4703$$

SOURCE: Data from U.S. Dept. Agr. Tech. Bull. 919, 1946, by W. T. Federer.

Table A.12 can be used quite satisfactorily. For strain 416, Table 10.1, $r = .527$ and we find

$$n - 3 = 47 \qquad s_z = \sqrt{\frac{1}{47}} = .146$$

$$.5 \ln \frac{1+r}{1-r} = .5 \ln \frac{1.527}{.473} = .5 \ln 3.23$$

$$= .5(1.172) = .586$$

$$\text{CL} = .59 \pm 1.96(.146)$$

$$= .30, .88$$

Hence the 95% confidence interval for ρ is (.30, .71).

To test the null hypothesis that ρ equals some specified value other than zero, transform to approximate normality as in the previous paragraph and use the normal test. If the hypothesis is that $\rho = 0$, compute

$$t = \frac{r}{\sqrt{(1-r^2)/(n-2)}}$$

and compare with Student's t for $n - 2$ degrees of freedom. The square of t equals F from the analysis of variance for regression and, consequently, the procedure is equivalent to testing the hypothesis that $\beta = 0$. This test cannot be used for testing the hypothesis $\rho = k \neq 0$, that is, ρ equals a constant other than zero.

Table A.13 gives values of r at the 95 and 99% levels of significance for a number of values of the degrees of freedom.

The distinction between significance and meaningfulness is not a serious problem with small samples but can be so with large samples. This is especially true of correlation since many large bivariate samples are collected in certain types of studies. In large samples, small values of r may be significant. However, if the percentage reduction in total sum of squares is small for either variable considered as dependent, then the correlation may be quite useless.

Exercise 10.4.1 Set confidence intervals on one of the correlation coefficients computed in Exercise 10.2.1. Use the chart by David (Table A.11) and a transformation to z and back to r. Do the results differ appreciably?

Exercise 10.4.2 For the correlation coefficient obtained in Exercise 10.2.2, test the null hypothesis that $\rho = 0$.

10.5 Homogeneity of correlation coefficients.

The test of a difference between two population values of ρ is simple. The two r's are converted to z's and the appropriate large sample normal test of Chap. 5 is applied. For tests of hypotheses about one or two ρ's, z values must be found but no reconversion is required; reconversion is required only in the case of confidence limits. An example of the test of homogeneity of two r's is given in Table 10.2.

TABLE 10.2

TEST OF THE HYPOTHESIS $\rho_1 = \rho_2$ FOR THE CORRELATION OF RUBBER CONTENT OF BRANCH ON DIAMETER OF WOOD IN GUAYULE

Strain	Number of plants	r	z	$1/(n-3)$
109	22	0.310	0.32055	0.0526
130	21	0.542	0.60699	0.0556
			Difference = 0.28644	sum = 0.1082

$$z = \frac{0.28644}{\sqrt{0.1082}} = \frac{0.28644}{0.329} = 0.87 \ ns$$

SOURCE: Data from U.S. Dept. Agr. Tech. Bull. 919, 1946, by W. T. Federer.

LINEAR CORRELATION

For the test shown in Table 10.2, a difference is compared with its standard deviation, the square root of the sum of the variances, and the result compared with values tabulated in a normal table. The probability of a larger difference arising by chance when there is no difference is about 38% for these two r values. Note that 0.310 is not significant whereas 0.542 is highly significant.

It is sometimes desired to test the homogeneity of several correlation coefficients and obtain a single coefficient if they seem to be homogeneous. For example, measurements on two characteristics of a crop or type of animal may be available for several strains or breeds. The strain or breed variances may not be homogeneous so that pooling sums of products and calculating a single correlation coefficient is not valid. The test of homogeneity and the method of pooling are illustrated in Table 10.1.

The z's are approximately normally distributed. Since the variances, $1/(n_i - 3)$, will generally be unequal, a weighted mean $\bar{z}_w$ is used in the computations. The test criterion is χ^2 with degrees of freedom equal to one less than the number of r's. The criterion χ^2 was originally defined by Eq. (3.9). In our present example, $z_i - \bar{z}_w$ replaces X_i and zero, the mean of the population of all possible deviations, replaces μ_i. The standard deviation σ_i is replaced by $1/\sqrt{n_i - 3}$. Hence Eq. (3.9) becomes

$$\chi^2 = \sum_i \left(\frac{z_i - \bar{z}_w}{1/\sqrt{n_i - 3}}\right)^2 = \sum_i (n_i - 3)(z_i - \bar{z}_w)^2$$

Since only $k - 1$ deviations are independent, χ^2 has $k - 1$ degrees of freedom only. Here χ^2 is not significant so we conclude that the correlation coefficients are homogeneous. Converting $\bar{z}_w$ back to r gives a pooled value of the several coefficients; pooled $r = 0.438$. To set confidence limits on the parameter being estimated, place an interval about $\bar{z}_w$ and convert to values of r. The appropriate standard deviation is $1/\sqrt{\Sigma(n_i - 3)}$. Since $n_i - 3$ is the amount of information in z_i, the amount of information in $\bar{z}_w$ is $\Sigma(n_i - 3)$ and the reciprocal of this is the variance appropriate to $\bar{z}_w$.

There is a small bias in z which may be serious if many correlations are averaged. Since only three are involved here, we convert without hesitation. The bias equals

$$\frac{\rho}{2(n-1)}$$

and is positive. Since ρ is unknown, the bias cannot be eliminated. It has been suggested that the average r obtained from $\bar{z}_w$ be used for ρ to compute the bias for each z. These biases are subtracted from their respective z's and a new $\bar{z}_w$ computed. This is converted to an adjusted r which should be more accurate than the unadjusted. It is not appropriate to compute a new χ^2.

Exercise 10.5.1 To show the effect of skewness on the distribution of r, assume that three sample r's of Table 10.1 are increased by 0.4 to 0.762, 0.819, and 0.927. Test the homogeneity and compare the two χ^2 values and the two probabilities of getting a larger χ^2.

10.6 Intraclass correlation. Sometimes a correlation coefficient is desired where there is no meaningful criterion for assigning one member of the pair to one variable rather than the other. In such cases, we obtain a value of the coefficient from the calculation of certain variances. The resulting coefficient is called an *intraclass correlation* and is computed by Eq. (10.4) when there are n observations per class.

$$r_I = \frac{\text{MS(among classes)} - \text{MS(within classes)}}{\text{MS(among classes)} + (n-1)\text{MS(within classes)}} \qquad (10.4)$$

The same result is also obtained from the equation

$$r_I = \frac{\widehat{\sigma_\tau^2}}{\widehat{\sigma_\tau^2} + \widehat{\sigma^2}} \qquad (10.5)$$

where $\widehat{\sigma^2}$ and $\widehat{\sigma_\tau^2}$ are estimates of the corresponding components of variance in an among- and within-classes analysis. This definition of a sample quantity is based on the definition of the population *intraclass correlation*, namely

$$\rho = \frac{\sigma_\tau^2}{\sigma_\tau^2 + \sigma^2}$$

where σ_τ^2 is the variation in the population of class means. The first paragraph of this section constitutes a justification of the term intraclass correlation rather than a definition. The definition is applicable even when the number of observations varies from class to class. It is of interest to note that this is the second time variances have been used in determining correlation. The ratio of the reduction in sum of squares attributable to an independent variable to a total sum of squares is the square of a correlation coefficient; hence, the ratio of variances is a multiple of r^2.

The value of r_I may go as high as $+1$ but can never fall below $-1/(n-1)$, the case when MS(among classes) $= 0$. The product-moment correlation is an *interclass correlation*. The intraclass correlation is a measure of fraternal resemblance when this term is meaningful.

For the data of Table 10.3,

$$r_I = \frac{16{,}320.5 - 2{,}199.2}{16{,}320.5 + 13(2{,}199.2)} = .314$$

$$= \frac{1{,}008.7}{1{,}008.7 + 2{,}199.2} = .314$$

TABLE 10.3

ANALYSIS OF VARIANCE OF GAIN IN WEIGHT OF TWO LOTS OF HOLSTEIN HEIFERS

Source	df	MS	MS estimates	Component
Treatments	1	16,320.5	$\sigma^2 + 14\sigma_\tau^2$	$\sigma_\tau^2 = 1{,}008.7$
Error	26	2,199.2	σ^2	$\sigma^2 = 2{,}199.2$

A test of significance of an intraclass correlation is now obvious. Test for the presence of σ_r^2 in the analysis of variance; the criterion is F and the numerator and the denominator are the among- and within-classes variances. One usually performs a test of significance before computing r_I.

References

10.1 David, F. N.: *Tables of the Ordinates and Probability Integral of the Distribution of the Correlation Coefficient in Small Samples*, Cambridge University Press, 1938.

10.2 Fisher, R. A.: "On the 'probable error' of a coefficient of correlation deduced from a small sample," *Metron*, **1:** 3–32 (1921).

10.3 Kabat, C., I. O. Buss, and R. K. Meyer: "The use of ovulated follicles in determining eggs laid by ring-necked pheasant," *J. Wildlife Management*, **12:** 399–416 (1948).

Chapter 11

ANALYSIS OF VARIANCE III: FACTORIAL EXPERIMENTS

11.1 Introduction. In this chapter, we deal with factorial experiments, experiments especially designed to give a set of independent comparisons of a type dependent on the choice of treatments.

The topic of independent comparisons based on single degrees of freedom is discussed. The discussion covers equally spaced treatments and regression. Tukey's test for nonadditivity is given.

11.2 Factorial experiments. A *factor* is a kind of treatment and, in factorial experiments, any factor will supply several treatments. For example, if diet is a factor in an experiment, then several diets will be used; if baking temperature is a factor, then baking will be done at several temperatures.

The concept of the factorial experiment can be illustrated by an example. Consider an experiment to evaluate the yielding abilities of several soybean varieties. In a single-factor approach, all variables other than variety are held as uniform as possible, that is, one particular level of each of the other factors is chosen. Suppose that a second factor, distance between rows, is of interest. A two-factor experiment can be planned in which the treatments consist of all combinations of varieties and the chosen row spacings, that is, each variety is present at all row spacings. In a single-factor experiment, all varieties would be planted at only one row spacing or one variety would be planted for all row spacings. In soils, an experiment could be designed to compare all combinations of several rates of phosphorus and potassium fertilizers. In an animal feeding experiment, factors under consideration might be several amounts and kinds of protein supplement.

The term *level* refers to the several treatments within any factor. It is derived from some of the earliest factorial experiments. These dealt with soil fertility where combinations of several amounts, or levels, of different fertilizers were the treatments. Today the word has a more general meaning, implying a particular amount or state of a factor. Thus, if five varieties of a crop are compared using three different management practices, the experiment is called a 5×3 factorial experiment with five levels of the variety factor and three levels of the management factor. The number of factors and levels which may be compared in a single experiment is limited only by practical considerations.

Factorial experiments are used in practically all fields of research. They are of great value in exploratory work where little is known concerning the optimum levels of the factors, or even which ones are important. Consider a new crop for which several promising varieties are available but little is known regarding a suitable date and rate of planting. A three-factor experiment is indicated. If the single-factor approach is used, one date and rate of planting are selected and a varietal experiment is conducted. However, the variety that does best at the chosen date and rate may not be the best for some other date and rate. Any other single factor must involve only one variety. At other times, the experimenter may be primarily interested in the interaction between factors, that is, he may wish to know if the differences in response to the levels of one factor are similar or different at different levels of another factor or factors. Where considerable information is available, the best approach may be to compare only a very limited number of combinations of several factors at specific levels.

Thus, we see that the scope of an experiment, or the population concerning which inferences can be made, often can be increased by the use of a factorial experiment. It is particularly important to do this where information is wanted on some factor for which recommendations are to be made over a wide range of conditions.

Notation and definitions. Systems of notation used with factorial experiments are generally similar but differ enough that the reader has to check carefully as he uses any new reference. We follow a notation similar in most respects to that suggested by Yates (11.18). Capital letters are used to refer to *factors*; for example, if an experiment involves several sprays to control insects and these are applied by several methods, we can denote the spray factor by A and the method factor by B. Combinations of lower-case letters and numerical subscripts are used to denote *treatment combinations* and *means;* for example, a_1b_3 refers to the treatment combination consisting of the lowest level of A and the third level of B, and to the corresponding treatment mean. (*Warning*. Most writers denote successively increasing levels by the subscripts 0, 1, 2, etc.)

For a two-factor experiment with two levels of each factor, that is, for a 2×2 or 2^2 factorial, any of the following notations is adequate. The first and third are readily extended to many factors and levels whereas the second is readily extended to many factors but to only two levels of each.

Factor				A				
		Complete form			Abbreviated forms†			
	Level	a_1	a_2	a_1	a_2	a_1	a_2	
B	b_1	a_1b_1	a_2b_1	(1)	a	00	10	
	b_2	a_1b_2	a_2b_2	b	ab	01	11	

† Not used elsewhere in this text.

The three degrees of freedom and sum of squares for the variance among the four treatment means in a 2^2 factorial can be partitioned into single independent degrees of freedom and corresponding sums of squares whose general interpretation is meaningful and relatively simple. For the general factorial experiment, degrees of freedom and sums of squares are partitioned to give independent and meaningful sets of contrasts but not necessarily with single degrees of freedom. The principles involved in the partitioning are best illustrated by the use of a table. Table 11.1 is an illustration for a 2^2 factorial.

In Table 11.1, let the numbers be the average or mean measured responses to the treatment combinations indicated by the row and column headings; the means are for all replicates.

The four differences, $a_2 - a_1$ at each level of B and $b_2 - b_1$ at each level of A, are termed *simple effects*. They are not included in the usual summarization of a factorial experiment but are useful in interpreting the summary as

TABLE 11.1

ILLUSTRATION OF SIMPLE EFFECTS, MAIN EFFECTS, AND INTERACTIONS

I

Factor			A = kind			
	Level	a_1	a_2		Mean	$a_2 - a_1$
B = rate	b_1	30	32		31	2
	b_2	36	44		40	8
	Mean	33	38		35.5	5
	$b_2 - b_1$	6	12		9	

II

Factor			A = kind			
	Level	a_1	a_2		Mean	$a_2 - a_1$
B = rate	b_1	30	32		31	2
	b_2	36	26		31	−10
	Mean	33	29		31	−4
	$b_2 - b_1$	6	−6		0	

III

Factor			A = kind			
	Level	a_1	a_2		Mean	$a_2 - a_1$
B = rate	b_1	30	32		31	2
	b_2	36	38		37	2
	Mean	33	35		34	2
	$b_2 - b_1$	6	6		6	

ANALYSIS OF VARIANCE III: FACTORIAL EXPERIMENTS

well as in themselves. For the data under I, the simple effect of A at the first level of B is 2; under II, the simple effect of B at the second level of A is -6.

When simple effects are averaged, the results are termed *main effects*. These are denoted by capital letters, as also are factors. The main effect of factor A for the data under I is 5; the main effect of factor B for the data under III is 6. In general for the 2^2 factorial, A and B are given by Eqs. (11.1) and (11.2).

$$A = \tfrac{1}{2}[(a_2b_2 - a_1b_2) + (a_2b_1 - a_1b_1)] \qquad (11.1)$$
$$B = \tfrac{1}{2}[(a_2b_2 - a_2b_1) + (a_1b_2 - a_1b_1)] \qquad (11.2)$$

Main effects are seen to be computed on a per-unit basis. Equations (11.1) and (11.2) are easily extended to 2^n factorials.

Main effects are averages over a variety of conditions just as are any other treatment means. For a factorial experiment in a randomized complete-block or Latin-square design, the variety of conditions exists within blocks as well as among blocks; thus factor A is replicated within every block since it is present at both levels for each level of factor B. We have hidden replication. Averaging implies that the differences, that is, the simple effects, vary only because of chance from level to level of the other factor or factors. This is, in fact, a hypothesis that is usually subjected to a test of significance when treatments are factorially arranged; the hypothesis is that of no interaction between factors.

For the data under I and II, the simple effects for both kind and rate differ. Under III, the simple effects for A are the same, as are the simple effects for B; here, they also equal the corresponding main effect. When simple effects for a factor differ by more than can be attributed to chance, this differential response is termed an *interaction* of the two factors. The relation is a symmetric one; that is, the interaction of A with B is the same as that of B with A. From Table 11.1 you will see that the difference between the simple effects of A equals that for B in all three cases; it would be impossible to construct a table otherwise. In our notation, the interaction of A and B is defined in Eq. (11.3).

$$AB = \tfrac{1}{2}[(a_2b_2 - a_1b_2) - (a_2b_1 - a_1b_1)] \qquad (11.3)$$

The value $\tfrac{1}{2}$ is used so that interaction, like the main effects, is on a per-unit basis. For the data under I,

$$AB = \tfrac{1}{2}(8 - 2) \text{ in terms of simple effects of } A$$
$$= \tfrac{1}{2}(12 - 6) \text{ in terms of simple effects of } B$$
$$= 3$$

For the data under II, we find

$$AB = \tfrac{1}{2}(26 - 36 - 32 + 30) = -6$$

and under III,

$$AB = \tfrac{1}{2}(38 - 36 - 32 + 30) = 0$$

Notice that the interaction is also one-half the difference between the sums of the two diagonals of the 2×2 table, which is one-half the difference between the sums of the treatments where both A and B are present at the higher and lower levels and of the treatments where only one is present at the higher level. This is always true of the 2^2 factorial.

Interaction measures the failure of the A effect, or the response to A, to be the same for each level of B or, conversely, the failure of the B effect to be the same for each level of A. Under I, the simple effects for kind are 2 and 8 while the main effect is 5. Interaction may be defined as a measure of the departure of the simple effects from an additive law or model based on main effects only.

Under I, Table 11.1, the response to A or the increase from a_1 to a_2 is greater for b_2 than for b_1, that is, there has been a change in the magnitude of the

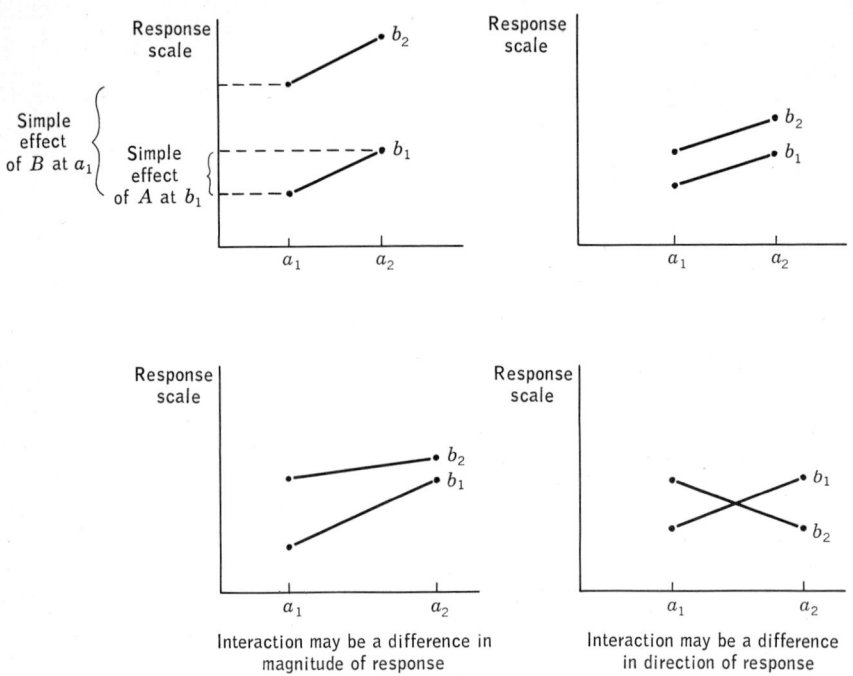

FIG. 11.1 Illustration of interaction

increase. Under II, the response to A is an increase in the presence of b_1 and a decrease in the presence of b_2; there has been a change in the direction of the increase. In terms of treatment means presented in a two-way table, sufficiently large changes in the magnitudes of the differences between treatment means in a column (or row), as one goes from column to column (or row to row), constitute an interaction. Also, changes in the rank of any treatment mean for a column (or row), as one changes columns (or rows), may constitute an interaction.

Figure 11.1 illustrates graphically what is meant by an interaction. The presence or absence of main effects tells us nothing about the presence or absence of interaction. The presence or absence of interaction tells us nothing about the presence or absence of main effects but does tell us something about the homogeneity of simple effects.

A significant interaction is one that is too large to be explained on the basis of chance and the null hypothesis of no interaction. With a significant interaction, the factors are not independent of one another; the simple effects of a factor differ and the magnitude of any simple effect depends upon the level of the other factor of the interaction term. Where factors interact, a single-factor experiment will lead to disconnected and possibly misleading information.

If the interaction is nonsignificant, it is concluded that the factors under consideration act independently of each other; the simple effects of a factor are the same for all levels of the other factors, within chance variation as measured by experimental error. Where factors are independent, the factorial experiment saves considerable time and effort. This is so since the simple effects are equal to the corresponding main effects and a main effect, in a factorial experiment, is estimated as accurately as if the entire experiment had been devoted to that factor.

In the analysis of a factorial experiment, it is correct to partition the treatment degrees of freedom and sum of squares into the components attributable to main effects and interactions even when the over-all F test of no differences among treatments is not significant; main effects and interaction comparisons are planned comparisons. It is easy to visualize a situation where one factor, say B, does neither good nor harm in itself nor has any effect on A, hence contributes no more to the treatment sum of squares than can be attributed to chance; a significant response to A might well be lost in an over-all test of significance such as F. Thus in a 2^2 factorial with 3 degrees of freedom for treatments, the sum of squares for the real A effect with its 1 degree of freedom may easily be lost when averaged with the sums of squares for the nonreal B and AB with their 2 degrees of freedom. Calculation of the sum of squares for treatments is more often used as part of a computational procedure than to supply the numerator of an F test.

All units of the factorial experiment are involved in measuring any one main effect or interaction. This is apparent from Eqs. (11.1), (11.2), and (11.3). Thus it appears that, in a factorial experiment as opposed to several single-factor experiments, all units are devoted to measuring A, and in turn to B, and to AB, and so on when there are more factors; nothing is lost either in replication or in measuring main effects; something is gained in that any one factor is measured at various levels of other factors and, in terms of treatment comparisons, in that we are able to measure interactions and test hypotheses concerning them.

The results of a factorial experiment lend themselves to a relatively simple explanation because of the variety and nature of treatment comparisons. If the factors are largely independent, the table of treatment means and analysis of variance summarize the data well. When the factors are not independent, the data require a detailed study with the possibility of further experimentation. Here the difficulty lies in the complex nature of the situation and not in the factorial approach to it. The factorial experiment has indicated the complexity, a fact that might well have been missed had a single-factor approach been used.

11.3 The 2 × 2 factorial experiment, an example. Wilkinson (11.17) reports the results of an experiment to study the influence of time of bleeding, factor A, and diethylstilbestrol (an estrogenic compound), factor B, on plasma phospholipid in lambs. Five lambs were assigned at random to each of four treatment groups; treatment combinations are for morning and afternoon times of bleeding with and without diethylstilbestrol treatment. The data are shown in Table 11.2.

TABLE 11.2

THE INFLUENCE OF TIME OF BLEEDING AND DIETHYLSTILBESTROL ON PHOSPHOLIPID IN LAMBS

Treatment groups

	$a_1 = $ AM		$a_2 = $ PM		Totals
	$a_1b_1 = $ control 1	$a_1b_2 = $ treated 1	$a_2b_1 = $ control 2	$a_2b_2 = $ treated 2	
	8.53	17.53	39.14	32.00	
	20.53	21.07	26.20	23.80	
	12.53	20.80	31.33	28.87	
	14.00	17.33	45.80	25.06	
	10.80	20.07	40.20	29.33	
ΣX	66.39	96.80	182.67	139.06	484.92
ΣX^2	963.88	1,887.02	6,913.63	3,912.17	13,676.70
$\bar{x}$	13.28	19.36	36.53	27.81	24.25

Treatment totals

Factor		$A = $ time		
	Level	$(a_1) = $ AM	$(a_2) = $ PM	Totals
$B = $ estrogen	$(b_1) = $ control	66.39	182.67	249.06
	$(b_2) = $ treated	96.80	139.06	235.86
	Totals	163.19	321.73	484.92

The table of treatment totals is used in computing sums of squares for testing hypotheses concerning main effects and interactions. Treatment totals are distinguished from treatment means by parentheses about the treatment combination symbol. Thus (a_2b_2) is the sum over all replicates of the observations made on treatment combination a_2b_2 while (a_1) is the sum over all replicates of observations made on treatment combinations a_1b_1 and a_1b_2. For example, for the factorial effect of A, we have the total

$$(A) = [(a_2b_2) - (a_1b_2) + (a_2b_1) - (a_1b_1)] \qquad (11.4)$$

The computing proceeds as follows. Let r, a, and b represent the number of replicates (number of observations per treatment combination), levels of A and levels of B. Then the number of treatments is ab. The design is completely random.

ANALYSIS OF VARIANCE III: FACTORIAL EXPERIMENTS

Step 1. Compute the analysis of variance without regard to the factorial arrangement of treatments for the experimental design used. We obtain

$$
\begin{aligned}
\text{Correction term} = C &= 11{,}757.37 \\
\text{Total SS} &= 1{,}919.33 \\
\text{Treatment SS} &= 1{,}539.41 \\
\text{Error SS} &= 379.92
\end{aligned}
$$

Step 2. From the table of treatment totals compute the sums of squares for main effects and interaction as follows:

$$\text{SS}(A) = \frac{\sum_i (a_i)^2}{rb} - C$$

$$= \frac{163.19^2 + 321.73^2}{5(2)} - \frac{(484.92)^2}{5(4)} = 1{,}256.75$$

or, for the 2 × 2 factorial

$$\text{SS}(A) = \frac{(A)^2}{2rb} \tag{11.5}$$

$$= \frac{(321.73 - 163.19)^2}{2(5)2}$$

$$\text{SS}(B) = \frac{\sum_j (b_j)^2}{ra} - C$$

$$= \frac{249.06^2 + 235.86^2}{5(2)} - C = 8.71$$

or, for the 2 × 2 factorial,

$$\text{SS}(B) = \frac{(B)^2}{2ra} \tag{11.6}$$

$$= \frac{(249.06 - 235.86)^2}{2(5)2}$$

$$\text{SS}(AB) = \text{trt SS} - \text{SS}(A) - \text{SS}(B)$$
$$= 1{,}539.41 - 1{,}256.75 - 8.71 = 273.95$$

or, for the 2 × 2 factorial,

$$\text{SS}(AB) = \frac{(AB)^2}{rab} \tag{11.7}$$

$$= \frac{[139.06 + 66.39 - (182.67 + 96.80)]^2}{5(2)2}$$

The results are transferred to an analysis of variance table such as Table 11.3 where the degrees of freedom for the general case are also shown.

TABLE 11.3

ANALYSIS OF VARIANCE FOR DATA OF TABLE 11.2

Source	df	SS	Mean square	F
Treatments	$(ab - 1 = 3)$	$(1{,}539.41)$		
A	$a - 1 = 1$	$1{,}256.75$	$1{,}256.75$	53^{**}
B	$b - 1 = 1$	8.71	8.71	<1
AB	$(a-1)(b-1) = 1$	273.95	273.95	11.5^{**}
Error	$ab(r-1) = 16$	379.92	23.75	
Total	$rab - 1 = 19$	$1{,}919.33$		

The significant interaction indicates that the factors are not independent; the difference between simple effects of A for the two levels of B is significant and, conversely, the difference in simple effects of B at the two levels of A is significant. In other words, the difference in measurements between times of bleeding differs for the control and treated groups or, the same thing, the difference in measurements between the treated and control animals differs for the two times of bleedings. Thus, any simple effect is dependent upon the level of the other factor in the experiment. We have

$$(AB) = [(a_2b_2) - (a_1b_2)] - [(a_2b_1) - (a_1b_1)]$$
$$= [139.06 - 96.80] - [182.67 - 66.39] = 74.02$$
$$= [(a_2b_2) - (a_2b_1)] - [(a_1b_2) - (a_1b_1)]$$
$$= [139.06 - 182.67] - [96.80 - 66.39] = 74.02$$

where parentheses indicate a total over replicates and brackets simply call attention to the simple effects referred to in the text. Figure 11.2 shows the nature of the interaction.

As a result of the significant AB interaction, the investigator may decide to examine the simple effects. Sums of squares for the simple effects are calculated as follows:

$$\text{SS}(A \text{ within } b_1) = \frac{(182.67 - 66.39)^2}{2 \times 5} = 1{,}352.10$$

$$\text{SS}(A \text{ within } b_2) = \frac{(139.06 - 96.80)^2}{2 \times 5} = 178.59$$

Note that the sum of these equals that for A and AB, that is

$$1{,}352.10 + 178.59 = 1{,}530.69 \text{ against } 1{,}530.70 = 1{,}256.75 + 273.95$$

ANALYSIS OF VARIANCE III: FACTORIAL EXPERIMENTS

Also

$$\text{SS }(B \text{ within } a_1) = \frac{(96.80 - 66.39)^2}{2 \times 5} = 92.48$$

$$\text{SS }(B \text{ within } a_2) = \frac{(139.06 - 182.67)^2}{2 \times 5} = 190.18$$

Here the sum equals that for B and AB, that is

$$92.48 + 190.18 = 282.66 = 8.71 + 273.95$$

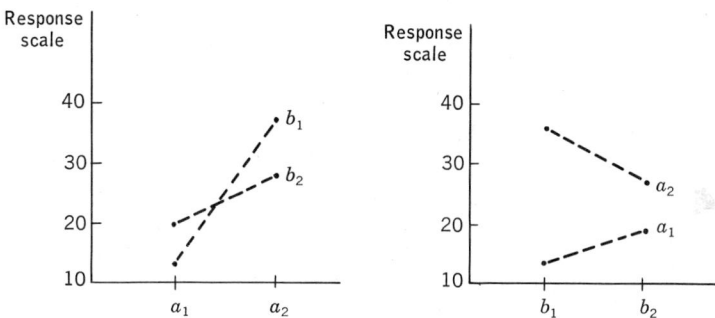

This interaction is seen to involve a change in direction as well as in magnitude of response

FIG. 11.2   Interaction for the data of Table 11.2

The results may be presented in an auxiliary treatment sums of squares table. For example:

Treatment comparison	df	Mean square	F
Between times within control	$a - 1 = 1$	1,352.10	57**
Between times within treated	$a - 1 = 1$	178.59	7.5*
Between estrogen levels, AM	$b - 1 = 1$	92.48	3.9
Between estrogen levels, PM	$b - 1 = 1$	190.18	8.0*

Where there is only one degree of freedom for each comparison, there seems little point in presenting sums of squares unless they are made part of a table such as Table 11.3. Notice that the auxiliary table contains 4 degrees of freedom for treatment comparisons whereas there are only 3 degrees of freedom among treatment means. Our four comparisons cannot be independent as are the main effects and interaction comparisons of Table 11.3. These F tests are essentially *lsd* comparisons but made after the significant interaction showed evidence of real differences attributable to treatments and evidence of the nature of these differences. Severe criticism of these tests would seem unjustified.

Exercise 11.3.1   A. Wojta of the Departments of Agricultural Engineering and Soils, University of Wisconsin, determined the draft, in 10-lb units, required to plow

TABLE 11.4

Number† of Plants Emerged for Three Legume Species A, Planted ½ in. Deep in Three Soil Types B, with Seed Treated and Not Treated with a Fungicide C

Species = A	Fungicide = C	Soil type = B			Total = $b_1 + b_2 + b_3$
		Silt loam = b_1	Sand = b_2	Clay = b_3	
Alfalfa = a_1	None = c_1	266 = $(a_1b_1c_1)$	286 = $(a_1b_2c_1)$	66 = $(a_1b_3c_1)$	618 = (a_1c_1)
	Treated = c_2	276 = $(a_1b_1c_2)$	271 = $(a_1b_2c_2)$	215 = $(a_1b_3c_2)$	762 = (a_1c_2)
	Total = $c_1 + c_2$	542 = (a_1b_1)	557 = (a_1b_2)	281 = (a_1b_3)	1,380 = (a_1)
Red clover = a_2	None = c_1	252 = $(a_2b_1c_1)$	289 = $(a_2b_2c_1)$	167 = $(a_2b_3c_1)$	708 = (a_2c_1)
	Treated = c_2	275 = $(a_2b_1c_2)$	292 = $(a_2b_2c_2)$	203 = $(a_2b_3c_2)$	770 = (a_2c_2)
	Total = $c_1 + c_2$	527 = (a_2b_1)	581 = (a_2b_2)	370 = (a_2b_3)	1,478 = (a_2)
Sweet clover = a_3	None = c_1	152 = $(a_3b_1c_1)$	197 = $(a_3b_2c_1)$	52 = $(a_3b_3c_1)$	401 = (a_3c_1)
	Treated = c_2	178 = $(a_3b_1c_2)$	219 = $(a_3b_2c_2)$	121 = $(a_3b_3c_2)$	518 = (a_3c_2)
	Total = $c_1 + c_2$	330 = (a_3b_1)	416 = (a_3b_2)	173 = (a_3b_3)	919 = (a_3)
Total = $a_1 + a_2 + a_3$	None = c_1	670 = (b_1c_1)	772 = (b_2c_1)	285 = (b_3c_1)	1,727 = (c_1)
	Treated = c_2	729 = (b_1c_2)	782 = (b_2c_2)	539 = (b_3c_2)	2,050 = (c_2)
	Total = $c_1 + c_2$	1,399 = (b_1)	1,554 = (b_2)	824 = (b_3)	3,777 = G

† Each value is a total of three replicates of 100 seeds each.

ANALYSIS OF VARIANCE III: FACTORIAL EXPERIMENTS

wet mud where the tractor traveled at 2 mph. The data are given below for one of three positions of the right wheel, two tire sizes of the left wheel, and two hitches. The original experiment was a $3 \times 2 \times 2$ factorial in a completely random design with three observations per treatment.

Hitch, in.	Right wheel straight	
	Left wheel tire	
	6.50 × 16	7.50 × 16
2	76, 76, 81	76, 76, 75
4	69, 74, 79	50, 63, 62

Compute the analysis of variance for these data; compute main effects and interaction sums of squares by Eqs. (11.5), (11.6), and (11.7). Interpret the results.

Exercise 11.3.2 The data of Table 7.8 are from a 2×3 factorial experiment with sampling. Partition the treatment sum of squares into main effects and interaction components. Note that Eqs. (11.5), (11.6), and (11.7) are not applicable to B and AB since each of B and AB involves 2 degrees of freedom. Interpret the results.

11.4 The $3 \times 3 \times 2$ or $3^2 \times 2$ factorial, an example. The data in Table 11.4 are the results of a greenhouse experiment conducted by Wagner (11.15) to determine the rate of emergence of seed of three species of legumes, treated and not treated with a fungicide, and planted in three soil types. (These data are from an experiment with a fourth factor, depth of planting, at three levels.) The layout was a randomized complete-block design. The data of Table 11.4 are in a $3 \times 3 \times 2$ or $3^2 \times 2$ factorial, each datum being a sum over blocks.

The raw data are arranged in a two-way table similar to Table 8.2 with treatments listed down the side and blocks across the top. The initial analysis of variance is then carried out as described in Sec. 8.3. In addition, the 18 treatment totals are entered in Table 11.4 and various subtotals are obtained. This table is necessary for the computation of the sums of squares of the main effects and interactions.

In the computations that follow, a, b, c, and r represent the number of levels of factors A, B, C, and the number of blocks or replicates. Previously described symbols are generalized. Thus in a three-factor experiment, $(a_1b_2c_1)$ is the total of the r observations on the treatment combination with A and C at their lowest levels and B at its second level; (a_1b_2) is the total of the rc observations made on units for which A was at the lowest level and B at the second level; and so on.

Step 1. Compute

$$\begin{aligned}\text{Correction term} = C &= 264{,}180.17\\ \text{Total SS} &= 35{,}597.67\\ \text{Block SS} &= 356.77\\ \text{Treatment SS} &= 32{,}041.50\\ \text{Error SS} &= 3{,}199.40\end{aligned}$$

Step 2. The treatment sum of squares is partitioned into components attributable to main effects and interactions. In most cases, these will involve more than single degrees of freedom. Definitions such as those of Eqs. (11.1), (11.2), and (11.3) will not be applicable, nor will the computing formulas given as Eqs. (11.5), (11.6), and (11.7). We must use standard sums of squares formulas. Thus,

$$\text{SS}(A) = \frac{\sum\limits_{i}(a_i)^2}{rbc} - C \tag{11.8}$$

$$= \frac{1,380^2 + 1,478^2 + 919^2}{3(3)2} - C = 9,900.11$$

$$\text{SS}(B) = \frac{\sum\limits_{j}(b_j)^2}{rac} - C$$

$$= \frac{1,399^2 + 1,554^2 + 824^2}{3(3)2} - C = 16,436.11$$

$$\text{SS}(C) = \frac{\sum\limits_{k}(c_k)^2}{rab} - C = \frac{[(c_2) - (c_1)]^2}{2rab}$$

$$= \frac{1,727^2 + 2,050^2}{3(3)3} - C = \frac{(2,050 - 1,727)^2}{2(3)3(3)} = 1,932.02$$

$$\text{SS}(AB) = \frac{\sum\limits_{i,j}(a_i b_j)^2}{rc} - C - \text{SS}(A) - \text{SS}(B) \tag{11.9}$$

$$= \frac{542^2 + \cdots + 173^2}{3(2)} - C - (9,900.11 + 16,436.11) = 658.44$$

$$\text{SS}(AC) = \frac{\sum\limits_{i,k}(a_i c_k)^2}{rb} - C - \text{SS}(A) - \text{SS}(C)$$

$$= \frac{618^2 + \cdots + 518^2}{3(3)} - C - (9,900.11 + 1,932.02) = 194.03$$

$$\text{SS}(BC) = \frac{\sum\limits_{j,k}(b_j c_k)^2}{ra} - C - \text{SS}(B) - \text{SS}(C)$$

$$= \frac{670^2 + \cdots + 539^2}{3(3)} - C - (16,436.11 + 1,932.02) = 1,851.14$$

$$\text{SS}(ABC) = \frac{\sum\limits_{i,j,k}(a_i b_j c_k)^2}{r} - C - \text{SS}(A) - \text{SS}(B) - \text{SS}(C) - \text{SS}(AB)$$

$$- \text{SS}(AC) - \text{SS}(BC) \tag{11.10}$$

$$= \frac{266^2 + \cdots + 121^2}{3} - C - (9,900.11 + 16,436.11 + 1,932.02$$

$$+ 658.44 + 194.03 + 1,851.14) = 1,069.65$$

ANALYSIS OF VARIANCE III: FACTORIAL EXPERIMENTS

Observe that sums of squares for interactions AB, AC, and BC are no more than residuals from two-way tables. Similarly, SS(ABC) is a residual from a three-way table. The computer may decide to construct the two-way tables or to mark appropriate totals distinctively, as with colored pencils.

The results are presented in an analysis of variance table such as Table 11.5. Here we have not shown the total sum of squares for treatments though this would have been quite correct.

TABLE 11.5

ANALYSIS OF VARIANCE OF DATA IN TABLE 11.4

Source	df	SS	Mean square	F
Blocks	$(r-1) = 2$	356.77	178.39	1.90
A = species	$a-1 = 2$	9,900.11	4,500.06	47.82**
B = soil type	$b-1 = 2$	16,436.11	8,218.06	87.33**
C = fungicide	$c-1 = 1$	1,932.02	1,932.02	20.53**
AB	$(a-1)(b-1) = 4$	658.44	164.61	1.75
AC	$(a-1)(c-1) = 2$	194.03	97.02	1.03
BC	$(b-1)(c-1) = 2$	1,851.14	925.57	9.84**
ABC	$(a-1)(b-1)(c-1) = 4$	1,069.65	267.41	2.84*
Error	$(r-1)(abc-1) = 34$	3,199.40	94.10	
Total	$abcr-1 = 53$	35,597.67		

Interactions involving two factors are called *two-factor* or *first-order interactions*, for example, AB, AC, BC. Interactions with three factors are *three-factor* or *second-order interactions*.

For these data, the analysis of variance shows that all three main effects and interactions BC and ABC are significant. The significant BC interaction implies that the differences between responses to C vary with the level of B, where responses are measured over all levels of A. Alternately, the differences among responses to levels of B vary for the two levels of C, where responses are again measured as totals over all levels of A. Specifically, the differences in emergence rates, when averaged over all species, between treated and untreated seed are not the same for the three soil types; or the differences among emergence rates of seed grown in three soil types are not the same for treated as for untreated seed.

The significant ABC interaction is more difficult to interpret. It can be considered in three ways; namely, as an interaction of the interaction AB with factor C, of the interaction AC with factor B, or of the interaction BC with factor A. Here the BC interaction is not consistent for the levels of A, and so on. The way to handle this will depend upon which approach is most meaningful and upon the significance of the two-factor interactions.

For these data, since the BC and ABC interactions are significant, it seems logical to begin by examining the BC interaction. Since C is at two levels only, let us begin by considering the simple effects of C at the various levels of B. Apparently, the simple effects of C are not homogeneous for the three soil types. Hence let us examine the two-way table, contained in Table 11.4, used in computing the BC interaction. This is presented as Table 11.6 where the responses to fungicide are compared for the various soil types. The difference 254 stands out immediately.

TABLE 11.6

EXAMINATION OF THE BC INTERACTION FOR THE DATA OF TABLE 11.4

		Fungicide		
		c_1	c_2	$c_2 - c_1$
Soil type	b_1	670	729	59
	b_2	772	782	10
	b_3	285	539	254
	$b_1 + b_2 + b_3$	1,727	2,050	

$$C \text{ within } b_1 \text{ SS} = \frac{[(b_1 c_2) - (b_1 c_1)]^2}{2ra}$$

$$= \frac{(729 - 670)^2}{2(3)3} = 193.39 \text{ ns}$$

$$C \text{ within } b_2 \text{ SS} = \frac{[(b_2 c_2) - (b_2 c_1)]^2}{2ra}$$

$$= \frac{(782 - 772)^2}{2(3)3} = 5.56 \text{ ns}$$

$$C \text{ within } b_3 \text{ SS} = \frac{[(b_3 c_2) - (b_3 c_1)]^2}{2ra}$$

$$= \frac{(539 - 285)^2}{2(3)3} = 3,584.22**$$

Each sum of squares has a single degree of freedom and their sum is the same as for the sum for C and BC, that is, $193.39 + 5.56 + 3,584.22 = 3,783.17 = 1,932.02 + 1,851.14$. We find that the difference in emergence rates of treated and untreated seed, averaged over three species, is not significant for silt loam or for sand but is for clay.

Because of the difficulties of interpreting significant interactions in the analysis of variance, we have chosen to examine certain simple effects at the expense of a loss of replication; each of the simple effects of C is measured over only a single level of B.

The significant ABC interaction implies that the BC interaction differs with the level of A. We take this point of view in looking at ABC since BC is significant. From having looked at the BC interaction, it seems reasonable to conclude that the difficulty is tied to C on clay soil and we proceed directly to an examination of the simple effects of C for the clay soil type b_3 at the various levels of A. A glance at the three soil type times fungicide tables contained within Table 11.4 seems to justify this approach. Thus, we look at the AC interaction on clay soil. The appropriate two-way table and computations are shown in Table 11.7.

TABLE 11.7
EXAMINATION OF AN AC INTERACTION, THAT FOR b_3, FOR THE DATA OF TABLE 11.4

		Clay $= b_3$		
		c_1	c_2	$c_2 - c_1$
Alfalfa	$= a_1$	66	215	149
Red clover	$= a_2$	167	203	36
Sweet clover	$= a_3$	52	121	69

$$C \text{ within } a_1 \text{ for } b_3 \ \ SS = \frac{[(a_1 b_3 c_2) - (a_1 b_3 c_1)]^2}{2r}$$

$$= \frac{(215 - 66)^2}{2(3)} = 3{,}700.17**$$

$$C \text{ within } a_2 \text{ for } b_3 \ \ SS = \frac{[(a_2 b_3 c_2) - (a_2 b_3 c_1)]^2}{2r}$$

$$= \frac{(203 - 167)^2}{2(3)} = 216.00 \ ns$$

$$C \text{ within } a_3 \text{ for } b_3 \ \ SS = \frac{[(a_3 b_3 c_2) - (a_3 b_3 c_1)]^2}{2r}$$

$$= \frac{(121 - 52)^2}{2(3)} = 793.50**$$

Some of the more important conclusions to be drawn from this experiment follow. No difference was found between the emergence rates of treated seed c_2 and untreated seed c_1, when averaged over all species for silt loam b_1, and sand soils b_2; however, the difference was significant in favor of treated seed for the clay soil b_3. Since the three-factor interaction was significant, a further analysis was made and indicated that for clay soil, treated seed of alfalfa a_1 and sweet clover a_3 emerged better than untreated seed whereas no difference was found for red clover a_2. The differences between emergence rates for treated and untreated seed, for each of the three species, were not significant for silt loam and sand soils.

Exercise 11.4.1 Sketch the BC interaction of Table 11.6 and the BC interactions for each level of A. This will add to your appreciation of the nature of a three-factor interaction.

Table 11.8
Average Values of Mean Squares for Factorial Experiments: The Three-factor Experiment

Source	df	Average value of mean square	
		Model I (fixed)	Model II (random)
Blocks	$r - 1$	$\sigma^2 + abc\Sigma\rho^2/(r-1)$	$\sigma^2 + abc\sigma_\rho^2$
A	$a - 1$	$\sigma^2 + rbc\Sigma\alpha^2/(a-1)$	$\sigma^2 + r\sigma_{ABC}^2 + rc\sigma_{AB}^2 + rb\sigma_{AC}^2 + rbc\sigma_A^2$
B	$b - 1$	$\sigma^2 + rac\Sigma\beta^2/(b-1)$	$\sigma^2 + r\sigma_{ABC}^2 + rc\sigma_{AB}^2 + ra\sigma_{BC}^2 + rac\sigma_B^2$
C	$c - 1$	$\sigma^2 + rab\Sigma\gamma^2/(c-1)$	$\sigma^2 + r\sigma_{ABC}^2 + rb\sigma_{AC}^2 + ra\sigma_{BC}^2 + rab\sigma_C^2$
AB	$(a-1)(b-1)$	$\sigma^2 + rc\Sigma(\alpha\beta)^2/(a-1)(b-1)$	$\sigma^2 + r\sigma_{ABC}^2 + rc\sigma_{AB}^2$
AC	$(a-1)(c-1)$	$\sigma^2 + rb\Sigma(\alpha\gamma)^2/(a-1)(c-1)$	$\sigma^2 + r\sigma_{ABC}^2 + rb\sigma_{AC}^2$
BC	$(b-1)(c-1)$	$\sigma^2 + ra\Sigma(\beta\gamma)^2/(b-1)(c-1)$	$\sigma^2 + r\sigma_{ABC}^2 + ra\sigma_{BC}^2$
ABC	$(a-1)(b-1)(c-1)$	$\sigma^2 + r\Sigma(\alpha\beta\gamma)^2/(a-1)(b-1)(c-1)$	$\sigma^2 + r\sigma_{ABC}^2$
Error	$(r-1)(abc-1)$	σ^2	σ^2

Mixed model; A and B fixed, C random

Source		
Blocks	$\sigma^2 + abc\sigma_\rho^2$	
A	$\sigma^2 + rb\dfrac{a}{a-1}\sigma_{AC}^2 + rbc\Sigma\alpha^2/(a-1)$	
B	$\sigma^2 + ra\dfrac{b}{b-1}\sigma_{BC}^2 + rac\Sigma\beta^2/(b-1)$	
C	$\sigma^2 + rab\sigma_C^2$	
AB	$\sigma^2 + r\dfrac{a}{a-1}\dfrac{b}{b-1}\sigma_{ABC}^2 + rc\Sigma(\alpha\beta)^2/(a-1)(b-1)$	
AC	$\sigma^2 + rb\dfrac{a}{a-1}\sigma_{AC}^2$	
BC	$\sigma^2 + ra\dfrac{b}{b-1}\sigma_{BC}^2$	
ABC	$\sigma^2 + r\dfrac{a}{a-1}\dfrac{b}{b-1}\sigma_{ABC}^2$	

Exercise 11.4.2 The following data and those of Exercise 11.3.1 constitute all the data for a 3 × 2 × 2 factorial experiment. Compute the analysis of variance for the 3 × 2 × 2 experiment. Interpret the results.

Hitch, in.	Right wheel toe in 1.29°		Right wheel castered	
	Left wheel tire		Left wheel tire	
	6.50 × 16	7.50 × 16	6.50 × 16	7.50 × 16
2	65, 78, 68	42, 56, 35	74, 85, 79	74, 69, 74
4	77, 60, 82	48, 50, 48	60, 88, 88	66, 62, 57

11.5 Linear models for factorial experiments.

Linear models have been discussed throughout the text. If we look upon the randomized complete-block design as a two-factor experiment, then the linear model has been discussed for some factorial experiments.

Two fundamentally different classes of problems have been raised. Class I problems involve the *fixed effects model*, Model I. Class II problems involve the *random effects model*, Model II.

Many sets of data present a mixture of the two classes of problems and so we have the *mixed model*. Still other models are possible. In any case, the computations will be the same regardless of the model, though the choice of error terms and the type of inference will vary. See also Scheffé (11.10) and Wilk and Kempthorne (11.16).

The average values of the mean squares in a three-factor experiment, in a randomized complete-block design, are given in Table 11.8. Capital letters refer to effects, that is, main effects or interactions; lower-case letters refer to the numbers of levels of the effects designated by the corresponding capital letters. The error variance is σ^2; other variances have subscripts which relate them to the effect concerned. Greek letters refer to the individual components used to describe any particular observation; these are used in average values where effects are fixed since the ones in the experiment constitute the complete population. Subscripts, on Greek letters or combinations of letters, are omitted as a matter of convenience in presenting the table. However, the complete mathematical description for any observation is as follows.

$$X_{ijkl} = \mu + \rho_i + \alpha_j + \beta_k + \gamma_l + (\alpha\beta)_{jk} + (\alpha\gamma)_{jl} + (\beta\gamma)_{kl} + (\alpha\beta\gamma)_{jkl} + \epsilon_{ijkl}$$

It is readily seen from Table 11.8 that, for the fixed model, the error variance is an appropriate term for testing hypotheses about any source of variation in the analysis of variance. However, as we have seen in our examples, a significant interaction may cause us to lose interest in tests of hypotheses concerning main effects and to become interested in other tests such as those of simple effects. Such a shift in emphasis is more likely to lead to a satisfactory interpretation of the data.

For the random model the choice of a suitable error term, when all sources of variation are real, is more difficult when hypotheses concerning main effects are to be tested. Table 11.8 shows that error is appropriate for testing the three-factor interaction; if σ^2_{ABC} is real, the ABC mean square is appropriate for testing the two-factor interactions. For tests of main effects, some pooling of mean squares is necessary. For example, the sum of the mean squares for ABC and C differs from the sum for AC and BC only in that the former contains $rab\sigma_C^2$ as an additional component. Thus a test of σ_C^2 is implied. For the distribution of such a test criterion and the appropriate degrees of freedom, the reader is referred to Satterthwaite (11.9) and Cochran (11.3).

Let us consider how average values are obtained for the random model. A convenient rule for doing so is given by Crump (11.4) and rules for more general situations, including mixed models, are given by Schultz (11.11).

Rule 1. For the random model, any effect will have, in the average value of its mean square, a linear combination of σ^2 and those variances, but no others, whose subscripts contain all the letters of the effect. For example, the average value of the mean square for AC will include σ^2, σ_{ABC}^2, and σ_{AC}^2. The coefficients of the variances are: 1 for σ^2 and, for any other variance, the product of the number of replicates (blocks in Table 11.8) and all the small letters corresponding to the capital letters not in the subscript. For example, for AC we have σ^2, $r\sigma_{ABC}^2$, and $rb\sigma_{AC}^2$. Notice that the complete set of letters used for factors appears with each variance other than σ^2, either as a coefficient (lower case) or a subscript (capital).

Rule 2. For the mixed model, begin by finding average values of mean squares for the random model and then delete certain variances and replace others by mean squares of population effects. The component with the same subscript as the name of the effect is always present in the effect mean square. In the mean square for any effect, consider any other component. In any subscript, ignore any letter which is used in naming the effect; if any other letter of a subscript corresponds to a fixed effect, cross out the variance component. For example, in Table 11.8 for the mixed model and opposite A, we have to consider σ_{ABC}^2, σ_{AB}^2, and σ_{AC}^2; in each subscript, ignore A; for σ_{ABC}^2 and σ_{AB}^2, B is fixed so both variances are crossed out; for σ_{AC}^2, C is random so σ_{AC}^2 is not crossed out; finally, since A is fixed, σ_A^2 is replaced by $\Sigma \alpha^2/(a-1)$. Again, for AC, we look at B only; B is fixed so we cross out σ_{ABC}^2; since C is random, AC is also random and σ_{AC}^2 is left as a variance.

Rule 3. (Mixed model only). After application of rule 2, if any variance left in an average value has, in its subscript, one or more letters corresponding to fixed effects, then the coefficient of the variance requires a factor for each fixed effect. The factor is the ratio of the number of levels of the fixed effect to one less than the number of levels. For example, in Table 11.8 for the mixed model and opposite A, σ_{AC}^2 has A as a fixed effect. Consequently, the coefficient requires the factor $a/(a-1)$. [Some workers do not use this factor; for example, Schultz (11.11)].

Divisors like $a-1$ occur with finite populations, as when an effect is fixed. Thus they also occur with $\Sigma \alpha^2$, a sum of a quantities. Consider the set of $(\alpha\gamma)$'s, A fixed and C random, that have been sampled in the mixed model of Table 11.8. The set is of the form

$$\begin{array}{cccc} (\alpha\gamma)_{11} & (\alpha\gamma)_{12} & \cdots & (\alpha\gamma)_{1j} & \cdots \\ (\alpha\gamma)_{21} & (\alpha\gamma)_{22} & \cdots & (\alpha\gamma)_{2j} & \cdots \\ \cdots\cdots\cdots\cdots\cdots\cdots\cdots\cdots\cdots \\ (\alpha\gamma)_{a1} & (\alpha\gamma)_{a2} & \cdots & (\alpha\gamma)_{aj} & \cdots \end{array}$$

ANALYSIS OF VARIANCE III: FACTORIAL EXPERIMENTS

The sampling procedure calls for a random sample of columns across all rows. Now it is customary to measure effects as deviations. For a row of this table, the deviations will be measured from a population mean chosen to be zero so that the sum of the deviations will be zero. Since we obtain a sample of columns, the deviations which get into our sample will not likely sum to zero in any row. For a column of this table, the deviations are from a finite population of a deviations whose mean is zero in all cases. Since we obtain complete columns, the deviations which get into our sample must equal zero in any column. It is this latter restriction on the population of deviations which results in the divisor $(a - 1)$. When the divisor is $(a - 1)$, then the numerator will involve $a\sum_1^a$ or $a\sigma^2$, where a multiple subscript on σ^2 will include A and at least one other letter corresponding to a random effect. The same is true for other fixed effects.

The rules given in this section are also applicable when a factorial experiment includes sampling, that is, when we have both a factorial and a nested or sampling experiment. Table 11.9 consists of an example. New notation has been introduced to handle this new situation: a subscript for a variance may contain letters in and letters not in parentheses; letters in parentheses indicate the position in the hierarchy at which the component arises. For example, $\sigma^2_{D(C)(AB)}$ is the variance of D within C within AB. Letters in parentheses are not involved in the application of rule 2 as it applies to deletions.

To illustrate the use of the rules with a nested classification, consider the average value of the mean square for A in Table 11.9 for the random model. An A total includes variation due to subsamples so has the component $\sigma^2_{D(C)(AB)}$, variation due to samples so has $\sigma^2_{C(AB)}$, variation due to AB so has σ^2_{AB}, variation due to A so has σ^2_A, but no variation due to B since every level of B appears in each A total. The appropriate coefficient of any σ^2 is composed of the letters (lower-case) that do not appear in the subscript of the coefficient. Hence, the average value is $\sigma^2_{D(C)(AB)} + d\sigma^2_{C(AB)} + cd\sigma^2_{AB} + bcd\sigma^2_A$. Thus far, only rule 1 has been used.

If the model calls for a fixed A effect only, then all other letters refer to random effects so that no component in the average value of the mean square for A will be deleted and the only change will be to replace σ^2_A by $\Sigma\alpha^2/(a-1)$.

If the model calls for both A and B fixed, then in the average value of the mean square for A, we delete σ^2_{AB} because B refers to a fixed effect and is not in parentheses. We obtain the value given in Table 11.9.

If the model calls for A fixed and B random, then there will be a σ^2_{AB} in the average value for B since B is random and appears in the subscript AB. Rule 3 also applies.

In some areas of investigation, equal subclass numbers are the exception rather than the rule. Methods for obtaining average values of mean squares in such cases are given by Henderson (11.8).

Exercise 11.5.1 Decide upon an appropriate model for the data of Tables 7.8, 11.2, 11.4, and Exercise 11.4.2. Make out a table of the average values of the mean squares as called for by your models.

11.6 Single degree of freedom comparisons. An F test with more than one degree of freedom for the numerator mean square is an average test

TABLE 11.9

AVERAGE VALUES OF MEAN SQUARES FOR A FACTORIAL EXPERIMENT INVOLVING SAMPLING

Source†	df	Average value of mean square, random model
A	$a-1$	$\sigma^2_{D(C\mid AB)} + d\sigma^2_{C(AB)} + cd\sigma^2_{AB} + bcd\sigma^2_A$
B	$b-1$	$\sigma^2_{D(C\mid AB)} + d\sigma^2_{C(AB)} + cd\sigma^2_{AB} + acd\sigma^2_B$
AB	$(a-1)(b-1)$	$\sigma^2_{D(C\mid AB)} + d\sigma^2_{C(AB)} + cd\sigma^2_{AB}$
C in AB	$(c-1)ab$	$\sigma^2_{D(C\mid AB)} + d\sigma^2_{C(AB)}$
D in C in AB	$(d-1)abc$	$\sigma^2_{D(C\mid AB)}$

Mixed model

Source	A fixed	A and B fixed
A	$\sigma^2_{D(C\mid AB)} + d\sigma^2_{C(AB)} + cd\,\dfrac{a}{a-1}\sigma^2_{AB} + bcd\,\Sigma\alpha^2/(a-1)$	$\sigma^2_{D(C\mid AB)} + d\sigma^2_{C(AB)} + bcd\,\Sigma\alpha^2/(a-1)$
B	$\sigma^2_{D(C\mid AB)} + d\sigma^2_{C(AB)} + acd\sigma^2_B$	$\sigma^2_{D(C\mid AB)} + d\sigma^2_{C(AB)} + acd\,\Sigma\beta^2/(b-1)$
AB	$\sigma^2_{D(C\mid AB)} + d\sigma^2_{C(AB)} + cd\,\dfrac{a}{a-1}\sigma^2_{AB}$	$\sigma^2_{D(C\mid AB)} + d\sigma^2_{C(AB)} + cd\,\Sigma(\alpha\beta)^2/(a-1)(b-1)$
C in AB	$\sigma^2_{D(C\mid AB)} + d\sigma^2_{C(AB)}$	$\sigma^2_{D(C\mid AB)} + d\sigma^2_{C(AB)}$
D in C in AB	$\sigma^2_{D(C\mid AB)}$	$\sigma^2_{D(C\mid AB)}$

† For example, A might refer to treatment and B to observer. Each observer is required to make an observation at each of a number of times C. (There is to be no learning process or trend in time that would imply that first observations are more alike than any other set, for example, than would be a set involving several times.) Finally, D might imply a set of subsamples observed within each time.

214

ANALYSIS OF VARIANCE III: FACTORIAL EXPERIMENTS

of as many independent comparisons as there are degrees of freedom. If only one of the comparisons involves a real difference and if this difference should be averaged with a number of nonreal differences, then a test of this average might fail to detect the real difference. It is for this reason that we plan meaningful comparisons, preferably independent ones. The factorial experiment is an example of such planning which results in independent comparisons. This section and Sec. 11.8 are concerned with planned, independent comparisons, where each comparison is based on a single degree of freedom.

In Sec. 11.3, Eqs. (11.5), (11.6), and (11.7) give short-cut computing formulas for sums of squares of single degree of freedom comparisons. The procedure consists of obtaining a linear function of the treatment totals as in Eq. (11.4), squaring the result, and dividing by an appropriate divisor. We now examine the procedure in more general terms.

Let $T_1, \ldots, T_t$ be t treatment totals based on the same numbers of observations. The linear function given by Eq. (11.11)

$$Q = \sum_i c_i T_i \quad \text{with} \quad \sum_i c_i = 0 \qquad (11.11)$$

is called a *comparison* or *contrast* among the T's; it has one degree of freedom. The c's are numerical constants. The restriction that $\Sigma c_i = 0$ is an essential one and it is generally arranged to have the c's be integers. For example, Eq. (11.4) is a comparison because $\Sigma c_i = 0$; $c_1 = 1$, $c_2 = -1$, $c_3 = 1$, and $c_4 = -1$. (Since means are easily changed to totals, means may also be used in defining comparisons.)

The sum of squares attributable to such a comparison is computed by

$$Q \text{ SS} = \frac{Q^2}{Kr} \quad \text{for } K = \Sigma c_i^2 \qquad (11.12)$$

and is on a per-observation basis, r being the number of observations in each total. For Eq. (11.4), $K = 1^2 + (-1)^2 + 1^2 + (-1)^2 = 4$. The main effects and interaction comparisons of Sec. 11.3, as illustrated by Eq. (11.4), all have two $+1$'s and two -1's as c's. Hence the corresponding sums of squares all have divisors $4r$; this can be checked in Eqs. (11.5), (11.6), and (11.7).

Again, we may wish to compare the mean of two treatments with the mean of three other treatments. Using totals, we make the equivalent comparison

$$Q = 3(T_1 + T_2) - 2(T_3 + T_4 + T_5)$$

Hence, the appropriate sum of squares for testing Q is

$$\frac{Q^2}{Kr} = \frac{[3(T_1 + T_2) - 2(T_3 + T_4 + T_5)]^2}{[2(3)^2 + 3(-2)^2]r}$$

This serves as the numerator for an F test. The alternative is to compute

$$\frac{(T_1 + T_2)^2}{2r} + \frac{(T_3 + T_4 + T_5)^2}{3r} - \frac{(T_1 + \cdots + T_5)^2}{5r}$$

Many experiments involving more than two treatments are designed for a set of single degree of freedom comparisons among treatments. Thus a treatment sum of squares with $t - 1$ degrees of freedom may be partitioned to give $t - 1$ independent comparisons, each based upon one degree of freedom. Each of these separate comparisons can be tested individually by experimental error. The 2^2 factorial is an example of such an experiment.

If $t - 1$ comparisons are made among t treatments and if the comparisons are independent of each other, then they are called an *orthogonal* set. The sum of the sums of squares for the $t - 1$ comparisons in an orthogonal set add to the total sum of squares for treatments. It is possible to partition a sum of squares with $t - 1$ degrees of freedom to give many different orthogonal sets. In any given experiment, ordinarily only one set is selected and this is done at the planning stage. Where only certain members of an orthogonal set have meaning to the investigator, only those comparisons of interest are made.

Orthogonality is defined for any two comparisons in terms of the coefficients, the c's. Consider two comparisons involving the same treatment totals, for example, $Q_1 = \Sigma c_{1i} T_i$ and $Q_2 = \Sigma c_{2i} T_i$. Comparisons Q_1 and Q_2 are orthogonal if

$$\Sigma c_{1i} c_{2i} = 0 \qquad (11.13)$$

that is, if the sum of the products of the coefficients of any two comparisons is zero. For example, an experiment with two treatments and a check may be planned to compare the two treatments and to compare their mean against check. Then

$$Q_1 = T_1 - T_2 \quad \text{and} \quad Q_2 = T_1 + T_2 - 2T_3$$

where $c_{11} = 1$, $c_{12} = -1$, and $c_{13} = 0$, $c_{21} = 1$, $c_{22} = 1$, and $c_{23} = -2$, and

$$\Sigma c_{1i} c_{2i} = 1(1) + (-1)1 + 0(-2) = 0$$

The comparisons are independent. If the plan had been to compare each treatment with check, then

$$Q_1 = T_1 - T_3 \quad \text{and} \quad Q_2 = T_2 - T_3$$

and
$$\Sigma c_{1i} c_{2i} = 1(0) + 0(1) + (-1)(-1) = 1$$

The comparisons are not independent.

Independence or orthogonality is seen to be directly related to covariance and correlation. Since $\Sigma c_i = 0$, then Σc_i^2 and $\Sigma c_{1i} c_{2i}$ are sums of products as in variances and covariances. In the second example above, the correlation between comparisons is

$$\frac{\Sigma c_{1i} c_{2i}}{\sqrt{\Sigma c_{1i}^2 \Sigma c_{2i}^2}} = \frac{1}{\sqrt{2(2)}} = \frac{1}{2}$$

If a set of $t - 1$ nonorthogonal comparisons are made, the sum of the $t - 1$ sums of squares will not add to the treatment sum of squares.

With an orthogonal set of treatment comparisons, it is possible to partition error sums of squares and degrees of freedom into $t - 1$ components, a component being associated with each of the treatment comparisons. This is a useful procedure when the error term may be heterogeneous. An example follows of an orthogonal set of comparisons and the partitioning of the error term into components.

Hoppe recorded observations intended to compare seven seed fungicides and an untreated check for emergence of corn seedlings infected with *Diplodia spp.* The experiment was conducted in a greenhouse with six blocks in a

TABLE 11.10

GREENHOUSE STAND FROM CORN SEED INFECTED WITH *Diplodia spp.* FOLLOWING SEED TREATMENT WITH VARIOUS FUNGICIDES

Block	Treatment								Block totals
	A	B	C	D	E	F	G	H	
1	8	16	14	10	8	8	7	12	83
2	8	19	16	11	7	8	6	19	94
3	9	24	14	12	1	3	6	9	78
4	7	22	13	8	1	3	6	11	71
5	7	19	14	7	3	3	4	9	66
6	5	19	13	3	2	7	4	5	58
Treatment totals	44	119	84	51	22	32	33	65	$G = 450$

Symbol	Treatment
A	untreated check
B and C	mercuric fungicides
D and H	nonmercuric fungicides, company I
E, F, and G	nonmercuric fungicides, company II, where F and G are newer formulations of E

Analysis of variance

Source	df	SS	Mean square	F
Blocks	5	102.50	20.50	
Treatments	7	1,210.58	172.94	29.92**
Error	35	202.17	5.78	
Total	47	1,515.25		

SOURCE: Data used with permission of P. E. Hoppe, University of Wisconsin, Madison, Wis.

randomized complete-block design. Each experimental unit consisted of 25 seeds. The data and a treatment code are given in Table 11.10 along with the conventional analysis of variance.

It is desired to partition the sum of squares and seven degrees of freedom for treatments to give an orthogonal set of comparisons. The comparisons of interest are between the means of: the check and the seven fungicides, mercuric and nonmercuric fungicides, the two mercuric fungicides, fungicides of companies I and II, fungicides of company I, the original and the new formulations of company II, and the new formulations of company II.

Table 11.11 shows the treatment comparisons, coefficients, divisors, and sums of squares. Integers, rather than the fractions called for by means, are used as coefficients for the comparisons. The sum of the coefficients in any row is zero and the sum of cross products of the coefficients in any two rows is zero; thus we have comparisons according to our definition and they are independent.

TABLE 11.11

PERTINENT INFORMATION FOR SEVEN ORTHOGONAL COMPARISONS

Treatment	A	B	C	D	E	F	G	H	Q	Kr	SS
Treatment total, T_i	44	119	84	51	22	32	33	65			
Comparison and no.											
1 A vs. rest	−7	+1	+1	+1	+1	+1	+1	+1	98	56(6)	28.58*
2 BC vs. DEFGH	0	+5	+5	−2	−2	−2	−2	−2	609	70(6)	883.05**
3 B vs. C	0	+1	−1	0	0	0	0	0	35	2(6)	102.08**
4 DH vs. EFG	0	0	0	+3	−2	−2	−2	+3	174	30(6)	168.20**
5 D vs. H	0	0	0	+1	0	0	0	−1	−14	2(6)	16.33
6 E vs. FG	0	0	0	0	+2	−1	−1	0	−21	6(6)	12.25
7 F vs. G	0	0	0	0	0	+1	−1	0	−1	2(6)	0.08
Total											1,210.57

With regard to the coefficients of Table 11.11, all signs in any line or lines can be changed without affecting any sum of squares. For comparison 1, we have $-7(44) + 1(119 + 84 + 51 + 22 + 32 + 33 + 65) = (-308 + 406) = 98$, the difference between seven times the total of treatment A and the total of the other seven treatment totals or, alternatively, $7r$ times the difference between the mean of the check and the mean of the seven fungicides. For this comparison, $K = (-7)^2 + (1)^2 + \cdots + (1)^2 = 56$.

Each sum of squares has a single degree of freedom and the sum of the sums of squares equals the treatment sum of squares because we have an orthogonal set of comparisons. Each sum of squares is tested against experimental error and compared with $F(1, n_e)$ where n_e is the number of degrees of freedom in error.

The sum of squares for error, with 35 degrees of freedom, can be partitioned into seven components with 5 degrees of freedom each, one component for each of the seven independent comparisons. The procedure is to obtain Q's for each of the comparisons of Table 11.11, for each block. These are shown in Table 11.12. For comparison 1, block 1, $Q = -7(8) + 1(16 + 14 + 10 + 8 + 8 + 7 + 12) = 19$; for comparison 2, block 1, $Q = 5(16 + 14) - 2(10 + 8 + 8 + 7 + 12) = 60$; and so on. The totals for each comparison within blocks are given in Table 11.12. The total, over blocks, of any of these comparisons is the total given in Table 11.11 for the same comparison; we have simply rearranged the arithmetic.

Any comparison within a block is unaffected by the general level of the block if the randomized complete-block model is valid. Thus, if we add 10 to every observation in a block, the comparison for that block is unchanged.

ANALYSIS OF VARIANCE III: FACTORIAL EXPERIMENTS

This is so because, for any comparison, $\Sigma c_i = 0$. Consequently, the variance among the block values of any comparison should be a suitable variance for testing a hypothesis about the mean of the comparison, that is, about the total.

TABLE 11.12

DIFFERENCES FOR SEVEN COMPARISONS BY BLOCKS, AND ERROR SUMS OF SQUARES FOR THESE COMPARISONS

Block	Comparison						
	1	2	3	4	5	6	7
1	19	60	2	20	−2	1	1
2	30	73	3	48	−8	0	2
3	6	128	10	43	3	−7	−3
4	15	117	9	37	−3	−7	−3
5	10	113	5	28	−2	−1	−1
6	18	118	6	−2	−2	−7	3
Total	98	609	35	174	−14	21	−1
Divisor	56	70	2	30	2	6	2

	Error components						
Comparison no.	1	2	3	4	5	6	7
Sum of squares among blocks	34.75	938.50	127.50	223.67	47.00	24.83	16.50
Correction term = Q^2/Kr	28.58	883.05	102.08	168.20	16.33	12.25	0.08
Error component	6.17	55.45	25.42	55.47	30.67	12.58	16.42

For example, for comparison 1, we have the sum of squares

$$\frac{19^2 + 30^2 + \cdots + 18^2}{56} - \frac{98^2}{6(56)} = 34.75 - 28.58 = 6.17 \text{ with 5 } df$$

Thus for any comparison that is to be tested by its own component of error, we begin by synthesizing six observations as linear combinations of the observations in any block. Block differences are not a source of variation among the synthesized observations. We then compute a variance among the six observations and use it to test the null hypothesis that the mean of the population of such observations is zero.

These component sums of squares are all computed similarly and are shown in Table 11.12. The divisor K is used throughout to give components on a per-observation basis. The sum of the seven components is 202.18 as compared with an error term of 202.17 obtained by subtraction in the initial analysis. The small difference is due to rounding. Bartlett's chi-square test of homogeneity, Sec. 17.3, gives a value of 7.95 with 6 degrees of freedom, a value to be exceeded in the random sampling of variances and when the null hypothesis of homogeneous variance is true, with probability between .30 and .20. It is concluded that the components of error variance are not heterogeneous.

Exercise 11.6.1 How many sums of products of coefficients in Table 11.11 must be checked to be sure the seven comparisons are independent?

Exercise 11.6.2 Check several pairs of the comparisons in Table 11.11 for independence.

Exercise 11.6.3 Compute the sum of squares among the six values given in Table 11.12 under comparison 2. Test the null hypothesis that the mean of the population from which these observations came is zero.

Compare the resulting F value with that obtained by dividing the mean square for comparison 2 (Table 11.11 or appropriate correction term in Table 11.12) by the mean square of the corresponding component of the error term. (The results should differ by rounding errors only.)

11.7 n-way classifications and factorial experiments; response surfaces. Table 11.4 contains sums. These sums are classified in a three-way system. They provide all the necessary material for computing sums of squares for main effects and interactions (Table 11.5). To compute block and error sums of squares, the individual observations are required.

In general, n-way classifications of data, not necessarily sums as in Table 11.4, are common. We now see that they may be analyzed, from a purely computational point of view, as we analyzed the treatment totals of Table 11.4. The question of tests of significance is another matter. For example, an investigator of seed emergence rates who must work in the field will not likely randomize all $3 \times 3 \times 2$ experiment combinations but only 3×2 of them; his soils will be at different locations. Suppose he regards soils as blocks so that each treatment combination appears only once on each soil type. He will probably analyze his data as follows:

Source of variation	df
Blocks (soil types, B)	2
Treatments:	5
Species, A	2
Fungicides, C	1
Species $\times$ fungicides, AC	2
Residual (= error)	10
Total	17

We have already seen that the 10 degrees of freedom for residual are those associated with interactions AB, BC, and ABC with 4, 2, and 4 degrees of freedom; also, AB is not significant, BC is highly significant, and ABC is significant. In other words, the residual or error variance would be an average of, probably, nonhomogeneous components. The conclusions drawn from the above analysis may well not be valid.

This illustrates that the choice of regarding data in an n-way classification as an n-factor experiment with one block or as a randomized block experiment with fewer than n factors is not always as simple as looking at the randomization scheme. Essentially, it is a problem of recognizing potential sources of variation and including them in the model. In many instances, such data are analyzed using the factorial approach since, at worst, the partitioning of the total sum of squares is excessive and, consequently, somewhat meaningless. It is also at the expense of degrees of freedom in error which leads to a less precise estimate of error variance. Here, the n-factor interaction is likely to be used as the error term, especially if it seems to have no meaningful interpretation. In any case, its significance cannot be tested. In the species-soil-fungicide experiment, the three-factor interaction had a clear explanation.

Often, n-way classifications and factorial experiments, in which the levels of any factor refer to measured amounts of a treatment such as fertilizer, insecticide, cooking temperature, or diet component, may be considered as experiments planned to determine the nature of a *response surface*. In particular, one is presumably looking for a value at or near a maximum since most responses are not purely linear. Main effects and interactions can be fairly easily interpreted in terms of such surfaces. Much of today's interest in response surfaces centers about the research of Box (11.2), work beyond the scope of this text.

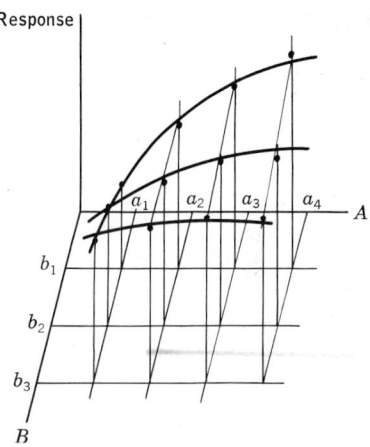

FIG. 11.3 Possible response surface for a two-factor experiment

Consider Fig. 11.3, which represents the results of a 4 × 3 factorial experiment. Plotted points are for the means of the 12 treatment combinations and the indicated surface is intended to fit these points reasonably well. Four lines indicate the responses to three amounts of B at four levels of A; these are comparable to simple effects. Here they are shown to be straight but with slopes that vary according to the level of A. It would appear that the response to B is real and mostly linear, and that this linear response is not homogeneous over the different levels of A; the slopes of the regression lines vary. Computationally, it is possible to examine the linear regression of response on B or the *linear component* of B at each level of A and to test the hypothesis of homogeneity. The latter would be a test of the interaction between the linear component of B and factor A. Such an interaction would seem to exist for the surface in Fig. 11.3. If the first of the four lines tipped up, the fourth tipped down, and the other two were intermediate, then we might find that the main effect B showed no significance, since it is based on B averages over all levels of A, while the interaction of the linear component of B and factor A was significant, since an interaction measures homogeneity of response.

Factor A shows a tendency toward curvilinearity which a test of significance might detect. Since the curvatures are fairly slight, the linear component of A would likely be significant also. Neither of these components appears to be unaffected by the level of B so we might expect the interactions of the linear component of A with B and of the curvilinear component of A with B to be significant since neither component appears to be homogeneous over levels of B. Further partitioning of the six degrees of freedom and sum of squares for this interaction is possible; six single, independent degrees of freedom and sums of squares can be obtained, tested, and interpreted.

Since a set of meaningful comparisons is to involve regression, the next section deals with regression components of a treatment sum of squares and with their homogeneity.

11.8 Individual degrees of freedom; equally spaced treatments. Many experiments are planned to determine the nature of a response curve or surface where the levels of a factor refer to increasing amounts of the factor. The standard regression analysis of Chap. 9 is applicable as far as it goes. However, when there are equal increments between successive levels of a factor, a simple coding device may be used to accomplish the analysis with much less effort. In addition, simple methods have been developed for dealing with nonlinear regression and the homogeneity of the various components of regression. We refer, in particular, to the use of coefficients of *orthogonal polynomials* or coefficients for orthogonal comparisons in regression. Orthogonal polynomials are regression equations such that each is associated with a power of the independent variable, for example, with x, x^2, or x^3, and all are independent of one another or are orthogonal. This permits an independent computation of any contribution according to the degree of the independent variable and an independent test of the contribution. Each sum of squares is the additional reduction due to fitting a curve of one degree higher. Computing time for curve fitting by means of orthogonal polynomials is much less than that for the usual regression analysis procedure, especially where a degree higher than the first is involved.

Table 11.13 gives the coefficients and divisors for orthogonal regression comparisons for up to six equally spaced treatments. Coefficients and divisors for up to $n = 52$ and $n = 104$ treatments are given by Fisher and Yates (11.6) and by Anderson and Houseman (11.1), respectively. These go as far as the fifth-degree polynomial. These orthogonal polynomials are not applicable to unequally spaced treatments. For unequally spaced treatments see Sec. 16.8 and reference (16.6).

With three levels of a factor, there are two degrees of freedom which can be partitioned into one associated with the linear response and one for the quadratic or second-degree response. Since a second or higher degree curve will always go through three points, it may be more appropriate to refer to the latter as being associated with nonlinear response. For four levels, an additional degree of freedom is available for estimating the cubic response; again, it might be better to refer to the latter as deviations from quadratic response; and so on. If more than four levels are present, usually only the linear and quadratic and, sometimes, the cubic responses are of interest. It may be desirable to calculate the sum of squares for each comparison of the set in order to check the work.

TABLE 11.13
Coefficients and Divisors for Orthogonal Comparisons in Regression: Equally Spaced Treatments

No. of treatments	Degree of polynomial	Treatment totals						Divisor $= \Sigma c_i^2$
		T_1	T_2	T_3	T_4	T_5	T_6	
2	1	−1	+1					2
3	1	−1	0	+1				2
	2	+1	−2	+1				6
4	1	−3	−1	+1	+3			20
	2	+1	−1	−1	+1			4
	3	−1	+3	−3	+1			20
5	1	−2	−1	0	+1	+2		10
	2	+2	−1	−2	−1	+2		14
	3	−1	+2	0	−2	+1		10
	4	+1	−4	+6	−4	+1		70
6	1	−5	−3	−1	+1	+3	+5	70
	2	+5	−1	−4	−4	−1	+5	84
	3	−5	+7	+4	−4	−7	+5	180
	4	+1	−3	+2	+2	−3	+1	28
	5	−1	+5	−10	+10	−5	+1	252

Since the individual comparisons are independent, the sum of their sums of squares equals the sum of squares for the factor concerned. Each of the individual sums of squares is tested by means of the error term, the null hypothesis being that the population mean for the comparison is zero. If only the linear effect is significant, we conclude that the increase in response between successive levels of the factor is constant within random variation of the order of experimental error. The response may be negative or positive depending upon whether it decreases or increases with additional increments of the factor. A significant quadratic effect indicates that a parabola fits the data better, that is, accounts for significantly more variation among treatment means, than a straight line; in other words, the increase or decrease for each additional increment is not constant but changes progressively. In planning experiments in which the levels of one or more factors involve equally spaced increments, it is desirable to have the highest level above the point where the maximum response is anticipated.

The use of orthogonal polynomials will be illustrated with an experiment on soybeans performed by Lambert. The effect on seed yield of five row spacings differing by increments of 6 in. was studied using Ottawa Mandarin soybeans in six blocks of a randomized complete-block design. The data, analysis of variance, and application of orthogonal regression comparisons are given in Table 11.14.

The last part of Table 11.14 illustrates the use of orthogonal polynomials in partitioning the treatment (row spacings) sum of squares into linear, quadratic, cubic, and quartic components. The procedure is a particular application of

Table 11.14
Yield of Ottawa Mandarin Soybeans Grown at Rosemount, Minn., 1951, in Bushels per Acre

Block	Row spacing, in.					Block totals
	18	24	30	36	42	
1	33.6	31.1	33.0	28.4	31.4	157.5
2	37.1	34.5	29.5	29.9	28.3	159.3
3	34.1	30.5	29.2	31.6	28.9	154.3
4	34.6	32.7	30.7	32.3	28.6	158.9
5	35.4	30.7	30.7	28.1	29.6†	154.5
6	36.1	30.3	27.9	26.9	33.4	154.6
Treatment totals	210.9	189.8	181.0	177.2	180.2	939.1
Means	35.2	31.6	30.2	29.5	30.0	

† Estimated value. See also degrees of freedom for error and total.

Analysis of variance

Source	df	SS	MS
Block	5	5.41	
Row spacing	4	125.66	31.42**
Error	19	73.92	3.89
Total	28	204.99	

Partition of row spacings SS by use of orthogonal polynomials

Effect	Row spacings (in.) and yields (bu/acre)					Q	Kr	SS	F
	18 210.9	24 189.8	30 181.0	36 177.2	42 180.2				
Linear	−2	−1	0	+1	+2	−74.0	10(6)	91.27	23.46**
Quadratic	+2	−1	−2	−1	+2	53.2	14(6)	33.69	8.66**
Cubic	−1	+2	0	−2	+1	−5.5	10(6)	0.50	<1
Quartic	+1	−4	+6	−4	+1	9.1	70(6)	0.20	<1
Total								125.66	

SOURCE: Data used with permission of J. W. Lambert, University of Minnesota, St. Paul, Minn.

single degree of freedom, orthogonal comparisons as discussed in Sec. 11.6. Thus, for the linear component, Eq. (11.11) gives

$$Q = -2(210.9) - 1(189.8) + 0(181.0) + 1(177.2) + 2(180.2) = -74.0$$

The sum of squares for this comparison is, from Eq. (11.12),

$$\frac{Q^2}{Kr} = \frac{(-74.0)^2}{[(-2)^2 + (-1)^2 + 0^2 + 1^2 + 2^2]6} = 91.27$$

Values of K are given in Table 11.13 for each comparison. Each sum of squares has one degree of freedom so is also a mean square; F values are obtained by dividing each mean square by the error mean square.

The analysis shows highly significant linear and quadratic effects for the row spacing treatments. On the average, yield decreases as distance between rows increases. The linear component is the portion of the sum of squares attributable to the linear regression of yield on spacing. The quadratic component measures the additional improvement due to fitting the second-order polynomial. It shows that the decrease in yield becomes less for each increment or increase in row spacing. The relation between row spacing and average yield in bushels per acre is depicted in Fig. 11.4.

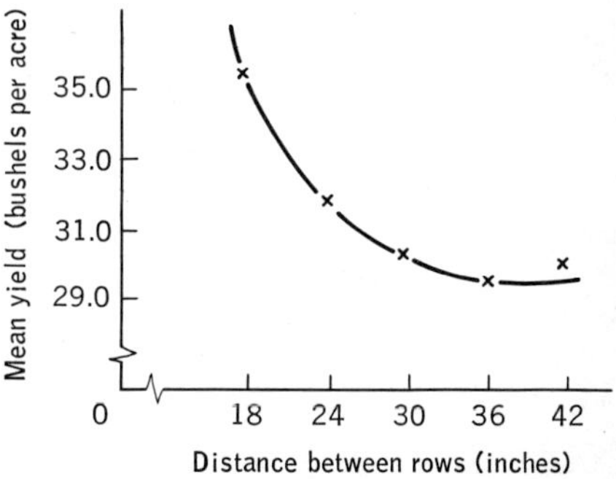

FIG. 11.4 Relation between seed yield and row spacing of Ottawa Mandarin soybeans

The coefficients for the linear component of the treatment sum of squares are clearly coded values of the row spacings. The coding is given by $x' = (X - 30)/6$, where X is row spacing in inches. Hence Q is a sum of cross products, as in a covariance. The regression coefficient is given by Eq. (11.14).

$$b = \frac{\Sigma xy}{\Sigma x^2} = \frac{Q}{K} \tag{11.14}$$

In our example, $Q/K = -74.0/10 = -7.4$ bu (per acre for a six-plot total) per unit of distance (6 in. of row spacing). The normal way to express this is to decode it. We then have $b = -7.4/6(6) = -.2$ bu per acre decrease in yield per inch increase in row width.

Another application of orthogonal polynomials is now shown for the data of Table 11.15. These are square roots of the number of quack-grass shoots per square foot, 52 days after spraying with maleic hydrazide. The experiment was conducted by Zick (11.19) at Madison, Wis., and involved two factors, namely, maleic hydrazide in applications of 0, 4, and 8 lb per acre, called rates, and days delay in cultivation after spraying, referred to as days. The data are used to illustrate how orthogonal polynomials can be used in partitioning the sums of squares for rates and rates times days into linear and quadratic

Table 11.15
Square Root of the Number of Quack-grass Shoots per Square Foot 52 Days After Spraying with Maleic Hydrazide

Days delay in cultivation	Maleic hydrazide per acre, lb	Blocks				Total
		1	2	3	4	
3	0	15.7	14.6	16.5	14.7	61.5
	4	9.8	14.6	11.9	12.4	48.7
	8	7.9	10.3	9.7	9.6	37.5
10	0	18.0	17.4	15.1	14.4	64.9
	4	13.6	10.6	11.8	13.3	49.3
	8	8.8	8.2	11.3	11.2	39.5
Total		73.8	75.7	76.3	75.6	301.4

Analysis of variance

Source	df	SS	Mean square
Blocks	3	0.58	0.19
Rates	2	153.66	76.83**
Days	1	1.50	1.50
Rates × days	2	0.50	0.25
Error	15	39.38	2.63
Total	23		

Partition of the sums of squares for rates and rates times days into simple and main effects for the computation of interactions

Rate	Simple effects				Main effects	
	3 days		10 days		3 + 10 days	
	Linear	Quadratic	Linear	Quadratic	Linear	Quadratic
0	−1(61.5)	+1(61.5)	−1(64.9)	+1(64.9)	−1(126.4)	+1(126.4)
4	0(48.7)	−2(48.7)	0(49.3)	−2(49.3)	0(98.0)	−2(98.0)
8	+1(37.5)	+1(37.5)	+1(39.5)	+1(39.5)	+1(77.0)	+1(77.0)
Sum	−24.0	1.6	−25.4	5.8	−49.4	7.4
Divisor	2(4)	6(4)	2(4)	6(4)	2(8)	6(8)
SS	72.0	0.11	80.64	1.40	152.52	1.14

components. In these data, since the interaction sum of squares is less than error, there is little point in partitioning it for other than illustrative purposes.

The procedure is to partition the sum of squares of rates within each day, and of rates over both days, into linear and quadratic components as shown at the bottom of Table 11.15. The linear component of interaction, called *rates linear times days*, measures the failure of the simple linear effects to be alike; it is found by subtracting the linear sum of squares for the main effect of rates from the sum of the simple linear components for rates within days; thus we have $72.0 + 80.64 - 152.52 = 0.12$. Since a main effect, when defined in terms of treatment totals, is the sum of the simple effects, this computation is seen to measure the variability of the simple linear effects of rate. The quadratic component for interaction is obtained similarly and is $0.11 + 1.40 - 1.14 = 0.37$. The sum of the two components for rates and that of the two interaction components equal the sums of squares for rates and rates times days, respectively.

Since no interaction is significant, interest lies only in the main effects. The sum of squares for the linear effect of rates is highly significant while that for the quadratic effect is less than error. Days do not appear to be a source of variation. It can be concluded that the decrease in the square root of the number of quack-grass shoots is the same for an increment of maleic hydrazide whether cultivation is delayed 3 or 10 days. Had the linear component of interaction been significant, it would indicate that the decrease in the square root of shoot number for each increment in rate differed for the two days; in other words, two regression coefficients of response on rate would differ. It would then be necessary to examine the simple effects, that is, the linear components or linear regression coefficients, within each of the two days.

When yield shows a linear response to a factor A and when this response is not homogeneous over levels of another factor B, then we have an (A linear) times B or $A_L B$ interaction. It is sometimes desired to examine the nature of this interaction by determining whether the linear regression coefficients show a linear response to B, and so on. This component of interaction would be labeled an A-linear times B-linear or $A_L B_L$ interaction. Like other interactions, this one is symmetrical in A_L and B_L; that is, we can consider this interaction as dealing with the linear regression coefficients of response to B and their linearity over levels of A. Similarly, we might consider $A_Q B_L$, $A_L B_Q$, $A_Q B_Q$, and other such interactions.

The computation of sums of squares for such interactions is straightforward. For the $A_L B_L$ interaction, we obtain Q values for A_L (or B_L) within each level of B (or A), then apply the set of coefficients for the linear orthogonal polynomial to these Q values. The resulting quantity is squared, divided by the product of the number of observations in each total, and the sum of squares for each of the linear comparisons (these will differ if the two factors have different numbers of levels). Tables of coefficients for such single degree of freedom interaction components are easily constructed. Federer (11.5) gives such a table for a 3×3 factorial experiment. Snedecor (11.12) works an example in detail. In the next section, we work an example of such an interaction where the spacing is not equal and the interaction has a special meaning.

TABLE 11.16

ONE DEGREE OF FREEDOM FOR NONADDITIVITY: AN EXAMPLE

(See also Table 8.2)

Treatment (stage when inoculated)	Block				Decoded treatment means	$\bar{x}_{i.} - \bar{x}_{..}$
	1	2	3	4		
Seedling	4.4	5.9	6.0	4.1	35.10	−0.43
Early bloom	3.3	1.9	4.9	7.1	34.30	−1.23
Full bloom	4.4	4.0	4.5	3.1	34.00	−1.53
Full bloom (1/100)	6.8	6.6	7.0	6.4	36.70	1.17
Ripening	6.3	4.9	5.9	7.1	36.05	0.52
Uninoculated	6.4	7.3	7.7	6.7	37.03	1.50
Decoded block means	35.27	35.10	36.00	35.75	35.53	
$\bar{x}_{.j} - \bar{x}_{..}$	−0.26	−0.43	0.47	0.22		0.00
$Q_j = \sum_i (\bar{x}_{i.} - \bar{x}_{..}) X_{ij}$	8.149	10.226	7.316	5.991		

$$Q = \sum_j (\bar{x}_{.j} - \bar{x}_{..}) Q_j = \sum_j (\bar{x}_{.j} - \bar{x}_{..}) \sum_i (\bar{x}_{i.} - \bar{x}_{..}) X_{ij}$$

$$= \sum_{i,j} (\bar{x}_{.j} - \bar{x}_{..})(\bar{x}_{i.} - \bar{x}_{..}) X_{ij}$$

$$= -1.759$$

$$\frac{Q^2}{Kr} = \frac{Q^2}{\sum_i (\bar{x}_{i.} - \bar{x}_{..})^2 \sum_j (\bar{x}_{.j} - \bar{x}_{..})^2}, \text{ since } r = 1$$

$$= \frac{(-1.759)^2}{7.928(.5218)} = .7483 \text{ with 1 } df$$

Analysis of variance

Source of variation	df	SS	MS	F
Blocks	$r - 1 = 3$	3.14	1.05	
Treatments	$t - 1 = 5$	31.65	6.33	4.83**
Error	$(r - 1)(t - 1) = 15$	19.72	1.31	
Additivity	1	.75	.75	<1
Residual	14	18.97	1.36	
Total	$rt - 1 = 23$	54.51		

ANALYSIS OF VARIANCE III: FACTORIAL EXPERIMENTS

Exercise 11.8.1 From Table 11.14, obtain treatment means and row spacings. Using this information, compute the regression coefficient of yield on row spacing by the methods of Chap. 9. Show that your value is the same as that found in this section by decoding the b of the linear orthogonal polynomial.

Exercise 11.8.2 The simple linear effects of Table 11.15 are sums of products. If we divide by $\Sigma x^2 = 2$, they become regression coefficients of yield totals for the four blocks on the coded rates. Compute the variation between these two regression coefficients.

The variance of totals is $n\sigma^2$ where σ^2 is on a per-unit or per-plot basis. Hence the variance of the totals used in computing the b's is $4\sigma^2$. The variance of a regression coefficient is $\sigma^2_{y \cdot x}/\Sigma x^2$. Here, $\Sigma x^2 = 2$. Hence the variance between our regression coefficients is an estimate of $4\sigma^2/2 = 2\sigma^2$. Show that the variance you have just computed is double that computed in the text for rates linear times days.

11.9 A single degree of freedom for nonadditivity.

Tukey (11.13) gives a method for isolating a sum of squares with one degree of freedom from error for the purpose of testing nonadditivity. The method originally proposed is applicable to two-way classifications or randomized complete-block designs and is illustrated below. A method (11.14) has also been given for Latin-square designs.

We now compute a sum of squares for nonadditivity with one degree of freedom for the data of Table 8.2. The data and computations are shown in Table 11.16. The first step is to measure deviations from treatment and block means. Next we compute the Q_j values, as shown in Eq. (11.11), at the bottom of the first part of the table. Thus

$$Q_j = \sum_i (\bar{x}_{i\cdot} - \bar{x}_{\cdot\cdot}) X_{ij} \qquad j = 1, \ldots, b(=4) \qquad (11.15)$$

In words, multiply each deviation of a treatment mean from the over-all mean by the corresponding value in block 1 and add. The data supply $c_i = (\bar{x}_{i\cdot} - \bar{x}_{\cdot\cdot})$; X_{ij} is a total of one observation. Repeat for all blocks and so obtain the Q_j's. Notice that each computation is equivalent to computing the numerator of a regression coefficient; we compute the regression of the response of the individuals in any block on the deviations of the treatment means from the over-all mean. We need only square each Q_j and divide by $K = \sum_i (\bar{x}_{i\cdot} - \bar{x}_{\cdot\cdot})^2$ to have four treatment-linear or T_L sums of squares.

The next computation involves measuring the homogeneity of regression coefficients; in particular, we see if they show a linear trend. The result may be looked upon as a treatment-linear times block-linear or $T_L B_L$ interaction. For coefficients, we use the deviations of the block means from the over-all mean, that is, the $(\bar{x}_{\cdot j} - \bar{x}_{\cdot\cdot})$'s. The computation is

$$Q = \sum_j (\bar{x}_{\cdot j} - \bar{x}_{\cdot\cdot}) Q_j \qquad (11.16)$$

Equations (11.14) and (11.15) may be written as

$$Q = \sum_{i,j} (\bar{x}_{i\cdot} - \bar{x}_{\cdot\cdot})(\bar{x}_{\cdot j} - \bar{x}_{\cdot\cdot}) X_{ij} \qquad (11.17)$$

The equation is seen to be symmetric in i and j.

Finally, we compute the sum of squares attributable to nonadditivity by Eq. (11.18).

$$\frac{Q^2}{Kr} = \frac{Q^2}{\Sigma(\bar{x}_{i\cdot} - \bar{x}_{\cdot\cdot})^2 \Sigma(\bar{x}_{\cdot j} - \bar{x}_{\cdot\cdot})^2} \text{ with 1 } df \qquad (11.18)$$

We have seen that this may be interpreted as a $T_L B_L$ interaction where the data supply the coefficients; ordinarily, in a two-factor experiment, we would arrange for equal spacing of the factors and rely on a table of orthogonal polynomial coefficients.

The nature of the nonadditivity being investigated may be seen by rewriting Eq. (11.17) in an equivalent form, namely, as Eq. (11.19).

$$Q = \sum_{i,j} (\bar{x}_{i\cdot} - \bar{x}_{\cdot\cdot})(\bar{x}_{\cdot j} - \bar{x}_{\cdot\cdot})(X_{ij} - \bar{x}_{i\cdot} - \bar{x}_{\cdot j} + \bar{x}_{\cdot\cdot}) \qquad (11.19)$$

Now $\bar{x}_{i\cdot} - \bar{x}_{\cdot\cdot}$ and $\bar{x}_{\cdot j} - \bar{x}_{\cdot\cdot}$ are estimates of τ_i and β_j, respectively. Hence $(\bar{x}_{i\cdot} - \bar{x}_{\cdot\cdot})(\bar{x}_{\cdot j} - \bar{x}_{\cdot\cdot})$ is an estimate of the block-treatment contribution to be expected in the i,jth cell if block and treatment effects, that is, main effects, are multiplicative instead of additive. Also, $X_{ij} - \bar{x}_{i\cdot} - \bar{x}_{\cdot j} + \bar{x}_{\cdot\cdot}$ is an estimate of the error component in the i,jth cell when the assumption of a linear additive model is valid. Hence, Q is the numerator of the sample coefficient of regression of error on the product of main effects. Now

$$\sum_i (\bar{x}_{i\cdot} - \bar{x}_{\cdot\cdot})^2 \sum_j (\bar{x}_{\cdot j} - \bar{x}_{\cdot\cdot})^2 = \sum_{i,j} [(\bar{x}_{i\cdot} - \bar{x}_{\cdot\cdot})(\bar{x}_{\cdot j} - \bar{x}_{\cdot\cdot})]^2$$

so that finally, the sum of squares for nonadditivity is that part of the customary residual sum of squares which can be attributed to this regression. When the mean square for nonadditivity is significant and not due to a few aberrant observations, a transformation is required. Harter and Lum (11.7) present the idea of regression and nonadditivity for a two-factor experiment.

Exercise 11.9.1 The data of Table 11.15 are transformed data. Compute the sum of squares attributable to nonadditivity with one degree of freedom. Has the transformation resulted in data for which the additive model still does not apply?

References

11.1 Anderson, R. L., and E. E. Houseman: "Tables of orthogonal polynomial values extended to $N = 104$," *Iowa Agr. Expt. Sta. Research Bull.* #297, 1942.

11.2 Box, G. E. P., and J. S. Hunter: "Experimental designs for exploring response surfaces," in *Experimental Designs in Industry*, edited by V. Chew, pp. 138–190, John Wiley & Sons, Inc., New York, 1958.

11.3 Cochran, W. G.: "Testing a linear relation among variances," *Biometrics*, **7:** 17–32 (1951).

11.4 Crump, S. L.: "The estimation of variance components in analysis of variance," *Biometrics Bull.*, **2:** 7–11 (1946).

11.5 Federer, W. T.: *Experimental Design*, The Macmillan Company, New York, 1955.

11.6 Fisher, R. A., and F. Yates: *Statistical Tables for Biological, Agricultural and Medical Research*, 5th ed., Hafner Publishing Company, New York, 1957.

11.7 Harter, H. L., and M. D. Lum: "A note on Tukey's one degree of freedom for non-additivity," Abstract 474, *Biometrics*, **14:** 136–137 (1958).

11.8 Henderson, C. R.: "Estimation of variance and covariance components," *Biometrics*, **9**: 226–252 (1953).

11.9 Satterthwaite, F. E.: "An approximate distribution of estimates of variance components," *Biometrics Bull.*, **2**: 110–114 (1946).

11.10 Scheffé, H.: "Statistical methods for evaluation of several sets of constants and several sources of variability," *Chem. Eng. Prog.*, **50**: 200–205 (1950).

11.11 Schultz, E. F., Jr.: "Rules of thumb for determining expectations of mean squares in analysis of variance," *Biometrics*, **11**: 123–135 (1955).

11.12 Snedecor, G. W.: *Statistical Methods*, 5th ed., Iowa State College Press, Ames, Iowa, 1956.

11.13 Tukey, J. W.: "One degree of freedom for non-additivity," *Biometrics*, **5**: 232–242 (1949).

11.14 Tukey, J. W.: "Reply to 'Query 113,'" *Biometrics*, **11**: 111–113 (1955).

11.15 Wagner, R. E.: "Effects of depth of planting and type of soil on the emergence of small-seeded grasses and legumes," M.Sc. Thesis, University of Wisconsin, Madison, 1943.

11.16 Wilk, M. B., and O. Kempthorne: "Fixed, mixed, and random models," *J. Am. Stat. Assoc.*, **50**: 1144–1167 (1955).

11.17 Wilkinson, W. S.: "Influence of diethylstilbestrol on feed digestibility and on blood and liver composition of lambs," Ph.D. Thesis, University of Wisconsin, Madison, 1954.

11.18 Yates, F.: "The principles of orthogonality and confounding in replicated experiments," *J. Agr. Sci.*, **23**: 108–145 (1933).

11.19 Zick, W.: "The influence of various factors upon the effectiveness of maleic hydrazide in controlling quack grass, *Agropyron repens*," Ph.D. Thesis, University of Wisconsin, Madison, 1956.

Chapter 12

ANALYSIS OF VARIANCE IV: SPLIT-PLOT DESIGNS AND ANALYSIS

12.1 Introduction. In our previous discussion of factorial experiments (Chap. 11), it was assumed that the set of all treatment combinations was to be applied to the experimental units according to the randomization procedure appropriate to the completely random, randomized complete-block, or Latin-square design. However, other randomization procedures are possible. One of the alternate randomizations gives rise to the split-plot design, which is a special kind of incomplete block design. The split-plot design and some of its applications are the subject of this chapter.

12.2 Split-plot designs. Split-plot designs are frequently used for factorial experiments. Such designs may incorporate one or more of the completely random, randomized complete-block, or Latin-square designs. The underlying principle is this: *whole plots* or *whole units*, to which levels of one or more factors are applied, are divided into *subplots* or *subunits* to which levels of one or more additional factors are applied. Thus each whole unit becomes a block for the subunit treatments. For example, consider an experiment to test factor A at four levels in three blocks of a randomized complete-block design. A second factor B, at two levels, can be superimposed by dividing each A unit into two subunits and assigning the two B treatments to these subunits. Here the A units are the whole units and the B units are the subunits.

After randomization, the layout may be as follows.

Block 1	Block 2	Block 3
a_4b_2 | a_1b_2 | a_2b_1 | a_3b_2	a_2b_1 | a_1b_2 | a_4b_1 | a_3b_1	a_1b_1 | a_2b_2 | a_4b_2 | a_3b_1
a_4b_1 | a_1b_1 | a_2b_2 | a_3b_1	a_2b_2 | a_1b_1 | a_4b_2 | a_3b_2	a_1b_2 | a_2b_1 | a_4b_1 | a_3b_2

Notice that the randomization is a two-stage one. We first randomize levels of factor A over the whole units; we then randomize levels of factor B over the subunits, two per whole unit. Each whole-unit plot may be considered as a block as far as factor B is concerned but only as an *incomplete block* as far as the full set of treatments is concerned. For this reason, split-plot designs may be called incomplete block designs.

The split-plot design is desirable in the following situations.

1. It may be used when the treatments associated with the levels of one or more of the factors require larger amounts of experimental material in an experimental unit than do treatments for other factors. This is common in field, laboratory, industrial, and social experimentation. For example, in a field experiment one of the factors could be methods of land preparation or application of a fertilizer, both usually requiring large experimental units or plots. The other factor could be varieties, which can be compared using smaller plots. Another example is the experiment designed to compare the keeping qualities of ice cream made from different formulas and stored at different temperatures. The procedure for a single replicate would be to manufacture a large batch by each formula, whole units, and then divide each batch for separate storage at the different temperatures, the subunits.

2. The design may be used when an additional factor is to be incorporated in an experiment to increase its scope. For example, suppose that the major purpose of an experiment is to compare the effect of several fungicides as protectants against infection from a disease. To increase the scope of the experiment, several varieties are included which are known to differ in their resistance to the disease. Here, the varieties could be arranged in whole units and the seed protectants in subunits.

3. From previous information, it may be known that larger differences can be expected among the levels of certain factors than among the levels of others. In this case, treatment combinations for the factors where large differences are expected could be assigned at random to the whole units simply as a matter of convenience.

4. The design is used where greater precision is desired for comparisons among certain factors than for others. This is essentially the same as the third situation but the reasons can be different.

In summary, since in split-plot experiments variation among subunits is expected to be less than among whole units, the factors which require smaller amounts of experimental material, or which are of major importance, or which are expected to exhibit smaller differences, or for which greater precision is desired for any reason, are assigned to the subunits.

The form of the analysis of variance for a two-factor split-plot experiment will now be discussed for a randomized complete-block design. Let r equal the number of blocks, a the number of levels of A or whole units per block, and b the number of levels of B or subunits per whole unit. Suppose $r = 3$, $a = 4$, and $b = 2$. The whole units comprise $ar = 12$ units. The 11 degrees of freedom *among whole units* are partitioned into 2 degrees of freedom for blocks, 3 degrees of freedom for the main effect A, and 6 degrees of freedom for an experimental error applicable to whole-unit comparisons. Within each whole unit there is 1 degree of freedom associated with the variation among subunits within a whole unit, giving a total of 12 degrees of freedom *within whole units* for the experiment. These 12 degrees of freedom are partitioned into 1 degree of freedom for the main effect B, 3 degrees of freedom for the interaction AB, and 8 degrees of freedom for an experimental error applicable to subunit comparisons.

The partition of the degrees of freedom for a split-plot design in which the whole units are arranged completely at random, in randomized complete blocks, and in a Latin square is given in Table 12.1. Factor A, applied to whole units, has a levels and factor B, applied to subunits, has b levels. Factor B may be applied to the subunits in arrangements other than that used here.

Notice that the degrees of freedom for error (b) may be obtained by multiplying $(b - 1)$ by the sum of the degrees of freedom for all sources, other

TABLE 12.1

PARTITION OF DEGREES OF FREEDOM FOR A SPLIT-PLOT DESIGN WITH DIFFERENT WHOLE-UNIT ARRANGEMENTS

Completely random (r replicates)		Randomized complete blocks (r replicates = blocks)		Latin square (r replicates = side of square)	
Source	df	Source	df	Source	df
Whole-unit analyses					
				Rows	$a - 1$
		Replicates	$r - 1$	Columns	$a - 1$
A	$a - 1$	A	$a - 1$	A	$a - 1$
Error (a)	$a(r - 1)$	Error (a)	$(a - 1)(r - 1)$	Error (a)	$(a - 1)(a - 2)$
Whole-unit total	$ar - 1$	Whole-unit total	$ar - 1$	Whole-unit total	$a^2 - 1$
Subunit analyses					
B	$b - 1$	B	$b - 1$	B	$b - 1$
AB	$(a - 1)(b - 1)$	AB	$(a - 1)(b - 1)$	AB	$(a - 1)(b - 1)$
Error (b)	$a(r - 1)(b - 1)$	Error (b)	$a(r - 1)(b - 1)$	Error (b)	$a(a - 1)(b - 1)$
Subunit total	$ar(b - 1)$	Subunit total	$ar(b - 1)$	Subunit total	$a^2(b - 1)$
Total	$abr - 1$	Total	$abr - 1$	Total	$a^2 b - 1$

than A, in the whole-unit analysis. For the randomized block design, this implies that blocks do not interact with factor B and, for the Latin-square design, that neither rows nor columns interact with factor B. If there is reason to doubt this assumption, error (b) should be further partitioned into components according to a more complete model. Any doubt as to the model must be examined in the light of the nature of the experimental material and past experience with it.

The whole-unit error, conveniently designated as E_a, is usually larger than the subunit error, designated as E_b. This is because the observations on subunits of the same whole unit tend to be positively correlated and thus react more alike than subunits from different whole units. E_a cannot be less than E_b except by chance and if this happens, it is proper to consider both E_a and E_b as estimates of the same σ^2 and consequently the two sums of squares can be pooled and divided by the pooled degrees of freedom to obtain an estimate of σ^2. (If E_a is significantly less than E_b, one should seriously consider the unlikely possibility of a negative intraclass correlation.)

If a factorial experiment is not laid out in a split-plot design, then the design that is used has a certain over-all precision applicable to treatment means. Relative to this experiment, the split-plot design should give increased precision for subunit comparisons but at the cost of lower precision for the whole-unit comparisons, since the over-all precision is not likely to be changed. Standard deviations or standard errors appropriate for comparisons among different means are given in Table 12.2. In the first three cases, divisors are the numbers of subunits in a mean; in the last case, r is the number of subunits in a treatment mean but rb is the correct divisor. The factor b is the sum of the weights used in obtaining a pooled error variance.

Comparisons of two A means, at the same or different levels of B, involve both main effect A and interaction AB, that is, they are both whole- and sub-unit comparisons; it is appropriate to use a weighted average of E_a and E_b as given in Table 12.2. The weights are $b-1$ and 1, their sum is b, so b appears in the divisor. For such comparisons, the ratio of the treatment difference to its standard error does not follow Student's t distribution. The approximation of Sec. 5.8 may be adapted to obtain a value for any significance level. Let t_a and t_b be the tabulated t values, at the chosen level of significance, corresponding to the degrees of freedom in E_a and E_b. Then

$$t' = \frac{(b-1)E_b t_b + E_a t_a}{(b-1)E_b + E_a} \tag{12.1}$$

is the value, at the chosen level of significance, with which we compare our sample t. Thus t' corresponds to a tabulated t. It will lie between t_a and t_b.

Many variants of the split-plot design are in common use. One of these involves dividing each subunit into c subsubunits for the inclusion of a third factor C at c levels. Levels of the C factor are assigned at random to the subsubunits. Partition of the degrees of freedom is exactly as in Table 12.1 with the addition of a subsubunit analysis. This is

Source	df
C	$c-1$
AC	$(a-1)(c-1)$
BC	$(b-1)(c-1)$
ABC	$(a-1)(b-1)(c-1)$
Error (c)	$ab(r-1)(c-1)$
Subsubunit total	$abr(c-1)$
Total	$abcr-1$

Calculation of sums of squares is on a subsubunit basis. Divisors are the numbers of subsubunits in any total being squared. The sum of squares for error (c), E_c, is obtained by subtracting from the total sum of squares for the subsubunits, the sum of all other sums of squares. This design is commonly referred to as a split-split-plot design. For other variants of the split-plot design, the reader is referred to Cochran and Cox (12.2) and Federer (12.3).

It is not necessary to have an additional split for each factor. If three factors are involved the AB combinations may be assigned to the whole units and the

TABLE 12.2

STANDARD ERRORS FOR A SPLIT-PLOT DESIGN

Difference between	Measured as†	Example	Standard errors
Two A means	$a_i - a_j$	$a_1 - a_2$	$\sqrt{\dfrac{2E_a}{rb}}$
Two B means	$b_i - b_j$	$b_1 - b_2$	$\sqrt{\dfrac{2E_b}{ra}}$
Two B means at the same level of A	$a_i b_j - a_i b_k$	$a_1 b_1 - a_1 b_2$	$\sqrt{\dfrac{2E_b}{r}}$
Two A means at the i. same level of B or ii. different levels of B (any two treatment means)	$a_i b_j - a_k b_j$ $a_i b_j - a_k b_l$	$a_1 b_1 - a_2 b_1$ $a_1 b_2 - a_2 b_1$	$\sqrt{\dfrac{2[(b-1)E_b + E_a]}{rb}}$

† All means are measured on a subunit basis. This is implied in the computational procedures.

levels of the C factor to the subunits, or the levels of A to the whole units and the BC combinations to the subunits.

12.3 An example of a split plot. An experiment performed by D. C. Arny at the University of Wisconsin compared the yields of four lots of oats for three chemical seed treatments and an untreated check. Two of the seed lots were Vicland, designated by Vicland (1) when infected with *H. Victoriae* and by Vicland (2) when not. The other two seed lots were samples of Clinton and Branch oats which are resistant to *H. Victoriae*. The seed lots, factor A, were assigned at random to the whole plots within each block; the seed protectants, factor B, were assigned at random to the subplots within each whole plot. The whole-plot design was a randomized complete block design of four blocks. Yields in bushels per acre are given in Table 12.3.

The analysis of variance of the data is computed on a subunit basis, the unit on which response is measured. It proceeds as follows. Let X_{ijk} denote the yield in the ith block from the subunit receiving the jth level of factor A and the kth level of factor B. Then $X_{i..}$ is the total for the ith block, the sum of ab subunit observations; $X_{.j.}$ is the total for all subunits receiving factor A at the jth level, the sum of rb observations; $X_{..k}$ is the total for all subunits receiving factor B at the kth level, the sum of ra observations; $X_{ij.}$ is a whole-unit total, the sum of b observations; etc.

Step 1. Find the correction term and total sum of squares.

$$\text{Correction term} = \frac{X_{...}^2}{rab} = \frac{3{,}379.8^2}{64} = 178{,}485.13$$

$$\text{Total SS (subunits)} = \sum_{i,j,k} X_{ijk}^2 - C$$

$$= 42.9^2 + \cdots + 47.4^2 - C = 7{,}797.39$$

TABLE 12.3
YIELDS OF OATS, IN BUSHELS PER ACRE

Seed lot, A	Blocks	Treatment, B				Totals
		Check	Ceresan M	Panogen	Agrox	
Vicland (1)	1	42.9	53.8	49.5	44.4	190.6
	2	41.6	58.5	53.8	41.8	195.7
	3	28.9	43.9	40.7	28.3	141.8
	4	30.8	46.3	39.4	34.7	151.2
Totals		144.2	202.5	183.4	149.2	679.3
Vicland (2)	1	53.3	57.6	59.8	64.1	234.8
	2	69.6	69.6	65.8	57.4	262.4
	3	45.4	42.4	41.4	44.1	173.3
	4	35.1	51.9	45.4	51.6	184.0
Totals		203.4	221.5	212.4	217.2	854.5
Clinton	1	62.3	63.4	64.5	63.6	253.8
	2	58.5	50.4	46.1	56.1	211.1
	3	44.6	45.0	62.6	52.7	204.9
	4	50.3	46.7	50.3	51.8	199.1
Totals		215.7	205.5	223.5	224.2	868.9
Branch	1	75.4	70.3	68.8	71.6	286.1
	2	65.6	67.3	65.3	69.4	267.6
	3	54.0	57.6	45.6	56.6	213.8
	4	52.7	58.5	51.0	47.4	209.6
Totals		247.7	253.7	230.7	245.0	977.1
Treatment totals		811.0	883.2	850.0	835.6	3,379.8

Block	Totals
1	965.3
2	936.8
3	733.8
4	743.9

SOURCE: Data used with permission of D. C. Arny, University of Wisconsin, Madison, Wis.

Step 2. Complete the whole-unit analysis.

$$\text{Whole-unit SS} = \frac{\sum\limits_{i,j} X_{ij\cdot}^2}{b} - C$$

$$= \frac{190.6^2 + \cdots + 209.6^2}{4} - C = 6{,}309.19$$

$$\text{Block SS} = \frac{\sum\limits_{i} X_{i\cdot\cdot}^2}{ab} - C$$

$$= \frac{965.3^2 + \cdots + 743.9^2}{4(4)} - C = 2{,}842.87$$

$$\text{SS } (A = \text{seed lots}) = \frac{\sum\limits_{j} X_{\cdot j \cdot}^2}{rb} - C$$

$$= \frac{679.3^2 + \cdots + 977.1^2}{4(4)} - C = 2{,}848.02$$

$$\begin{aligned}\text{Error } (a) \text{ SS} &= \text{Whole-unit SS} - \text{block SS} - \text{SS}(A) \\ &= 6{,}309.19 - (2{,}842.87 + 2{,}848.02) = 618.30\end{aligned}$$

Step 3. Complete the subunit analysis.

$$\text{SS } (B = \text{seed treatments}) = \frac{\sum\limits_{k} X_{\cdot\cdot k}^2}{ra} - C$$

$$= \frac{811.0^2 + \cdots + 835.6^2}{4(4)} - C = 170.53$$

TABLE 12.4
ANALYSIS OF VARIANCE FOR DATA OF TABLE 12.3
(On a subunit basis)

Source of variation	df	SS	Mean square	F
Blocks	3	2,842.87	947.62	
Seed lots, factor A	3	2,848.02	949.34	13.82**
Error (a)	9	618.30	68.70	
Seed treatments, factor B	3	170.53	56.84	2.80
Interaction, AB	9	586.47	65.16	3.21**
Error (b)	36	731.20	20.31	
Total	63	7,797.39		

Coeff. of variability: $\text{CV}(a) = \dfrac{\sqrt{68.70}}{52.8}(100) = 15.7\%$; $\text{CV}(b) = \dfrac{\sqrt{20.31}}{52.8}(100) = 8.5\%$

$$SS(AB) = \frac{\sum\limits_{j,k} X_{jk}^2}{r} - C - SS(A) - SS(B)$$

$$= \frac{144.2^2 + \cdots + 245.0^2}{4} - C$$

$$- (2{,}848.02 + 170.53)$$

$$= 586.47$$

Error (b) = total SS (subunits) − whole-unit SS − SS(B) − SS(AB)
    = 7,797.39 − 6,309.19 − 170.53 − 586.47 = 731.20

The sums of squares are entered in an analysis of variance table such as Table 12.4, which is for the data under discussion. For the fixed model, the F value for seed lots requires E_a in the denominator; those for seed treatments and interaction require E_b. The mean square for seed lots is highly significant, that for seed treatments is just short of the 5% level, and that for interaction is highly significant. For the random model, the choice of denominator for the several F tests may not be as straightforward. The reader is referred to Table 12.10.

TABLE 12.5

MEAN YIELDS AND STANDARD ERRORS

Mean yields of oats, in bushels per acre, for data of Table 12.3

Seed lots	Seed treatment				Seed lot means
	Check	Ceresan M	Panogen	Agrox	
Vicland (1)	36.1	50.6	45.9	37.3	42.5
Vicland (2)	50.9	55.4	53.1	54.3	53.4
Clinton	53.9	51.4	55.9	56.1	54.3
Branch	61.9	63.4	57.7	61.3	61.1
Seed treatment means	50.7	55.2	53.1	52.2	52.8

Standard errors

Difference between	Standard error as bushels per acre	df
Two seed lot means, factor A	$\sqrt{\dfrac{2(68.70)}{4(4)}} = 2.93$	9
Two seed treatment means, factor B	$\sqrt{\dfrac{2(20.31)}{4(4)}} = 1.59$	36
Two seed treatment means in same seed lot	$\sqrt{\dfrac{2(20.31)}{4}} = 3.19$	36
Two seed lot means for same seed treatment	$\sqrt{\dfrac{2[(3)20.31 + 68.70]}{4(4)}} = 4.02$	—

Since interaction is significant, differences in the responses among seed lots vary over seed treatments in a way that chance and the null hypothesis cannot easily explain; it is important to examine the simple effects. The simple effects of most interest are those among the four seed treatments within each seed lot. For the desired comparisons, treatment means and standard errors, as in Table 12.2, are given in Table 12.5.

To compute a t value, corresponding to a tabulated $t_{.05}$, for comparing two seed lots means with the same seed treatment, we use Eq. (12.1). Tabulated t's for 9 and 36 degrees of freedom are 2.262 and 2.028, respectively. Hence, $t_{.05}$ for the comparison is

$$t' = \frac{3(20.31)2.028 + 68.70(2.262)}{3(20.31) + 68.70} = 2.152$$

Such a t' value always lies between the two t values used in computing it, since it is a weighted mean of these tabulated t's. Knowing this will often obviate the necessity for computing t'.

For comparisons involving the check with each of the seed protectants within lots, Dunnett's procedure gives

$$t_{.05}(36\ df)s_{\bar{d}} = 2.14(3.19) = 6.8\ \text{bu/acre}, \quad \text{one-sided test}$$
$$t_{.01}(36\ df)s_{\bar{d}} = 2.84(3.19) = 9.1\ \text{bu/acre}, \quad \text{one-sided test}$$

where t is from Dunnett's tables, Table A.9. The probability applies to the joint set of statements, that is, a set of three, rather than to each individual statement.

We conclude that for Vicland (1) the increase in yield over check is highly significant for Ceresan M and Panogen but not for Agrox. No significant differences are found for the other seed lots.

Exercise 12.3.1 J. W. Lambert, at the University of Minnesota, in 1951 compared the effect of five row spacings on the yield of two soybean varieties. The design was a split plot with varieties as whole-plot treatments in a randomized complete-block design; row spacings were applied to subplots. The yield in bushels per acre for six blocks is given in the accompanying table.

Row spacing, in.	Variety†	Block					
		1	2	3	4	5	6
18	OM	33.6	37.1	34.1	34.6	35.4	36.1
	B	28.0	25.5	28.3	29.4	27.3	28.3
24	OM	31.1	34.5	30.5	32.7	30.7	30.3
	B	23.7	26.2	27.0	25.8	26.8	23.8
30	OM	33.0	29.5	29.2	30.7	30.7	27.9
	B	23.5	26.8	24.9	23.3	21.4	22.0
36	OM	28.4	29.9	31.6	32.3	28.1	26.9
	B	25.0	25.3	25.6	26.4	24.6	24.5
42	OM	31.4	28.3	28.9	28.6	18.5	33.4
	B	25.7	23.2	23.4	25.6	24.5	22.9

† OM = Ottawa Mandarin; B = Blackhawk

Write out the analysis of variance. Perform the computations. Interpret the data. Calculate the CV's for whole and subunits.

Exercise 12.3.2 Partition the sum of squares for row spacings and the variety times row-spacing interaction into linear, quadratic, and deviations from quadratic components. Use orthogonal polynomial coefficients where possible. Interpret the data further.

12.4 Missing data in split-plot designs. Formulas for estimating missing observations in the split-plot design are given by Anderson (12.1). Consider the case where a *single subunit* is missing and the treatment is $a_j b_k$. Let Y represent the missing subunit observation, W be the total of the observed subunits in the whole unit from which the observation is missing, $(a_j b_k)$ be the total of observed subunits that received the same treatment $a_j b_k$, and (a_j) be the total of observed subunits that received the jth level of A. Then the estimate of the missing value is given by Eq. (12.2).

$$Y = \frac{rW + b(a_j b_k) - (a_j)}{(r-1)(b-1)} \qquad (12.2)$$

For example, suppose that in Table 12.3 the value of the check in block 1 for Vicland (1), namely, 42.9, is missing. Then

$$W = 190.6 - 42.9 = 147.7$$
$$(a_j b_k) = 144.2 - 42.9 = 101.3$$
$$(a_j) = 679.3 - 42.9 = 636.4$$

and $$Y = \frac{4(147.7) + 4(101.3) - 636.4}{3(3)} = \frac{359.6}{9} = 40.0$$

If several values are missing, each in different whole-unit treatments, estimate the missing values within each whole-unit treatment as described above. If more than one subunit is missing in a whole-unit treatment, make repeated use of the above equation.

The analysis of variance is carried out after the missing value or values have been inserted. One degree of freedom is subtracted from error (b) for each missing subunit. The estimate of E_b is unbiased; however, the mean squares for treatments and E_a are biased upward. If only a few values are missing, these biases can be ignored. Procedures for obtaining unbiased estimates are given by Anderson (12.1), as are procedures for estimating a missing whole unit. The reader is also referred to Khargonkar (12.4).

Formulas for estimating the standard errors of differences between two means where missing values are involved are given by Cochran and Cox (12.2) and are reproduced in Table 12.6.

Where only one missing value occurs, the factor f in Table 12.6 is $(r-1)(b-1)/2$ for comparisons involving a mean with the missing value and another mean. However, if more than one observation is missing, f depends

TABLE 12.6

STANDARD ERRORS FOR THE SPLIT-PLOT DESIGN WITH MISSING DATA

Comparison	Measured as	Standard error
Difference between two A means	$a_i - a_j$	$\sqrt{\dfrac{2(E_a + fE_b)}{rb}}$
Difference between two B means	$b_i - b_j$	$\sqrt{\dfrac{2E_b(1 + fb/a)}{ra}}$
Difference between two B means at the same level of A	$a_i b_j - a_i b_k$	$\sqrt{\dfrac{2E_b(1 + fb/a)}{r}}$
Difference between two A means i. at the same level of B ii. at different levels of B	$a_i b_j - a_k b_j$ $a_i b_j - a_k b_l$	$\sqrt{\dfrac{2E_a + 2E_b[(b-1) + fb^2]}{rb}}$

upon the location of the missing subunits. The following approximation is correct for certain cases but tends to be slightly large for others.

$$f = \frac{k}{2(r-d)(b-k+c-1)}$$

where k, c, and d refer only to missing observations for the two means being compared; in particular

$k =$ number of missing observations
$c =$ number of blocks containing missing observations
$d =$ number of observations in the subunit treatment $a_j b_k$ that is most affected

Exercise 12.4.1 For the data of Exercise 12.3.1, assume that the values for variety OM, row spacing 18, block 1, and for variety B, row spacing 30, block 5, and for variety OM, row spacing 42, block 5 are missing.

Compute missing values for these observations. Set up a table such as Table 12.6, putting in numerical values except for E_a and E_b, which should come from a new analysis.

12.5 Split plots in time. The split-plot designs previously discussed are frequently referred to as split plots in space, since each whole unit is subdivided into distinct subunits. In some experiments, successive observations are made on the same whole unit over a period of time. For example, with a forage crop such as alfalfa, data on forage yield are usually obtained two or more times a year over a period of years. Such data are analogous to those from a split-plot design in many respects, and their analysis is often conducted as such and referred to as a split plot in time.

In conducting the analysis, suppose that yields are obtained on each plot for b cuttings of a alfalfa varieties in a randomized complete-block design of r blocks. Let X_{ijk} represent the observation in the ith block on the jth variety where the kth cutting was made.

Step 1. Conduct an analysis of variance for each cutting, that is, an analysis for the X_{ij1}'s, for the X_{ij2}'s, and so on.

Step 2. Prepare a two-way table of block times variety totals over all cuttings, that is, a table of $X_{ij.}$'s (see Table 12.7). These correspond to whole-unit totals in the split-plot design.

Step 3. From the table of totals, compute the whole-unit analysis on a subunit basis as in Table 12.8, that is, use divisors based on number of subunits.

TABLE 12.7
WHOLE-PLOT TOTALS FOR THE ANALYSIS OF A SPLIT PLOT IN TIME

Block	Variety			Block totals
	1 ... j ... a			
1	$X_{11.}$ ... $X_{1j.}$ ... $X_{1a.}$			$X_{1..}$
.	. . . .			.
i	$X_{i1.}$ ... $X_{ij.}$ ... $X_{ia.}$			$X_{i..}$
.	. . . .			.
r	$X_{r1.}$ ... $X_{rj.}$ ... $X_{ra.}$			$X_{r..}$
Variety totals	$X_{.1.}$ ... $X_{.j.}$ ... $X_{.a.}$			$X_{...}$

Step 4. Perform the subunit analysis as in Table 12.8. Error (b) of previous examples is partitioned into two components in this case.

The totals necessary for the calculation of the sums of squares for B, AB, and BR are contained in the individual analyses of cuttings. Thus, SS(B) requires the grand totals from the individual cuttings analyses, SS(AB) requires the variety totals from the individual cuttings analyses, and SS(BR) requires the block totals from the individual cuttings analyses.

Certain relationships exist between the individual cuttings analyses and the combined analysis. These serve as checks on the computations or as computing procedures. SS(B) is the sum of the individual correction terms less C, the over-all correction term. The sums of the degrees of freedom and sums of squares for A and AB equal those for A in the individual cuttings. This implies that SS(AB) can be obtained by subtracting SS(A) in the combined analysis from the sum of the SS(A)'s in the individual cutting analyses. Similarly, the sums of the degrees of freedom and sums of squares for R and RB in the combined analysis equal those for R in the individual cuttings. Also, the total SS equals the sum of the total SS's for the individual analyses plus SS(B).

Several differences exist between the analyses of a split plot in time and a split plot in space. In particular, the cuttings times blocks SS is usually not included as a part of E_b in the split plot in time analysis, since there is often a pronounced interaction. For example, if the blocks are different areas on a sloping field, moisture or other conditions may favor higher yields in certain blocks for one cutting but can be the reverse for other cuttings. This

TABLE 12.8

ANALYSIS OF VARIANCE FOR THE SPLIT PLOT IN TIME

Source	df	SS
Blocks, R	$r-1$	$\dfrac{\sum_i X_{i\cdot\cdot}^2}{ab} - C$
Varieties, A	$a-1$	$\dfrac{\sum_j X_{\cdot j\cdot}^2}{rb} - C$
Error (a), AR	$(r-1)(a-1)$	$\dfrac{\sum_{i,j} X_{ij\cdot}^2}{b} - C - SS(R) - SS(A)$
Whole units	$ar-1$	$\dfrac{\sum_{i,j} X_{ij\cdot}^2}{b} - C$
Cuttings, B	$b-1$	$\dfrac{\sum_k X_{\cdot\cdot k}^2}{ra} - C$
Cuttings $\times$ blocks, BR	$(r-1)(b-1)$	$\dfrac{\sum_{i,k} X_{i\cdot k}^2}{a} - C - SS(R) - SS(B)$
Cuttings $\times$ varieties, AB	$(a-1)(b-1)$	$\dfrac{\sum_{j,k} X_{\cdot jk}^2}{r} - C - SS(A) - SS(B)$
Error (b)	$(r-1)(a-1)(b-1)$	$\sum_{i,j,k} X_{ijk}^2 - C -$ whole units SS $- SS(B) - SS(BR) - SS(AB)$
Total	$rab-1$	$\sum_{i,j,k} X_{ijk}^2 - C$

partitioning amounts to recognizing an additional source of variance, namely cuttings times blocks, putting a term for the source into the model, and computing the analysis accordingly.

Another difference is that the standard errors for various comparisons among treatment means are not always the same for both analyses. (Compare Tables 12.2 and 12.9.) For the fixed model, the determination of F values is the same for the two analyses in that E_a is the appropriate divisor for whole-unit comparisons and E_b for subunits. Likewise, the standard error for comparisons among two A means (varieties) is $\sqrt{2E_a/rb}$ or $\sqrt{2bE_a/r}$ for forage crops where yields are reported as average totals. However, to compare differences among two A means at the same level of B (cuttings) for the split plot in time, the appropriate standard error is $\sqrt{2E/r}$, where E is the error mean square for the level of B under consideration, that is, the error from the appropriate cutting analysis of step 1. Likewise, to compare two B means at the same level of A,

TABLE 12.9
STANDARD ERRORS FOR A SPLIT-SPLIT-PLOT IN TIME

Treatment comparison	Standard error
Difference between two A means (Varieties over all years and cuttings)	$\sqrt{\dfrac{2cE_a}{rb}}$ †
Difference between two A means at the same level of B (Varieties over all cuttings within a year)†	$\sqrt{\dfrac{2cE_{a(j)}}{r}}$
Difference between two A means at the same level of B and C (Varieties for a specified cutting and year)	$\sqrt{\dfrac{2E_{(jk)}}{r}}$
Difference between two C means at the same level of A and B (Cuttings in same year and variety)	$\sqrt{\dfrac{2(E_{(jk)} + E_{(jl)})}{2r}}$

† For a perennial crop, yield is usually expressed as total of all cuttings within a year. When a mean is actually used, c goes in denominator rather than numerator.

the standard error is $\sqrt{2(E_{(1)} + E_{(2)})/2r}$, where $E_{(1)}$ and $E_{(2)}$ are the error mean squares for the two levels of B. Here the 2 in the denominator is associated with averaging $E_{(1)}$ and $E_{(2)}$. This particular average is used because the error mean squares are not necessarily estimates of a common σ^2. Big differences in average yields from cutting to cutting and heterogeneous variances are common. This is part of the reason why it is desirable to perform a separate analysis of each cutting first, and then perform the combined analysis over all cuttings. It is usually more convenient to make separate analyses and use these in obtaining the combined analysis than to reverse the process.

If several years' results involving several cuttings each year for a number of varieties are to be analyzed, the procedure is basically the same as for a split-split-plot design. Varieties are analogous to whole units, years within varieties to subunits, and cuttings within years to subsubunits. Usually individual analyses are made for each cutting and for each year over all cuttings, in addition to the combined analysis. The formulas of Table 12.9 are suggested for comparisons between two means. Since the problem of heterogeneity and correlation of error variance is very likely to arise when an experiment is conducted over a period of years, other methods of analysis have been proposed. Steel (12.5) has proposed a multivariate analysis, a procedure which has also been used by Tukey (12.6) in pooling the results of a group of experiments.

In Table 12.9, r, a, b, and c refer to the numbers of blocks, varieties, years, and cuttings, respectively. It is assumed that there is the same number of cuttings per year and that all first cuttings have more in common than a first in one year and a not-first in another, and so on. E_a, $E_{a(j)}$, $E_{(jk)}$ refer to error mean squares for the whole-unit analysis of the combined analysis, for the whole-unit analysis in the combined analysis for the jth year, and for the analysis of the kth cutting in the jth year, respectively.

12.6 The split-plot model. Let

$$X_{ijk} = \mu + \rho_i + \alpha_j + \delta_{ij} + \beta_k + (\alpha\beta)_{jk} + \epsilon_{ijk} \qquad (12.3)$$

represent the observation in the ith block of a randomized complete-block design, on the jth whole unit and the kth subunit. Let $i = 1, \ldots, r$ blocks, $j = 1, \ldots, a$ whole-unit treatments, and $k = 1, \ldots, b$ subunit treatments. Let δ_{ij} and ϵ_{ijk} be normally and independently distributed about zero mean with σ_δ^2 as the common variance of the δ's, the whole-unit random components, and with σ_ϵ^2 as the common variance of the ϵ's, the subunit random components. This representation serves for the analysis of variance of the randomized complete-block design of Table 12.1 and for the example of Sec. 12.3.

The model may be fixed, random, or mixed. If either the α's or the β's are random, then the $(\alpha\beta)$'s are random. [The component $(\alpha\beta)$ is not intended to imply a product; compare $\alpha_j\beta_k$.] Average values of the mean squares are given in Table 12.10. From this table it is clear what null hypotheses can be

TABLE 12.10
AVERAGE VALUES OF THE MEAN SQUARES FOR A SPLIT-PLOT MODEL IN A RANDOMIZED COMPLETE-BLOCK DESIGN

Source of variation	df	Model I Fixed effects	Model II Random effects
Blocks	$r - 1$	$\sigma_\epsilon^2 + b\sigma_\delta^2 + ab\sigma_\rho^2$	$\sigma_\epsilon^2 + b\sigma_\delta^2 + ab\sigma_\rho^2$
A	$a - 1$	$\sigma_\epsilon^2 + b\sigma_\delta^2 + rb \dfrac{\sum_j \alpha_j^2}{a - 1}$	$\sigma_\epsilon^2 + b\sigma_\delta^2 + r\sigma_{\alpha\beta}^2 + rb\sigma_\alpha^2$
Error a	$(r - 1)(a - 1)$	$\sigma_\epsilon^2 + b\sigma_\delta^2$	$\sigma_\epsilon^2 + b\sigma_\delta^2$
B	$b - 1$	$\sigma_\epsilon^2 + ra \dfrac{\sum_k \beta_k^2}{b - 1}$	$\sigma_\epsilon^2 + r\sigma_{\alpha\beta}^2 + ra\sigma_\beta^2$
AB	$(a - 1)(b - 1)$	$\sigma_\epsilon^2 + r \dfrac{\sum_{i,k}(\alpha\beta)_{jk}^2}{(a - 1)(b - 1)}$	$\sigma_\epsilon^2 + r\sigma_{\alpha\beta}^2$
Error b	$a(b - 1)(r - 1)$	σ_ϵ^2	σ_ϵ^2

Mixed model	
A random, B fixed	A fixed, B random
$\sigma_\epsilon^2 + b\sigma_\delta^2 + ab\sigma_\rho^2$	$\sigma_\epsilon^2 + b\sigma_\delta^2 + ab\sigma_\rho^2$
$\sigma_\epsilon^2 + b\sigma_\delta^2 + rb\sigma_\alpha^2$	$\sigma_\epsilon^2 + b\sigma_\delta^2 + r\dfrac{a}{a-1}\sigma_{\alpha\beta}^2 + rb\dfrac{\Sigma\alpha^2}{a-1}$
$\sigma_\epsilon^2 + b\sigma_\delta^2$	$\sigma_\epsilon^2 + b\sigma_\delta^2$
$\sigma_\epsilon^2 + r\dfrac{b}{b-1}\sigma_{\alpha\beta}^2 + ra\dfrac{\Sigma\beta^2}{b-1}$	$\sigma_\epsilon^2 + ra\sigma_\beta^2$
$\sigma_\epsilon^2 + r\dfrac{b}{b-1}\sigma_{\alpha\beta}^2$	$\sigma_\epsilon^2 + r\dfrac{a}{a-1}\sigma_{\alpha\beta}^2$
σ_ϵ^2	σ_ϵ^2

ANALYSIS OF VARIANCE IV: SPLIT-PLOT DESIGNS AND ANALYSIS

tested by errors (a) and (b). For the random and mixed models where interactions are real, it is necessary to synthesize an error in several cases. For the procedure, see references (11.3) and (11.9).

In Sec. 12.5, an observation was represented by

$$X_{ijk} = \mu + \rho_i + \alpha_j + \delta_{ij} + \beta_k + (\alpha\beta)_{jk} + (\rho\beta)_{ik} + \epsilon_{ijk} \quad (12.4)$$

A component for the block times cuttings interaction has been included. Let us assume that the ρ's and α's are random and that the β's are fixed. This is a mixed model and, again, some of the tests require a synthetic error.

Exercise 12.6.1 For the experiment that led to Table 12.11, assume that both A and B are fixed. Use the rules of Chap. 11 to obtain average values for the mean squares in this case. (Note that AB will also be fixed.)

TABLE 12.11

AVERAGE VALUES OF THE MEAN SQUARES FOR A POSSIBLE MODEL FOR THE DATA OF TABLE 12.8

Model: ρ's, α's random, β's fixed

Sources of variation	df	Average value of mean square
Blocks, R	$r-1$	$\sigma_\epsilon^2 + b\sigma_\delta^2 + ab\sigma_\rho^2$
Varieties, A	$a-1$	$\sigma_\epsilon^2 + b\sigma_\delta^2 + rb\sigma_\alpha^2$
Error a	$(r-1)(a-1)$	$\sigma_\epsilon^2 + b\sigma_\delta^2$
Cuttings, B	$b-1$	$\sigma_\epsilon^2 + r\dfrac{b}{b-1}\sigma_{\alpha\beta}^2 + a\dfrac{b}{b-1}\sigma_{\rho\beta}^2 + ra\dfrac{\Sigma\beta^2}{b-1}$
AB	$(a-1)(b-1)$	$\sigma_\epsilon^2 + r\dfrac{b}{b-1}\sigma_{\alpha\beta}^2$
RB	$(r-1)(b-1)$	$\sigma_\epsilon^2 + a\dfrac{b}{b-1}\sigma_{\rho\beta}^2$
Error b	$(r-1)(a-1)(b-1)$	σ_ϵ^2

12.7 Split plots in time and space. An example of a split plot in both space and time is the experiment with a perennial crop laid out in a split-plot design. Consider a pasture experiment to evaluate different fertilizer treatments, factor A, arranged in whole units with several management practices, factor B, as subunits, and conducted without re-randomization for a number of years, factor C. Data from such analyses are usually analyzed each year as described for the split plot in Sec. 12.2. A combined analysis of the data can be made to determine the average treatment responses over all years and whether these responses are consistent from year to year. The form of the analysis is given in Table 12.12, where X_{ijkm} represents the observation made in the ith block on the subunit receiving the treatment consisting of the jth level of factor A and the kth level of factor B in the mth year, factor C. Then $X_{i...}$ is a block total over all subunits and years; it is a total of abc observations. Also $X_{.jk.}$ is the total for the j,kth treatment combination over all blocks and years; it is a total of rc observations. Other totals are given similarly.

Table 12.12
Analysis of Variance for a Split Plot in Time and Space

Reference no.	Source	df	SS
I	Blocks, R	$r-1$	$\dfrac{\sum_i X_{i\cdots}^2}{abc} - C$
I	Fertilizers, A	$a-1$	$\dfrac{\sum_j X_{\cdot j\cdot\cdot}^2}{rbc} - C$
I	Error (a)	$(r-1)(a-1)$	$\dfrac{\sum_{i,j} X_{ij\cdot\cdot}^2}{bc} - C - \text{SS}(R) - \text{SS}(A)$
	Subtotal I	$ra-1$	$\dfrac{\sum_{i,j} X_{ij\cdot\cdot}^2}{bc} - C$
II	Years, C	$c-1$	$\dfrac{\sum_m X_{\cdots m}^2}{rab} - C$
II	RC	$(r-1)(c-1)$	$\dfrac{\sum_{i,m} X_{i\cdot\cdot m}^2}{ab} - C - \text{SS}(R) - \text{SS}(C)$
II	AC	$(a-1)(c-1)$	$\dfrac{\sum_{j,m} X_{\cdot j\cdot m}^2}{rb} - C - \text{SS}(A) - \text{SS}(C)$
II	E_b	$(r-1)(a-1)(c-1)$	$\dfrac{\sum_{i,j,m} X_{ij\cdot m}^2}{b} - C - \text{Subtotal I SS} - \text{SS}(C) - \text{SS}(RC) - \text{SS}(AC)$
	Subtotal I + II	$rac-1$	$\dfrac{\sum_{i,j,m} X_{ij\cdot m}^2}{b} - C$
III	Management practices, B	$b-1$	$\dfrac{\sum_k X_{\cdot\cdot k\cdot}^2}{rac} - C$
III	AB	$(a-1)(b-1)$	$\dfrac{\sum_{j,k} X_{\cdot jk\cdot}^2}{rc} - C - \text{SS}(A) - \text{SS}(B)$
III	E_c	$(r-1)a(b-1)$	$\dfrac{\sum_{i,j,k} X_{ijk\cdot}^2}{c} - C - \text{SS}(R) - \text{SS}(A) - \text{SS}(E_a) - \text{SS}(B) - \text{SS}(AB)$
	Subtotal I + III	$rab-1$	$\dfrac{\sum_{i,j,k} X_{ijk\cdot}^2}{c} - C$
IV	BC	$(b-1)(c-1)$	$\dfrac{\sum_{k,m} X_{\cdot\cdot km}^2}{ra} - C - \text{SS}(B) - \text{SS}(C)$
IV	ABC	$(a-1)(b-1)(c-1)$	$\dfrac{\sum_{j,k,m} X_{\cdot jkm}^2}{r} - C - \text{SS}(A) - \text{SS}(B) - \text{SS}(C) - \text{SS}(AB) - \text{SS}(AC) - \text{SS}(BC)$
IV	E_d	$(r-1)a(b-1)(c-1)$	$\sum_{i,j,k,m} X_{ijkm}^2 - C - \text{Subtotal I + II} - \text{SS}(B) - \text{SS}(AB) - \text{SS}(E_c) - \text{SS}(BC) - \text{SS}(ABC)$
	Grand total	$rabc-1$	$\sum_{i,j,k,m} X_{ijkm}^2 - C$

Parts I and III in Table 12.12 give the partition of the degrees of freedom and sums of squares for the portion of the analysis concerned with the average treatment responses over all years. The procedure is identical with that described for a split-plot design in Secs. 12.2 and 12.3 except for the additional constant c in the divisors; totals used in the calculations are for c years. Parts II and IV are extensions of the analyses described in Sec. 12.5.

Standard errors for comparisons between treatment means for all years are the same as those in Table 12.2 except that c is included in the divisor since the means being compared cover c years. The same precautions which are discussed in Sec. 12.5 apply to comparisons among treatments within and between individual years. Since the complete analysis and interpretation of experiments involving split units in both space and time may be complex, it is suggested that the reader seek the advice of a competent statistician.

12.8 Series of similar experiments. Many agricultural experiments are conducted at more than one location and for several years. This practice is very common in varietal evaluation of different crops. The purpose is to obtain information that permits recommendations for future years over a wide area. Both locations and years can be considered as broad types of replication, locations being replications or samples of the area for which information is desired and years being replications or samples of future years.

A series of similar experiments may be conducted to determine the effect of different external environmental conditions, such as different day lengths, on a response such as the growth of plants or the iodine number of flax; or to determine whether differently manufactured tires behave similarly under varying driving conditions. The repetition of similar experiments under different conditions is essential to provide variation in the external conditions under study. Again, an experiment to evaluate a product by biologic assay or a procedure for chemical determinations may be conducted at several laboratories; the object might be to ascertain the accuracy at the several laboratories and whether or not the same conclusions are reached.

The procedure for the analysis of such data varies with the objectives. Preliminary analyses are usually the same. However, the final analyses differ and are often quite complex, the details being beyond the scope of this text. The reader is referred to Cochran and Cox (12.2), Federer (12.3), Steel (12.5), and Tukey (12.6), where other references may also be found. Some of the more important points that concern the analyses of a series of agricultural experiments will be discussed here.

Treatment effects, in agricultural experiments repeated at several locations over a period of years, are usually fixed; that is, they are not a random set of treatments but are selected in advance by the experimenter as those most likely to succeed. Location and year effects are usually considered to be random, locations being a random set of possible locations and years being a random set of future years. These assumptions in actual practice are rarely, if ever, fulfilled. In most situations, the locations used are not selected at random but are experimental stations or fields permanently located in the area for which information is desired. Such locations are assumed to be at least representative of particular soil types or areas. The same locations are used year after year for the duration of the experiment. With annual crops, different fields at the selected experiment stations are generally used. Finally, a sample of several successive years will not always be representative of future years.

Error variances often differ considerably from test to test. When the pooled error term is used in the denominator of F, such heterogeneity tends to invalidate the F test for comparisons involving the interaction of treatments with locations and with years. The result is that the F test tends to produce too many significant results, and, in any one experiment, the stated probability level may be quite incorrect if the heterogeneity is extreme. A conservative approach in testing interactions involving treatments is to compare the calculated F with tabulated F for $(t-1)$ and n degrees of freedom where t is the number of treatments and n is the number of degrees of freedom for error in a single test, rather than with tabulated F for the degrees of freedom in the interaction and pooled error. The true numbers of degrees of freedom for the distribution of such a ratio lie somewhere between the two extremes. If this test indicates significance there need be few doubts; the difficulty comes in deciding about F values that are between the extremes.

Heterogeneity of interaction frequently complicates the interpretation of data from similar experiments. This results when certain treatments differ widely from trial to trial while others do not. Such heterogeneity invalidates the F test of treatments against interaction in a way similar to that described for testing interaction against pooled error. In some cases, a subdivision of the treatment comparisons and interactions into components of special interest may be helpful.

In the interpretation of data from varietal trials conducted at several places for a period of years, it is very helpful to study carefully the varietal means at the different locations. That is, analyses should be made for each station for each year and for each station over the several years. An interaction of varieties times locations is usually expected, especially if the locations or varieties differ widely. Such an interaction does not mean that all varieties react differently at all stations but usually that only certain varieties at certain stations do so. Even then, it may be only that the magnitudes of differences change and not that the order or ranking changes. A particular variety could be at or near the top for all locations in spite of a very large variety times location interaction. In this case, other things being equal, the single variety could be recommended over the area covered by the test. If, however, a variety is good in performance at certain locations but not at others, different recommendations could be made for different parts of the area.

Another point concerns the variability that exists within varieties from test to test. Two varieties may give about the same performance at each of several locations. However, one may vary considerably from year to year while the other may be quite consistent. If it were possible to predict, in advance of a year and with reasonable accuracy, the weather conditions likely to be encountered, then specific recommendations could be made as to what variety to plant. Since this is not fully possible at present, the grower who is interested in reasonably uniform yield each season is to be advised to plant the consistent variety.

References

12.1 Anderson, R. L.: "Missing-plot techniques," *Biometrics Bull.*, **2:** 41–47 (1946).
12.2 Cochran, W. G., and G. M. Cox: *Experimental Designs*, 2d ed., John Wiley & Sons, Inc., New York, 1957.
12.3 Federer, W. T.: *Experimental Design*, The Macmillan Company, New York, 1955.
12.4 Khargonkar, S. A.: "The estimation of missing plot values in split-plot and strip trials," *J. Ind. Soc. Agr. Stat.*, **1:** 147–161 (1948).
12.5 Steel, R. G. D.: "An analysis of perennial crop data," *Biometrics*, **11:** 201–212 (1955).
12.6 Tukey, J. W.: "Diadic anova," *Human Biol.*, **21:** 65–110 (1949).

Chapter 13

ANALYSIS OF VARIANCE V: UNEQUAL SUBCLASS NUMBERS

13.1 Introduction. Not all research workers are fortunate enough to obtain equal numbers of observations for the various subclasses of interest. This chapter is concerned with unequal numbers of observations in the subclasses, and in particular, with disproportionate subclass numbers.

13.2 Disproportionate subclass numbers; general. In certain fields of research, it is not always possible to have an equal number of observations on all treatment combinations; for example, it is not possible to control the number of offspring from bred sows yet such sows may have received treatments intended to affect some character of their offspring. When disproportionality occurs, the analysis of the data becomes somewhat involved. Such data are considered to be from a completely random design with one or more criteria of classification, not of the nested type. Where many observations are missing from replicated designs, the same analytic difficulties arise and the procedures to be discussed are applicable and may be necessary to give a valid analysis.

Data, such that there are *unequal numbers* of observations on the various treatments, can be divided into three main classes.

First, data may be in a one-way classification with unequal numbers of observations on each treatment. This is a single-factor experiment. Models and analyses were discussed in Chap. 7. The error sum of squares is the sum of the within-treatment sums of squares. The treatment sum of squares is defined as $\sum_i n_i(\bar{x}_{i.} - \bar{x}_{..})^2$ and computed as $\sum_i (X_{i.}^2/n_i) - X_{..}^2/n_{..}$. It is a weighted sum of squares of the deviations from the general or over-all mean $\bar{x}$, which is a weighted average of treatment means. The weights are essentially the reciprocals of the variances σ^2/n_i, where σ^2 is unknown but constant for all treatments.

Second, there is the two-way classification with *proportional subclass numbers*. Consider the case of a factorial experiment where the model contains an interaction component. An illustration is given in Table 13.1. The analysis presents no particular difficulties. The error sum of squares is the sum of the within-treatment sums of squares. The sum of squares for a main effect is computed from appropriate marginal totals, as for a single-factor experiment with unequal numbers; this is a weighted sum of squares of deviations. The interaction sum of squares is obtained by subtracting the main effects sums of squares from the weighted sum of squares among cell totals, that is, as

$$\sum_{i,j} \frac{X_{ij.}^2}{n_{ij}} - \frac{X_{...}^2}{n_{..}} - \mathrm{SS}(A) - \mathrm{SS}(B) \tag{13.1}$$

where $X_{ij.}$ represents a cell total of n_{ij} observations, or as

$$\frac{(a_1b_1)^2}{3} + \cdots + \frac{(a_2b_3)^2}{12} - \frac{X_{...}^2}{39} - \mathrm{SS}(A) - \mathrm{SS}(B)$$

for the illustration.

TABLE 13.1

PROPORTIONAL SUBCLASS NUMBERS ANALYSIS

Model: $X_{ijk} = \mu + \alpha_i + \beta_j + (\alpha\beta)_{ij} + \epsilon_{ijk}$

Factor	Number of observations				Analysis of variance	
	B			A totals	Source	df
	Level	b_1 b_2 b_3				
A	a_1	3 4 6		13	A	1
	a_2	6 8 12		26	B	2
					AB	2
					Error	33
B totals		9 12 18		39	Total	38

In developing an analysis for any set of data, it is necessary to estimate the various unknown parameters in the model. This is as true for balanced data, such as come from a randomized complete-block experiment, as it is for unbalanced data such as those with disproportionate subclass numbers. (Once the analysis has been developed, we may never consciously use the estimates.) In order to obtain a unique set of estimates, it is generally necessary to impose restrictions on them. These restrictions are usually related to the model. Thus, for the randomized complete-block design, the mathematical expression for an observation is $X_{ij} = \mu + \beta_i + \tau_j + \epsilon_{ij}$ and we impose a restriction that $\Sigma \hat{\tau}_j = 0$ where $\hat{\tau}_j$ is our estimate of τ_j; in the population, either $\Sigma \tau_j = 0$ (fixed effects) or $\mu_\tau = 0$ (random effects). The restriction $\Sigma \hat{\beta}_i = 0$ is also imposed. For the illustration of Table 13.1, the analysis outlined imposes the restrictions that weighted sums of $\hat{\alpha}$'s, of $\hat{\beta}$'s, and of $\widehat{(\alpha\beta)}$'s be zero, the weights being essentially the proportions. In our illustration, $\hat{\alpha}_1 + 2\hat{\alpha}_2 = 0$, $3\hat{\beta}_1 + 4\hat{\beta}_2 + 6\hat{\beta}_3 = 0$, $\widehat{(\alpha\beta)}_{11} + 2\widehat{(\alpha\beta)}_{21} = 0$, $\widehat{(\alpha\beta)}_{12} + 2\widehat{(\alpha\beta)}_{22} = 0$, $\widehat{(\alpha\beta)}_{13} + 2\widehat{(\alpha\beta)}_{23} = 0$, $3\widehat{(\alpha\beta)}_{11} + 4\widehat{(\alpha\beta)}_{12} + 6\widehat{(\alpha\beta)}_{13} = 0$, and $3\widehat{(\alpha\beta)}_{21} + 4\widehat{(\alpha\beta)}_{22} + 6\widehat{(\alpha\beta)}_{23} = 0$. For these restrictions the experimental mean, a *weighted* mean of the cell means, is an unbiased estimate of μ. Examination of the treatment totals and means of Table 13.2 shows that this is a reasonable set of restrictions from a purely computational point of view. For example, to estimate the variance among, or the sum of squares of, the β's,

TABLE 13.2

TREATMENT TOTALS AND MEANS FOR ILLUSTRATION OF TABLE 13.1

Factor	Level	b_1	B b_2	b_3	A totals	A means
A	a_1	$3\mu + 3\alpha_1 + 3\beta_1$ $+ 3(\alpha\beta)_{11} + \epsilon_{11\cdot}$	$4\mu + 4\alpha_1 + 4\beta_2$ $+ 4(\alpha\beta)_{12} + \epsilon_{12\cdot}$	$6\mu + 6\alpha_1 + 6\beta_3$ $+ 6(\alpha\beta)_{13} + \epsilon_{13\cdot}$	$13\mu + 13\alpha_1 + 3\beta_1 + 4\beta_2$ $+ 6\beta_3 + 3(\alpha\beta)_{11} + 4(\alpha\beta)_{12}$ $+ 6(\alpha\beta)_{13} + \epsilon_{1\cdot\cdot}$	$\mu + \alpha_1 + 3/13\beta_1 + 4/13\beta_2 + 6/13\beta_3$ $+ 3/13(\alpha\beta)_{11} + 4/13(\alpha\beta)_{12} + 6/13(\alpha\beta)_{13}$ $+ \bar{\epsilon}_{1\cdot\cdot}$
	a_2	$6\mu + 6\alpha_2 + 6\beta_1$ $+ 6(\alpha\beta)_{21} + \epsilon_{21\cdot}$	$8\mu + 8\alpha_2 + 8\beta_2$ $+ 8(\alpha\beta)_{22} + \epsilon_{22\cdot}$	$12\mu + 12\alpha_2 + 12\beta_3$ $+ 12(\alpha\beta)_{23} + \epsilon_{23\cdot}$	$26\mu + 26\alpha_2 + 6\beta_1 + 8\beta_2$ $+ 12\beta_3 + 6(\alpha\beta)_{21} + 8(\alpha\beta)_{22}$ $+ 12(\alpha\beta)_{23} + \epsilon_{2\cdot\cdot}$	$\mu + \alpha_2 + 3/13\beta_1 + 4/13\beta_2 + 6/13\beta_3$ $+ 3/13(\alpha\beta)_{21} + 4/13(\alpha\beta)_{22} + 6/13(\alpha\beta)_{23}$ $+ \bar{\epsilon}_{2\cdot\cdot}$
B totals		$9\mu + 3\alpha_1 + 6\alpha_2$ $+ 9\beta_1 + 3(\alpha\beta)_{11}$ $+ 6(\alpha\beta)_{21} + \epsilon_{\cdot 1\cdot}$	$12\mu + 4\alpha_1 + 8\alpha_2$ $+ 12\beta_2 + 4(\alpha\beta)_{12}$ $+ 8(\alpha\beta)_{22} + \epsilon_{\cdot 2\cdot}$	$18\mu + 6\alpha_1 + 12\alpha_2$ $+ 18\beta_3 + 6(\alpha\beta)_{13}$ $+ 12(\alpha\beta)_{23} + \epsilon_{\cdot 3\cdot}$	$39\mu + 13\alpha_1 + 26\alpha_2 + 9\beta_1$ $+ 12\beta_2 + 18\beta_3 + 3(\alpha\beta)_{11}$ $+ 4(\alpha\beta)_{12} + 6(\alpha\beta)_{13} + 6(\alpha\beta)_{21}$ $+ 8(\alpha\beta)_{22} + 12(\alpha\beta)_{23} + \epsilon_{\cdot\cdot\cdot}$	
B means		$\mu + 1/3\alpha_1 + 2/3\alpha_2 + \beta_1$ $+ 1/3(\alpha\beta)_{11} + 2/3(\alpha\beta)_{21}$ $+ \bar{\epsilon}_{\cdot 1\cdot}$	$\mu + 1/3\alpha_1 + 2/3\alpha_2 + \beta_2$ $+ 1/3(\alpha\beta)_{12} + 2/3(\alpha\beta)_{22}$ $+ \bar{\epsilon}_{\cdot 2\cdot}$	$\mu + 1/3\alpha_1 + 2/3\alpha_2 + \beta_3$ $+ 1/3(\alpha\beta)_{13} + 2/3(\alpha\beta)_{23}$ $+ \bar{\epsilon}_{\cdot 3\cdot}$		$\mu + 1/3\alpha_1 + 2/3\alpha_2 + 3/13\beta_1 + 4/13\beta_2 + 6/13\beta_3$ $+ 3/39(\alpha\beta)_{11} + 4/39(\alpha\beta)_{12} + 6/39(\alpha\beta)_{13}$ $+ 6/39(\alpha\beta)_{21} + 8/39(\alpha\beta)_{22} + 12/39(\alpha\beta)_{23} + \bar{\epsilon}_{\cdot\cdot\cdot}$

Unweighted mean $= 1/6[(\mu + \alpha_1 + \beta_1 + (\alpha\beta)_{11} + \bar{\epsilon}_{11\cdot}) + (\mu + \alpha_1 + \beta_2 + (\alpha\beta)_{12} + \bar{\epsilon}_{12\cdot}) + (\mu + \alpha_1 + \beta_3 + (\alpha\beta)_{13} + \bar{\epsilon}_{13\cdot}) + (\mu + \alpha_2 + \beta_1 + (\alpha\beta)_{21} + \bar{\epsilon}_{21\cdot})$
$+ (\mu + \alpha_2 + \beta_2 + (\alpha\beta)_{22} + \bar{\epsilon}_{22\cdot}) + (\mu + \alpha_2 + \beta_3 + (\alpha\beta)_{23} + \bar{\epsilon}_{23\cdot})] = \mu + [3\Sigma\alpha_i + 2\Sigma\beta_j + \Sigma(\alpha\beta)_{ij} + (\Sigma\bar{\epsilon}_{ij\cdot})]/6$

we must arrange to get rid of the $(\alpha\beta)$'s in the B means since they differ from mean to mean and, consequently, would contribute to the variance among the B means. On the other hand, the restrictions imply the relative frequencies with which various α's, β's, and $(\alpha\beta)$'s are present in their respective populations. This is information about the model and should not depend upon the computing procedure. Rather, the computing procedure should depend upon the model.

TABLE 13.3

NUMBERS OF OBSERVATIONS AND ROW AND COLUMN MEANS FOR A DISPROPORTIONATE SUBCLASS NUMBERS ILLUSTRATION

Model: $X_{ijk} = \mu + \alpha_i + \beta_j + \epsilon_{ijk}$, no interaction

		Numbers				A means
		b_1	b_2	b_3	Total	
Numbers	a_1	3	5	12	20	$\mu + \alpha_1 + 3/20\beta_1 + 5/20\beta_2 + 12/20\beta_3 + \bar{\epsilon}_{1\cdot\cdot}$
	a_2	6	7	8	21	$\mu + \alpha_2 + 6/21\beta_1 + 7/21\beta_2 + 8/21\beta_3 + \bar{\epsilon}_{2\cdot\cdot}$
	Total	9	12	20	41	
B means		$\mu + 3/9\alpha_1 + 6/9\alpha_2 + \beta_1 + \bar{\epsilon}_{\cdot 1 \cdot}$	$\mu + 5/12\alpha_1 + 7/12\alpha_2 + \beta_2 + \bar{\epsilon}_{\cdot 2 \cdot}$	$\mu + 12/20\alpha_1 + 8/20\alpha_2 + \beta_3 + \bar{\epsilon}_{\cdot 3 \cdot}$		

Other restrictions may result in an analysis of a different form. Consider the alternative set of restrictions $\Sigma\hat{\alpha}_i = 0$, $\Sigma\hat{\beta}_j = 0$, $\sum_i \widehat{(\alpha\beta)}_{ij} = 0$ for all j, and $\sum_j \widehat{(\alpha\beta)}_{ij} = 0$ for all i. Now the *unweighted* mean of the cell means is an unbiased estimate of μ, as can be seen from Table 13.2. For these restrictions, the analysis is different and more involved than that outlined at the beginning of this section.

Now suppose that the model does not include an interaction. By crossing out the $(\alpha\beta)$'s in Table 13.2, we see that all B means contain the same weighted sum of the α's. Hence they contribute nothing to the variance among the B means. Any reasonable restrictions on the $\hat{\alpha}$'s would not affect our estimate of the variance among, or the sum of squares of, the β's.

Thus it becomes clear that the choice of a set of restrictions, on the estimates of the various effects, has a real effect on the analysis of variance when there is interaction and should be made on the basis of assumptions which are a part of the model.

Third, subclass numbers classified in more than one direction may be disproportionate. Consider Table 13.3. It is apparent that we cannot compute a variance among the weighted A means that will be free from the B effect. No restriction on the β's can eliminate this difficulty since the ratio of β_1 to β_2 varies with the A level. One possibility is to obtain row and column means as unweighted averages of cell means and use the restriction that $\Sigma\hat{\beta}_j = 0$, provided there is at least one entry in every cell. This raises the question of efficiency since the means have different variances. A similar difficulty arises in connection with a sum of squares for factor B.

The inability to obtain a sum of squares for A or for B that is free of the other factor is summed up by saying that the data are *nonorthogonal*. In other words, we cannot use the familiar property of additivity of sums of squares. As a result, special methods of analysis must be used. While there is, essentially, only one basic method, a number of computing procedures are available. For two-way tables, these vary according to the number of levels of each effect and whether or not interaction is included in the model.

The basic method is the least-squares method, which amounts to this: Suppose we wish to test the null hypothesis that a certain effect or interaction is not a real source of variation. For example, in a two-way classification we might wish to test the null hypothesis of no interaction. First we include interaction with the other sources of variation called for by the model, then we estimate all the components, that is, the α's, β's, and $(\alpha\beta)$'s, in such a way that the residual sum of squares is a minimum. Thus we obtain the reduction in the total sum of squares attributable to all sources of variation including interaction. Secondly, we drop interaction out of the model and again proceed to minimize the residual sum of squares. Now we obtain the reduction in the total sum of squares attributable to all sources of variation other than interaction. Finally, the difference between the two reductions (the first is always greater) gives the *additional reduction* attributable to interaction or *interaction adjusted for other effects*. This sum of squares is tested against the error mean square obtained in the first case. The analysis of variance table may be written as shown below.

Source	df†	SS
All, including interaction	f	Compute
All but interaction	$f - k$	Compute
Additional due to interaction	k	By subtraction
Error	e	Total minus all
Total	$f + e$	Compute

† df column refers to no particular case.

The analysis of variance for data with disproportionate subclass numbers and the estimation of effects are inextricably linked. However, consider estimation alone for the moment. This raises the question of whether or not to weight cell means when combining them.

Consider the use of unweighted cell means. Their use in computing means is easily justified on grounds other than computational. For example, suppose the average monthly income of a sample of 200 men in a certain age group is $500 and of a sample of 100 men in another age group is $300. The unweighted average, $(500 + 300)/2 = \$400$, is an excellent estimate of the mean salary if the numbers of men in the two age groups are about equal. However, if the numbers in these age groups are about proportional to the numbers in the sample, then the weighted mean, $[2(500) + (300)]/3 = \$433$, is preferable.

It is possible to construct sets of data such that unweighted means and their differences are the same for all sets while weighted means and their differences change. As an example, consider the data of Table 13.4 where set 1 shows no interaction, set 3 shows pronounced interaction, and set 2 is intermediate. Here we see the effect of interaction on a main effect and the need of a clear definition of terms along with some advance information of sources of variation, in particular, information relative to interaction.

TABLE 13.4

UNWEIGHTED AND WEIGHTED MEANS AND DIFFERENCES

Set	Level of A and B	Number of observations		Totals		Means		Un-weighted mean	Differ-ence	Weighted mean	Differ-ence
		b_1	b_2	b_1	b_2	b_1	b_2				
1	a_1	2	6	6	30	3	5	4	6	4.5	5.83
	a_2	4	8	36	88	9	11	10		10.33	
2	a_1	2	6	6	30	3	5	4	6	4.5	7.83
	a_2	4	8	12	136	3	17	10		12.33	
3	a_1	2	6	2	42	1	7	4	6	5.5	2.17
	a_2	4	8	68	24	17	3	10		7.67	

In general, subclass means are unbiased estimates of corresponding population means, and variation within subclasses is an unbiased estimate of the common population variance. The problem is to find unbiased estimates of the main effects and interactions and of the sums of squares or variances attributable to them.

13.3 Disproportionate subclass numbers; the method of fitting constants. The method of fitting constants is general. It is applicable, though not computationally simple, to data from well-designed and conducted experiments as well as to data where good design has been an impossibility. In particular, it is the only method applicable where some possible categories contain no observations, as when there are vacant cells in an $r \times c$ table. It may also be used with a covariate.

Consider the analysis of disproportionate data in an $r \times c$ table. The mathematical description with interaction is given by Eq. (13.2).

$$X_{ijk} = \mu + \alpha_i + \beta_j + (\alpha\beta)_{ij} + \epsilon_{ijk} \qquad (13.2)$$

The k subscript refers to observations within each cell. The usual restrictions are $\sum_i \hat{\alpha}_i = 0$, $\sum_j \hat{\beta}_j = 0$, $\sum_i \widehat{(\alpha\beta)}_{ij} = 0$ for each j and $\sum_j \widehat{(\alpha\beta)}_{ij} = 0$ for each i. (Other restrictions may be used if they appear to be more convenient.) The method of fitting constants requires us to estimate μ, the α's, the β's, and the $(\alpha\beta)$'s, and to compute the reduction in the total sum of squares attributable to them, the estimates being such that the residual sum of squares is a minimum.

The description of an observation without interaction is

$$X_{ijk} = \mu + \alpha_i + \beta_j + \epsilon_{ijk} \qquad (13.3)$$

with restrictions $\Sigma \hat{\alpha}_i = 0 = \Sigma \hat{\beta}_j$. Again we estimate the α's and β's, obtaining different values than for the preceding model, and compute the reduction in the total sum of squares.

Since the method of fitting constants requires considerably more effort when interaction is present, it is desirable to have prior knowledge about interaction. If this is not the case, it may be possible to make a decision concerning interaction by comparing successive differences between means in several rows or columns; cell means are unbiased estimates of the corresponding population means. The within-cells variance is an unbiased estimate of the error variance and is appropriate for such comparisons.

If interaction is present but Eq. (13.3) has been the description, then any main effect sum of squares is only an approximation, increasingly poor as the disproportion increases; however, an efficient estimate of the interaction sum of squares can be obtained with little additional effort. If there is no interaction but Eq. (13.2) has been the description, any main effect sum of squares is unbiased but inefficiently estimated. The experimenter will probably wish to recompute his analysis if the evidence is against interaction, using Eq. (13.3). Unbiased main effect sums of squares will again be obtained; they will differ from the previous set but will be more efficient.

An example of the method of fitting constants is now illustrated for a model without interaction; see Eq. (13.3).

Table 13.5 gives numbers of cows and total times for the interval in days to

TABLE 13.5

INTERVAL IN DAYS TO FIRST HEAT FOLLOWING PARTURITION OF OUTBRED HOLSTEIN-FRIESIAN COWS HAVING NORMAL CALVINGS, 1950–1953

Season = factor A	Parturition = factor B						Totals	
	Second		Third		Fourth			
	n_{i1}	$(b_1) = X_{i1}$	n_{i2}	$(b_2) = X_{i2}$	n_{i3}	$(b_3) = X_{i3}$	$n_i.$	$X_i..$
Winter, n_{1j} and $(a_1) = X_{1j}$.	26	987	26	1024	27	962	79	2,973
Spring, n_{2j} and $(a_2) = X_{2j}$.	30	997	22	676	28	873	80	2,546
Summer, n_{3j} and $(a_3) = X_{3j}$.	17	463	18	463	18	558	53	1,484
Autumn, n_{4j} and $(a_4) = X_{4j}$.	15	321	11	331	14	401	40	1,053
Totals, $n_{.j}$ and $X_{.j}$.	88	2,768	77	2,494	87	2,794	252	8,056

first heat following parturition of outbred Holstein-Friesian cows having normal calvings in the period 1950–1953; data were obtained by Buch et al. (13.1). The A factor is season and the B factor refers to the second, third, and fourth parturitions. The within-cell sum of squares is 59,186.40 with 240 degrees of freedom.

First, we compute a preliminary analysis. This is an among-cells and within-cells analysis plus A-ignoring-B and B-ignoring-A sums of squares. The results are shown in Table 13.6. For the among-cells sum of squares, consider the data to be in a one-way table with rc categories. Section 7.9 gives the computing procedure. Thus,

$$\text{Among-cells SS} = \frac{(987)^2}{26} + \frac{(997)^2}{30} + \cdots + \frac{(401)^2}{14} - C = 5,794.11$$

TABLE 13.6
PRELIMINARY ANALYSIS OF THE DATA IN TABLE 13.5

Source	df	SS	MS
Among cells	11	5,794.11	526.74
A, ignoring B	3	4,645.07	
B, ignoring A	2	38.77	
Within cells	240	59,186.64	246.61
Total		64,980.75	

The A-ignoring-B sum of squares is computed as though factor B were not available in classifying the data. Again Sec. 7.9 applies.

$$\text{SS }(A, \text{ ignoring } B) = \frac{(2{,}973)^2}{79} + \cdots + \frac{(1{,}053)^2}{40} - C = 4{,}645.07$$

The B-ignoring-A sum of squares is computed similarly. The within-cells sum of squares may be computed directly or by subtracting the among-cells sum of squares from the total sum of squares. The within-cells mean square is appropriate for testing the among-cells mean square.

Next, a set of simultaneous equations is set up, one equation for each parameter being estimated. The parameters are the general mean μ, the set of row (seasons) deviations, namely, the α's, and the set of column (parturition) deviations, namely, the β's. The equations are:

To estimate	Equation
μ	$n_{..}\hat{\mu} + n_1.\hat{\alpha}_1 + n_2.\hat{\alpha}_2 + n_3.\hat{\alpha}_3 + n_4.\hat{\alpha}_4 + n_{.1}\hat{\beta}_1 + n_{.2}\hat{\beta}_2 + n_{.3}\hat{\beta}_3 = X_{...}$
α_1	$n_1.\hat{\mu} + n_1.\hat{\alpha}_1 \hspace{3.5cm} + n_{11}\hat{\beta}_1 + n_{12}\hat{\beta}_2 + n_{13}\hat{\beta}_3 = X_{1..}$
α_2	$n_2.\hat{\mu} \hspace{1.2cm} + n_2.\hat{\alpha}_2 \hspace{2.3cm} + n_{21}\hat{\beta}_1 + n_{22}\hat{\beta}_2 + n_{23}\hat{\beta}_3 = X_{2..}$
α_3	$n_3.\hat{\mu} \hspace{2.4cm} + n_3.\hat{\alpha}_3 \hspace{1.1cm} + n_{31}\hat{\beta}_1 + n_{32}\hat{\beta}_2 + n_{33}\hat{\beta}_3 = X_{3..}$
α_4	$n_4.\hat{\mu} \hspace{3.6cm} + n_4.\hat{\alpha}_4 + n_{41}\hat{\beta}_1 + n_{42}\hat{\beta}_2 + n_{43}\hat{\beta}_3 = X_{4..}$
β_1	$n_{.1}\hat{\mu} + n_{11}\hat{\alpha}_1 + n_{21}\hat{\alpha}_2 + n_{31}\hat{\alpha}_3 + n_{41}\hat{\alpha}_4 + n_{.1}\hat{\beta}_1 \hspace{3cm} = X_{.1.}$
β_2	$n_{.2}\hat{\mu} + n_{12}\hat{\alpha}_1 + n_{22}\hat{\alpha}_2 + n_{32}\hat{\alpha}_3 + n_{42}\hat{\alpha}_4 \hspace{1.2cm} + n_{.2}\hat{\beta}_2 \hspace{1.5cm} = X_{.2.}$
β_3	$n_{.3}\hat{\mu} + n_{13}\hat{\alpha}_1 + n_{23}\hat{\alpha}_2 + n_{33}\hat{\alpha}_3 + n_{43}\hat{\alpha}_4 \hspace{3cm} + n_{.3}\hat{\beta}_3 = X_{.3.}$

The ^'s are used to indicate that we can obtain only estimates of the parameters. If you add the α equations, you find they give the μ equation; similarly for the β equations. For this reason we cannot obtain a unique solution of the equations so will need auxiliary equations. These are the usual restrictions.

$$\Sigma \hat{\alpha}_i = 0 = \Sigma \hat{\beta}_j$$

A solution is obtained by eliminating the larger set of constants. In our example, solve the α equations for the $\hat{\alpha}$'s in terms of $\hat{\mu}$ and the $\hat{\beta}$'s. Substitute these values in the β equations. Solve for the $\hat{\beta}$'s, using $\Sigma \hat{\beta}_j = 0$. Substitute these solutions in the α equations and solve for the $(\hat{\mu} + \hat{\alpha}_i)$'s. Since $\Sigma(\hat{\mu} + \hat{\alpha}_i) = r\hat{\mu}$, $\hat{\mu}$ is easily obtained, as are the $\hat{\alpha}_i$'s. The reduction in the sum of squares due to fitting the constants is

$$\hat{\mu} X_{...} + \Sigma \hat{\alpha}_i X_{i..} + \Sigma \hat{\beta}_j X_{.j.}$$

This reduction includes the usual correction term, which must be subtracted from the above reduction to give the sum of squares attributable to constants other than the mean. The notation a_i and b_j is also commonly used for $\hat{\alpha}_i$ and $\hat{\beta}_j$. We shall use a_i and b_j in the remainder of this section. (To avoid confusion, do not use a_i and b_j if a and b are being used to represent the numbers of levels of the two factors.) Now, we write the *reduction in sum of squares due to fitting constants* as

$$\hat{\mu}X... + \Sigma a_i X_{i..} + \Sigma b_j X_{.j.} \tag{13.4}$$

A particular method of solving the linear equations where only row and column effects are included in the model follows. The method is applicable even if some cells contain no observations. The data are arranged as in Table 13.7, with the factor having fewer levels used for columns. This lessens the

TABLE 13.7

DATA OF TABLE 13.5 ARRANGED FOR METHOD OF FITTING CONSTANTS

		n_{ij} values			$n_{i.}$	$n_{ij}/n_{i.}$	$n_{i1}/n_{i.}$	$n_{i2}/n_{i.}$	$n_{i3}/n_{i.}$	$X_{i..}$
		n_{i1}	n_{i2}	n_{i3}						
n_{ij} values	n_{1j}	26	26	27	79	$n_{1j}/n_{1.}$	0.329114	0.329114	0.341772	2,973
	n_{2j}	30	22	28	80	$n_{2j}/n_{2.}$	0.375000	0.275000	0.350000	2,546
	n_{3j}	17	18	18	53	$n_{3j}/n_{3.}$	0.320755	0.339623	0.339623	1,484
	n_{4j}	15	11	14	40	$n_{4j}/n_{4.}$	0.375000	0.275000	0.350000	1,053
$n_{.j}$		88	77	87	252					
$X_{.j.}$		2,768	2,494	2,794			$X... = 8,056$			

work. The first part of the method is essentially an elimination of the a_i's from the β equations.

To obtain the $n_{ij}/n_{i.}$, we have, for example

$$\frac{n_{11}}{n_{1.}} = \frac{26}{79} = 0.329114 \quad \text{and} \quad \frac{n_{23}}{n_{2.}} = \frac{28}{80} = 0.350000$$

Here we are using six decimal places to minimize rounding errors. Note that $\sum_j (n_{ij}/n_{i.}) = 1$ within rounding errors for each value of i. For example, for $i = 3$,

$$\Sigma \frac{n_{3j}}{n_{3.}} = 0.320755 + 0.339623 + 0.339623 = 1.000001$$

This provides a check on the computing.

Define c_i and c_{ij} by Eqs. (13.5) to (13.7).

$$c_1 = \sum_i \left(\frac{n_{i1}}{n_{i.}}\right) X_{i..} - X_{.1.} \qquad c_2 = \sum_i \left(\frac{n_{i2}}{n_{i.}}\right) X_{i..} - X_{.2.}$$

$$c_3 = \sum_i \left(\frac{n_{i3}}{n_{i.}}\right) X_{i..} - X_{.3.} \tag{13.5}$$

$$c_{11} = \sum_i \left(\frac{n_{i1}}{n_{i.}}\right) n_{i1} - n_{.1} \qquad c_{22} = \sum_i \left(\frac{n_{i2}}{n_{i.}}\right) n_{i2} - n_{.2} \qquad c_{33} = \sum_i \left(\frac{n_{i3}}{n_{i.}}\right) n_{i3} - n_{.3} \tag{13.6}$$

$$c_{12} = \sum_i \left(\frac{n_{i1}}{n_{i.}}\right) n_{i2} \qquad c_{13} = \sum_i \left(\frac{n_{i1}}{n_{i.}}\right) n_{i3} \qquad c_{23} = \sum_i \left(\frac{n_{i2}}{n_{i.}}\right) n_{i3} \tag{13.7}$$

ANALYSIS OF VARIANCE V: UNEQUAL SUBCLASS NUMBERS

These values are easily obtained from Table 13.7 by use of a calculating machine. Note that Eq. (13.5) is a set of c_i's, Eq. (13.6) is a set of c_{ii}'s and Eq. (13.7) is a set of c_{ij}'s where i is less than j. We obtain

$$c_1 = 0.329114(2{,}973) + 0.375000(2{,}546) + 0.320755(1{,}484)$$
$$+ 0.375000(1{,}053) - 2{,}768 = 36.0813$$

Similarly $c_2 = -21.8185$ and $c_3 = -14.2613$. Their sums should be zero within rounding errors. Here $\Sigma c_i = .0015$. Also

$$c_{11} = 0.329114(26) + 0.375000(30) + 0.320755(17)$$
$$+ 0.375000(15) - 88 = -57.1152$$

$c_{22} = -53.2548$ and $c_{33} = -56.9589$

Finally

$$c_{12} = 0.329114(26) + 0.375000(22) + 0.320755(18) + 0.375000(11)$$
$$= 26.7055$$

$c_{13} = 30.4097$ and $c_{23} = 26.5493$

The sums $c_{11} + c_{12} + c_{13}$, $c_{12} + c_{22} + c_{23}$, and $c_{13} + c_{23} + c_{33}$ should also equal zero within rounding errors. Here they equal .0000, .0000, and .0001, respectively.

Simultaneous equations are now solved for the b's, where b is now replacing $\hat{\beta}$. The b's may be regarded as deviations, adjusted for disproportionality, of the column means from the general mean. The equations follow.

$$c_1 = c_{11}b_1 + c_{12}b_2 + c_{13}b_3$$
$$c_2 = c_{12}b_1 + c_{22}b_2 + c_{23}b_3$$
$$c_3 = c_{13}b_1 + c_{23}b_2 + c_{33}b_3$$

For our example,

$$36.0813 = -57.1152b_1 + 26.7055b_2 + 30.4097b_3$$
$$-21.8185 = 26.7055b_1 - 53.2548b_2 + 26.5493b_3$$
$$-14.2613 = 30.4097b_1 + 26.5493b_2 - 56.9589b_3$$

Since $\Sigma b_j = 0$, $b_3 = -(b_1 + b_2)$. Substituting for b_3, we obtain

$$c_1 = b_1(c_{11} - c_{13}) + b_2(c_{12} - c_{13})$$
$$c_2 = b_1(c_{12} - c_{23}) + b_2(c_{22} - c_{23})$$

or $\quad 36.0813 = (-57.1152 - 30.4097)b_1 + (26.7055 - 30.4097)b_2$
$$= -87.5249b_1 - 3.7042b_2$$

and $\quad -21.8185 = (26.7055 - 26.5493)b_1 + (-53.2548 - 26.5493)b_2$
$$= 0.1562b_1 - 79.8041b_2$$

We can eliminate b_1 by multiplying the first equation by 0.1562, the second equation by -87.5249, and subtracting. This gives

$$b_2 = \frac{-1904.0261}{-6985.4245} = 0.272571$$

From the first of the two equations above,
$$b_1 = \frac{36.0813 + 3.7042(0.272571)}{-87.5249} = -0.423776$$
and finally
$$b_3 = -(b_1 + b_2) = -(-0.423776 + 0.272571) = 0.151205$$

The original α equations are now easily solved for the values $\hat{\mu} + \hat{a}_i = \hat{\mu} + a_i$. They are

$$\hat{\mu} + a_1 = \frac{X_{1..}}{n_{1.}} - \frac{n_{11}b_1}{n_{1.}} - \frac{n_{12}b_2}{n_{1.}} - \frac{n_{13}b_3}{n_{1.}}$$

$$\hat{\mu} + a_2 = \frac{X_{2..}}{n_{2.}} - \frac{n_{21}b_1}{n_{2.}} - \frac{n_{22}b_2}{n_{2.}} - \frac{n_{23}b_3}{n_{2.}}$$

$$\hat{\mu} + a_3 = \frac{X_{3..}}{n_{3.}} - \frac{n_{31}b_1}{n_{3.}} - \frac{n_{32}b_2}{n_{3.}} - \frac{n_{33}b_3}{n_{3.}}$$

$$\hat{\mu} + a_4 = \frac{X_{4..}}{n_{4.}} - \frac{n_{41}b_1}{n_{4.}} - \frac{n_{42}b_2}{n_{4.}} - \frac{n_{43}b_3}{n_{4.}}$$

We obtain
$$\hat{\mu} + a_1 = \frac{2{,}973}{79} - 0.329114(-0.423776) - 0.329114(0.272571)$$
$$- 0.341772(0.151205) = 37.6310$$
$$\hat{\mu} + a_2 = 31.8560$$
$$\hat{\mu} + a_3 = 27.9920$$
$$\hat{\mu} + a_4 = 26.3560$$

Now $\Sigma(\hat{\mu} + a_i) = 4\hat{\mu} = 123.835$ since $\Sigma a_i = 0$ and $\hat{\mu} = 30.9588$. Note that $\hat{\mu}$ does not equal $\bar{x} = 8{,}056/252 = 31.9683$; $\bar{x}$ would be our estimate of μ only if we were to ignore A and B, that is, if our description of an observation were $X_{ijk} = \mu + \epsilon_{ijk}$.

Finally, the reduction in sum of squares due to fitting the constants is found by Eq. (13.4) or the equivalent alternative given in Eq. (13.8). The second expression in Eq. (13.8) should be less subject to rounding errors.

$$\hat{\mu}X_{...} + \Sigma a_i X_{i..} + \Sigma b_j X_{.j.} = \Sigma(\hat{\mu} + a_i)X_{i..} + \Sigma b_j X_{.j.} \qquad (13.8)$$
$$= [37.6310(2{,}973) + 31.8560(2{,}546)$$
$$+ 27.9920(1{,}484) + 26.3560(1{,}053)]$$
$$+ [-.423776(2{,}768) + .272571(2{,}494)$$
$$+ .151120(2{,}794)] = 262{,}275.34 - 70.75$$
$$= 262{,}204.59$$

To find the *additional* (to the mean) *reduction attributable to A and B*, subtract the correction term. We find
$$262{,}204.59 - (8{,}056)^2/252 = 4{,}668.34$$

If there is no interaction, that is, if we have the correct model, the a's and b's are unbiased estimates of the corresponding parameters. Unbiased estimates of population cell means may be calculated as
$$\hat{\mu}_{ij} = \hat{\mu} + a_i + b_j$$

These are shown in Table 13.8 along with the means for the various levels of the main effects. These will differ from the less efficient, observed cell means, which are also unbiased estimates of population cell means.

So far, we have computed the sum of squares attributable to both main effects. To test a hypothesis about any single main effect, we compute the sum of squares attributable to the other main effect, subtract this from that due to both, and thus obtain the additional reduction due to the main effect in which we are interested. The result is often called an *adjusted sum of squares*.

TABLE 13.8

ADJUSTED TREATMENT AND MAIN-EFFECT MEANS, INTERACTION NEGLIGIBLE, FOR THE DATA OF TABLE 13.5

Season	Constants	Parturition			Season means
		Second $b_1 = -0.4238$	Third $b_2 = 0.2726$	Fourth $b_3 = 0.1511$	
Winter	$a_1 = 6.6722$	37.2072	37.9036	37.7821	37.6310
Spring	$a_2 = 0.8972$	31.4322	32.1286	32.0071	31.8560
Summer	$a_3 = -2.9668$	27.5682	28.2646	28.1431	27.9920
Autumn	$a_4 = -4.6028$	25.9322	26.6286	26.5071	26.3560
Parturition means		30.5350	31.2316	31.1100	$30.9588 = \hat{\mu}$

Suppose we wish to test the null hypothesis that there is no row or A effect. Under this null hypothesis, namely $H_0: \alpha_i = 0, i = 1, 2, 3, 4$, the appropriate mathematical description is

$$X_{ijk} = \mu + \beta_j + \epsilon_{ijk}$$

Again we could fit constants. This time $\hat{\mu} = (\bar{x}_{.1.} + \bar{x}_{.2.} + \bar{x}_{.3.})/3 \neq \bar{x}_{...}$. The reduction attributable to the constants equals the weighted sum of squares among column or parturition means plus the correction term. We call the weighted sum of squares *an unadjusted sum of squares for B*; it is 38.77; that for A is 4,645.07 (see Table 13.6). The final analysis of variance is given in Table 13.9. The adjusted sums of squares are appropriate for tests of significance.

When interaction is included in the model, that is, when

$$X_{ijk} = \mu + \alpha_i + \beta_j + (\alpha\beta)_{ij} + \epsilon_{ijk}$$

there are more parameters, more equations, and more arithmetic. Restrictions for this model are given at the beginning of the section. The basic procedure is the same though we will obtain different estimates of the α's and β's than before. This procedure, fitting constants, will rarely be used when interaction is present if there is at least one observation in each cell and if there is no covariate. Instead, the method of weighted squares of means is used since it gives unbiased estimates and sums of squares for main effects, though not for interaction. This method is discussed in the next section.

TABLE 13.9

FINAL ANALYSIS OF VARIANCE FOR DATA OF TABLE 13.5

Source	df	SS	MS	F
Factors A and B	5	4,668.34		
A adjusted for B (i.e., additional attributable to A)	3	(4,668.34 − 38.77) = 4,629.57	1,543.19	6.26**
B adjusted for A (i.e., additional attributable to B)	2	(4,668.34 − 4,645.07) = 23.27	11.64	ns
Error	240	59,186.64	246.61	

The above model requires a mean for each cell and the appropriate description is

$$X_{ijk} = \mu + (\widehat{\alpha\beta})'_{ij} + \epsilon_{ijk}$$

where $\sum_{i,j} (\widehat{\alpha\beta})'_{ij} = 0$ but the two sets of restrictions on the $(\widehat{\alpha\beta})$'s do not apply. With this restriction, $(rc - 1) = [(r - 1) + (c - 1) + (r - 1)(c - 1)]$ degrees of freedom, rather than $(r - 1)(c - 1)$, are accounted for. Again, we could fit constants but this is not necessary since the sum of squares among cell totals, that is, $\Sigma(X_{ij.}^2/n_{ij})$, gives the same reduction as fitting constants, including μ. Hence

$$\sum_{i,j} \frac{X_{ij.}^2}{n_{ij}} - C$$

is the additional reduction attributable to factors A and B and their interaction AB. This has already been computed for the preliminary analysis. Thus with little effort, we have the material for a valid test of the null hypothesis about interaction. This is shown in Table 13.10. The numbers are for the example under discussion.

TABLE 13.10

TEST OF INTERACTION FOR THE DATA OF TABLE 13.5

Source	df	SS	MS	F
Main effects and interaction (among cell means)	11	5,794.11		
Main effects only (fitting constants)	5	4,668.34		
Interaction adjusted for main effects	6	1,125.77	187.63	<1
Error	240	59,186.64	246.61	

When interaction is significant, the previous work of this section does not give valid tests of main effects. To test a null hypothesis for a main effect, it is necessary to fit constants for the other main effect and for the interaction with the double set of restrictions. The difference between the reduction thus obtained and that for main effects plus interaction, computed as "among-cell means," is appropriate for testing the null hypothesis about the main effect. The process must be repeated for the other main effect. For this reason, the method of fitting constants is not generally used to test main effects when there is a real interaction. Instead, the method of weighted squares of means is used. For our example, interaction was not present, the method used is the recommended one, and the tests of main effects were valid.

Exercise 13.3.1 The following data were obtained by Carroll et al. (13.2). Assays were made by injecting day-old cockerels with a suspension prepared from pituitary tissue of cattle. Data were recorded as the thyroid weights, in milligrams, of control and experimental cockerels, as shown in the accompanying table. The within-cells mean square is 1.384.

Assay		Control	Normal pituitary	Dwarf pituitary
A	n_{1j}	11	10	7
	$\bar{x}_{1j.}$	3.33	6.30	3.69
B	n_{2j}	20	20	20
	$\bar{x}_{2j.}$	3.88	5.76	4.33
C	n_{3j}	21	20	20
	$\bar{x}_{3j.}$	3.59	5.57	4.19

How many degrees of freedom are associated with the within-cells variance? Analyze the data by the method of fitting constants, assuming no interaction. Compute and test the interaction variance. Do you now wish to accept or revise your analysis?

13.4 Disproportionate subclass numbers; the method of weighted squares of means. The method of weighted squares of means, devised by Yates (13.5), is applicable to the analysis of data in a two-way table *where there is interaction*. It gives unbiased estimates and tests of main effects if the experimenter considers them to be meaningful, that is, if he can specify populations for which unweighted row and column means are estimates of meaningful parameters. It provides no information about interaction unless one of the criteria of classification contains only two categories. If there is no interaction, estimates and tests of significance are inefficient but unbiased. The method requires that every cell contain at least one observation.

The data in Table 13.11 are coded mean lengths, in days, of gestation periods according to sex of calf, factor B, and calving sequence of dam, factor A, reported by Jafar et al. (13.3). A preliminary analysis of variance is shown below.

Source	df	SS	MS
Treatments	9	678.62	75.40
Error	374	8,003.13	21.40
Total	383	8,681.75	

TABLE 13.11

Mean Length (Days) of Gestation Period According to Sex (B) of Calf and Calving Sequence (A) of Dam; Computations for the Analysis of Variance. (Individual Observations Were Coded by Subtracting 270 from Each.)

Calving sequence, A		Sex, B		$\sum_j \left(\frac{1}{n_{ij}}\right)$	$w_i = 1/\sum_j \left(\frac{1}{n_{ij}}\right)$ †	Unweighted mean $= (\bar{x}_{i1\cdot} + \bar{x}_{i2\cdot})/2$	$\frac{w_i(\bar{x}_{i1\cdot} + \bar{x}_{i2\cdot})}{2}$	Unweighted $d_i = \bar{x}_{i1\cdot} - \bar{x}_{i2\cdot}$	$w_i d_i$
		Male, b_1	Female, b_2						
1 a_1	n_{1j} $1/n_{1j}$ $\bar{x}_{1j\cdot}$	54 0.018519 8.722	52 0.019231 6.462	0.037750	26.49057	7.592	201.116	2.260	59.869
2 a_2	n_{2j} $1/n_{2j}$ $\bar{x}_{2j\cdot}$	37 0.027027 9.649	34 0.029412 6.500	0.056439	17.71831	8.074	143.058	3.149	55.795
3 a_3	n_{3j} $1/n_{3j}$ $\bar{x}_{3j\cdot}$	37 0.027027 9.622	37 0.027027 8.973	0.054054	18.50000	9.297	171.995	0.649	12.007
4 a_4	n_{4j} $1/n_{4j}$ $\bar{x}_{4j\cdot}$	28 0.035714 9.679	28 0.035714 5.750	0.071428	14.00000	7.714	107.996	3.929	55.006
5 a_5	n_{5j} $1/n_{5j}$ $\bar{x}_{5j\cdot}$	46 0.021739 8.261	31 0.032258 8.581	0.053997	18.51948	8.421	155.953	−0.320	−5.926
$\sum_i (1/n_{ij})$		0.130026	0.143642	0.273668	95.22836 $= \Sigma w_i$		780.118 $= \sum_i w_i(\bar{x}_{i1\cdot} + \bar{x}_{i2\cdot})/2$		176.751 $= \Sigma w_i d_i$
$v_j = 1/(\sum_i \frac{1}{n_{ij}})$		7.69077	6.96175	14.65252 $= \Sigma v_j$					
Unweighted mean $= (\sum_i \bar{x}_{ij\cdot})/5$		9.187	7.253			8.220			
$v_j \sum_i \bar{x}_{ij\cdot}/5$		70.655	50.494	121.149 $= \sum_j v_j \sum_i \bar{x}_{ij\cdot}/5$					

† Example: $w_1 = 1/[(1/n_{11} + 1/n_{12})] = 1/(\frac{1}{54} + \frac{1}{52})$ $= 1/.037750 = [54(52)]/(54 + 52) = 26.49057$. The procedure in brackets, that is, $[54(52)]/(54 + 52)$, is applicable for a w_i with two n_{ij}'s and more desirable when such is the case since it is less subject to rounding errors.

ANALYSIS OF VARIANCE V: UNEQUAL SUBCLASS NUMBERS

The within-cells mean square is 21.4 with 374 degrees of freedom. The wide variation in unweighted sex differences for the five sequences, $-.320$ to 3.929, suggests a significant interaction. An approximate, and perhaps generous, standard deviation of the difference between two treatment means is $\sqrt{2(21.40)/30} = \sqrt{1.43} = 1.2$ days. This should add weight to the investigator's assumption of interaction. An exact test is available by the method of the previous section; a simpler test, applicable only when at least one criterion has but two categories, is shown later.

The computing procedure is based on a theorem which states that the variance of a sum of independent observations is the sum of the variances. For example, in the case of means, the variance of $\bar{x}_{11\cdot} + \bar{x}_{12\cdot}$ is $\sigma^2/54 + \sigma^2/52$ and the variance of the unweighted mean $(\bar{x}_{11\cdot} + \bar{x}_{12\cdot})/2$ is $(1/2^2)(1/54 + 1/52)\sigma^2$; also, the variance of the unweighted sum of column 1 means is $(1/54 + 1/37 + 1/37 + 1/28 + 1/46)\sigma^2 = 0.130026\sigma^2$ and of the unweighted mean is $(1/5^2)(0.130026)\sigma^2$. The numbers 2 and 5 in the divisors are the numbers of row and column means in the unweighted means.

The use of unweighted means is dictated by the assumed presence of interaction. For example, the unweighted mean for column 1 is $(\sum_i \bar{x}_{i1\cdot})/r = \mu + (\Sigma\alpha_i)/5 + \beta_1 + [\sum_i (\alpha\beta)_{i1}]/5 + (\sum_i \bar{\epsilon}_{i1\cdot})/5$ and for column 2 mean is $(\sum_i \bar{x}_{i2\cdot})/r = \mu + (\Sigma\alpha_i)/5 + \beta_2 + [\sum_i (\alpha\beta)_{i2}]/5 + (\sum_i \bar{\epsilon}_{i2\cdot})/5$. A variance among these unweighted means is attributable to β's and ϵ's only. However, if weighted means were used, a variance among them would be attributable to α's, β's, $(\alpha\beta)$'s, and ϵ's since disproportionality would lead to $\sum_i n_{i1}\alpha_i$ and $\sum_i n_{i1}(\alpha\beta)_{i1}$ in $\bar{x}_{\cdot 1\cdot}$, and $\sum_i n_{i2}\alpha_i$ and $\sum_i n_{i2}(\alpha\beta)_{i2}$ in $\bar{x}_{\cdot 2\cdot}$; similarly for the row means. Thus it would be impossible to compute a variance among the column means that would be free from the α and $\alpha\beta$ effects; similarly for a variance among row means.

The appropriate numerator variance for testing the null hypothesis of no differences among the unweighted row means is a weighted variance with sum of squares

$$c^2 \Sigma w_i \left(\frac{\bar{x}_{i1\cdot} + \bar{x}_{i2\cdot}}{2} - \bar{x}_w\right)^2 = c^2 \left\{ \Sigma w_i \left(\frac{\bar{x}_{i1\cdot} + \bar{x}_{i2\cdot}}{2}\right)^2 - \frac{[\Sigma w_i(\bar{x}_{i1\cdot} + \bar{x}_{i2\cdot})/2]^2}{\Sigma w_i} \right\}$$

(13.9)

where $\bar{x}_w = [\Sigma w_i(\bar{x}_{i1\cdot} + \bar{x}_{i2\cdot})/2]/\Sigma w_i$, c is the number of columns (here $c = 2$), and the w_i's are the reciprocals of the coefficients of the corresponding variances (see Table 13.11). Reciprocals are used because an observation with a low variance has relatively high precision so should receive a greater weight. We have used w_i and v_j for weights and have not developed additional notation for unweighted means. For factor A,

$$A \text{ SS} = c^2 \left\{ \Sigma w_i \left(\frac{\bar{x}_{i1\cdot} + \bar{x}_{i2\cdot}}{2}\right)^2 - \frac{[\Sigma w_i(\bar{x}_{i1\cdot} + \bar{x}_{i2\cdot})/2]^2}{\Sigma w_i} \right\}$$

$$= 2^2 \left[6,427.322 - \frac{(780.118)^2}{95.22836} \right] = 146.144 \text{ with } 4 \, df$$

The individual weighted squares of means are computed by multiplying $(\bar{x}_{i1.} + \bar{x}_{i2.})/2$ by $[w_i(\bar{x}_{i1.} + \bar{x}_{i2.})]/2$ since a column of each of these quantities is available. For factor B, where $r =$ number of rows,

$$B\text{ SS} = r^2 \left\{ \sum_i v_j \left(\frac{\sum_i \bar{x}_{ij.}}{5} \right)^2 - \frac{[\sum_j v_j(\sum_i \bar{x}_{ij.}/5)]^2}{\sum_j v_j} \right\} \quad (13.10)$$

$$= 5^2 \left[1,015.340 - \frac{(121.149)^2}{14.65252} \right] = 341.60 \text{ with } 1 \text{ } df$$

Notice that Eq. (13.10) is of exactly the same form as the right-hand side of Eq. (13.9). In other words, $(\bar{x}_{i1.} + \bar{x}_{i2.})/2 = (\sum_j \bar{x}_{ij.})/2$ or $(\sum_j \bar{x}_{ij.})/c$ where c, the number of columns, will be greater than 2 in the general case. Also $(\sum_i \bar{x}_{ij.})/5 = (\sum_i \bar{x}_{ij.})/r$ where r is used in the general case to indicate the number of rows.

For the case where at least one of the factors has only two categories, interaction may also be measured, as in Eq. (13.11). For this, we need the last two columns of Table 13.11.

$$\text{Interaction SS} = \Sigma w_i d_i^2 - \frac{(\Sigma w_i d_i)^2}{\Sigma w_i} \quad (13.11)$$

$$= 536.810 - \frac{(176.751)^2}{95.22836} = 208.747$$

It measures the variance of differences so is an interaction in the two-factor case. The use of differences in computing interaction has been illustrated in several other instances.

We are now able to complete the analysis of variance table, Table 13.12.

TABLE 13.12

COMPLETED ANALYSIS OF VARIANCE FOR DATA OF TABLE 13.11

Source	df	SS	MS
A	4	146.14	36.54
B	1	341.60	341.60**
AB	4	208.75	52.19*
Error	374	8,003.13	21.40

The AB interaction is specially marked since it is available only when at least one effect is at two levels. The procedure for main effects is applicable for any $r \times c$ table with no missing cells.

Since interaction is significant, estimates of simple effects as well as main effects are obtained from unweighted means. The reasons are clear from the discussion given with the computing procedure. If there is no interaction, these estimates are still valid but are inefficient and, therefore, are not recommended.

ANALYSIS OF VARIANCE V: UNEQUAL SUBCLASS NUMBERS

Exercise 13.4.1 Show the algebraic expressions for the means of the observations in Table 13.11. (These expressions will be in terms of parameters and means of ϵ's.) From this table, it is apparent why weighted means are not used for the computation of main effects and their sums of squares. Show why differences are appropriate for the computation of a valid interaction sum of squares.

Exercise 13.4.2 The data in the accompanying table appeared in "Query 100" in *Biometrics* (13.4). The difference in response to litter size for mating PP appears to be different from the other responses. The data are weights, in grams, of foetal membranes in swine at the twenty-fifth day of gestation. The error mean square is 7.73.

Analyze the data by the method of squares of means and test interaction.

	Litter size and mean			
Mating	Not more than 9		More than 9	
	n_{i1}	$\bar{x}_{i1.}$	n_{i2}	$\bar{x}_{i2.}$
CC	2	12.700	4	9.525
PC	4	8.575	1	5.100
CP	2	10.600	3	7.200
PP	3	8.233	3	8.400

13.5 Disproportionate subclass numbers; methods for $r \times 2$ tables. While the method of fitting constants is always applicable to $r \times 2$ tables, shorter computational procedures are available. Some such procedures will now be discussed. These give the same results in terms of tests of significance but do not always yield estimates of the effects.

A preliminary analysis of variance consisting of among- and within-cells sums of squares and unadjusted main effects sums of squares is usually performed. Next a test of interaction is carried out. This is illustrated in the following examples. The procedure used to complete the analysis depends upon whether or not there is evidence of interaction. When interaction is present, the method of weighted squares of means, given in Sec. 13.4, is used. It may also be used if interaction is not present.

The data in Table 13.13, from C. Lemke, Wisconsin Conservation Department, are average brood sizes of spring released and wild hen pheasants. The preliminary analysis of variance is also given. Sums of squares for main effects must be adjusted to eliminate any contribution from the other effect and from interaction, if present; that is, we must find an additional reduction (in the unexplained variance) which can be attributed to the main effect under test.

Consider the column headed d_i in Table 13.13. For the ith row, d_i is

$$d_i = (\mu + \alpha_i + \beta_1 + (\alpha\beta)_{i1} + \bar{\epsilon}_{i1.}) - (\mu + \alpha_i + \beta_2 + (\alpha\beta)_{i2} + \bar{\epsilon}_{i2.})$$
$$= (\beta_1 - \beta_2) + [(\alpha\beta)_{i1} - (\alpha\beta)_{i2}] + (\bar{\epsilon}_{i1.} - \bar{\epsilon}_{i2.}) \qquad (13.12)$$

A variance of these measures interaction, that is, the failure of the differences to vary no more than experimental error. This is seen from the fact that the

Table 13.13

Average Brood Size of Spring Released and Wild Pheasant Hens at Four Locations; Computations for the Analysis of Variance

Loca-tion	A Spring released, b_1		B Wild, b_2		$w_i = \dfrac{n_{i1}n_{i2}}{n_{i1}+n_{i2}}$	d_i	$w_i d_i$	$\bar{x}_{i1\cdot} + \bar{x}_{i2\cdot}$	$w_i(\bar{x}_{i1\cdot} + \bar{x}_{i2\cdot})$
	n_{i1}	$\bar{x}_{i1\cdot}$	n_{i2}	$\bar{x}_{i2\cdot}$					
a_1	5	3.600	8	7.000	3.0769	3.400	10.4615	10.600	32.6151
a_2	41	7.634	29	8.483	16.9857	0.849	14.4209	16.117	273.7585
a_3	29	7.103	27	7.556	13.9821	0.453	6.3339	14.659	204.9636
a_4	3	6.667	15	9.333	2.5000	2.666	6.6650	16.000	40.0000
Sums					36.5447		37.8813		551.3372

$$\Sigma w_i d_i^2 = 68.4506 \qquad \Sigma w_i(\bar{x}_{i1\cdot} + \bar{x}_{i2\cdot})^2 = 8{,}402.4472$$

Preliminary analysis of variance

Source	df	SS	MS	F
Treatment	7	159.80	22.83	2.47*
A	3	91.36	30.45	
B	1	42.14	42.14	
Error	149	1,379.31	9.26	
Total	156	1,539.11		

Completed analysis of variance

Source	df	SS	MS	F
A, adjusted for B	3	84.61	28.20	2.96
B, adjusted for A	1	39.27	39.27	4.24*
AB, adjusted for A and B	3	29.18	9.73	1.05
Error	149	1,379.31	9.26	
Total	156	1,539.11		

only quantities in the expression that vary with i are $(\alpha\beta)_{i1} - (\alpha\beta)_{i2}$ and $(\bar{\epsilon}_{i1\cdot} - \bar{\epsilon}_{i2\cdot})$. Thus, the interaction sum of squares adjusted for main effects is

$$\text{Interaction SS, adjusted for } A \text{ and } B = \Sigma w_i d_i^2 - \frac{(\Sigma w_i d_i)^2}{\Sigma w_i}$$

$$= 68.4506 - 39.27 = 29.18 \text{ with 3 } df$$

This is Eq. (13.11) again. Interaction is not significant.

The weights $w_i = (n_{i1}n_{i2})/(n_{i1} + n_{i2})$, are the reciprocals of $(1/n_{i1} + 1/n_{i2})$ as described in the previous section and defined in Table 13.11.

If there is no interaction, then either a weighted or an unweighted sum of the differences is suitable for calculation of the main effect B sum of squares. Efficiency dictates a weighted sum. The sum of squares for B adjusted for A is given by Eq. (13.13).

$$\text{SS } (B, \text{ adjusted for } A) = \frac{(\Sigma w_i d_i)^2}{\Sigma w_i} \tag{13.13}$$

$$= (37.8813)^2/36.5447 = 39.27^* \text{ with 1 } df$$

ANALYSIS OF VARIANCE V: UNEQUAL SUBCLASS NUMBERS

This is not a valid estimate of the true B sum of squares if there is a real interaction. It can be seen from Eq. (13.12) that the $[(\alpha\beta)_{i1} - (\alpha\beta)_{i2}]$ terms do not disappear from a weighted sum of d_i's; they disappear from an unweighted sum because the restrictions $\Sigma(\widehat{\alpha\beta})_{i1} = 0 = \Sigma(\widehat{\alpha\beta})_{i2}$, imposed upon the estimates, imply similar restrictions on the $(\alpha\beta)$'s. Equation (13.13) reduces to the usual Q type formula (Sec. 11.6) when the numbers of observations are the same for all means.

In the absence of interaction, an unbiased estimate of the difference between the B means is given by

$$\frac{\Sigma w_i d_i}{\Sigma w_i} \quad (13.14)$$

This gives $37.8813/36.5447 = 1.04$ birds. This difference is significant; $F = 39.27/9.26 = 4.24^*$.

The weighted sum of squares for the A effect is

$$\text{SS}(A, \text{ adjusted for } B) = \Sigma w_i (\bar{x}_{i1\cdot} + \bar{x}_{i2\cdot})^2 - \frac{[\Sigma w_i (\bar{x}_{i1\cdot} + \bar{x}_{i2\cdot})]^2}{\Sigma w_i}$$

$$= 8{,}402.4472 - \frac{(551.3372)^2}{36.5447} = 84.61 \text{ with 3 } df$$

This computing formula is equivalent to the right side of Eq. (13.9). The sum of squares is an unbiased estimate of the true A sum of squares.

Unbiased estimates of the differences between A means are obtained by the method of fitting constants, which requires quite extensive calculations. Instead, an approximation involving two levels of A at a time is generally used. The procedure is illustrated for the first two levels of A in Table 13.14. When there is no real interaction, a weighted mean difference gives a valid

TABLE 13.14

APPROXIMATE METHOD OF OBTAINING AND TESTING THE DIFFERENCE BETWEEN TWO A MEANS TAKEN FROM AN $r \times 2$ TABLE

Type of release, B	Location, A				w_i	d_i	$w_i d_i$
	a_1		a_2				
	n_{i1}	$\bar{x}_{i1}$	n_{i2}	$\bar{x}_{i2}$			
b_1	5	3.600	41	7.634	4.4565	4.034	17.9775
b_2	8	7.000	29	8.483	6.2703	1.483	9.2989
Sums					10.7268		27.2764

$$\text{Weighted mean difference} = \frac{\Sigma w_i d_i}{\Sigma w_i} = \frac{27.2764}{10.7268} = 2.54 \text{ birds}$$

$$\text{Standard error of the mean difference} = \sqrt{\frac{E_e}{\Sigma w_i}} = \sqrt{\frac{9.26}{10.7268}} = 0.93 \text{ birds}$$

$$t = \frac{2.54}{0.93} = 2.73^{**} \text{ compared with } t_{.01} \text{ for 149 } df$$

estimate of the mean of the population of differences and is more efficient than an unweighted mean difference. A test of significance is applied but the investigator is cautioned against testing differences suggested by the data. Only planned-for comparisons can be compared logically by this procedure.

The completed analysis of variance is also shown in Table 13.13.

A significant interaction is evidence that differences among the levels of A are dependent upon the level of B. Before testing main effects, one must be convinced that such tests are meaningful, that is, that the populations specified by the marginal totals and means are meaningful. For calculating variances, use weighted squares of means; the method of differences is no longer applicable to the two-category classification. To estimate means and differences, use unweighted cell means. An example was given in the previous section.

13.6 Disproportionate subclass numbers; 2×2 tables. While the general methods are still applicable, special computing procedures have been developed for 2×2 tables. These methods depend upon the presence or absence of interaction although the initial steps are the same.

The data in Table 13.15 are ages at puberty of inbred and outbred Holstein heifers that did and did not have scours. These data are used to illustrate the procedure when there is *no real interaction*.

TABLE 13.15

AGE AT PUBERTY OF INBRED AND OUTBRED HOLSTEIN HEIFERS WITH AND WITHOUT SCOURS

	Inbred, b_1			Outbred, b_2			Totals	
	n_{i1}	$X_{i1\cdot}$	$\bar{x}_{i1\cdot}$	n_{i2}	$X_{i2\cdot}$	$\bar{x}_{i2\cdot}$	$n_{i\cdot}$	$X_{i\cdot\cdot}$
Scours, a_1	17	6,991	411.24	24	9,609	400.38	41	16,600
No scours, a_2	10	4,176	417.60	16	5,774	360.88	26	9,950
Totals	27	11,167	(413.59 $= \bar{x}_{\cdot 1 \cdot}$)	40	15,383	(384.60 $= \bar{x}_{\cdot 2 \cdot}$)	67	26,550

SOURCE: Data courtesy of L. E. Casida, University of Wisconsin, Madison, Wis.

Preliminary analysis of variance

Source	df	SS	MS	F
Treatments	3	28,806.34	9,602.11	3.56*
A, unadjusted	1	7,831.23	7,831.23	
B, unadjusted	1	13,572.87	13,572.87	
Error	63	169,775.83	2,694.85	
Total	66	198,582.17		

First, a preliminary analysis of variance is made. This consists of among- and within-treatments sums of squares plus unadjusted sums of squares for A and B, obtained as in Sec. 7.9. It gives an unbiased estimate of the error sum of squares, that is, the within-cells sum of squares. The remainder of the method is that used in the previous section for an $r \times 2$ table and involves a correction for disproportion.

The four treatment means with respective numbers of observations are entered in Table 13.16. Weights for differences between B means at each

TABLE 13.16

BASIC CALCULATIONS TO OBTAIN ADJUSTED SUMS OF SQUARES FOR MAIN EFFECTS AND INTERACTION; DATA OF TABLE 13.15

	Inbred, b_1		Outbred, b_2		w_i	d_i	$w_i d_i$	$w_i d_i^2$
	n_{i1}	$\bar{x}_{i1\cdot}$	n_{i2}	$\bar{x}_{i2\cdot}$				
Scours, a_1	17	411.24	24	400.38	9.951	10.86	108.0679	1,173.6174
No scours, a_2	10	417.60	16	360.88	6.154	56.72	349.0549	19,798.3939
Sums					16.105		457.1228	20,972.0113
					$= \Sigma w_i$		$= \Sigma w_i d_i$	$= \Sigma w_i d_i^2$

SS (AB, adjusted for A and B), from Eq. (13.11) $= \Sigma w_i d_i^2 - \dfrac{(\Sigma w_i d_i)^2}{\Sigma w_i}$

$= 20{,}972.01 - 12{,}974.93$

$= 7{,}997.08$ with 1 df

SS (B, adjusted for A), from Eq. (13.13) $= \dfrac{(\Sigma w_i d_i)^2}{\Sigma w_i}$

$= \dfrac{(457.1228)^2}{16.105} = 12{,}974.93$

Correction for disproportion $=$ SS (B, unadjusted) $-$ SS (B, adjusted)

$= 13{,}572.87 - 12{,}974.93 = 597.94$

SS (A, adjusted for B) $=$ SS (A, unadjusted) $-$ correction for disproportion

$= 7{,}831.23 - 597.94 = 7{,}233.29$

Completed analysis of variance

Source	df	Mean square	F
A, adjusted for B	1	7,233.29	2.68
B, adjusted for A	1	12,974.93	4.81*
AB, adjusted	1	7,997.08	2.97
Error	63	2,694.85	

level of A are computed as usual (see Table 13.11). Individual means are not weighted; only the differences are. The remaining calculations are shown in the table. To minimize rounding errors in all calculations, carry d_i and w_i to several extra decimal places.

Interaction is a measure of the variability of the differences for one factor at the two levels of the other. Since it is not significant, $F = 2.97$ for 1 and 63 degrees of freedom, main effects require no adjustment for interaction. Adjusted sums of squares for the main effects may be calculated as illustrated, beginning with B, or beginning with A. The correction for disproportion is the same in either case and one usually calculates it only once. The completed analysis is shown in Table 13.16.

Since there is no interaction, efficiency dictates the weighted mean difference for computing an unbiased estimate of the mean difference due to breeding. From Eq. (13.14), we obtain

$$\frac{\Sigma w_i d_i}{\Sigma w_i} = \frac{457.1228}{16.105} = 28.4 \text{ days}$$

This differs from both the difference between weighted means, namely, $\bar{x}_{.1.} - \bar{x}_{.2.} = 413.59 - 384.60 = 28.99$ days, and the difference between unweighted means, namely $[(\bar{x}_{11.} + \bar{x}_{21.})/2 - (\bar{x}_{12.} + \bar{x}_{22.})/2] = 414.4 - 380.6 = 33.8$ days. Equation (13.14) gives 21.3 days as the weighted mean difference for A. The analysis of variance supplies tests of these differences.

When interaction is present, the method of weighted squares of means is used for the analysis of variance and effects are measured by unweighted means. The presence or absence of interaction in the true model may be concluded on the basis of a test of significance or past experience.

The data in Table 13.17 are average ages at puberty of Holstein heifers from two families, O and M, which had or did not have scours. The preliminary analysis of variance is also shown. The procedure of the previous example provides a test of significance for interaction, also shown in the table. Since the evidence is for an interaction, the method of weighted squares of means may be used to estimate mean squares for main effects. An alternative procedure is to use the harmonic mean of all subclass n_{ij}'s; the calculations are less involved in this case which is illustrated below. Computation of interaction is not necessary if it is assumed to exist or obviously is significant. Here it is shown to be significant.

Obtain the harmonic mean, as shown in Eq. (2.4), of the subclass numbers, namely

$$\frac{1}{n_0} = \frac{1}{4}\left(\frac{1}{5} + \frac{1}{5} + \frac{1}{8} + \frac{1}{3}\right) = 0.2145, \quad n_0 = 4.662$$

We now use differences between unweighted means to find

$$\text{SS }(A, \text{ adjusted for } B \text{ and } AB) = \frac{n_0[(\bar{x}_{11.} + \bar{x}_{12.}) - (\bar{x}_{21.} + \bar{x}_{22.})]^2}{4}$$

$$= \frac{4.662[(393.80 + 422.50) - (438.40 + 351.67)]^2}{4} = 801.88$$

$$\text{SS }(B, \text{ adjusted for } A \text{ and } AB) = \frac{n_0[(\bar{x}_{11.} + \bar{x}_{21.}) - (\bar{x}_{12.} + \bar{x}_{22.})]^2}{4}$$

$$= \frac{4.662[(393.80 + 438.40) - (422.50 + 351.67)]^2}{4} = 3{,}924.80$$

$$\text{SS }(AB, \text{ adjusted for } A \text{ and } B) = \frac{n_0[(\bar{x}_{11.} + \bar{x}_{22.}) - (\bar{x}_{12.} + \bar{x}_{21.})]^2}{4}$$

$$= \frac{4.662[(393.80 + 351.67) - (422.50 + 438.40)]^2}{4} = 15{,}529.22$$

TABLE 13.17
AGE AT PUBERTY OF HOLSTEIN HEIFERS FROM TWO FAMILIES, WITH AND WITHOUT SCOURS

	Family, B				w_i	d_i	$w_i d_i$	$w_i d_i^2$
	$O = b_1$		$M = b_2$					
	n_{i1}	$\bar{x}_{i1}$	n_{i2}	$\bar{x}_{i2}$				
Scours, a_1	5	393.80	8	422.50	3.077	−28.70	−88.3099	2,534.4941
No scours, a_2	5	438.40	3	351.67	1.875	86.73	162.6188	14,103.9285
Sums					4.952		74.3089	16,638.4226

SOURCE: Data courtesy of L. E. Casida, University of Wisconsin, Madison, Wis.

Preliminary analysis of variance

Source	df	SS	MS
Treatments	3	16,794.00	5,598.00
Error	17	52,664.67	3,097.92
Total	20	69,458.67	

$$\text{SS } (AB, \text{ adjusted for } A \text{ and } B) = \Sigma w_i d_i^2 - \frac{(\Sigma w_i d_i)^2}{\Sigma w_i}$$

$$= 16,638.42 - 1,115.07 = 15,523.35$$

$$F = \frac{15,523.35}{3,097.92} = 5.01, 1 \text{ and } 17 \text{ } df$$

SS(AB) was previously computed as 15,523.35; the difference between the two is attributable to rounding errors. Observe that the above sums of squares are closely related to the Q type comparisons and their sums of squares as developed in Sec. 11.6; the only real difference is the use of the multiplier n_0. The completed analysis is given in Table 13.18.

TABLE 13.18
COMPLETED ANALYSIS OF VARIANCE FOR DATA OF TABLE 13.17

Source	df	Mean square	F
A, adjusted for B and AB	1	801.88	0.26
B, adjusted for A and AB	1	3,924.80	1.27
AB, adjusted for A and B	1	15,529.22	5.01*
Error	17	3,097.92	

Since interaction is present, unweighted means and differences between them are used to obtain unbiased estimates of main effects and interactions.

These are shown in the computations giving mean squares. Because of the significant interaction, interest is likely to be in the simple effects.

Exercise 13.6.1 J. L. Adams, University of Wisconsin, compared the effect of adding diethylstilbestrol to the ration, on gain in weight for two breeds of turkeys. The within subclass mean square is 0.1192.
Analyze the data shown in the accompanying table.

	Breed I		Breed II	
	n_{i1}	$\bar{x}_{i1\cdot}$	n_{i2}	$\bar{x}_{i2\cdot}$
Control	18	3.8902	28	2.4599
Treated	15	3.6037	27	2.1794

References

13.1 Buch, N. C., W. J. Tyler, and L. E. Casida: "Postpartum estrus and involution of the uterus in an experimental herd of Holstein-Friesian cows," *J. Dairy Sci.*, **38:** 73–79 (1955).

13.2 Carroll, F. D., P. W. Gregory, and W. C. Rollins: "Thyrotropic-hormone deficiency in homozygous dwarf beef cattle," *J. Animal Sci.*, **10:** 916–921 (1951).

13.3 Jafar, S. M., A. B. Chapman, and L. E. Casida: "Causes of variation in length of gestation in dairy cattle," *J. Am. Sci.*, **9:** 593–601 (1950).

13.4 "Query 100," *Biometrics*, **9:** 253–255 (1953).

13.5 Yates, F.: "The analysis of multiple classifications with unequal numbers in the different subclasses," *J. Am. Stat. Assoc.*, **29:** 51–66 (1934).

Chapter 14

MULTIPLE AND PARTIAL REGRESSION AND CORRELATION

14.1 Introduction. Chapter 9 dealt with linear regression where values of the dependent variable must be drawn at random. Chapter 10 was concerned with the measurement of the strength of the linear association or the *total* or *simple correlation* between two variables where pairs of observations must be drawn at random. Multiple regression and correlation, discussed in this chapter, involve linear relations among more than two variables.

The simple correlation may not be what is desired in situations where the dependent variable is influenced by several independent variables. Consider the data reported by Drapala (14.6) where the following total correlations are for 152 F_2 plants of Sudan grass and Sudan grass-sorghum hybrids.

$$\text{Green yield and coarseness} \quad r_{yc} = +0.554$$

$$\text{Green yield and height} \quad r_{yh} = +0.636$$

$$\text{Coarseness and height} \quad r_{ch} = +0.786$$

From the simple correlation r_{yc} we cannot conclude that the relationship between yield and coarseness will be the same if a different set of values of height is considered. The correlation between yield and coarseness at a single value of height, called a *partial correlation*, is $+0.113$ and is nonsignificant. Thus when height is held constant, that is, for plants of a given height, there is no apparent relationship between yield and coarseness. The simple correlation between coarseness and yield appears to be largely a reflection of the close relation between coarseness and height which is also closely related to yield. That is, most of the causes producing correlation between height and coarseness appear to be the same as those producing correlation between height and yield. Partial linear correlation and regression are discussed within the framework of multiple linear correlation and regression.

When interest is primarily in estimation or prediction of values of one character from knowledge of several other characters, multiple regression and correlation give the combined effect of several variables on the one of primary concern. For the data referred to, the *multiple correlation* of height and coarseness with yield is 0.642.

14.2 The linear equation and its interpretation in more than two dimensions.
Some of the ideas and terms of the introduction will be clearer if related to Fig. 14.1, which shows a plane in a three-dimensional space.

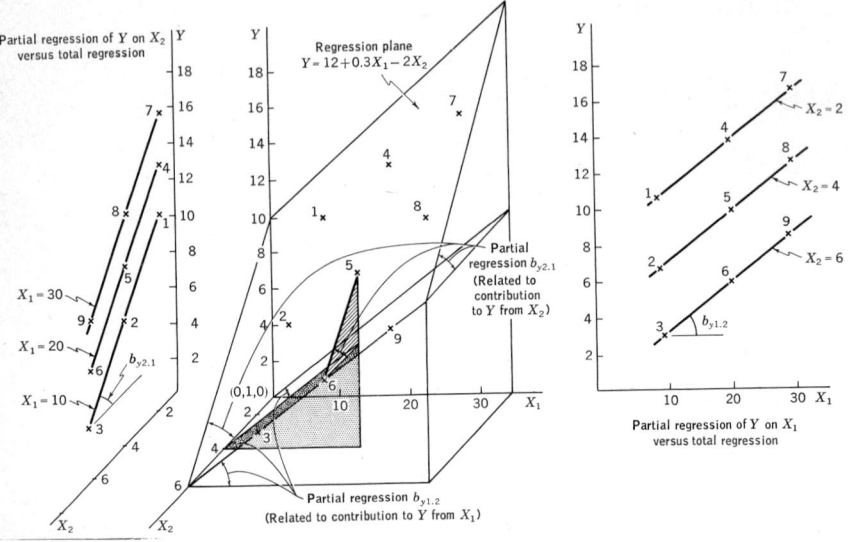

FIG. 14.1 Multiple regression surface

The equation of a plane is

$$Y = a + b_1 X_1 + b_2 X_2 \tag{14.1}$$

The ideas and arithmetic are easily generalized to include more than two independent variables. The plotted points in Fig. 14.1 have the coordinates

Point No.	X_1	X_2	Y
1	10	2	11
2	10	4	7
3	10	6	3
4	20	2	14
5	20	4	10
6	20	6	6
7	30	2	17
8	30	4	13
9	30	6	9

To construct a physical model of such a surface, cut two identical right-angled triangles and place them parallel to the X_1, Y plane, and with the right angles at the same X_1 value; see one at the front of the figure with corner at $X_2 = 6$ and base perpendicular to the X_2 axis and the other at the back on the X_1 axis with $X_2 = 1$. (The axes meet at 0,1,0 for illustrative convenience.)

Now cut two other identical right-angled triangles, not necessarily the same as the first pair, and place them parallel to the X_2,Y plane, resting on the previous two triangles and with their right angles at the same X_2 value; see one on the X_2,Y plane and the other resting on the rectangle parallel to the X_2,Y plane at $X_1 = 35$. The plane resting on the latter triangles is the desired plane. In Fig. 14.1, the equation of the plane is $y = 12 + 0.3X_1 - 2X_2$.

From the method of constructing this plane, it is clear that all lines on it which are parallel to the Y,X_1 plane have the same slope, that of the first pair of triangles. Similarly, all lines on the constructed plane which are parallel to the Y,X_2 plane have the same slope, that of the second pair of triangles and probably different from that of the other pair.

A point on the plane has three components:

1. A base value or Y reference point. In Fig. 14.1, this could be the X_1,X_2 plane shown. In Eq. (14.1), it is the value $Y = a$. In the illustration, $a = 12$.

2. A contribution due to X_1 alone. In Fig. 14.1, it is a height from the X_1,X_2 plane to the plane on the first pair of triangles and is measured, in our figure, as the product of the distance of X_1 from the X_2 axis by the slope, $b_{Y1\cdot 2}$, of the plane resting on the first two triangles (see point 5). This contribution is seen to be independent of X_2. In Eq. (14.1), the contribution is b_1X_1. In the illustration, $b_1X_1 = 0.3X$.

3. A contribution due to X_2 alone. In Fig. 14.1, it is a height from the plane resting on the first two triangles to the plane resting on the second two triangles and is measured as the product of the distance of X_2 from $X_2 = 6$ by the slope, $b_{Y2\cdot 1}$, of the plane resting on the second two triangles (see point 5). This contribution is independent of X_1. In Eq. (14.1), the contribution is b_2X_2. In the illustration, $b_2X_2 = -2X_2$.

14.3 Partial, total, and multiple linear regression. Any sample regression equation determines or supplies an estimate of a population mean. A multiple linear regression equation has more than one independent variable, all to the first power. For a population, parameters are concerned and a regression equation can be written in the form

$$\mu_{Y\cdot 12\cdots k} = \mu_{Y\cdot \overline{12\cdots k}} + \beta_1 x_1 + \cdots + \beta_k x_k \tag{14.2}$$

where $\mu_{Y\cdot 12\cdots k}$ represents the mean Y for the population of Y's at a fixed set of $X_1, \ldots, X_k$ and $\mu_{Y\cdot \overline{12\cdots k}}$ represents the population mean when the fixed set is $(\bar{x}_1, \ldots, \bar{x}_k)$. In practice β_i is written as $\beta_{yi\cdot 1\cdots i-1, i+1\cdots k}$ and may be read as the *regression of Y on X_i for fixed values of the other variables* or as the *partial regression of Y on X_i*. When such an equation arises from a sample, we write

$$\hat{Y} = \hat{\mu}_{Y\cdot 1\cdots k} = \bar{y} + b_1 x_1 + \cdots + b_k x_k \tag{14.3}$$

where $b_i = b_{yi\cdot 1\cdots i-1, i+1\cdots k}$. The estimate of $\mu_{Y\cdot 1\cdots k}$ is $\bar{y} + b_1 x_1 + \cdots + b_k x_k$. Consequently, we customarily write the mathematical description as Eq. (14.4).

$$Y_i = \mu + \beta_1 x_{1i} + \cdots + \beta_k x_{ki} + \epsilon_i \tag{14.4}$$

For a sample, an observed point will rarely lie on the regression plane but will be above or below it. Thus to the components of any point on a plane as listed in Sec. 14.2, add a random component to give a sample point. In other words, the illustration used there gives values of $\mu_{Y \cdot 1 \cdots k}$.

In Fig. 14.1, $b_{y1 \cdot 2}$ and $b_{y2 \cdot 1}$ are slopes or angles of the triangles used to construct the regression plane. As a slope, $b_{y1 \cdot 2}$ measures the increase in Y per unit of X_1 for any value of X_2. This contribution is independent of X_2 because of the nature of a plane. This can be seen in the right-hand portion of Fig. 14.1 where Y, X_1 values are plotted for the different values of X_2. The same applies to $b_{y2 \cdot 1}$ and the left-hand portion of Fig. 14.1. Values of $b_{y1 \cdot 2}$ and $b_{y2 \cdot 1}$ are called *partial regression coefficients*.

From the right-hand part of Fig. 14.1, the distinction between total or simple and partial regression is evident. The figure was constructed so that all points fell on the regression plane. A total regression, say of Y on X_1, ignores the observed X_2 and is concerned with the nine points in the X_1, Y plane. Clearly the total regression of Y on X does not account for much variation in Y. Also this regression can be greatly changed without affecting the partial regression. For example, if $X_2 = 4$ had been observed for $X_1 = 20, 30$, and 40, and $X_2 = 2$ for $X_1 = 30, 40$, and 50, and if the Y's were still on the given regression plane, the total regression of Y on X_1 would seem quite different. The right-hand part of Fig. 14.1 may be altered to illustrate this point.

Similar comments apply to total and partial correlation. The *partial correlation* of Y on X_1, denoted by $r_{y1 \cdot 2}$, is a measure of the association of Y and X_1 at $X_2 = 2$, at $X_2 = 4$, and at $X_2 = 6$. Clearly this is perfect correlation. The total or simple correlation between Y and X_1 ignores X_2 and is seen, from the right-hand part of Fig. 14.1, to be rather low. The example in the introduction was of the opposite type, with a high total correlation masking a low partial correlation between yield and coarseness for fixed height.

A *multiple correlation* coefficient measures the closeness of association between the observed Y values and a function of the independent values. In sample situations, the function is that which determines the regression plane. The multiple correlation coefficient is denoted by $R_{y \cdot 12 \cdots k}$.

14.4 The sample multiple linear regression equation. By definition, a sample regression equation supplies estimates of population means. In practice, it may also be used to predict events. The contribution to an observation due solely to regression is defined as

$$\hat{y} = \hat{Y} - \bar{y}$$
$$= b_1 x_1 + b_2 x_2 + \cdots + b_k x_k \tag{14.5}$$

where b_i is the partial regression of y on X_i. The latter expression comes directly from Eq. (14.3). A deviation from regression is given by Eq. (14.6).

$$d = Y - \hat{Y}$$
$$= y - \hat{y}$$
$$= y - b_1 x_1 - \cdots - b_k x_k \tag{14.6}$$

The regression analysis, based on least squares, is such that

$$\Sigma(Y - \hat{Y}) = 0$$

and $\Sigma(Y - \hat{Y})^2 = \text{minimum}$ (14.7)

The computational procedure requires the solution of a set of equations in the b's. These equations are easily obtained from Eq. (14.8).

$$b_1 x_1 + \cdots + b_k x_k = y \tag{14.8}$$

This is obtained from Eq. (14.6) by setting $d = 0$. To get the set of equations, multiply Eq. (14.8) by x_1 and then sum over all observations to obtain an equation associated with b_1. Repeat the process with x_2 to obtain an equation associated with b_2, and so on. Thus we obtain

$$\begin{aligned}
x_1: & \quad b_1 \Sigma x_1^2 + b_2 \Sigma x_1 x_2 + \cdots + b_k \Sigma x_1 x_k = \Sigma x_1 y \\
x_2: & \quad b_1 \Sigma x_2 x_1 + b_2 \Sigma x_2^2 + \cdots + b_k \Sigma x_2 x_k = \Sigma x_2 y \\
& \quad \cdots\cdots\cdots\cdots\cdots\cdots\cdots\cdots\cdots\cdots\cdots\cdots\cdots \\
x_k: & \quad b_1 \Sigma x_k x_1 + b_2 \Sigma x_k x_2 + \cdots + b_k \Sigma x_k^2 = \Sigma x_k y
\end{aligned} \tag{14.9}$$

These are called *normal equations*.

Solution, by direct methods, of Eqs. (14.9) for the b's becomes increasingly time-consuming as the number of independent variables increases. Consequently a tabular general procedure, called the abbreviated Doolittle method, will be illustrated in Sec. 14.10 for a problem with three independent variables. Regression and correlation aspects of the problem will be discussed in Secs. 14.7 and 14.8 for a problem with two independent variables where the arithmetic is less involved. The latter problem is begun in Sec. 14.5.

Exercise 14.4.1 Compute $\Sigma x_1 x_2$, $\Sigma x_1 y$, $\Sigma x_2 y$ for the illustration in Sec. 14.2. Note the value of $\Sigma x_1 x_2$. It is this property of orthogonality that makes the solution of Eqs. (14.9) so simple and permits us to compute independent contributions, as in the case of orthogonal polynomials where X corresponds to the present X_1 and X^2 to the present X_2.

14.5 Multiple linear regression equation; three variables. Table 14.1 gives percentages of nitrogen X_1, chlorine X_2, and potassium X_3, and log of leaf burn in seconds, Y, for 30 samples of tobacco taken from farmers' fields. Of these four variables, the first two independent variables and the dependent one will be used in this section to determine a regression equation. The data for all four variables will be used to illustrate the procedure for any number of independent variables starting in Sec. 14.9.

To obtain the regression equation, proceed as follows:

1. Compute sums, sums of squares, and sums of cross products for the raw data. These are given at the foot of Table 14.1.
2. Adjust the sums of squares and cross products for the means. Adjusted values are shown in Table 14.2 with the simple correlation coefficients.

TABLE 14.1
Percentages of Nitrogen X_1, Chlorine X_2, Potassium X_3, and Log of Leaf Burn in Seconds Y, in Samples of Tobacco from Farmers' Fields

Sample No.	Nitrogen % X_1	Chlorine % X_2	Potassium % X_3	Log of leaf burn Y, sec
1	3.05	1.45	5.67	0.34
2	4.22	1.35	4.86	0.11
3	3.34	0.26	4.19	0.38
4	3.77	0.23	4.42	0.68
5	3.52	1.10	3.17	0.18
6	3.54	0.76	2.76	0.00
7	3.74	1.59	3.81	0.08
8	3.78	0.39	3.23	0.11
9	2.92	0.39	5.44	1.53
10	3.10	0.64	6.16	0.77
11	2.86	0.82	5.48	1.17
12	2.78	0.64	4.62	1.01
13	2.22	0.85	4.49	0.89
14	2.67	0.90	5.59	1.40
15	3.12	0.92	5.86	1.05
16	3.03	0.97	6.60	1.15
17	2.45	0.18	4.51	1.49
18	4.12	0.62	5.31	0.51
19	4.61	0.51	5.16	0.18
20	3.94	0.45	4.45	0.34
21	4.12	1.79	6.17	0.36
22	2.93	0.25	3.38	0.89
23	2.66	0.31	3.51	0.91
24	3.17	0.20	3.08	0.92
25	2.79	0.24	3.98	1.35
26	2.61	0.20	3.64	1.33
27	3.74	2.27	6.50	0.23
28	3.13	1.48	4.28	0.26
29	3.49	0.25	4.71	0.73
30	2.94	2.22	4.58	0.23
ΣX_i	98.36	24.23	139.61	20.58
$\bar{x}$	3.28	0.81	4.65	0.69
ΣX_i^2	332.3352	30.1907	682.7813	20.8074

$\Sigma X_i X_j$	$\Sigma X_1 X_2 = 81.5834$	$\Sigma X_1 X_3 = 459.4052$	$\Sigma X_1 Y = 61.6502$
	$\Sigma X_2 X_3 = 120.3950$	$\Sigma X_2 Y = 12.4103$	$\Sigma X_3 Y = 98.4408$

SOURCE: Data obtained through the courtesy of O. J. Attoe, University of Wisconsin, Madison, Wis.

The multiple regression equation is given by Eq. (14.1) or Eq. (14.3) for $k = 2$. Respectively,

$$\hat{Y} = a + b_1 X_1 + b_2 X_2$$

and

$$\hat{Y} = \bar{y} + b_1 x_1 + b_2 x_2 \qquad (14.10)$$

TABLE 14.2

SUMS OF PRODUCTS AND SIMPLE CORRELATIONS
FOR DATA OF TABLE 14.1

	X_1	X_2	Y
X_1	$\Sigma x_1^2 = 9.8455$	$\Sigma x_1 x_2 = 2.1413$ $r_{12} = 0.2094$	$\Sigma x_1 y = -5.8248$ $r_{y1} = -0.7177$
X_2		$\Sigma x_2^2 = 10.6209$	$\Sigma x_2 y = -4.2115$ $r_{y2} = -0.4998$
Y			$\Sigma y^2 = 6.6895$

The normal equations are a special case of Eqs. (14.9), namely, Eqs. (14.11).

$$\begin{aligned} x_1: & \quad b_1 \Sigma x_1^2 + b_2 \Sigma x_1 x_2 = \Sigma x_1 y \\ x_2: & \quad b_1 \Sigma x_1 x_2 + b_2 \Sigma x_2^2 = \Sigma x_2 y \end{aligned} \qquad (14.11)$$

3. Write out and solve the normal equations. We obtain

$$9.8455 b_1 + 2.1413 b_2 = -5.8248$$

$$2.1413 b_1 + 10.6209 b_2 = -4.2115$$

for which $\quad b_1 = -0.529 \quad$ and $\quad b_2 = -0.290$

where b_1 is the number of leaf burn units per percentage of nitrogen for fixed chlorine percentage. Interpret b_2 similarly.

4. Write the multiple regression equation. Here

$$Y = 2.66 - 0.529 X_1 - 0.290 X_2$$

From a log leaf burn in seconds of 2.66 as a starting point, log leaf burn in seconds decreases by .529 units for each per cent increase of nitrogen X_1, and by .290 units for each per cent increase of chlorine X_2.

Exercise 14.5.1 Solve the equations referred to in Exercise 14.4.1 and show that the regression equation of Sec. 14.2 is correct. Compare the total and partial regression coefficients. Explain this particular result.

Exercise 14.5.2 Birch (14.2) collected the accompanying data on maize in a study of phosphate response, base saturation, and silica relationship in acid soils. Percentage of response is measured as the difference between yield on plots receiving P and those not receiving P, divided by yield on plots receiving no P, and multiplied by 100. Hence, a correlation between Y and X_1 has been introduced by the computing procedure. BEC refers to base exchange capacity.

Compute the regression equation for the linear regression of Y on X_1 and X_3.

Y = response to phosphate, %	X_1 = Control yield, lb grain/acre	X_2 = saturation of BEC, %	X_3 = pH of soil
88	844	67	5.75
80	1,678	57	6.05
42	1,573	39	5.45
37	3,025	54	5.70
37	653	46	5.55
20	1,991	62	5.00
20	2,187	69	6.40
18	1,262	74	6.10
18	4,624	69	6.05
4	5,249	76	6.15
2	4,258	80	5.55
2	2,943	79	6.40
−2	5,092	82	6.55
−7	4,496	85	6.50

14.6 Standard partial regression coefficients. Standard partial regression coefficients, denoted by b'_i or $b'_{yi\cdot 1\cdots i-1, i+1\cdots k}$, are the partial regression coefficients when each variable is in standard measure, that is, is a deviation from the mean in units of its standard deviation. For example

$$x'_i = \frac{X_i - \bar{x}_i}{s_i}$$

is a standard deviate. The standard regression equation is

$$\hat{y}' = b'_1 x'_1 + \cdots + b'_k x'_k \tag{14.12}$$

or

$$\frac{\hat{Y} - \bar{y}}{s_y} = b'_1 \frac{X_1 - \bar{x}_1}{s_1} + \cdots + b'_k \frac{X_k - \bar{x}_k}{s_k} \tag{14.13}$$

A comparison of Eqs. (14.3) and (14.13) shows that

$$b'_i = b_i \frac{s_i}{s_y} \quad \text{and} \quad b_i = b'_i \frac{s_y}{s_i} \tag{14.14}$$

Since each b' is independent of the original units of measurement, a comparison of any two indicates the relative importance of the independent variables involved. Thus if b'_1 is twice the size of b'_2, then X_1 is approximately twice as important as X_2 in estimating or predicting Y.

If the partial regression coefficients, sometimes called the net partial regression coefficients, have been computed, then the standard partial regression coefficients may be computed using Eq. (14.14). For the example of Sec. 14.5,

$$b'_1 = b_1 \frac{s_1}{s_y} = -0.529 \frac{(0.5827)}{(0.4803)} = -0.642$$

$$b'_2 = b_2 \frac{s_2}{s_y} = -0.290 \frac{(0.6051)}{(0.4803)} = -0.365$$

When the standard equation is to be found directly from the original data, the necessary equations are Eqs. (14.9) with x_i and y replaced by x'_i and y'. Now $\Sigma x'^2_i/(n-1) = 1$ and $\Sigma x'_i x'_j/(n-1) = r_{ij}$, and so on. Thus for two independent variables, the equations are

$$b'_1 + r_{12} b'_2 = r_{y1}$$
$$r_{12} b'_1 + b'_2 = r_{y2} \quad (14.15)$$

Solving Eqs. (14.15) gives Eqs. (14.16).

$$b'_1 = \frac{r_{y1} - r_{y2} r_{12}}{1 - r^2_{12}}$$
$$b'_2 = \frac{r_{y2} - r_{y1} r_{12}}{1 - r^2_{12}} \quad (14.16)$$

For the example of Sec. 14.5, we obtain

$$b'_1 = \frac{-0.7177 - (-0.4998)(0.2094)}{1 - (0.2094)^2} = -0.641$$

$$b'_2 = \frac{-0.4998 - (-0.7177)(0.2094)}{1 - (0.2094)^2} = -0.366$$

We see from b'_1 and b'_2 that X_1 is almost twice as useful as X_2 in estimating or predicting Y. Differences in b''s computed from Eqs. (14.14) and (14.16) are attributable to rounding errors.

Exercise 14.6.1 Compute standard partial regression coefficients for the data of Exercise 14.5.2.

14.7 Partial and multiple correlation; three variables. Section 14.6 was a digression that gave us a method for comparing variables for their relative worth. In this section we return to the general problem and example of Sec. 14.5, and compute partial and multiple correlation coefficients.

Multiple and partial correlation coefficients are strictly applicable only when the total observation, that is $(Y_i, X_{1i}, \ldots, X_{ki})$, is random. However, regardless of the randomness of the observations, these correlation coefficients may be useful for computing and other reasons.

For three variables, the partial correlation coefficients used in conjunction with the partial regression equation are given in Eqs. (14.17).

$$r_{y1 \cdot 2} = \frac{r_{y1} - r_{y2} r_{12}}{\sqrt{(1 - r^2_{y2})(1 - r^2_{12})}}$$
$$r_{y2 \cdot 1} = \frac{r_{y2} - r_{y1} r_{12}}{\sqrt{(1 - r^2_{y1})(1 - r^2_{12})}} \quad (14.17)$$

For the data of Table 14.1, we find

$$r_{y1 \cdot 2} = \frac{-0.7177 - (-0.4998)(0.2094)}{\sqrt{[1 - (-0.4998)^2][1 - (0.2094)^2]}} = -0.7238$$

$$r_{y2 \cdot 1} = \frac{-0.4998 - (-0.7177)(0.2094)}{\sqrt{[1 - (-0.7177)^2][1 - (0.2094)^2]}} = -0.5133$$

Thus -0.7238 is the partial correlation between Y and X_1, that is, it is an estimate of the correlation to be expected between Y and X_1 if (Y,X_1) pairs were to be observed for a fixed X_2.

The multiple correlation coefficient, denoted by $R_{y\cdot 12}$ for three variables, measures the closeness with which the regression plane fits the observed points. That is, it is the correlation between the observed Y's and the regression Y's, the $\hat{Y}$'s. Thus $R_{y\cdot 12}$ measures the combined effect of the independent variables X_1 and X_2 on the dependent variable Y. It can be computed in several ways. In general

$$1 - R^2_{y\cdot 1\cdots k} = (1 - r^2_{y1})(1 - r^2_{y2\cdot 1}) \cdots (1 - r^2_{yk\cdot 1\cdots k-1}) \qquad (14.18)$$

For three variables, two independent and one dependent, we have

$$1 - R^2_{y\cdot 12} = (1 - r^2_{y1})(1 - r^2_{y2\cdot 1})$$

Also,
$$R^2_{y\cdot 1\cdots k} = r_{y1}b'_1 + r_{y2}b'_2 + \cdots + r_{yk}b'_k \qquad (14.19)$$

which, for three variables, becomes

$$R^2_{y\cdot 12} = r_{y1}b'_1 + r_{y2}b'_2$$

Again,
$$R_{y\cdot 1\cdots k} = \sqrt{\frac{\Sigma \hat{y}^2}{\Sigma y^2}} \qquad (14.20)$$

Equation (14.5) defines $\hat{y}$. It can be shown that

$$\Sigma \hat{y}^2 = \Sigma(\hat{Y} - \bar{y})^2$$
$$= b_1 \Sigma x_1 y + b_2 \Sigma x_2 y + \cdots + b_k \Sigma x_k y \qquad (14.21)$$
$$= R^2_{y\cdot 12\cdots k} \Sigma(Y - \bar{y})^2 \qquad (14.22)$$

from which Eq. (14.20) is obtained. Equation (14.21) gives the reduction in the total sum of squares Σy^2 attributable to the regression of Y on the set of X's. Equation (14.22) shows this reduction as a fraction of the total sum of squares.

Substituting in Eqs. (14.18), (14.19), and (14.20), we obtain the following values for our example.

From (14.18):

$$1 - R^2_{y\cdot 12} = [1 - (-0.7177)^2][1 - (-0.5133)^2] = 0.3571$$

and $\quad R_{y\cdot 12} = \sqrt{1 - 0.3571} = 0.802$

From (14.19):

$$R^2_{y\cdot 12} = (-0.7177)(-0.642) + (-0.4998)(-0.365) = 0.6428$$

and $\quad R_{y\cdot 12} = \sqrt{0.6432} = 0.802$

From (14.20):

$$R_{y\cdot 12} = \sqrt{\frac{(-0.529)(-5.8248) + (-0.290)(-4.2115)}{6.6895}}$$

$$= 0.802$$

The numerical value of the multiple correlation coefficient is always at least as large as that of any simple or partial correlation, a fact useful in detecting gross errors. The value of a multiple correlation coefficient lies between 0 and $+1$, whereas the simple and partial correlation coefficients can range from -1 through 0 to $+1$.

Exercise 14.7.1 Compute the partial correlation coefficients and the multiple correlation coefficient for the data of Exercise 14.5.2. Do you think the partial correlation coefficients are appropriate?

Exercise 14.7.2 Can you see how the partial correlation coefficient $r_{12\cdot y}$ would be computed?

14.8 Tests of significance; three variables. Partial regression and correlation coefficients are tested for significance by t or F tests. Multiple correlation coefficients are tested by F since more than one degree of freedom is associated with such coefficients; the same applies to multiple regression as a source of variation. Partial regression coefficients are not independent so that tests of significance are not orthogonal; a covariance exists. The same applies to partial correlations. A test of the additional reduction due to introducing a particular independent variable into a regression problem (in the manner of testing an additional reduction due to an added source of variation in Chap. 13) is equivalent to a test of a partial regression or correlation coefficient.

Table A.13 gives 5 and 1% values for partial and multiple correlation coefficients. For partial correlations, use the column headed one independent variable and the degrees of freedom for error in multiple regression. Tests of significance for corresponding partial correlations and partial regressions are essentially the same and only one need be performed. The same applies to the multiple correlation coefficient and multiple regression as a source of variation. The null hypothesis is that the various population correlation or regression coefficients are zero.

The regression analysis for variables X_1, X_2, and Y of Table 14.1 is given in Table 14.3. The highly significant F indicates that, in the total sum of squares of the dependent variable, the reduction due to the combined effect of X_1 and X_2 is not likely the result of chance. The *coefficient of determination*, 100 $R^2 = 64\%$, is the per cent reduction in the sum of squares of Y attributable to the combined effect of X_1 and X_2. The standard deviation is $s_{y\cdot 12} = \sqrt{0.0885} = 0.2975$ units of log leaf burn in seconds.

To determine if X_2 gives information about Y which is not given by X_1, partition the regression sum of squares into two parts:
1. That due to X_1 alone, namely,

$$r_{y1}^2 \Sigma y^2 = (-0.7177)^2(6.6895) = 3.4458, 1 \; df$$

2. The additional reduction due to X_2, namely,

$$R^2_{y \cdot 12}\Sigma y^2 - r^2_{y1}\Sigma y^2 = 4.3027 - 3.4458 = 0.8569, 1\ df$$

The test of significance is

$$F = \frac{\text{additional reduction mean square}}{\text{residual mean square}}$$

$$= \frac{0.8569}{0.0884} = 9.69^{**} \text{ with 1 and 27 } df$$

TABLE 14.3
MULTIPLE REGRESSION ANALYSIS
(n = number of multiple observations; k = number of independent variables)
1. Symbolic

Source of variation	df	Sum of squares	
		Definition	Calculation
Regression	k	$\Sigma(\hat{Y} - \bar{y})^2$	$b_1\Sigma x_1 y + \cdots + b_k\Sigma x_k y$ $= R^2_{y \cdot 1 \ldots k}\Sigma(Y - \bar{y})^2$
Error	$n - k - 1$	$\Sigma(Y - \hat{Y})^2$	Total SS $-$ regression SS $= (1 - R^2_{y \cdot 1 \ldots k})\Sigma(Y - \bar{y})^2$
Total	$n - 1$	$\Sigma(Y - \bar{y})^2$	$\Sigma Y^2 - \dfrac{(\Sigma Y)^2}{n}$

2. Numerical example

Source of variation	df	Sum of squares	MS	F
Regression	2	4.3027	2.1514	24.34**
Error	27	2.3868	0.0884	
Total	29	6.6895		

The quantity

$$t = \frac{r\sqrt{n - k - 1}}{\sqrt{1 - r^2}} \tag{14.23}$$

where r is a partial correlation coefficient, n is the number of multiple observations, and k is the number of independent variables, is distributed as t when $\rho = 0$. To test the null hypothesis $H_0: \rho_{y2 \cdot 1} = 0$, compute

$$t = \frac{r_{y2 \cdot 1}\sqrt{n - 3}}{\sqrt{1 - r^2_{y2 \cdot 1}}}$$

$$= \frac{-0.5133}{\sqrt{[1 - (-0.5133)^2]/27}} = -3.11^{**} \text{ with 27 } df$$

(Ignore sign of t for a two-tailed test.) This is seen to be the equivalent of the preceding F test of the additional reduction due to X_2. Note that $t^2 = 3.11^2 = 9.67$ is within rounding errors of $F = 9.69$.

A test of significance of a partial regression coefficient requires computations given in Sec. 14.12. The test is essentially the same as a test of a standard partial regression coefficient or a partial correlation coefficient. Standard errors of partial regression coefficients depend upon the unit of measurement. However, the standard errors of the standard partial regression coefficients are the same for three variables, namely,

$$s_{b'} = \sqrt{\frac{1 - R^2_{y \cdot 12}}{(1 - r^2_{12})(n - k)}} \qquad (14.24)$$

$$= \sqrt{\frac{1 - (0.802)^2}{[1 - (0.2094)^2](30 - 3)}} = 0.1175 \text{ units}$$

To test the null hypothesis $H_0: \beta'_{y1 \cdot 2} = 0$,

$$t = \frac{b'_{y1 \cdot 2}}{s_{b'}} = \frac{-0.642}{0.1175} = -5.46**$$

To test $H_0: \beta'_{y2 \cdot 1} = 0$,

$$t = \frac{b'_{y2 \cdot 1}}{s_{b'}} = \frac{-0.365}{0.1175} = -3.11**$$

Note that the latter t equals that for the test of $r_{y2 \cdot 1}$.

Exercise 14.8.1 Compute the regression equation for the regression of Y on X_1, X_2, and X_3. (This exercise and the following one may be delayed until Sec. 14.10 has been covered.)

Exercise 14.8.2 The variable X_3 is considered to be a good alternative to X_2 for assessing phosphate responses. Does X_2 give information about Y which is not given by X_3?

14.9 Multiple linear regression; computations for more than three variables. With more than two independent variables, the normal Eqs. (14.9) are customarily solved by use of a tabular method. We shall illustrate the *abbreviated Doolittle* or *Gauss-Doolittle* method. The basic Doolittle method was introduced in 1878 while Doolittle was an engineer with the United States Coast and Geodetic Survey. Dwyer (14.8) presents a number of modifications of the basic method. Computational methods are also given in Agriculture Handbook No. 94 (14.9) and in other places.

An outline of the procedure follows together with an example. The data are given in Table 14.1 for four variables. Anderson and Bancroft (14.1) recommend that data be coded so that each $\Sigma x_i x_j$ lies between 1 and 10. This was not done for the illustration since the data already approach this recommendation.

Write the sums of products and the partial regression coefficients for normal Eqs (14.9) in the following arrays:

$$A = \begin{bmatrix} \Sigma x_1^2 & \Sigma x_1 x_2 & \Sigma x_1 x_3 \\ \Sigma x_2 x_1 & \Sigma x_2^2 & \Sigma x_2 x_3 \\ \Sigma x_3 x_1 & \Sigma x_3 x_2 & \Sigma x_3^2 \end{bmatrix} = \begin{bmatrix} 9.8455 & 2.1413 & 1.6705 \\ 2.1413 & 10.6209 & 7.6367 \\ 1.6705 & 7.6367 & 33.0829 \end{bmatrix}$$

(14.25)

$$G = \begin{bmatrix} \Sigma x_1 y \\ \Sigma x_2 y \\ \Sigma x_3 y \end{bmatrix} = \begin{bmatrix} -5.8248 \\ -4.2115 \\ 2.6683 \end{bmatrix} \qquad B = \begin{bmatrix} b_1 \\ b_2 \\ b_3 \end{bmatrix}, \text{ unknowns}$$

These rectangular arrays are called *matrices*. When A is symmetric, as in a regression problem, one does not ordinarily show the numbers in the triangle when using the abbreviated Doolittle method.

The relation between matrices (14.25) and Eqs. (14.9) is such that Eqs. (14.9) are often written in matrix notation as Eq. (14.26).

$$AB = G \qquad (14.26)$$

Now just as $5b = 3$ is solved to give $b = (1/5)3$, so Eq. (14.26) is solved to give $B = (1/A)G$. However, we write Eq. (14.27).

$$B = A^{-1}G \qquad (14.27)$$

Call A^{-1} the *inverse of A* rather than the reciprocal of A. Actually, A^{-1} is another matrix of real numbers from which B, the set of partial regression coefficients, can be found. The abbreviated Doolittle method is a tabular procedure for finding the inverse of A or of *inverting A*.

Denote the elements of A by a_{ij} for $i,j = 1, 2, 3$, with i and j referring to rows and columns, respectively. Denote the elements of G by g_i, $i = 1, 2, 3$. For example, $a_{11} = \Sigma x_1^2 = 9.8455$, $a_{32} = a_{23} = \Sigma x_2 x_3 = 7.6367$, and $g_2 = \Sigma x_2 y = -4.2115$.

14.10 The abbreviated Doolittle method. This method consists of a *forward solution* and a *backward solution*. The forward solution permits us to find partial regression coefficients and the sum of squares attributable to regression. This is done without computing A^{-1} as required for Eq. (14.27). The backward solution completes the inversion of A and thus supplies constants necessary for computing the standard errors of the partial regression coefficients. The complete method is a short-cut procedure for use with a symmetric A matrix. Table 14.4 is an instruction code describing the following instructions.

Perform the *forward solution* according to the instructions:

1. In a table such as 14.4, enter the A and G matrices as defined in Eqs. (14.25). The check column consists of the sum of the observations in the row; this requires the full symmetric matrix though all need not be filled in. For example, $S_2 = a_{21} + a_{22} + a_{23} + g_2$, where $a_{21} = a_{12} = \Sigma x_1 x_2$, and so on.

2. Copy the first line from instruction 1, then divide each element by a_{11}. For reference, denote the results as indicated in Table 14.4. Thus $B_{13} = A_{13}/A_{11} = a_{13}/a_{11}$. Subscripts y and c, used in the last two columns of the table, do not refer to location within a matrix but to location in instruction 1 and, in addition, call attention to the source; thus B_{1y} comes from a cross product involving y, and B_{1c} comes from a check. Capital A's and B's are used to prevent confusion with the a_{ij}'s and the regression coefficients.

TABLE 14.4

ABBREVIATED DOOLITTLE METHOD FOR MULTIPLE REGRESSION

Instruction		A matrix			G matrix	Check
		x_1	x_2	x_3	y	
1	x_1	a_{11}	a_{12}	a_{13}	g_1	S_1
	x_2		a_{22}	a_{23}	g_2	S_2
	x_3			a_{33}	g_3	S_3
2	A_{1j} B_{1j}	A_{11} 1	A_{12} B_{12}	A_{13} B_{13}	A_{1y} B_{1y}	A_{1c} B_{1c}
3	A_{2j} B_{2j}		A_{22} 1	A_{23} B_{23}	A_{2y} B_{2y}	A_{2c} B_{2c}
4	A_{3j} B_{3j}			A_{33} 1	A_{3y} B_{3y}	A_{3c} B_{3c}
		End of forward solution				
	C_{1j} C_{2j} C_{3j}	C_{11}	C_{12} C_{22}	C_{13} C_{23} C_{33}		
		C_{i1}	C_{i2}	C_{i3}		
Instruction		7	6	5		

3. Compute the A entry for the jth column, $j = 2, 3, y, c$ as $A_{2j} = a_{2j} - A_{12}B_{1j} = a_{2j} - A_{1j}B_{12}$. $A_{12}B_{1j} = A_{1j}B_{12}$ within rounding errors and it is advisable to compute the one for which the numbers are more nearly equal. Compute the B entry by dividing each A by A_{22}. A_{2c} and B_{2c} are also row sums, giving a computational check.

4. Compute A entries as $A_{3j} = a_{3j} - (A_{13}B_{1j} + A_{23}B_{2j}) = a_{3j} - (A_{1j}B_{13} + A_{2j}B_{23})$. Again $A_{13}B_{1j} + A_{23}B_{2j} = A_{1j}B_{13} + A_{2j}B_{23}$ within rounding errors. Compute B entries by dividing each A by A_{33}.

This completes the forward solution for three independent variables. In generalization, for each additional independent variable the method requires an additional column under the A matrix heading, an additional line in instruction 1, and one new instruction requiring two lines of entries. Thus for four independent variables $A_{4j} = a_{4j} - (A_{14}B_{1j} + A_{24}B_{2j} + A_{34}B_{3j}) = a_{4j} - (A_{1j}B_{14} + A_{2j}B_{24} + A_{3j}B_{34})$, and so on.

To minimize errors due to rounding, carry as many significant digits as possible, preferably six or more after coding.

When the forward solution is complete, partial regression coefficients and the sum of squares attributable to multiple regression can be obtained from Eqs. (14.28).

$$b_3 = B_{3y}$$
$$b_2 = B_{2y} - B_{23}b_3 \qquad (14.28)$$
$$b_1 = B_{1y} - B_{12}b_2 - B_{13}b_3$$
$$\text{Regression SS} = b_1\Sigma x_1 y + b_2\Sigma x_2 y + b_3\Sigma x_3 y$$

Check: $b_1\Sigma x_1^2 + b_2\Sigma x_1 x_2 + b_3\Sigma x_1 x_3 = \Sigma x_1 y$

A check is supplied for the b's.

If it is desired to have standard errors for the partial regression coefficients or for an estimated population mean (Sec. 14.12), or to delete a variable (Sec. 14.14), we compute the C matrix by the *backward solution*. C is the inverse of A. The C_{ij}'s are

5. $C_{33} = 1/A_{33}$; $C_{23} = -B_{23}C_{33}$; $C_{13} = -B_{12}C_{23} - B_{13}C_{33}$

Check: $a_{13}C_{13} + a_{23}C_{23} + a_{33}C_{33} = 1$

6. $C_{32} = C_{23}$; $C_{22} = 1/A_{22} - B_{23}C_{23}$; $C_{12} = -B_{12}C_{22} - B_{13}C_{23}$

Check: $a_{12}C_{12} + a_{22}C_{22} + a_{32}C_{32} = 1$

7. $C_{31} = C_{13}$; $C_{21} = C_{12}$; $C_{11} = 1/A_{11} - B_{12}C_{12} - B_{13}C_{13}$

Check: $a_{11}C_{11} + a_{21}C_{21} + a_{31}C_{31} = 1$

These checks may be summarized as $\sum_i a_{ij}C_{ij} = 1, j = 1, 2, 3$. It can also be shown that $\sum_i a_{ij}C_{ik} = 0, j \neq k$, that is, that sums such as $\sum_i a_{i2}C_{i3}$ equal zero. Note that the second subscripts on a and C must be *different*; in the first set of checks, second subscripts are the same.

The partial regression coefficients can be calculated as $b_i = \sum_j (C_{ij}\Sigma x_j y)$.

Recall that $\Sigma x_i y$ is a sum of x_i by y products over an unwritten subscript. Computation of standard errors is left for Sec. 14.12.

In generalizing to more than three independent variables, each additional variable requires an additional computing instruction for the C_{ij}'s. For example, with four independent variables:

$$C_{44} = 1/A_{44} \qquad C_{34} = -B_{34}C_{44} \qquad C_{24} = -B_{23}C_{34} - B_{24}C_{44}$$
$$C_{14} = -B_{12}C_{24} - B_{13}C_{34} - B_{14}C_{44}$$
$$C_{33} = 1/A_{33} - B_{34}C_{43} \qquad C_{23} = -B_{23}C_{33} - B_{24}C_{43}$$
$$C_{13} = -B_{12}C_{23} - B_{13}C_{33} - B_{14}C_{43}$$

and so on. Recall that $C_{ij} = C_{ji}$ and notice the symmetry of the computing formulas. For example, consider C_{13}. Subscripts 1 and 3 are the first on B and the second on C; the center subscripts (second on B and first on C) are the same, begin one higher than the first on C and run to the last permissible. For example, they run 2,2, 3,3, and 4,4 for C_{13}. Thus it is seen that, for any computation, the B's in a row and the C's in a column are required. Only a slight modification is made for the C_{ii}'s.

Cramer (14.5) modifies the above procedure to permit the use of a uniform number of decimal places in the computations. The investigator who is continually faced with regression problems is referred to this computing procedure.

An example. The abbreviated Doolittle method just outlined will be illustrated using the data of Table 14.1. The computations follow in detail and are summarized in Table 14.5.

TABLE 14.5

ABBREVIATED DOOLITTLE METHOD FOR DATA OF TABLE 14.1

Instruction		A matrix			G matrix	Check
		x_1	x_2	x_3	y	
1	x_1	9.8455	2.1413	1.6705	−5.8248	7.8325
	x_2		10.6209	7.6367	−4.2115	16.1874
	x_3			33.0829	2.6683	45.0584
2	A_{1j}	9.8455	2.1413	1.6705	−5.8248	7.8325
	B_{1j}	1	0.217490	0.169671	−0.591621	0.795541
3	A_{2j}		10.155189	7.273383	−2.944662	14.483908
	B_{2j}		1	0.716223	−0.289966	1.426257
4	A_{3j}			27.590100	5.765637	33.355737
	B_{3j}			1	0.208975	1.208975
			End of forward solution			
	C_{1j}	0.106234	−0.021056	−0.000504		
	C_{2j}		0.117065	−0.025960		
	C_{3j}			0.036245		
		C_{i1}	C_{i2}	C_{i3}		
Instruction		7	6	5		

1. Enter the A and G matrix in a suitable table and compute sums for the check column (see Table 14.5).

2. Copy line 1 of instruction 1 as A_{1j}'s and divide each by $A_{11} = 9.8455$ to give the B_{1j}'s. For example, $B_{13} = 1.6705/9.8455 = 0.169671$. B_{1c} is found by division and checked as the sum of the other entries in the line. Such checks here and in later instructions may differ by rounding errors.

3. Compute A_{2j}'s as

$$A_{2j} = a_{2j} - \begin{Bmatrix} A_{12}B_{1j} \\ \text{or} \\ A_{1j}B_{12} \end{Bmatrix}, j = 2,3,y,c$$

For example

$$A_{2y} = -4.2115 - \begin{Bmatrix} (2.1413)(-0.591621) = -1.266838 \\ \text{or} \\ (-5.8248)(0.217490) = -1.266836 \end{Bmatrix} = -2.944664$$

Obtain B_{2j}'s by dividing each A_{2j} by A_{22}. A_{2c} and B_{2c} are found by formula and checked as sums of the other entries in the lines.

4. Compute A_{3j}'s as

$$A_{3j} = a_{3j} - \begin{Bmatrix} A_{13}B_{1j} + A_{23}B_{2j} \\ \text{or} \\ A_{1j}B_{13} + A_{2j}B_{23} \end{Bmatrix}, j = 3,y,c$$

For example,

$$A_{3c} = a_{3c} - \begin{Bmatrix} A_{13}B_{1c} + A_{23}B_{2c} \\ \text{or} \\ A_{1c}B_{13} + A_{2c}B_{23} \end{Bmatrix}$$

$$= 45.0584$$

$$- \begin{Bmatrix} (1.6705)(0.795540) + (7.273383)(1.426257) = 11.702663 \\ \text{or} \\ (7.8325)(0.169671) + (14.483908)(0.716223) = 11.702656 \end{Bmatrix}$$

$$= 33.355737$$

Obtain B_{3j}'s by dividing each A_{3j} by A_{33}. Again A_{3c} and B_{3c} supply checks on the computations since they are found by two methods.

This completes the forward solution. The partial regression coefficients and the reduction attributable to regression are obtained from Eqs. (14.28). Thus,

$$b_3 = B_{3y} = 0.208975$$

$$b_2 = B_{2y} - B_{23}b_3$$
$$= -0.289966 - (0.716223)(0.208975) = -0.439639$$

$$b_1 = B_{1y} - B_{12}b_2 - B_{13}b_3$$
$$= -0.591621 - (0.217490)(-0.439639) - (0.169671)(0.208975)$$
$$= -0.531461$$

Check: $b_1 \Sigma x_1^2 + b_2 \Sigma x_1 x_2 + b_3 \Sigma x_1 x_3 = \Sigma x_1 y$

$$(-0.5314961)(9.8455) + (-0.439639)(2.1413)$$
$$+ (0.208975)(1.6705) = -5.8248$$

Observed $\Sigma x_1 y = -5.8248$

$$\text{Regression SS} = b_1 \Sigma x_1 y + b_2 \Sigma x_2 y + b_3 \Sigma x_3 y$$
$$= (-0.531461)(-5.8248) + (-0.439639)(-4.2115)$$
$$+ (0.208975)(2.6683) = 5.5048$$

Continuing to the backward solution, obtain the C_{ij}'s by the following computations:

5. Compute the C_{i3}'s and check that $\Sigma a_{i3} C_{i3} = 1$. We obtain

$C_{33} = 1/A_{33} = 1/27.590100 = 0.036245$

$C_{23} = -B_{23} C_{33} = -(0.716223)(0.036245) = -0.025960$

$C_{13} = -B_{12} C_{23} - B_{13} C_{33}$
$= -(0.217490)(-0.025960) - (0.169671)(0.036245) = -0.000504$

Check: $\Sigma a_{i3} C_{i3} = (1.6705)(-0.000504) + (7.6367)(-0.025960)$
$+ (33.0829)(0.036245) = 0.999999$

6. Compute the C_{i2}'s and check that $\Sigma a_{i2} C_{i2} = 1$. We obtain

$C_{32} = C_{23} = -0.025960$

$C_{22} = 1/A_{22} - B_{23} C_{23} = 1/(10.155189) - (0.716223)(-0.025960)$
$= 0.117065$

$C_{12} = -B_{12} C_{22} - B_{13} C_{23}$
$= -(0.217490)(0.117065) - (0.169671)(-0.025960) = -0.021056$

Check: $\Sigma a_{i2} C_{i2} = (2.1413)(-0.021056) + (10.6209)(0.117065)$
$+ (7.6367)(-0.025960) = 1.000000$

7. Compute the C_{i1}'s and check that $\Sigma a_{i1} C_{i1} = 1$. We obtain

$C_{31} = C_{13} = -0.000504$

$C_{21} = C_{12} = -0.021056$

$C_{11} = 1/A_{11} - B_{12} C_{12} - B_{13} C_{13}$
$= 1/(9.8455) - (0.217490)(-0.021056) - (0.169671)(-0.000504)$
$= 0.106234$

Check: $\Sigma a_{i1} C_{i1} = (9.8455)(0.106234) + (2.1413)(-0.021056)$
$+ (1.6705)(-0.000504) = 0.999998$

This completes the backward solution.

The partial regressions, already found, can also be obtained using the C values. We have

$$b_1 = C_{11}\Sigma x_1 y + C_{12}\Sigma x_2 y + C_{13}\Sigma x_3 y$$
$$= (0.106234)(-5.8248) + (-0.021056)(-4.2115)$$
$$+ (-0.000504)(2.6683) = -0.531459$$

$$b_2 = C_{21}\Sigma x_1 y + C_{22}\Sigma x_2 y + C_{23}\Sigma x_3 y$$
$$= (-0.021056)(-5.8248) + (0.117065)(-4.2115)$$
$$+ (-0.025960)(2.6683) = -0.439641$$

$$b_3 = C_{31}\Sigma x_1 y + C_{32}\Sigma x_2 y + C_{33}\Sigma x_3 y$$
$$= (-0.000504)(-5.8248) + (-0.025960)(-4.2115)$$
$$+ (0.036245)(2.6683) = 0.208979$$

Check: $b_1 \Sigma x_1^2 + b_2 \Sigma x_1 x_2 + b_3 \Sigma x_1 x_3 = \Sigma x_1 y$

$$(-0.531459)(9.8455) + (-0.439641)(2.1413)$$
$$+ (0.208979)(1.6705) = -5.8248$$

Observed $\Sigma x_1 y = -5.8248$

The multiple regression equation is

$$\hat{Y} = \bar{y} + b_1(X_1 - \bar{x}_1) + b_2(X_2 - \bar{x}_2) + b_3(X_3 - \bar{x}_3)$$
$$= 0.69 - 0.5315 x_1 - 0.4396 x_2 + 0.2090 x_3$$
$$= 1.8175 - 0.5315 X_1 - 0.4396 X_2 + 0.2090 X_3$$

Exercise 14.10.1 See Exercise 14.8.1, which may be carried out by the abbreviated Doolittle method.

14.11 Test of significance of multiple regression. The reduction in sums of squares attributable to regression can be tested for significance by F. The test is shown in analysis of variance presentation in Table 14.6.

TABLE 14.6
TESTING THE SIGNIFICANCE OF REGRESSION

Source	Symbolic		Data of Table 14.1			
	df	SS	df	SS	MS	F
Regression on k variables	k	$\sum_i b_i(\Sigma x_i y)$	3	5.5048	1.8349	40.24**
Residual	$n - k - 1$	By subtraction	26	1.1847	0.0456	
Total	$n - 1$	Σy^2	29	6.6895		

MULTIPLE AND PARTIAL REGRESSION AND CORRELATION 297

The multiple correlation coefficient is

$$R_{y \cdot 123} = \sqrt{\frac{\Sigma \hat{y}^2}{\Sigma y^2}}$$

$$= \sqrt{\frac{5.5048}{6.6895}} = 0.907$$

Note that $R^2_{y \cdot 123}$ is the proportion of the total sum of squares attributable to regression, that is,

$$R^2 \Sigma y^2 = (0.907)^2 (6.6895)$$
$$= 5.5048, \text{ the reduction attributable to regression}$$

A test of the significance of R^2 is equivalent to the test in Table 14.6. Tabulated values for R for 5 and 1% significance levels are given in Table A.13.

Exercise 14.11.1 Compute $R^2_{y \cdot 123}$, the reduction attributable to regression, deviations from regression, and test the significance of $R^2_{y \cdot 123}$ for Exercise 14.10.1 (or Exercise 14.8.1).

14.12 Standard errors and tests of significance for partial regression coefficients. The estimate of the common σ, the standard deviation or standard error of estimate of the populations of Y values, is given by Eq. (14.29).

$$s_{y \cdot 1 \cdots k} = \sqrt{\frac{\Sigma(Y - \hat{Y})^2}{n - k - 1}} \tag{14.29}$$

k is the number of independent variables and $\Sigma(Y - \hat{Y})^2 = \Sigma y^2 - \sum_i b_i(\Sigma x_i y)$, as already seen. For the example,

$$s_{y \cdot 123} = \sqrt{\frac{6.6895 - 5.5048}{26}}$$

$$= \sqrt{\frac{1.1847}{26}} = 0.2135 \text{ units of log leaf burn, in seconds}$$

The standard error of b_i is, in general, given by Eq. (14.30).

$$s_{b_i} = \sqrt{C_{ii} s^2_{y \cdot 1 \cdots k}} \tag{14.30}$$

For the example,

$$s_{b_1} = \sqrt{C_{11} s^2_{y \cdot 123}} = \sqrt{(0.106234)(0.0456)} = 0.0696$$

$$s_{b_2} = \sqrt{C_{22} s^2_{y \cdot 123}} = \sqrt{(0.117065)(0.0456)} = 0.0731$$

$$s_{b_3} = \sqrt{C_{33} s^2_{y \cdot 123}} = \sqrt{(0.032645)(0.0456)} = 0.0386$$

The t test is used to test hypotheses about partial regression coefficients. To test $H_0: \beta_i = \beta_{i0}$, that is, that β_i has the value β_{i0},

$$t = \frac{b_i - \beta_{i0}}{s_{b_i}} \quad \text{with } n - k - 1 \text{ df} \tag{14.31}$$

To test the hypotheses $H_0: \beta_i = \beta_{i0} = 0$, $i = 1,2,3$, for the example

$$t = \frac{b_1}{s_{b_1}} = \frac{-0.531459}{0.0696} = -7.64**$$

$$t = \frac{b_2}{s_{b_2}} = \frac{-0.439641}{0.0731} = -6.01**$$

$$t = \frac{b_3}{s_{b_3}} = \frac{0.208979}{0.0386} = 5.41**$$

(Ignore the sign of t for two-tailed tests.) All t's have 26 degrees of freedom. These tests are not independent in the sense of orthogonal comparisons as discussed in Chap. 11.

The standard error of the difference between two partial regression coefficients is given by Eq. (14.32).

$$s_{(b_i - b_j)} = \sqrt{s_{y \cdot 1 \cdots k}^2 (C_{ii} + C_{jj} - 2C_{ij})} \tag{14.32}$$

The C_{ij} quantity is essentially a covariance; it takes care of the fact that b_i and b_j are not independent. For the example,

$$s_{(b_1 - b_2)} = \sqrt{s_{y \cdot 123}^2 (C_{11} + C_{22} - 2C_{12})}$$
$$= \sqrt{0.0456[0.106234 + 0.117065 - 2(-0.021056)]} = 0.1100$$

To test the null hypothesis $H_0: \beta_i = \beta_j$,

$$t = \frac{b_i - b_j}{s_{(b_i - b_j)}} \quad \text{with } n - k - 1 \text{ df} \tag{14.33}$$

To test $\beta_1 = \beta_2$ for the example,

$$t = \frac{b_1 - b_2}{s_{(b_1 - b_2)}}$$
$$= \frac{[-0.531459 - (-0.439641)]}{0.1100} = -0.84 \text{ with 26 df}$$

We conclude that there is no difference between β_1 and β_2, in other words, that the observed difference between b_1 and b_2 is attributable to chance and the null hypothesis. This test was performed without an attempt to justify it. In practice, one would not conduct such a test unless it was meaningful to the investigator.

Computation of confidence intervals for a β_i or a difference $\beta_i - \beta_j$ is implicit in Eqs. (14.31) and (14.33). Thus the 95% confidence interval on the partial regression coefficient β_i is computed by Eq. (14.34).

$$b_i \pm t_{.05}(n - k - 1)s_{b_i} \tag{14.34}$$

For b_1, we have

$$-0.5315 \pm (2.06)(0.0696) = (-0.3881, -0.6749)$$

Confidence intervals computed for several β's are not independent. Hence, the tabulated confidence coefficient does not truly apply to the two or more statements. An appropriate procedure is to compute a confidence region in a k-dimensional space. The interested reader is referred to Durand (14.7). The 95% confidence interval on the difference $\beta_i - \beta_j$ is

$$(b_i - b_j) \pm t_{.05}(n - k - 1)s_{(b_i - b_j)} \qquad (14.35)$$

The standard error of an estimated mean for the population of Y's for which the set of X's is (X_1, X_2, X_3) is

$$s_{\hat{Y}} = \sqrt{s_{y \cdot 123}^2 \left(\frac{1}{n} + C_{11}x_1^2 + C_{22}x_2^2 + C_{33}x_3^2 + 2C_{12}x_1x_2 + 2C_{13}x_1x_3 + 2C_{23}x_2x_3 \right)} \qquad (14.36)$$

Notice that each X is measured as a deviation x in this equation. The equation is easily generalized to k independent variables. If it is desired to predict a Y, then an additional unity is required in the bracket.

Exercise 14.12.1 Compute the standard error of $b_{y1 \cdot 23}$ for Exercise 14.10.1.

Exercise 14.12.2 Compute the standard error of $b_{y2 \cdot 13} - b_{y3 \cdot 12}$ and test H_0: $\beta_{y2 \cdot 13} - \beta_{y3 \cdot 12} = 0$. Does this seem like a justifiable test?

14.13 Standard partial regression coefficients. To determine the relative importance of the various independent variables in relation to the dependent variable, compare the standard partial regression coefficients; partial regressions are in the units of measurement whereas standard partial regressions are in the same units, standard deviations.

The standard partial regression coefficients are determined from the partials by Eq. (14.14). For our example,

$$b_1' = b_1 \frac{s_1}{s_y} = (-0.531459) \frac{0.5827}{0.4803} = -0.6448$$

$$b_2' = b_2 \frac{s_2}{s_y} = (-0.439641) \frac{0.6051}{0.4803} = -0.5539$$

$$b_3' = b_3 \frac{s_3}{s_y} = (0.208979) \frac{1.0682}{0.4803} = 0.4648$$

Exercise 14.13.1 Compute standard partial regression coefficients for Exercise 14.5.2.

14.14 Deletion and addition of an independent variable. The results of a multiple regression or correlation analysis may indicate that one or more of the independent variables contributes so little to variation in the dependent variable that it is advantageous to delete this variable from the analysis. Again, it may be that so little variation in the dependent variable is attributable to the independent variable that it becomes desirable to incorporate an additional variable. Procedures for the addition or deletion of an independent variable have been given by Cochran (14.3).

Suppose X_k is to be deleted. Compute C'_{ij} values as defined by Eq. (14.37) for $i,j = 1, \ldots, k-1$.

$$C'_{ij} = C_{ij} - \frac{C_{ik}C_{jk}}{C_{kk}} \qquad (14.37)$$

The C'_{ij}'s are the C_{ij}'s which would have been obtained if X_k had not been included originally.

When a single variable is to be deleted, Eq. (14.37) provides a faster solution than setting up and solving a reduced set of normal equations, except possibly when only one or two independent variables remain. When two independent variables are to be deleted, the above procedure is performed in two successive stages. This is probably faster than solving a reduced set of normal equations when there are five or more independent variables initially. If there are more than two variables, it is probably best to solve the reduced set of normal equations on the basis of amount of work and introduction of rounding errors.

The deletion procedure is now illustrated for the data of Table 14.1 and the analysis of Table 14.5. We delete X_3, which is already in the last position. (If the variable to be deleted is not in the final position, permute rows and columns of the A matrix to attain this and still have a symmetric matrix with variances in the diagonal. Corresponding permutations must be made in the C matrix.) The variable X_3 would not normally be deleted since b_3, like the other b's, is highly significant. The required C' values are C'_{11}, C'_{12}, and C'_{22}. We find

$$C'_{11} = C_{11} - \frac{C^2_{13}}{C_{33}} = 0.106234 - \frac{(-0.000504)^2}{0.036245} = 0.106227$$

$$C'_{12} = C_{12} - \frac{C_{13}C_{23}}{C_{33}} = -0.021056 - \frac{(-0.000504)(-0.025960)}{0.036245}$$
$$= -0.021417$$

$$C'_{22} = C_{22} - \frac{C^2_{23}}{C_{33}} = 0.117065 - \frac{(-0.025960)^2}{0.036245} = 0.098471$$

Check: $C'_{11}\Sigma x_1^2 + C'_{12}\Sigma x_1 x_2 = 1$

$$0.106227(9.8455) + (-0.021417)2.1413 = 0.999998$$

The new set of partial regression coefficients may be found by the equation

$$b_i = \sum_j C'_{ij} \Sigma x_j y \qquad (14.38)$$

Recall that $\Sigma x_j y$ is over an unnamed subscript and that $C_{ij} = C_{ji}$. This method of computing partial regression coefficients was illustrated following the completion of the backward solution for the example in Sec. 14.10. Thus

$$b_1 = C'_{11}\Sigma x_1 y + C'_{12}\Sigma x_2 y$$
$$= 0.106227(-5.8248) + (-0.021417)(-4.2115) = -0.528553$$

$$b_2 = C'_{12}\Sigma x_1 y + C'_{22}\Sigma x_2 y$$
$$= (0.021417)(-5.8248) + 0.098471(-4.2115) = -0.289961$$

Check: $b_1 \Sigma x_1^2 + b_2 \Sigma x_1 x_2 = \Sigma x_1 y$

$$(-0.528553)9.8455 + (-0.289961)2.1413 = -5.8249$$

$$\text{Observed } \Sigma x_1 y = -5.8248$$

The b values are, of course, the same as those found in Sec. 14.5.

To add a variable, say X_{k+1}, compute Σx_{k+1}^2, $\Sigma x_{k+1} y$, and $\Sigma x_i x_{k+1}$ for $i = 1, \ldots, k$. The new Gauss multipliers C_{ij}' are computed from Eq. (14.41) and the partial regression coefficients are computed from Eq. (14.42). Initial computations given in Eqs. (14.39) and (14.40) are also required.

$$C_{k+1,k+1}' = \frac{1}{\Sigma x_{k+1}^2 - \sum_{i,j} C_{ij}(\Sigma x_1 x_{k+1})(\Sigma x_j x_{k+1})} \tag{14.39}$$

$$C_{i,k+1}' = C_{k+1,k+1}'(C_{i1} \Sigma x_1 x_{k+1} + \cdots + C_{ik} \Sigma x_k x_{k+1}) \tag{14.40}$$

$$C_{ij}' = C_{ij} - \frac{C_{i,k+1}' C_{j,k+1}'}{C_{k+1,k+1}'} \tag{14.41}$$

$$b_i' = b_i + \frac{C_{i,k+1}' b_{k+1}'}{C_{k+1,k+1}'} \tag{14.42}$$

where b_{k+1}' is found from the C''s and Σxy's by Eq. (14.38).

Exercise 14.14.1 From Exercise 14.10.1, delete X_3.

Exercise 14.14.2 To Exercise 14.5.2, add X_2.

14.15 Partial correlation. Having discussed the problem of adding or deleting an independent variable in multiple regression, we return to partial correlation. An example with two independent variables is given in Eqs. (14.17). We now consider the general case.

For partial correlations to be valid, three criteria must be clearly in mind. They are:
 1. The sampled distribution must be a multivariate normal distribution.
 2. The total observation must be random.
 3. No distinction is made between dependent and independent variables.

The third criterion implies that the problem is not one of regression. (However, such correlations may be very useful in regression problems.) Hence computed partial correlation coefficients need not be limited to those for which one of the "not-fixed" variables is a dependent variable; for example, in Sec. 14.7, $r_{Y1 \cdot 2}$ and $r_{Y2 \cdot 1}$ were computed while $r_{12 \cdot Y}$ was not. With this in mind, label the variables as $X_1, X_2, \ldots, X_k$.

To obtain partial correlation coefficients,
 1. Compute the correlation matrix.
 2. Invert the correlation matrix, say by the abbreviated Doolittle method. Call the resulting element in the ith row and jth column C_{ij}' rather than C_{ij}.
 3. Any partial correlation may now be found by Eq. (14.43). Partial correlation coefficients are symmetric; for example, $r_{12 \cdot 3} = r_{21 \cdot 3}$. Partial regression coefficients are not.

$$r_{ij \cdot 1 \cdots i-1, i+1 \cdots j-1, j+1 \cdots k} = \frac{-C_{ij}'}{\sqrt{C_{ii}' C_{jj}'}} \tag{14.43}$$

To test the null hypothesis $H_0: \rho = 0$, where ρ is a population partial correlation coefficient, the t test is appropriate. For example, to test H_0: $\rho_{12\cdot3\cdots k} = 0$,

$$t = \frac{r_{12\cdot3\cdots k}}{\sqrt{1 - r_{12\cdot3\cdots k}^2}} \sqrt{n - k} \qquad (14.44)$$

with $n - k$ degrees of freedom, where k is now the total number of variables rather than the number of independent variables, the term independent variable now being inapplicable. The square of this quantity is distributed as F with $1, n - k$ degrees of freedom.

It is interesting that we are now able to compute any regression equation that may be desired. Thus standard partial regression coefficients may be computed by Eq. (14.45).

$$b'_{ij\cdot 1\cdots i-1, i+1\cdots j-1, j+1\cdots k} = \frac{-C'_{ij}}{C'_{ii}} \qquad (14.45)$$

These are not symmetric; in fact, the relation given by Eq. (14.46) holds.

$$r_{12\cdot3\cdots k} = \pm \sqrt{b'_{12\cdot3\cdots k} b'_{21\cdot3\cdots k}} \qquad (14.46)$$

Both b'''s will have the same sign and the common sign is given to the square root. Equation (14.46) is also valid for the b's.

The partial regression coefficients may be obtained from the standard partial regression coefficients by Eq. (14.14). The multiple correlation coefficient may then be obtained by an obvious modification of Eq. (14.19) or by computing the sum of squares attributable to regression as implied by one of Eqs. (14.28) and then applying Eq. (14.20). Computation of the residual sum of squares is given by

$$(1 - R_{i\cdot 1\cdots i-1, i+1\cdots k}^2) \Sigma x_i^2$$

or by subtracting the sum of squares attributable to regression from the total sum of squares for the dependent variable, here X_i.

The null hypothesis $H_0: \rho_{i\cdot 1\cdots i-1, i+1\cdots k}$ may be tested by F as given in Eq. (14.47).

$$F = \frac{R_{i\cdot 1\cdots i-1, i+1\cdots k}^2}{1 - R_{i\cdot 1\cdots i-1, i+1\cdots k}^2} \frac{n - k}{k - 1} \qquad (14.47)$$

with $k - 1, n - k$ degrees of freedom where k is the total number of variables.

Notice that a $k \times k$ array (matrix) of correlation coefficients has been inverted. It is this fact that makes it possible to compute any and all regression equations. If only one regression equation is required, then it is necessary only to invert a $(k - 1)(k - 1)$ array. The number of computations required for inverting a matrix increases rapidly with each additional variable; the relation is not a linear one.

Section 14.10 provided the method for inverting a symmetric matrix and completing the general regression problem. Since we are now interested in correlation, it seems appropriate to comment on the relation between the two computing procedures. To begin with, the a_{ij}'s of Table 14.4 are replaced by r_{ij}'s; recall that $r_{ii} = 1$. The G matrix may be replaced by a column of ones.

The forward solution now proceeds exactly as before with the check column still applicable. The backward solution follows in the same way. Again, all checks should be made.

Partial correlation coefficients may also be computed from three partial correlation coefficients of one order lower. *Order* refers to the number of "fixed" subscripts, that is, subscripts after the dot. The formula is essentially given in Eqs. (14.17). For four variables, it becomes Eq. (14.48), and so on.

$$r_{ij \cdot hk} = \frac{r_{ij \cdot k} - r_{ih \cdot k} r_{jh \cdot k}}{\sqrt{(1 - r_{ih \cdot k}^2)(1 - r_{jh \cdot k}^2)}} \quad (14.48)$$

Recall that symmetry exists for the subscripts on the left of the dot as well as those on the right of the dot. Where only a few, say not more than four or five, variables are involved, this procedure is probably no more time-consuming than the abbreviated Doolittle method for inverting a matrix.

Partial regression coefficients may be computed similarly by modifying Eqs. (14.16), which is for standard partial regression coefficients. This gives Eq. (14.49) for four variables. The interested reader is referred to Cowden (14.4). Recall that these coefficients are not symmetric in the two subscripts preceding the dot.

$$b_{ij \cdot hk} = \frac{b_{ij \cdot k} - b_{ih \cdot k} b_{hj \cdot k}}{1 - b_{jh \cdot k} b_{hj \cdot k}} \quad (14.49)$$

Exercise 14.15.1 Johnson and Hasler (14.10) studied several factors influencing rainbow trout production in three dystrophic lakes. The following simple correlation coefficients were obtained.

	X_2	X_3	X_4	X_5
X_1	0.2206	−0.3284	−0.0910	−0.2160
X_2		0.6448	−0.1566	−0.1079
X_3			0.0240	−0.2010
X_4				−0.7698

X_1 = instantaneous growth rate of age group I trout for the interval concerned
X_2 = index of zooplankton density (availability of food)
X_3 = total standing crop of trout (competition)
X_4 = temperature of water
X_5 = size of age group I trout

Invert the matrix of correlations by the abbreviated Doolittle method and compute $r_{12 \cdot 345}$, $r_{13 \cdot 245}$, $r_{14 \cdot 235}$, and $r_{15 \cdot 234}$.

References

14.1 Anderson, R. L., and T. A. Bancroft: *Statistical Theory in Research*, McGraw-Hill Book Company, Inc., New York, 1952.

14.2 Birch, H. F.: "Phosphate response, base saturation and silica relationships in acid soils," *J. Agr. Sci.*, **43:** 229–235 (1953).

14.3 Cochran, W. G.: "The omission or addition of an independent variate in multiple linear regression," *J. Roy. Stat. Soc. Suppl.*, **5:** 171–176 (1938).

14.4 Cowden, D. J.: "A procedure for computing regression coefficients," *J. Am. Stat. Assoc.*, **53:** 144–150 (1958).

14.5 Cramer, C. Y.: "Simplified computations for multiple regression," *Industrial Quality Control*, **8:** 8–11 (1957).

14.6 Drapala, W. J.: "Early generation parent-progeny relationships in spaced plantings of soybeans, medium red clover, barley, Sudan grass, and Sudan grass times sorghum segregates," Ph.D. Thesis, Library, University of Wisconsin, Madison, 1949.

14.7 Durand, D.: "Joint confidence regions for multiple regression coefficients," *J. Am. Stat. Assoc.*, **49:** 130–146 (1954).

14.8 Dwyer, P. S.: *Linear Computations*, John Wiley & Sons, Inc., New York, 1951.

14.9 Friedman, J., and R. J. Foote: "Computational methods for handling systems of simultaneous equations," *U.S. Dept. Agr. Handbook* 94, 1957.

14.10 Johnson, W. E., and A. D. Hasler: "Rainbow trout production in dystrophic lakes," *J. Wildlife Management*, **18:** 113–134 (1954).

Chapter 15

ANALYSIS OF COVARIANCE

15.1 Introduction. The analysis of covariance is concerned with two or more measured variables where no exact control has been exercised over measurable variables regarded as independent. It makes use of the concepts of both analysis of variance and of regression. This chapter deals with linear covariance. A linear relation is often a reasonably good approximation for a nonlinear relation provided the values of the independent variables do not cover too wide a range.

15.2 Uses of covariance analysis. The most important uses of covariance analysis are:

1. To assist in the interpretation of data, especially with regard to the nature of treatment effects.
2. To partition a total covariance or sum of cross products into component parts.
3. To control error and increase precision.
4. To adjust treatment means of the dependent variable for differences in sets of values of corresponding independent variables.
5. To estimate missing data.

These will now be covered in more detail.

1. *Interpretation of data.* Any arithmetic procedure and associated statistical technique are meant to assist in the interpretation of data. Thus, use 1 includes the other uses of covariance analysis. However, use 1 is intended to be more specific in that covariance analysis often aids the experimenter in understanding the principles underlying the results of an investigation. For example, it may be well known that certain treatments produce real effects on both the dependent and independent variables; a covariance analysis may aid in determining the manner by which this is brought about. A discussion of a case where an independent variable is influenced by treatments is given by DeLury (15.3).

2. *Partition of a total covariance.* As in the analysis of variance, a sum of products is partitioned. While this use supplies the title for our chapter, it is often an incidental, although necessary, part of an analysis of covariance.

A covariance from a replicated experiment is partitioned when it is desired to determine the relation between two or more measured variables when it is uninfluenced by other sources of variation. For example, consider the data from a randomized complete-block design with four blocks of 25 varieties of soybeans, where it is desired to determine the relationship between oil and protein content. The total sum of products of the 100 observations can be partitioned into components according to the sources of variation, namely blocks, varieties, and error. The components have 3, 24, and 72 degrees of freedom, respectively. If the several regression and correlation coefficients corresponding to these sources differ significantly, the total regression and correlation are heterogeneous and not interpretable. For this experiment, interest would be in the relationships for the variety mean and for the residual, the latter measuring the average relationship between the two observed variables within varieties after over-all block effects have been removed. There is often every reason to assume these two regressions will be different.

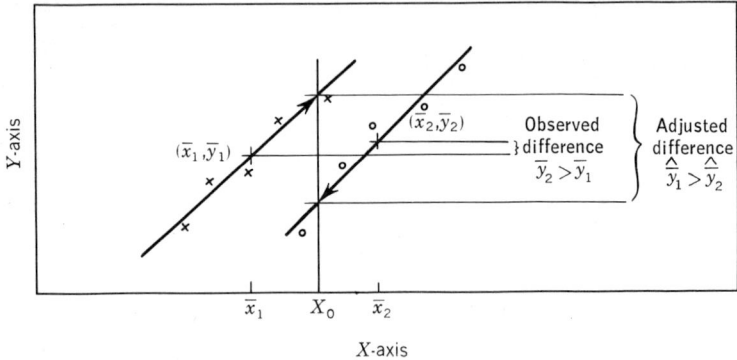

Fig. 15.1 Error control and adjustment of treatment means by covariance

3 and 4. Error control and adjusted treatment means. These two uses of covariance often appear together. The variance of a treatment mean is σ^2/n. Hence, to lower this variance, we have only two approaches. When covariance is used as a method of error control, that is, control of σ^2, it is in recognition of the fact that observed variation in the dependent variable Y is partly attributable to variation in the independent variable X. In turn, this implies that variation among treatment $\bar{y}$'s is affected by variation among treatment $\bar{x}$'s, and that, to be comparable, treatment $\bar{y}$'s should be adjusted to make them the best estimates of what they would have been if all treatment $\bar{x}$'s had been the same. Similarly if the prime object of covariance is to adjust treatment $\bar{y}$'s, it is in recognition of a regression situation which calls for a corresponding adjustment of error. In either case, it is necessary to measure the appropriate regression, independent of other sources of variation that may be called for by the model.

The general idea is apparent from Fig. 15.1 for two treatments. For each treatment, variation in X is seen to contribute to variation in Y. Hence a need to control error variance by use of the covariate is seen. At the same time, the distance between $\bar{x}_1$ and $\bar{x}_2$ can contribute greatly to the difference between $\bar{y}_1$ and $\bar{y}_2$. If the treatment $\bar{y}$'s had been observed from some common $\bar{x}$, say X_0, then they would be comparable. Thus the need for adjusting treatment means is apparent.

The use of covariance to control error is a means of increasing the precision with which treatment effects can be measured by removing, by regression, certain recognized environmental effects that cannot be or have not been controlled effectively by experimental design. For example, in a cattle feeding experiment to compare the effects of several rations on gain in weight, animals assigned to any one block will vary in initial weight. Now if initial weight is correlated with gain in weight, a portion of the experimental error for gain can be the result of differences in initial weight. By covariance analysis, the contribution to experimental error for gain, a contribution which can be attributed to differences in initial weight, may be computed and eliminated from experimental error for gain.

Forester (15.6) illustrates the application of covariance to error control on data from fields where the variability has been increased by previous experiments. Federer and Schlottfeldt (15.4) use covariance as a substitute for the use of blocks to control gradients in experimental material. Outhwaite and Rutherford (15.9) extend Federer and Schlottfeldt's results.

Covariance may be used primarily to adjust treatment means of the dependent variable for differences in the independent variable, that is, to adjust treatment $\bar{y}$'s by regression to be estimates of what they would have been had they had a common $\bar{x}$. Consider canning peas, a crop which increases rapidly in yield with increase in maturity. In a trial to evaluate yields of different varieties, it is difficult to harvest all at the same state of maturity. An analysis of yields unadjusted for differences in maturity may have little value. However, maturity can be used as a covariate when measured by a mechanical device, the tenderometer, which measures the pressure required to puncture the peas. A comparison of yields adjusted for maturity differences would be more meaningful than a comparison among unadjusted yields.

In field experiments, yields can be adjusted for differences in plot productivity as determined by uniformity trials. A uniformity trial measures yields from plots handled in a uniform manner prior to the performance of the main experiment. Love (15.7) concludes that with annual crops, the increased precision resulting from the use of uniformity data rarely pays; however, with long-lived perennials such as tree crops, there is often much to be gained.

In animal feeding experiments, differences among unadjusted treatment means may be due to differences in the nutritive value of the rations, to differences in the amounts consumed, or to both. If differences among mean gains in weight for the different rations are adjusted to a common food intake, the adjusted means indicate whether or not the rations differ in nutritive value. Here covariance is getting at the principles underlying the results of the investigation by supplying information on the way in which the treatments produce effects.

When covariance is used in testing adjusted treatment means, a joint use, it is important to know whether or not the independent variable is influenced by the treatments. If the independent variable is so influenced, the interpretation of the data is changed. This is so because the adjusted treatment means estimate the values expected when the treatment means for the independent variable are the same. Adjustment removes part of the treatment effects when means of the independent variable are affected by treatments. This does not mean that covariance should not be used in such cases but that care must be exercised in the interpretation of the data.

Covariance as a means of controlling error and adjusting treatment means is intended for use primarily when the independent variable measures environmental effects and is not influenced by treatments. In a fertilizer trial on sugar beets, the treatments may cause differences in stand. When stand, the independent variable, is influenced by treatments, the analysis of yield adjusted for stand differences removes part of the treatment effect and the experimenter may be misled in the interpretation of the data.

An analysis of covariance may supply additional useful information when treatments affect the independent variable. Again in the sugar beet problem, total yield is a function of average weight per beet and of stand. Now if stand is influenced by treatments, the analysis of covariance of yield adjusted for stand measures essentially the effects of the treatments on the average weight of the beets. A significant F value for yields adjusted for stand differences would indicate that treatments affect individual beet weights on the average.

For space-planted crops such as sugar beets, where variation in number of plants per plot is random and there is a correlation between number and yield, experimental error for yield will be increased by the covariate. To counter this increase and adjust yield means, an adjustment in proportion to the number of plants is sometimes practiced. This procedure is not recommended, because it usually results in an overcorrection for the plots with smallest stand since yields are rarely proportional to the number of plants per plot. In other words, the true regression of yield on stand for any treatment rarely goes through the origin. The analysis of covariance provides a more satisfactory and appropriate method of adjusting the experimental data.

Where it is doubtful if the independent variable is influenced by treatments, an F test may be helpful. However, in situations where real differences among treatments for the independent variable do occur but are not the direct effect of the treatments, adjustment is warranted. For example, consider a variety trial for which the seed of the various varieties or strains has been produced in different areas. Such seed may differ widely in germination, not because of inherent differences but as a result of the environment in which it was grown. Consequently, differences in stand may occur even if planting rate is controlled. In this situation, the use of covariance for both error control and yield adjustment is warranted. Thus the problem of whether or not covariance is applicable is one for which the experimenter must use careful judgment. Extreme care is also required in the interpretation of the differences among adjusted treatment means.

5. *Estimation of missing data.* Formulas given previously for estimating missing data result in a minimum residual sum of squares. However, the treatment sum of squares is biased upward. The use of covariance to estimate missing plots results in a minimum residual sum of squares and an unbiased treatment sum of squares. The covariance procedure is simple to carry out though more difficult to describe than previous procedures which required little more than a formula.

15.3 The model and assumptions for covariance. The assumptions for covariance are a combination of those for the analysis of variance and linear regression. The linear additive model for any given design is that for

ANALYSIS OF COVARIANCE

the analysis of variance plus an additional term for the concomitant or independent variable. Thus for the randomized complete-block design with one observation per cell, the mathematical description is given by Eq. (15.1).

$$Y_{ij} = \mu + \tau_i + \rho_j + \beta(X_{ij} - \bar{x}..) + \epsilon_{ij} \qquad (15.1)$$

The variable being analyzed, the dependent variable, is generally denoted by Y while the variable used in the control of error and adjustment of means, the independent variable or covariate, is denoted by X.

It is of interest to rewrite this expression in the following forms:

$$Y_{ij} - \beta(X_{ij} - \bar{x}..) = \mu + \tau_i + \rho_j + \epsilon_{ij}$$

$$Y_{ij} - \tau_i - \rho_j = \mu + \beta(X_{ij} - \bar{x}..) + \epsilon_{ij}$$

In the first case, we are stressing the experimental design aspects of the problem. We wish to carry out an analysis of variance of values which have been adjusted for regression on an independent variable. We are stressing use 4 (Sec. 15.2) though we obviously have uses 1 and 3 in mind.

In the second case, we are stressing the regression approach. We wish to measure the regression of Y on X without the interference of treatment and block effects. Use 2 is now uppermost.

The assumptions necessary for the valid use of covariance are:
1. The X's are fixed and measured without error.
2. The regression of Y on X after removal of block and treatment differences is linear and independent of treatments and blocks.
3. The residuals are normally and independently distributed with zero mean and a common variance.

Assumption 1 states that the X's are parameters associated with the means of the sampled Y populations. As such, they must be measured exactly. This assumption implies that, for the standard use of covariance, the treatments will not affect the X values since we may choose them for reasons of convenience. It has already been pointed out that covariance can be used where the X values are so affected but must be used with caution. Assumption 2 states that the effect of X on Y is to increase or decrease any Y, on the average, by a constant multiple of the deviation of the corresponding X from the mean of that variable for the whole experiment, that is, by $\beta(X_{ij} - \bar{x}..)$. The regression is assumed to be stable or homogeneous. Assumption 3 is the one on which the validity of the usual tests, t and F, depends. An analysis, as determined by the model, supplies a valid estimate of the common variance when there has been randomization. The assumption of normality is not necessary for estimating components of the variance of Y; randomization is necessary.

The residual variance is estimated on the basis of estimating values for μ, the τ_i's, the ρ_j's, and β, indicated by ^'s, such that Eq. (15.2) holds.

$$\sum_{i,j} [Y_{ij} - \hat{\mu} - \hat{\tau}_i - \hat{\rho}_j - \hat{\beta}(X_{ij} - \bar{x}..)]^2 = \text{minimum} \qquad (15.2)$$

The estimates of the parameters are termed least-squares estimates. For these estimates, Eq. (15.3) is correct, that is,

$$\sum_{i,j} [Y_{ij} - \hat{\mu} - \hat{\tau}_i - \hat{\rho}_j - \hat{\beta}(X_{ij} - \bar{x}..)] = 0 \qquad (15.3)$$

the sum of all the deviations is zero. It is not necessary to obtain estimates of the parameters in Eq. (15.2) any more than it was for the randomized complete-block analysis without covariance. Equations (15.4) to (15.6) define the estimates and give the residual variance.

$$\hat{\mu} = \bar{y}..$$
$$\hat{\tau}_i = t_i = \bar{y}_{i.} - \bar{y}.. - b(\bar{x}_{i.} - \bar{x}..)$$
$$\hat{\rho}_j = r_j = \bar{y}_{.j} - \bar{y}.. - b(\bar{x}_{.j} - \bar{x}..)$$
(15.4)

$$\hat{\beta} = b = \frac{E_{xy}}{E_{xx}} \tag{15.5}$$

$$\widehat{\sigma^2_{y \cdot x}} = s^2_{y \cdot x} = \frac{E_{yy} - [(E_{xy})^2/E_{xx}]}{f_e} \tag{15.6}$$

where E_{xx}, E_{xy}, and E_{yy} are sums of products for error, for example, E_{xx} is the error sum of squares for X, and f_e is error degrees of freedom. It is seen from the second of Eqs. (15.4) that, in order to estimate the treatment effect τ_i, the deviation of any treatment mean from the general mean must be adjusted by the quantity $b(\bar{x}_{i.} - \bar{x}..)$. This adjustment removes any effect that is attributable to the variable X. It is the adjusted treatment means that are comparable. Equation (15.6) is the computing formula for what amounts to the definition formula, namely, Eq. (15.2).

15.4 Testing adjusted treatment means. Table 15.1 gives the analysis of covariance for a randomized complete-block design (the only case

TABLE 15.1

TESTING ADJUSTED TREATMENT MEANS

The analysis of covariance for the randomized complete-block design

Source	df	Sums of products of			df	Adjusted Σy^2	MS
		x,x	x,y	y,y			
Total	$rt - 1$	Σx^2	Σxy	Σy^2			
Blocks	$r - 1$	R_{xx}	R_{xy}	R_{yy}			
Treatments	$t - 1$	T_{xx}	T_{xy}	T_{yy}			
Error	$(r-1)(t-1)$	E_{xx}	E_{xy}	E_{yy}	$(r-1)(t-1) - 1$	$E_{yy} - \dfrac{(E_{xy})^2}{E_{xx}}$	$s^2_{y \cdot x}$
Treatments + error	$r(t-1)$	S_{xx}	S_{xy}	S_{yy}	$r(t-1) - 1$	$S_{yy} - \dfrac{(S_{xy})^2}{S_{xx}}$	
Treatments adjusted					$t - 1$	$\left[S_{yy} - \dfrac{(S_{xy})^2}{S_{xx}}\right]$ $- \left[E_{yy} - \dfrac{(E_{xy})^2}{E_{xx}}\right]$	

given) and, at the same time, illustrates the general procedure. Notice the new notation with capital letters and paired subscripts to denote sums of products; a square is a particular kind of product.

The general procedure requires all three sums of products for treatments, and for error after adjustment for all other sources of variation included in the model. For the completely random design, there would be no block sums of products; for the Latin-square design, there would be sums of products for rows and columns instead of for blocks. From the treatment and error lines, a line for treatments plus error is obtained by addition. The sums of squares E_{yy} and S_{yy} are adjusted by subtracting the contributions due to linear regression. The difference between these adjusted sums of squares is the sum of squares for testing adjusted treatment means. To test the mean square for adjusted treatments, the appropriate error mean square is $s_{y \cdot x}^2$. Notice that the lines for treatments and error are the essentials from which the test of adjusted treatment means is constructed.

When it is desired to make a number of tests, as when a treatment sum of squares is partitioned into components, exact tests require a separate computation of treatment plus error and adjusted treatments for each comparison. An approximation to shorten the computations is illustrated in Sec. 15.10.

Exercise 15.4.1 Matrone et al. (15.8) compared the effect of three factors, each at two levels, on gain in weight of lambs. The factors were soybean hay A, from plots fertilized with KCa and PKCa, urea supplement B, and dicalcium supplement C. A randomized complete-block design of two replicates was used and food intake was determined for each lamb. The data are given in the accompanying table.

Ration	Block 1 Ewe lambs		Block 2 Wether lambs	
	X = intake	Y = gain	X = intake	Y = gain
(1)	209.3	11.2	286.9	27.2
a	252.4	26.1	302.1	30.6
b	241.5	13.2	246.8	15.4
c	259.1	24.4	273.2	24.0
ab	201.1	18.8	274.6	20.1
ac	287.5	31.0	276.3	24.4
bc	286.6	27.9	270.7	29.9
abc	255.7	20.8	253.0	20.8

Ignore the factorial nature of the experiment and carry out the computations outlined in Table 15.1. Compute b and $s_{y \cdot x}^2$. (Computational procedures are given in detail in the next section.) In terms of a statistical analysis of the data, how valuable are the decimal points?

15.5 Covariance in the randomized complete-block design. The form of the analysis of covariance in a randomized complete-block design is shown in Table 15.1. The procedure will be illustrated using the data of Table 15.2 from an experiment by J. L. Bowers and L. B. Permenter, Mississippi

State College, in which 11 varieties of lima beans were compared for ascorbic acid content. From previous experience it was known that increase in maturity resulted in decrease in vitamin C content. Since all varieties were not of the same maturity at harvest and since all plots of a given variety did not reach the same level of maturity on the same day, it was not possible to harvest all plots at the same stage of maturity. Hence the percentage of dry matter based on 100 grams of freshly harvested beans was observed as an index of maturity and used as a covariate.

TABLE 15.2

ASCORBIC ACID CONTENT† Y AND PERCENTAGE OF DRY MATTER‡ X FOR LIMA BEANS

Variety	Replication										Variety totals	
	1		2		3		4		5			
	X	Y	X	Y	X	Y	X	Y	X	Y	$X_{i.}$	$Y_{i.}$
1	34.0	93.0	33.4	94.8	34.7	91.7	38.9	80.8	36.1	80.2	177.1	440.5
2	39.6	47.3	39.8	51.5	51.2	33.3	52.0	27.2	56.2	20.6	238.8	179.9
3	31.7	81.4	30.1	109.0	33.8	71.6	39.6	57.5	47.8	30.1	183.0	349.6
4	37.7	66.9	38.2	74.1	40.3	64.7	39.4	69.3	41.3	63.2	196.9	338.2
5	24.9	119.5	24.0	128.5	24.9	125.6	23.5	129.0	25.1	126.2	122.4	628.8
6	30.3	106.6	29.1	111.4	31.7	99.0	28.3	126.1	34.2	95.6	153.6	538.7
7	32.7	106.1	33.8	107.2	34.8	97.5	35.4	86.0	37.8	88.8	174.5	485.6
8	34.5	61.5	31.5	83.4	31.1	93.9	36.1	69.0	38.5	46.9	171.7	354.7
9	31.4	80.5	30.5	106.5	34.6	76.7	30.9	91.8	36.8	68.2	164.2	423.7
10	21.2	149.2	25.3	151.6	23.5	170.1	24.8	155.2	24.6	146.1	119.4	772.2
11	30.8	78.7	26.4	116.9	33.2	71.8	33.5	70.3	43.8	40.9	167.7	378.6
Block totals $X_{.j}$ and $Y_{.j}$	348.8	990.7	342.1	1,134.9	373.8	995.9	382.4	962.2	422.2	806.8	1,869.3	4,890.5

† In milligrams per 100 grams dry weight.
‡ Based on 100 grams of freshly harvested beans.
SOURCE: Data used with permission of Bowers and Permenter, Mississippi State College, State College, Miss.

The calculation of sums of products for the randomized complete-block design is basically given in Eqs. (8.1), (8.2), (8.3), (8.4), (8.5), and (9.2). For the data of Table 15.2, the computations follow. Sums of products for total are

$$\Sigma x_{ij}^2 = \Sigma X_{ij}^2 - \frac{X_{..}^2}{rt} = 2{,}916.22$$

$$\Sigma y_{ij}^2 = 61{,}934.42$$

$$\Sigma x_{ij} y_{ij} = \Sigma X_{ij} Y_{ij} - \frac{X_{..}Y_{..}}{rt} = -12{,}226.14$$

Sums of products for blocks (replicates) are

$$R_{xx} = \frac{\sum\limits_{j} X_{.j}^2}{t} - \frac{X_{..}^2}{rt} = 367.85$$

$$R_{yy} = 4{,}968.94$$

$$R_{xy} = \frac{\sum\limits_{j} X_{.j} Y_{.j}}{t} - \frac{X_{..}Y_{..}}{rt} = -1{,}246.66$$

ANALYSIS OF COVARIANCE

Sums of products for treatments (varieties) are

$$T_{xx} = \frac{\sum_i X_{i\cdot}^2}{r} - \frac{X_{\cdot\cdot}^2}{rt} = 2{,}166.71$$

$$T_{yy} = 51{,}018.18$$

$$T_{xy} = \frac{\sum_i X_{i\cdot} Y_{i\cdot}}{r} - \frac{X_{\cdot\cdot} Y_{\cdot\cdot}}{rt} = -9{,}784.14$$

Sums of products for error are found by subtraction and are

$$E_{xx} = \Sigma x_{ij}^2 - R_{xx} - T_{xx} = 2{,}916.22 - 367.85 - 2{,}166.71 = 381.66$$
$$E_{yy} = \Sigma y_{ij}^2 - R_{yy} - T_{yy} = 5{,}947.30$$
$$E_{xy} = \Sigma x_{ij} y_{ij} - R_{xy} - T_{xy} = -12{,}226.14 - (-1{,}246.66) - (-9{,}784.14)$$
$$= -1{,}195.34$$

These results are entered in Table 15.3.

TABLE 15.3

ANALYSIS OF COVARIANCE OF DATA IN TABLE 15.2

Source of variation	df	Sum of products			Y adjusted for X			
		x,x	x,y	y,y	df	SS	MS	F
Total	54	2,916.22	−12,226.14	61,934.42				
Blocks	4	367.85	−1,246.66	4,968.94				
Treatments	10	2,166.71	−9,784.14	51,018.18				
Error	40	381.66	−1,195.34	5,947.30	39	2,203.56	56.50	
Treatments + error	50	2,548.37	−10,979.48	56,965.48	49	9,661.13		
Treatments adjusted					10	7,457.57	745.76	13.20**

These sums of products contain the material for the analyses of variance of X and Y as well as the analysis of covariance. Thus, to test the hypothesis of no differences among unadjusted variety means for ascorbic acid content Y,

$$F = \frac{T_{yy}/(t-1)}{E_{yy}/(r-1)(t-1)} = \frac{51{,}018.18/10}{5{,}947.30/40} = 34.31^{**}, \; 10 \text{ and } 40 \; df$$

We conclude that real differences exist among unadjusted variety means for ascorbic acid content.

To test the hypothesis of no differences among variety means for percentage of dry matter X at harvest,

$$F = \frac{T_{xx}/(t-1)}{E_{xx}/(r-1)(t-1)} = \frac{2{,}166.71/10}{381.66/40} = 22.71^{**}, \; 10 \text{ and } 40 \; df$$

We conclude that real differences exist among variety means for percentage dry matter in the beans at harvest.

The latter significant F illustrates a situation mentioned in Sec. 15.2, namely, significant differences may exist among treatment means for the independent variable and yet adjustment of the treatment means of the dependent variable is warranted. Differences in maturity as measured by percentage of dry matter are not treatment effects but occur because all varieties were not harvested at the same stage of maturity.

To test the hypothesis that no differences exist among the adjusted treatment means, it is necessary to calculate the error and treatment plus error sums of squares of Y adjusted for their respective regressions on the covariate X. The sum of squares for testing the hypothesis of no differences among adjusted treatment means is the difference between these adjusted sums of squares. The procedure was outlined in Table 15.1 and is illustrated in Table 15.3.

For error, the regression coefficient is given by Eq. (15.5) as

$$b_{yx} = \frac{E_{xy}}{E_{xx}} = \frac{-1{,}195.34}{381.66} = -3.13 \text{ mg ascorbic acid per } 1\% \text{ of dry matter}$$

The sum of squares of Y attributable to regression on X is

$$b_{yx} E_{xy} = \frac{(E_{xy})^2}{E_{xx}} = \frac{(-1{,}195.34)^2}{381.66} = 3{,}743.74 \text{ with 1 } df$$

The adjusted sum of squares is implied in Eq. (15.6) as

$$E_{yy} - \frac{(E_{xy})^2}{E_{xx}} = 5{,}947.30 - 3{,}743.74$$

$$= 2{,}203.56 \text{ with } (r-1)(t-1) - 1 = 39 \text{ } df$$

and the residual variance is

$$s_{y \cdot x}^2 = \frac{2{,}203.56}{39} = 56.50$$

For treatment plus error, the adjusted sum of squares is

$$S_{yy} - \frac{(S_{xy})^2}{S_{xx}} = 56{,}965.48 - \frac{(10{,}979.48)^2}{2{,}548.37}$$

$$= 9{,}661.13 \text{ with } r(t-1) - 1 = 49 \text{ } df$$

For adjusted treatment means, the sum of squares is the difference between the treatments plus error and the error sums of squares, that is,

$$\text{Adjusted } T_{yy} = 9{,}661.13 - 2{,}203.56 = 7{,}457.57 \text{ with } t - 1 = 10 \text{ } df$$

The adjusted sum of squares for error is used to estimate the variance within each and all populations of Y values. Homogeneity of variance is assumed. The adjusted sum of squares for treatments plus error is an estimate of the sum of squares to be expected for the two combined sources if all treatment $\bar{x}$'s were equal.

For the given adjustments to be both valid and useful, it is necessary for the linear regression of Y on X to be homogeneous for all treatments adjusted for block differences. Homogeneity of regression is, then, an assumption, as is homogeneity of error variance after adjustment for regression. There is no simple way of testing the homogeneity assumptions.

To test the hypothesis of no differences among treatment means for Y adjusted for the regression of Y on X,

$$F = \frac{\text{MS(adjusted treatment means)}}{s_{y \cdot x}^2} = \frac{745.76}{56.50} = 13.20^{**} \text{ with 10 and 39 } df$$

The highly significant F is evidence that real differences exist among the treatment means for Y when adjusted for X. If the unadjusted but not the adjusted treatment means had been significant, it would indicate that differences among the unadjusted means merely reflected differences in maturity and not in ascorbic acid content at a common maturity.

Exercise 15.5.1 Test adjusted treatment means for Exercise 15.4.1.

15.6 Adjustment of treatment means. The formula for adjusting treatment means is essentially given by Eq. (15.4), and is an application of the procedure given as Eq. (9.9). The basic idea is also illustrated in Fig. 15.1. In familiar notation, the equation for an adjusted treatment mean is given by Eq. (15.7).

$$\hat{\bar{y}}_{i \cdot} = \bar{y}_{i \cdot} - b_{yx}(\bar{x}_{i \cdot} - \bar{x}_{\cdot \cdot}) \tag{15.7}$$

$b_{yx} = b$ is the error regression coefficient. Adjusted treatment means are estimates of what the treatment means would be if all $\bar{x}_{i \cdot}$'s were at $\bar{x}_{\cdot \cdot}$. Adjusted treatment means are given in Table 15.4 for the data of Table 15.2. The

TABLE 15.4

ADJUSTMENT OF TREATMENT MEANS; DATA OF TABLE 15.2

Variety No.	Mean percentage of dry matter $\bar{x}_{i \cdot}$	Deviation $\bar{x}_{i \cdot} - \bar{x}_{\cdot \cdot}$	Adjustment $b_{yx}(\bar{x}_{i \cdot} - \bar{x}_{\cdot \cdot})$	Observed mean ascorbic acid content $\bar{y}_{i \cdot}$	Adjusted mean ascorbic acid content $\hat{\bar{y}}_{i \cdot} = \bar{y}_{i \cdot} - b_{yx}(\bar{x}_{i \cdot} - \bar{x}_{\cdot \cdot})$
1	35.42	1.43	−4.48	88.10 (5)†	92.58 (5)†
2	47.76	13.77	−43.10	35.98(11)	79.08 (8)
3	36.60	2.61	−8.17	69.92 (9)	78.09 (9)
4	39.38	5.39	−16.87	67.64(10)	84.51 (6)
5	24.48	−9.51	29.77	125.76 (2)	95.99 (4)
6	30.72	−3.27	10.24	107.74 (3)	97.50 (3)
7	34.90	0.91	−2.85	97.12 (4)	99.97 (2)
8	34.34	0.35	−1.10	70.94 (8)	72.04(11)
9	32.84	−1.15	3.60	84.74 (6)	81.14 (7)
10	23.88	−10.11	31.64	154.44 (1)	122.80 (1)
11	33.54	−0.45	1.41	75.72 (7)	74.31(10)
	$\bar{x} = 33.99$	$\Sigma = -0.03$‡	$\Sigma = +0.09$‡	$\bar{y}_{\cdot \cdot} = 88.92$‡	Mean $= 88.91$‡

† Ranks.
‡ Theoretically $\Sigma(\bar{x}_{i \cdot} - \bar{x}_{\cdot \cdot}) = 0$. Hence $\Sigma b_{yx}(\bar{x}_{i \cdot} - \bar{x}_{\cdot \cdot}) = 0$ and $\Sigma \bar{y}_{i \cdot} = \Sigma \hat{\bar{y}}_{i \cdot}$. In practice, rounding errors may appear.

standard error of an adjusted treatment mean is simply a modified Eq. (9.12). That is,

$$s_{\hat{y}_i.} = s_{y \cdot x}\sqrt{\frac{1}{r} + \frac{(\bar{x}_{i.} - \bar{x}_{..})^2}{E_{xx}}}$$

The standard error of the difference between two adjusted treatment means is given by Wishart (15.12) as Eq. 15.8.

$$s_{\bar{d}} = \sqrt{s_{y \cdot x}^2\left[\frac{2}{r} + \frac{(\bar{x}_{p.} - \bar{x}_{q.})^2}{E_{xx}}\right]} \qquad (15.8)$$

for the pth and qth means, p and q being specified values of i. For example, to compare varieties 7 and 10 (assuming a legitimate reason), we require

$$s_{\bar{d}} = \sqrt{56.50\left[\frac{2}{5} + \frac{(34.90 - 23.88)^2}{381.66}\right]} = 6.38 \text{ mg ascorbic acid}$$

The comparison is of the adjusted means $\hat{y}_{7.} = 99.97$ and $\hat{y}_{10.} = 122.78$, the difference obviously being significant.

Equation (15.8) necessitates a separate calculation for each comparison. This is probably justified if the experiment took an appreciable amount of time or was more than a preliminary experiment. However, Finney (15.5) suggests the following approximate $s_{\bar{d}}$ which utilizes an average in place of the separate $(\bar{x}_{p.} - \bar{x}_{q.})$'s required by Eq. (15.8). The formula is given in Eq. (15.9).

$$s_{\bar{d}} = \sqrt{\frac{2s_{y \cdot x}^2}{r}\left[1 + \frac{T_{xx}}{(t-1)E_{xx}}\right]} \qquad (15.9)$$

From Table 15.3,

$$s_{\bar{d}} = \sqrt{\frac{2(56.50)}{5}\left[1 + \frac{2{,}166.71}{10(381.66)}\right]} = 5.95 \text{ mg ascorbic acid}$$

for our example. For treatment means with unequal numbers, replace $2/r$ in Eqs. (15.8) and (15.9) by $(1/r_1 + 1/r_2)$. Cochran and Cox (15.2) state that Finney's approximation is usually close enough if error degrees of freedom exceed 20, since the contribution from sampling errors in b_{yx}, the adjustment factor, is small.

If variation among the $\bar{x}_{i.}$'s is nonrandom, for example, due to treatments, the approximate formula may lead to serious errors and should not be used. This would apply here also, since there were significant differences among treatment means for dry matter X.

Exercise 15.6.1 Compute adjusted treatment means for the eight treatments of Exercise 15.4.1.

Exercise 15.6.2 Compute the two means for main effect A. Adjust them. Compute the standard deviation appropriate to this comparison and test the null hypothesis of no A effect.

Exercise 15.6.3 Compute an approximate standard deviation for this comparison by Eq. (15.9). Compare with the exact standard deviation.

15.7 Increase in precision due to covariance. To test the effectiveness of covariance as a means of error control, a comparison is made of the variance of a treatment mean before and after adjustment for the independent

ANALYSIS OF COVARIANCE

variable X. The error mean square before adjustment is $5{,}947.30/40 = 148.68$ with 40 degrees of freedom and after adjustment is 56.50 with 39 degrees of freedom. It is necessary to adjust the latter value to allow for sampling error in the regression coefficient used in adjusting. The *effective error mean square* after adjustment for X is given by Eq. (15.10).

$$s_{y \cdot x}^2 \left[1 + \frac{T_{xx}}{(t-1)E_{xx}}\right] = 56.50 \left[1 + \frac{2{,}166.71}{10(381.66)}\right] = 88.58 \quad (15.10)$$

An estimate of the relative precision is $(148.68/88.58)100 = 168\%$. This indicates that 100 replicates with covariance are as effective as 168 without, a ratio of approximately $3:5$.

Exercise 15.7.1 Compute the effective mean square for Exercise 15.4.1. From the relative precision, conclude how many blocks would be required to give the same precision without using the covariate.

15.8 Partition of covariance. The term analysis of covariance implies a use not generally stressed, namely, the partitioning carried out in the cross-products column. This is extended in Table 15.5 to include regression and correlation coefficients.

TABLE 15.5
PARTITIONING FOR A COVARIANCE; DATA OF TABLE 15.2

Source of covariation	df	Sum of products	b_{yx}	r
Total	54	$-12{,}226.14$	-4.19	-0.910
Blocks	4	$-1{,}246.66$	-3.39	-0.922
Treatments	10	$-9{,}784.14$	-4.52	-0.931
Error	40	$-1{,}195.34$	-3.13	-0.793

When the null hypothesis of no treatment differences is true, treatments and error provide independent estimates of common regression and correlation coefficients. If a test of adjusted treatment means shows significance, then it may be presumed that treatment and error regressions differ. If the test does not show significance, then there may or may not be a treatment regression. If there is a treatment regression then it is the same as error regression. In rare instances, it may be desired to test the homogeneity of the regression coefficients. The appropriate numerator sum of squares is given by Eq. (15.11).

$$\text{Treatment versus error regression SS} = \frac{T_{xy}^2}{T_{xx}} + \frac{E_{xy}^2}{E_{xx}} - \frac{(T_{xy} + E_{xy})^2}{T_{xx} + E_{xx}} \quad (15.11)$$

$$= \frac{(-9{,}784.14)^2}{2{,}166.71}$$

$$+ \frac{(-1{,}195.34)^2}{381.66} - \frac{(-10{,}979.48)^2}{2{,}548.37}$$

$$= 621.31$$

The appropriate variance for testing the null hypothesis of homogeneity is generally considered to be the error variance. [The use of deviations from treatment regression for estimating the variance of the treatment regression coefficient and of error variance for the error regression coefficient has been suggested. These variances will often differ and require a test procedure like that of Sec. 5.8; divisors for Eqs. (5.13) and (5.14) will be T_{xx} and E_{xx} rather than n_1 and n_2.] For our example,

$$F = \frac{621.31}{56.50} = 11.00^{**} \text{ with 1 and 39 } df$$

The test of homogeneity of treatment and error regressions is sometimes presented in an alternate form. The following argument may be used to explain and justify the procedure.

1. Compute the sum of squares for adjusted treatment means. This may be considered to be the sum of squares of deviations of treatment means from a common regression line, a moving average. Since the slope of the regression line is obtained from error, a degree of freedom is not partitioned from those available for treatment means.

2. Compute the sum of squares of deviations of treatment means from their own regression. This will have $t - 2$ degrees of freedom since the means supply the estimate of the regression coefficient.

3. Subtract the latter sum of squares from the former. If the regressions are the same, the result should be no larger than one would expect from random sampling. If the result cannot reasonably be attributed to chance and the null hypothesis, then we conclude that the regressions are not homogeneous.

For our example,

$$\text{Deviations from treatment regression} = T_{yy} - \frac{T_{xy}^2}{T_{xx}}$$

$$= 51{,}018.08 - \frac{(-9{,}784.14)^2}{2{,}166.71} = 6{,}836.27 \text{ with } t - 2 = 9 \text{ } df$$

Treatment versus error regression SS

$$= \text{adj } T_{yy} - \text{deviations from treatment regression}$$
$$= 7{,}457.57 \text{ (10 } df\text{)} - 6{,}836.27 \text{ (9 } df\text{)} = 621.30 \text{ with 1 } df$$

Again, $F = 621.30/56.50 = 11.00^{**}$ with 1 and 39 degrees of freedom. Since differences among adjusted treatment means were significant, this result was to be expected and the test would not normally be made.

A comparison of correlation coefficients may be made using the appropriate procedure of Sec. 10.5. This procedure indicates no difference between correlation coefficients. That this can happen, that is, that regression coefficients may be very different while correlation coefficients are the same, can be seen from the relation

$$r = b_{yx} \frac{s_x}{s_y}$$

Two r's may be very much alike in spite of very different b_{yx}'s provided the ratio s_x/s_y counters these differences.

ANALYSIS OF COVARIANCE

15.9 Homogeneity of regression coefficients.

Where the experimental design is a completely random one, the regression of Y on X can be computed for each treatment. In this case, the usual assumption of homogeneity of the regression coefficients can be posed as a null hypothesis and tested by an appropriate F test in an analysis of covariance. (For other designs, we are not aware of methods presently available for testing homogeneity of regression coefficients.)

The procedure follows and is summarized in Table 15.6.

1. Compute Σx^2, Σxy, and Σy^2 for each treatment.
2. From the above and for each treatment, compute the sum of squares of Y attributable to regression on X (not shown in Table 15.6), and the residual sum of squares after this regression has been accounted for. This is called the "reduced SS" in Table 15.6. The reduction is given by $(\sum_j x_{ij} y_{ij})^2 / \sum_j x_{ij}^2$ for each value of i; the residual is found by subtracting this from $\sum_j y_{ij}^2$.
3. Total the reduced sums of squares to give the residual sum of squares of deviation from individual regression lines. This is denoted by A in Table 15.6 and has $\Sigma(n_i - 2) = \Sigma n_i - 2t$ degrees of freedom.
4. Total the Σx^2, Σxy, and Σy^2 over all treatments.
5. From the results in step 4, compute the sum of squares of Y attributable to regression on X, and the residual sum of squares. This residual sum of squares is of deviations from individual regression lines having a common regression coefficient. This is denoted by B in Table 15.6 and has $\Sigma(n_i - 1) - 1 = \Sigma n_i - t - 1$ degrees of freedom.
6. Compute $B - A$. This is the amount of the total sum of squares in Y that can be attributed to differences among the regression coefficients. It is necessarily positive since we can always do better by computing several coefficients rather than one.

The quantity $B - A$ can also be found by totaling the individual reductions attributable to regression and, from this total, subtracting the reduction based on the residual or error. Each regression reduction had 1 degree of freedom, hence A has t degrees of freedom, and the difference $B - A$ has $t - 1$ degrees of freedom. The appropriate error term is the mean square of the deviations from the individual regressions.

Exercise 15.9.1 Test the null hypothesis of homogeneity of regression coefficients for the data of Exercise 9.9.1. Compare the resulting F with the value t^2, where t is that obtained in Exercise 9.9.1.

15.10 Covariance where the treatment sum of squares is partitioned.

The application of covariance analysis to a factorial experiment is illustrated for the data in Table 15.7, from Wishart (15.13). The experiment was a 3×2 factorial in a randomized complete-block design of five replicates. The treatments were three rations, factor A, and two sexes, factor B. The procedure is applicable whenever a treatment sum of squares is to be partitioned.

The analysis is presented in Table 15.8. The computational procedure is the same as in Sec. 15.5 except that the treatment sum of squares is partitioned into main effects and interaction components as in Chap. 11. For easy

TABLE 15.6
Homogeneity of Within-treatment Regressions for a Completely Random Design

Treatment	df	Σx^2	Σxy	Σy^2	df	Reduced SS
1	$n_1 - 1$	Σx_{1j}^2	$\Sigma x_{1j}y_{1j}$	Σy_{1j}^2	$n_1 - 2$	$\Sigma y_{1j}'^2$
2	$n_2 - 1$	Σx_{2j}^2	$\Sigma x_{2j}y_{2j}$	Σy_{2j}^2	$n_2 - 2$	$\Sigma y_{2j}'^2$
.	.	.	.	.	.	.
t	$n_t - 1$	Σx_{tj}^2	$\Sigma x_{tj}y_{tj}$	Σy_{tj}^2	$n_t - 2$	$\Sigma y_{tj}'^2$
Residuals from individual regressions					$\Sigma n_i - 2t$	$\sum_i (\sum_j y_{ij}'^2) = A$
Totals for single regression	$\Sigma n_i - t$	$\sum_i (\sum_j x_{ij}^2)$	$\sum_i (\sum_j x_{ij}y_{ij})$	$\sum_i (\sum_j y_{ij}^2)$	$\Sigma n_i - t - 1$	$\sum_i (\sum_j y_{ij}^2) - \dfrac{[\sum_i (\sum_j x_{ij}y_{ij})]^2}{\sum_i (\sum_j x_{ij}^2)} = B$
Difference for homogeneity of regressions					$t - 1$	$B - A$

To test homogeneity of regression, $F = \dfrac{(B-A)/(t-1)}{A/(\Sigma n_i - 2t)}$ with $t - 1$ and $\Sigma n_i - 2t$ df.

TABLE 15.7
INITIAL WEIGHTS X AND GAINS IN WEIGHT Y IN POUNDS FOR BACON PIGS IN A FEEDING TRIAL

Pens (blocks)	Sex	Rations						Totals	
		a_1		a_2		a_3		X	Y
		X	Y	X	Y	X	Y		
1	$M = b_1$	38	9.52	39	8.51	48	9.11	269	56.83
	$F = b_2$	48	9.94	48	10.00	48	9.75		
2	M	35	8.21	38	9.95	37	8.50	202	54.04
	F	32	9.48	32	9.24	28	8.66		
3	M	41	9.32	46	8.43	42	8.90	238	52.94
	F	35	9.32	41	9.34	33	7.63		
4	M	48	10.56	40	8.86	42	9.51	272	59.88
	F	46	10.90	46	9.68	50	10.37		
5	M	43	10.42	40	9.20	40	8.76	222	55.44
	F	32	8.82	37	9.67	30	8.57		
Totals	M	205	48.03	203	44.95	209	44.78	617	137.76
	F	193	48.46	204	47.93	189	44.98	586	141.37
	M + F	398	96.49	407	92.88	398	89.76	1,203	279.13

reference, write $T_{xy} = A_{xy} + B_{xy} + AB_{xy}$ to denote the partitioning of the sum of cross products for treatments into components for rations A, sex B, and interaction AB. A similar notation is used with T_{xx} and T_{yy}. To partition the treatment sum of cross products, compute

$$A_{xy} = \frac{\sum_i (a_i)_x (a_i)_y}{rb} - \frac{(\Sigma X_{ij})(\Sigma Y_{ij})}{rt}$$

$$= \frac{398(96.49) + 407(92.88) + 398(89.76)}{5(2)} - \frac{1,203(279.13)}{5(6)}$$

$$= -0.147$$

$$B_{xy} = \frac{\Sigma (b_i)_x (b_i)_y}{ra} - \frac{(\Sigma X_{ij})(\Sigma Y_{ij})}{rt}$$

$$= \frac{617(137.76) + 586(141.37)}{5(3)} - \frac{1,203(279.13)}{5(6)} = -3.730$$

$$AB_{xy} = T_{xy} - A_{xy} - B_{xy}$$

$$= \frac{205(48.03) + \cdots + 189(44.98)}{5} - \frac{1,203(279.13)}{5(6)} - (-0.147)$$

$$- (-3.730) = 3.112$$

TABLE 15.8

ANALYSIS OF COVARIANCE FOR THE DATA IN TABLE 15.7

Source of variation	df	Sums of products			df	Y adjusted for X		
		x^2	xy	y^2		SS	MS	F
Total	29	1,108.70	78.507	16.3453				
Blocks (pens)	4	605.87	39.905	4.8518				
Rations	2	5.40	−0.147	2.2686				
Sex	1	32.03	−3.730	0.4344				
Rations × sex	2	22.47	3.112	0.4761				
Error	20	442.93	39.367	8.3144	19	4.8155	0.2534	
Rations + error	22	448.33	39.220	10.5830	21	7.1520		
Difference for testing adjusted ration means					2	2.3365	1.1683	4.61*
Sex + error	21	474.96	35.637	8.7488	20	6.0749		
Difference for testing adjusted sex means					1	1.2594	1.2594	4.97*
Ration × sex + error	22	465.40	42.479	8.7905	21	4.9133		
Difference for testing adjusted ration × sex interaction					2	0.0978	0.0489	0.19

where $(a_i)_x$ is the total of X values for level i of factor A, and so on, with $t = ab$. The directly computed sums of products are entered in Table 15.8 and the residuals obtained by subtracting the block, sex, ration, and ration times sex components from the total sum of products. To test the null hypotheses of no differences among levels of factor A, levels of factor B, and of no AB interaction, after adjustment for the concomitant variable, the procedure is basically that of Tables 15.1 and 15.3. However, each hypothesis is tested separately as shown in Table 15.8. Thus, the adjusted sum of squares for testing the null hypothesis about rations is the difference between the adjusted sum of squares for rations plus error and the adjusted sum of squares for error. For example, the adjusted sum of squares for error is

$$E_{yy} - \frac{(E_{xy})^2}{E_{xx}} = 8.3144 - \frac{(39.367)^2}{442.93} = 4.8155 \text{ with } (r-1)(t-1) - 1$$
$$= 19 \text{ df}$$

The adjusted sum of squares for rations plus error is

$$(E_{yy} + A_{yy}) - \frac{(E_{xy} + A_{xy})^2}{E_{xx} + A_{xx}}$$
$$= (8.3144 + 2.2686) - \frac{[(39.367) + (-0.147)]^2}{442.93 + 5.40}$$
$$= 7.1520 \text{ with } (a-1) + (r-1)(t-1) - 1 = 21 \text{ df}$$

The difference

$$\left[E_{yy} + A_{yy} - \frac{(E_{xy} + A_{xy})^2}{E_{xx} + A_{xx}} \right] - \left[E_{yy} - \frac{(E_{xy})^2}{E_{xx}} \right] = 7.1520 - 4.8155$$

$$= 2.3365 \text{ with } a - 1 = 2 \text{ df}$$

is the adjusted sum of squares for rations.

Adjusted sums of squares for sex and rations times sex are obtained similarly. F tests of the three null hypotheses are made with the adjusted error mean square in the denominator. For example, to test the null hypothesis that there are no differences among ration means after adjustment to a common $\bar{x}$, $F = 1.1683/0.2534$ with 2 and 19 degrees of freedom.

In an experiment where the treatment sum of squares is partitioned to test several null hypotheses, the above procedure is time-consuming but justifiable for most experiments. Cochran and Cox (15.2) give the following time-saving approximation. Prepare an analysis of adjusted Y's, that is, of $(Y - bx)$'s where $b = E_{xy}/E_{xx}$, from the preliminary analysis of covariance. This is illustrated in Table 15.9, using the covariance analysis in Table 15.8. The

TABLE 15.9
ANALYSIS OF VARIANCE OF $(Y - bx)$ FOR THE DATA OF TABLES 15.7 AND 15.8

Source of variation	df	Sum of squares	Mean square	F	F†
Rations	2	2.3374	1.1687	4.61*	4.61*
Sex	1	1.3507	1.3507	5.33*	4.97*
Rations × sex	2	0.1004	0.0502	0.20	0.19
Error	19	4.8155	0.2534		

† F from Table 15.8.

analysis of $Y - bx$ is accomplished by computing $\Sigma y^2 - 2b\Sigma xy + b^2 \Sigma x^2$ for error and each source of variation for which it is desired to test a null hypothesis; in all cases, $b = E_{xy}/E_{xx} = 39.367/442.93 = 0.0889$ lb of gain per pound of initial weight. For example, for sex,

$$0.4344 - 2(0.0889)(-3.730) + (0.0889)^2(32.03) = 1.3507$$

This approximate method gives the correct residual mean square within rounding errors; sums of squares for main effects and interactions, and consequently F values, are larger than those obtained using the exact method of Table 15.7. Overestimation of F is seldom great when variation in X is random and should be checked by the exact procedure only when just significant. The approximate procedure should not be used when X is influenced by treatments or is otherwise larger than due to chance, since the F values may be considerably in error in such cases.

Adjusted treatment means are obtained as in Table 15.4.

Exercise 15.10.1 Compute exact tests of adjusted treatment means for each of the main effects and interactions for the data of Exercise 15.4.1. State what the null hypothesis is, in each case.

Exercise 15.10.2 Compute the adjusted treatment means involved in the main effects comparisons. What is the value of the standard deviation applicable to one of these means?

Exercise 15.10.3 Compute tests of main effects and interactions using Cochran and Cox's approximate procedure. Compare the resulting F values with those found for Exercise 15.10.1.

15.11 Estimation of missing observations by covariance.

To illustrate this use of covariance, consider the data of Table 15.10, from Tucker et al. (15.11). Although all observations are available, assume that the one in the upper left corner is missing.

TABLE 15.10

MEAN ASCORBIC ACID CONTENT OF THREE 2-GRAM SAMPLES OF TURNIP GREENS, IN MILLIGRAMS PER 100 GRAMS DRY WEIGHT

Block (day)	1	2	3	4	5	Totals
Treatment						
A		887	897	850	975	3,609
B	857	1,189	918	968	909	4,841
C	917	1,072	975	930	954	4,848
Totals	1,774	3,148	2,790	2,748	2,838	13,298

The procedure, which gives an unbiased estimate of both treatment and error sums of squares, is:

1. Set $Y = 0$ for the missing plot.
2. Define a covariate as $X = 0$ for an observed Y, $X = +1$ (or -1) for $Y = 0$.
3. Carry out the analysis of covariance (see Table 15.11).
4. Compute $b = E_{xy}/E_{xx}$ and change sign to estimate the missing value.

The missing plot value, $-b$, is essentially an adjustment to the so-called observation $Y = 0$, to give an estimate of the Y that would have been obtained if X had been 0 instead of 1.

Missing plot Eq. (8.8) gives a value of 800 for a residual mean square of 5,816 (the same results as given by the covariance procedure) and an adjusted treatment sum of squares of 12,646 which is biased. Equation (8.9) may be used to compute the bias when Eq. (8.8) is used. The analysis of covariance procedure leads directly to an unbiased test of adjusted treatment means.

On application, the covariance method for missing plots is seen to be convenient and simple. The technique can be extended to handle several missing plots by introducing a new independent variate for each missing plot and using multiple covariance as illustrated in the next section. An example of the use of covariance for a missing plot along with the usual type of covariate is given by Bartlett (15.1).

Exercise 15.11.1 Apply the covariance technique to the missing plot problem given as Exercise 8.5.1. Obtain the missing plot value, error mean square, and treatment mean square. Compare these results with those of Exercise 8.5.1.

TABLE 15.11
ANALYSIS OF COVARIANCE AS AN ALTERNATIVE TO MISSING PLOT EQUATION FOR DATA OF TABLE 15.10

Source	df	Sums of products			df	Adjusted Σy^2	MS
		x,x	x,y	y,y			
Total	14	$1 - \frac{1}{15} = \frac{14}{15} = \frac{df}{n}$	-886.53	$945{,}296$			
Blocks	4	$\frac{1}{3} - \frac{1}{15} = \frac{4}{15} = \frac{df}{n}$	-295.20	$359{,}823$			
Treatments	2	$\frac{1}{5} - \frac{1}{15} = \frac{2}{15} = \frac{df}{n}$	-164.73	$203{,}533$			
Residual	8	By subtraction $= \frac{8}{15} = \frac{df}{n}$	-426.60	$381{,}940$	7	$40{,}713$	$5{,}816$
Treatments + residual	10	$\frac{10}{15} = \frac{df}{n}$	-591.33	$585{,}473$	9	$60{,}966$	
Treatments adjusted					2	$20{,}253$	$10{,}126$

$$\text{Missing plot} = -b = \frac{426.60}{8/15} = 800$$

15.12 Covariance with two independent variables. Sometimes a dependent variable is affected by two or more independent variables. Multiple covariance analysis is a method of analyzing such data. An example of such data is given in Table 15.12 with two independent variables X_1 and X_2 measuring initial weight of and forage (Ladino clover) consumed by guinea pigs, and a dependent variable Y measuring gain in weight. (It can be shown [see Steel (15.10)] that final weight as an alternative to gain in weight leads to essentially the same covariance analysis, when initial weight is a covariate, in that adjusted sums of squares for the two analyses are identical.)

Initial weight is unaffected by the treatments and is included to give error control and to adjust treatment means. Initial weight was largely the basis for assigning the animals to the three blocks but there was still enough variation remaining to advise its measurement. The amount of forage consumed is affected by treatments and was introduced to assist in interpreting the data. Differences among treatment means for gain in weight unadjusted for consumption could be due to both or either of nutritive value and consumption of forage. Thus the comparison among treatment means for gain adjusted for consumption gives information on the nutritive values of the treatments.

The computational procedure follows:

1. Obtain sums of products as in Secs. 15.5 and 15.10. (This is a factorial experiment with fertilizers and soils as factors.) The results are given in Table 15.13.

Table 15.12

Initial Weight X_1, Forage Consumed X_2, and Gain in Weight Y, All in Grams, from a Feeding Trial with Guinea Pigs

Soil treatment:	Unfertilized			Fertilized		
Block	X_1	X_2	Y	X_1	X_2	Y
Miami silt loam						
1	220	1,155	224	222	1,326	237
2	246	1,423	289	268	1,559	265
3	262	1,576	280	314	1,528	256
Total	728	4,154	793	804	4,413	758
Mean	242.7	1,384.7	264.3	268.0	1,471.0	252.7
Plainfield fine sand						
1	198	1,092	118	205	1,154	82
2	266	1,703	191	236	1,250	117
3	335	1,546	115	268	1,667	117
Total	799	4,341	424	709	4,071	316
Mean	266.3	1,447.0	141.3	236.3	1,357.0	105.3
Almena silt loam						
1	213	1,573	242	188	1,381	184
2	236	1,730	270	259	1,363	129
3	288	1,593	198	300	1,564	212
Total	737	4,896	710	747	4,308	525
Mean	245.7	1,632.0	236.7	249.0	1,436.0	175.0
Carlisle peat						
1	256	1,532	241	202	1,375	239
2	278	1,220	185	216	1,170	207
3	283	1,232	185	225	1,273	227
Total	817	3,984	611	643	3,818	673
Mean	272.3	1,328.0	203.7	214.3	1,272.7	224.3

$\Sigma X_1 = 5,984$ $\Sigma X_2 = 33,985$ $\Sigma Y = 4,810$
$\bar{x}_1 = 249.3$ $\bar{x}_2 = 1,416.0$ $\bar{y} = 200.4$
$\Sigma X_1^2 = 1,526,422$ $\Sigma X_2^2 = 48,971,371$ $\Sigma Y^2 = 1,045,898$
$\Sigma X_1 X_2 = 8,555,357$ $\Sigma X_1 Y = 1,199,664$ $\Sigma X_2 Y = 6,904,945$

Source: Data obtained through the courtesy of W. Wedin, formerly of University of Wisconsin, Madison, Wis.

ANALYSIS OF COVARIANCE

TABLE 15.13
SUM OF PRODUCTS FOR THE MULTIPLE COVARIANCE ANALYSIS OF THE DATA IN TABLE 15.12

Source of variation	df	Sums of products					
		Σx_1^2	Σx_2^2	$\Sigma x_1 x_2$	$\Sigma x_1 y$	$\Sigma x_2 y$	Σy^2
Total	23	34,411	847,195	81,764	371	93,785	81,894
Blocks	2	20,397	122,438	49,815	917	2,835	496
Soil types	3	480	165,833	3,006	612	5,008	57,176
Fertilizers	1	1,320	24,384	5,674	1,973	8,479	2,948
Interaction	3	6,055	61,163	5,491	−2,902	11,284	5,545
Error	14	6,159	473,377	17,778	−229	66,179	15,729

2. Calculate the necessary sets of partial regression coefficients for multiple regression equations in the form of Eq. (15.12). The procedure of Sec. 14.2 is applicable.

$$\hat{Y} = \bar{y} + b_{y1\cdot 2}(X_1 - \bar{x}_1) + b_{y2\cdot 1}(X_2 - \bar{x}_2) \qquad (15.12)$$

The necessary sets of coefficients for this two-factor experiment involve sums of products (squares and cross products) for error, error plus soil types, error plus fertilizers, and error plus interaction. The equations necessary to solve for the partial regression coefficients are

$$(\Sigma x_1^2)b_{y1\cdot 2} + (\Sigma x_1 x_2)b_{y2\cdot 1} = \Sigma x_1 y$$
$$(\Sigma x_1 x_2)b_{y1\cdot 2} + (\Sigma x_2^2)b_{y2\cdot 1} = \Sigma x_2 y \qquad (15.13)$$

Thus we have

2a. For *error* the equations are

$$6,159 b_{y1\cdot 2} + 17,778 b_{y2\cdot 1} = -229$$
$$17,778 b_{y1\cdot 2} + 473,377 b_{y2\cdot 1} = 66,179$$

Their solution is $b_{y1\cdot 2} = -0.4944$ and $b_{y2\cdot 1} = 0.1584$.

2b. For *soil types plus error*, the equations are

$$6,639 b_{y1\cdot 2} + 20,784 b_{y2\cdot 1} = 383$$
$$20,784 b_{y1\cdot 2} + 639,210 b_{y2\cdot 1} = 71,187$$

Their solution is $b_{y1\cdot 2} = -0.3240$ and $b_{y2\cdot 1} = 0.1219$.

2c. For *fertilizers plus error*, the equations are

$$7,479 b_{y1\cdot 2} + 23,452 b_{y2\cdot 1} = 1,744$$
$$23,452 b_{y1\cdot 2} + 497,761 b_{y2\cdot 1} = 74,658$$

Their solution is $b_{y1\cdot 2} = -0.2782$ and $b_{y2\cdot 1} = 0.1631$.

2d. For *interaction plus error*, the equations are

$$12,214 b_{y1\cdot 2} + 23,269 b_{y2\cdot 1} = -3,131$$
$$23,269 b_{y1\cdot 2} + 534,540 b_{y2\cdot 1} = 77,463$$

Their solution is $b_{y1\cdot 2} = -0.5806$ and $b_{y2\cdot 1} = 0.1702$.

3. Compute the necessary sums of squares of Y adjusted for X_1 and X_2 from formula (15.14).

$$\Sigma y^2 - b_{y1 \cdot 2}\Sigma x_1 y - b_{y2 \cdot 1}\Sigma x_2 y \qquad (15.14)$$

For *error*, we have

$15,729 - (-0.4944)(-229) - 0.1584(66,179) = 5,133.0$ with $14 - 2 = 12\,df$

The adjusted sums of squares for soil types plus error, fertilizers plus error, and interaction plus error are similarly computed. These values are shown in Table 15.14.

4. Compute sums of squares for adjusted treatment means as differences between the adjusted sum of squares for error computed at step 3 and the appropriate error plus treatment sum of squares computed at the same step. For example, the adjusted sum of squares for *soil types* is

$$64{,}351.4 - 5{,}133.0 = 59{,}218.4 \text{ with } 15 - 12 = 3\,df$$

The other adjusted sums of squares for use in testing the various null hypotheses are also given in Table 15.14.

TABLE 15.14

COMPUTATION OF ADJUSTED TREATMENT SUMS OF SQUARES FOR THE MULTIPLE COVARIANCE ANALYSIS BEGUN IN TABLE 15.13

Source of variation		Sums of squares				Y adjusted for X_1 and X_2		
	df	Σy^2	$b_{y1\cdot2}\,\Sigma x_1 y$	$b_{y2\cdot1}\,\Sigma x_2 y$	df	SS	MS	F
Error	14	15,729	113.2	10,482.8	12	5,133.0	427.8	
Soil types + error	17	72,905	−124.1	8,677.7	15	64,351.4		
Difference for testing adjusted soil types SS					3	59,218.4	19,739.5	46.14**
Fertilizers + error	15	18,677	−485.2	12,176.7	13	6,985.5		
Difference for testing adjusted fertilizer SS					1	1,852.5	1,852.5	4.33
Interaction + error	17	21,274	1,817.9	13,184.2	15	6,271.9		
Difference for testing adjusted interaction SS					3	1,138.9	379.6	0.89

5. Compute F values as required. To test adjusted soil-type means

$$F = \frac{19{,}739.5}{427.8} = 46.14** \text{ with 3 and 12 } df$$

The other F values are also shown in Table 15.14.

Adjusted treatment means for gain in weight are computed from Eq. (15.15).

$$\hat{\bar{y}}_{i\cdot} = \bar{y}_{i\cdot} - b_{y1\cdot 2}(\bar{x}_{1i\cdot} - \bar{x}_{1\cdot\cdot}) - b_{y2\cdot 1}(\bar{x}_{2i\cdot} - \bar{x}_{2\cdot\cdot}) \qquad (15.15)$$

where i refers to treatment for the $4(2) = 8$ treatment combinations, the numerical subscript refers to the corresponding independent variable, and $\bar{x}_{1\cdot\cdot}$ and $\bar{x}_{2\cdot\cdot}$ to general means. For example, the adjusted mean for gain in weight of animals fed forage from fertilized Miami silt loam is

$$252.7 - (-0.4944)(268.0 - 249.3) - (0.1584)(1{,}471 - 1{,}416)$$
$$= 253.2 \text{ grams}$$

All adjusted treatment means for gain in weight are given in Table 15.15. The feeding trial lasted 55 days and the *average* daily gains for both adjusted and unadjusted means are also given.

TABLE 15.15

ADJUSTED TREATMENT MEANS FOR DATA OF TABLE 15.12

Treatment		$\bar{y}_{i\cdot}$	$b_{y1\cdot 2}(\bar{x}_{1i\cdot} - \bar{x}_{1\cdot\cdot})$	$b_{y2\cdot 1}(\bar{x}_{2i\cdot} - \bar{x}_{2\cdot\cdot})$	$\hat{\bar{y}}_{i\cdot}$†	Average daily gain‡	
						Adjusted	Unadjusted
Miami silt loam	F	252.7	−9.2	+8.7	253.2	4.60	4.59
	Not F	264.3	+3.3	−5.0	266.0	4.84	4.81
Plainfield fine sand	F	105.3	+6.4	−9.3	108.2	1.97	1.92
	Not F	141.3	−8.4	+4.9	144.8	2.63	2.57
Almena silt loam	F	175.0	+0.1	+3.2	171.7	3.12	3.18
	Not F	236.7	+1.8	+34.2	200.7	3.65	4.30
Carlisle peat	F	224.3	+17.3	−22.7	229.7	4.18	4.08
	Not F	203.7	−11.4	−13.9	229.0	4.16	3.70
Total		1,603.3	0.1	0.1	1,603.3		

† $\hat{\bar{y}}_{i\cdot} = \bar{y}_{i\cdot} - b_{y1\cdot 2}(\bar{x}_{1i\cdot} - \bar{x}_{1\cdot\cdot}) - b_{y2\cdot 1}(\bar{x}_{2i\cdot} - \bar{x}_{2\cdot\cdot})$.
‡ The adjusted and unadjusted daily gains are obtained by dividing columns $\hat{\bar{y}}_{i\cdot}$ and $\bar{y}_{i\cdot}$ by 55, the number of days the animals were on trial.

The standard error of a difference between two adjusted treatment means is given in Eq. (15.16) for treatments p and q.

$$s_{\bar{d}} = \sqrt{s_{y\cdot 12}^2 \left[\frac{2}{r} + \frac{(\bar{x}_{1p\cdot} - \bar{x}_{1q\cdot})^2 E_{22} - 2(\bar{x}_{1p\cdot} - \bar{x}_{1q\cdot})(\bar{x}_{2p\cdot} - \bar{x}_{2q\cdot}) E_{12} + (\bar{x}_{2p\cdot} - \bar{x}_{2q\cdot})^2 E_{11}}{E_{11}E_{22} - E_{12}^2} \right]}$$
(15.16)

where $E_{12} = E_{x_1 x_2}$, and so on.

An approximate formula corresponding to Eq. (15.9) for a single covariate is given by Eq. (15.17).

$$s_{\bar{d}} = \sqrt{\frac{2}{r} s_{y\cdot 12}^2 \left[1 + \frac{T_{11}E_{22} - 2T_{12}E_{12} + T_{22}E_{11}}{(t-1)(E_{11}E_{22} - E_{12}^2)} \right]} \qquad (15.17)$$

T_{12} is the treatment sum of cross products of X_1 and X_2 with treatments including soil types, fertilizers, and interaction, and T_{11}, T_{22} are correspondingly interpreted. As illustration of Eq. (15.16), the standard deviation of the difference in response to fertilized and unfertilized forage grown on Plainfield sand is

$$s_{\bar{d}} = \sqrt{427.8\left[\frac{2}{3} + \frac{(236.3 - 266.3)^2 473{,}377 - 2(236.3 - 266.3)(1{,}357.0 - 1{,}447.0)17{,}778 + (1{,}357.0 - 1{,}447.0)^2 6{,}159}{6{,}159(473{,}377) - (17{,}778)^2}\right]}$$

$$= 18.65 \text{ grams}$$

By Eq. (15.17) we obtain an approximate standard deviation of

$$s_{\bar{d}} = \sqrt{\frac{2(427.8)}{3}\left\{1 + \frac{7{,}855(473{,}377) - 2(14{,}171)(17{,}778) + 251{,}380(6{,}159)}{(8-1)[(6{,}159)(473{,}377) - (17{,}778)^2]}\right\}}$$

$$= 18.97 \text{ grams}$$

Equation (15.16) gives a different standard deviation for each comparison whereas (15.17) gives the same, a compromise which gives too large a value for some comparisons and too small a value for others. When variation among the treatment means for one or more of the independent variables is larger than normally attributable to chance, as when influenced by treatments, serious error may result from applying Eq. (15.17).

Exercise 15.12.1 The accompanying table, furnished through the courtesy of C. R. Weber, Iowa State College, is a randomized block experiment with four replications. Eleven strains of soybeans were planted. The data and definition of the variables follow:

X_1 = maturity, measured in days later than the variety Hawkeye.
X_2 = lodging, measured on a scale from 1 to 5.
Y = infection by stem canker measured as a percentage of stalks infected.

Strain	Block 1			Block 2			Block 3			Block 4		
	X_1	X_2	Y	X_1	X_2	Y	X_1	X_2	Y	X_1	X_2	Y
Lincoln	9	3.0	19.3	10	2.0	29.2	12	3.0	1.0	9	2.5	6.4
A7–6102	10	3.0	10.1	10	2.0	34.7	9	2.0	14.0	9	3.0	5.6
A7–6323	10	2.5	13.1	9	1.5	59.3	12	2.5	1.1	10	2.5	8.1
A7–6520	8	2.0	15.6	5	2.0	49.0	8	2.0	17.4	6	2.0	11.7
A7–6905	12	2.5	4.3	11	1.0	48.2	13	3.0	6.3	10	2.5	6.7
C–739	4	2.0	25.2	2	1.5	36.5	2	2.0	23.4	1	2.0	12.9
C–776	3	1.5	67.6	4	1.0	79.3	6	2.0	13.6	2	1.5	39.4
H–6150	7	2.0	35.1	8	2.0	40.0	7	2.0	24.7	7	2.0	4.8
L6–8477	8	2.0	14.0	8	1.5	30.2	10	1.5	7.2	7	2.0	8.9
L7–1287	9	2.5	3.3	9	2.0	35.8	13	3.0	1.1	9	3.0	2.0
Bav. Sp.	10	3.5	3.1	10	3.0	9.6	11	3.0	1.0	10	3.5	0.1

The principal objective was to learn whether maturity or lodging is more closely related to infection. This will be determined from the error multiple regression. Incidentally, test the hypothesis of no differences among adjusted means for the varieties.
What is the appropriate test to attain the principal objective?

ANALYSIS OF COVARIANCE

Exercise 15.12.2 What is the appropriate equation for computing adjusted treatment means?

Exercise 15.12.3 Compute an approximate standard deviation for the difference between any pair of adjusted treatment means. Is there any reason to believe this standard deviation might not be applicable in this case?

References

15.1 Bartlett, M. S.: "Some examples of statistical methods of research in agriculture and applied biology," *J. Roy. Stat. Soc. Suppl.*, **4:** 137–183 (1937).

15.2 Cochran, W. G., and G. M. Cox: *Experimental Designs*, 2d ed., John Wiley & Sons, New York, 1957.

15.3 DeLury, D. B.: "The analysis of covariance," *Biometrics*, **4:** 153–170 (1948).

15.4 Federer, W. T., and C. S. Schlottfeldt: "The use of covariance to control gradients in experiments," *Biometrics*, **10:** 282–290 (1954).

15.5 Finney, D. J.: "Standard errors of yields adjusted for regression on an independent measurement," *Biometrics Bull.*, **2:** 53–55 (1946).

15.6 Forester, H. C.: "Design of agronomic experiments for plots differentiated in fertility by past treatments," *Iowa Agr. Expt. Sta. Research Bull.* no. 226, 1937.

15.7 Love, H. H.: "Are uniformity trials useful?" *J. Am. Soc. Agron.*, **28:** 234–245 (1936).

15.8 Matrone, G., F. H. Smith, V. B. Weldon, W. W. Woodhouse, Jr., W. J. Peterson, and K. C. Beeson: "Effects of phosphate fertilization on the nutritive value of soybean forage for sheep and rabbits," *U.S. Dept. Agr. Tech. Bull.* 1086, 1954.

15.9 Outhwaite, A. D., and A. Rutherford: "Covariance analysis as alternative to stratification in the control of gradients," *Biometrics*, **11:** 431–440 (1955).

15.10 Steel, R. G. D.: "Which dependent variate? Y or $Y - X$?" *Mimeo Series* BU-54-M, Biometrics Unit, Cornell Univ., Ithaca, N.Y., 1954.

15.11 Tucker, H. P., J. T. Wakeley, and F. D. Cochran: "Effect of washing and removing excess moisture by wiping or by air current on the ascorbic acid content of turnip greens," *Southern Coop. Ser. Bull.*, 10, 54–56 (1951).

15.12 Wishart, J.: "Tests of significance in the analysis of covariance," *J. Roy. Stat. Soc. Suppl.*, **3:** 79–82 (1936).

15.13 Wishart, J.: "Growth-rate determinations in nutrition studies with the bacon pig and their analysis," *Biometrika*, **30:** 16–28 (1938).

15.14 *Biometrics*, **13:** 261–405, No. 3 (1957). (This issue consists of seven papers devoted to covariance.)

Chapter 16

NONLINEAR REGRESSION

16.1. Introduction. The simplest and most common type of curve fitting is for the straight line. However, when pairs of observations are plotted, they often fall on a curved line; biological or other theory may even call for a nonlinear regression of a specified form. This chapter considers nonlinear or curvilinear regression. In addition, there is a brief discussion of the construction and use of orthogonal polynomials.

16.2 Curvilinear regression. A relation between two variables may be approximately linear when studied over a limited range but markedly curvilinear when a broader range is considered. For example, the relation between maturity and yield of canning peas is usually linear over the maturity range acceptable to the canning trade. However as maturity increases, the rate of increase in yield decreases, that is, becomes nonlinear. Similarly, the amount of increase in yield at very immature stages is smaller than at less immature stages. A nonlinear curve is required to describe the relation over the entire range.

In addition to the desirability of using an appropriately descriptive curve, it is sound procedure to remove from error any component that measures nonlinear regression. Thus if an observation is properly described as $Y = \mu + \beta_1 x + \beta_2 x^2 + \epsilon$, then to use $Y = \mu + \beta x + \epsilon$ as the model is to assign a part of the variation among population means, namely, that associated with $\beta_2 x^2$, to the measurement of error. Clearly this inflates our measure of error.

The selection of the form of the regression equation which best expresses a nonlinear relation is not always a simple problem. There is practically no limit to the number of kinds of curves that can be expressed by mathematical equations. Among the possible equations, many may exist which do about equally well in terms of goodness of fit. Hence, in choosing the form of curve, it is desirable to have some theory relative to the form, the theory being provided by the subject matter. In addition, it may be well to consider the work involved in fitting the regression and whether confidence limits and tests of significance will be valid.

Such considerations lead to two types of solutions, which we will refer to as direct and transformation solutions. For the *direct* type of solution, we decide upon the curve to be fitted, then fit the observed data to this curve. This chapter deals more particularly with the direct type of solution as applied to polynomial curves. For the *transformation* type of solution, we decide upon the curve to be fitted, then transform the data before fitting the curve; transformation is intended to provide an easier fitting procedure and/or valid estimation

NONLINEAR REGRESSION

and testing procedures. For example, we may agree that the equation $Y = aX^b$ is based on sound biological reasoning. Then $\log Y = \log a + b \log X$ is a linear equation if the pair of observations is considered to be $(\log Y, \log X)$. The procedures of Chaps. 9 and 15 apply. With such data as these, it is not uncommon to find that assumptions concerning normality are more nearly appropriate on the transformed scale than on the original.

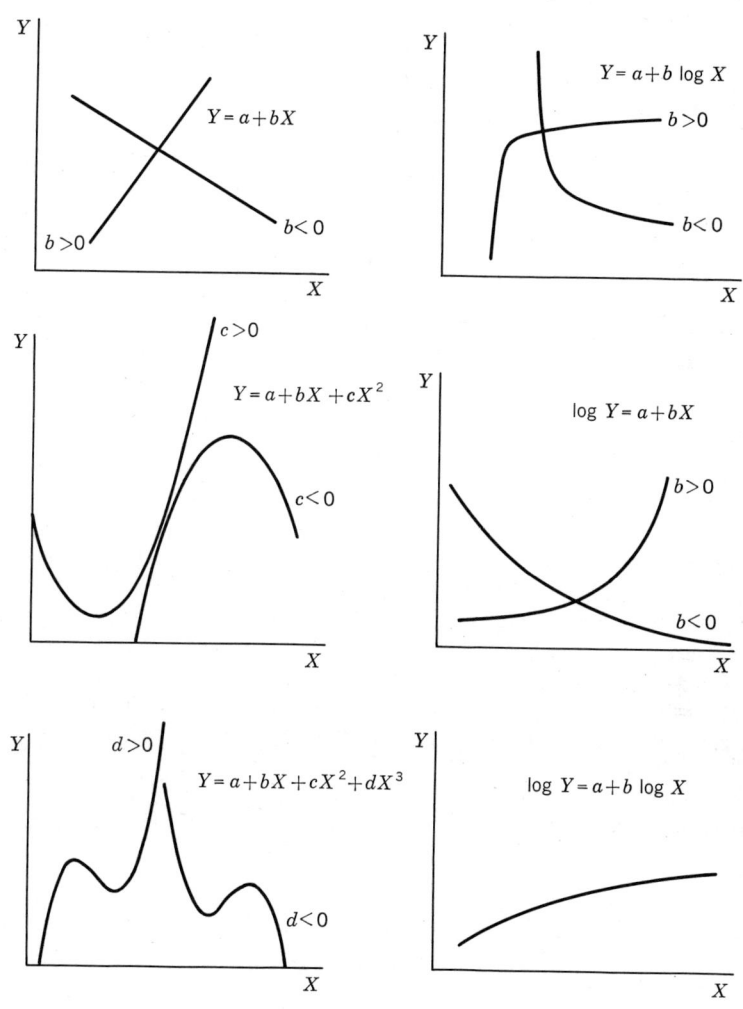

FIG. 16.1 General types of curves

We shall consider two general types of curves, polynomials and exponential or logarithmic curves. Examples follow for which general forms are shown in Fig. 16.1. For the exponential equations, e may be replaced by any other constant without affecting the form of the curve being fitted. For the equation $Y = aX^b$, integral values of the *exponent* b give special cases of polynomials.

However, this type of curve is more likely to be used when a fractional exponent is desired, as is often so in the field of economics.

	Polynomial	Exponential	Logarithmic
Linear:	$Y = a + bX$	$e^Y = aX^b$	$Y = a_1 + b \log X$
Quadratic:	$Y = a + bX + cX^2$	$Y = ab^X$	$\log Y = a_1 + b_1 X$
Cubic:	$Y = a + bX + cX^2 + dX^3$	$Y = aX^b$	$\log Y = a_1 + b \log X$

Polynomials may have peaks and depressions numbering one less than the highest exponent. For example, the bottom left illustration in Fig. 16.1 has one peak and one depression, two such places for a curve where the highest exponent is 3. Peaks are called *maxima*, depressions are called *minima*. Polynomials may have fewer maxima and minima than one less than the highest exponent. In fitting polynomial curves, the investigator is usually interested in some particular segment of the entire range represented by the equation.

The exponential or logarithmic curves, except those of the form $\log Y = a + b \log X$, are characterized by a flattening at one end of the range. For example, the curve $Y = \log X$ approaches closer and closer to $X = 0$ as Y takes on numerically larger and larger negative values; however, this curve never crosses the vertical line $X = 0$. Negative numbers do not have real logarithms.

16.3 Logarithmic or exponential curves. Logarithmic, or simply log, curves are linear when plotted on the proper logarithmic paper. Referring to Fig. 16.1, right-hand side, top to bottom, we have:

1. $e^Y = aX^b$ or $Y = a_1 + b \log X$. The points (X,Y) plot as a straight line on semilog paper where Y is plotted on the equal-interval scale and X on the log scale. Essentially, the semilog paper finds and plots the logs of X.

2. $Y = ab^X$ or $\log Y = a_1 + b_1 X$. The points (X,Y) plot as a straight line on semilog paper where Y is plotted on the log scale and X on the equal-interval scale (see Fig. 16.2 also).

3. $Y = aX^b$ or $\log Y = a_1 + b \log X$. This plots as a straight line on double-log paper. Here, both scales are logarithmic (see Fig. 16.3 also).

TABLE 16.1

HEIGHT ABOVE THE COTYLEDONS FOR GOLDEN ACRE CABBAGE MEASURED AT WEEKLY INTERVALS

Weeks after first observation X	Height, cm Y	Common logarithm of height Y
0	4.5	0.653
1	5.5	0.740
2	6.5	0.813
3	8.0	0.903
4	10.0	1.000
5	12.0	1.079
6	15.5	1.190
7	17.5	1.243

NONLINEAR REGRESSION

To determine whether data can be described by a log curve, it is often sufficient to plot the data on log paper. After a decision has been made as to the type of log curve, the observed values of X and/or Y are transformed to logarithms before the computations can be performed. The transformed data are then handled by the methods of Chaps. 9 and 15. The usual assumptions apply to the transformed data rather than to the original.

W. J. Drapala, Mississippi State College, has supplied us with two sets of data for illustration. The data in Table 16.1 are heights in centimeters above the cotyledons, at weekly intervals, of Golden Acre cabbage. The data are plotted in Fig. 16.2 with X on the fixed-interval scale and Y on the log as well as the fixed-interval scale. Notice how linear the set of points becomes when plotted on semilog paper.

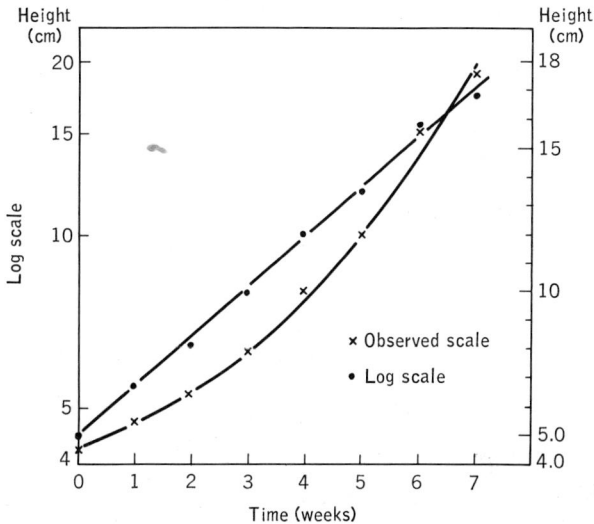

FIG. 16.2 Observed points plotted on fixed-interval and log scales

We now proceed with calculations on the transformed data, using the equation $\log \hat{Y} = a + bx$. For this equation

$$a = \frac{\sum_i \log Y_i}{n} = 0.953$$

$$b = \frac{\sum_i X_i(\log Y_i) - (\sum_i X_i)(\sum_i \log Y_i)/n}{\sum_i X_i^2 - (\sum_i X_i)^2/n} = 0.087 \text{ units on log scale per week}$$

The regression equation is

$$\log \hat{Y} = 0.953 + 0.087x = 0.649 + 0.087X$$

The data in Table 16.2 are pressures, in atmospheres, of oxygen gas kept at 25°C when made to occupy various volumes, measured in liters. The data are curvilinear when plotted on the equal-interval scale, nearly linear on the double log scale (see Fig. 16.3). The relation $\log \hat{Y} = a + b \log x$ is fitted. For this, we compute

$$a = \frac{\sum_i \log Y_i}{n} = 0.374$$

$$b = \frac{\sum_i (\log X_i)(\log Y_i) - (\sum_i \log X_i)(\sum_i \log Y_i)/n}{\sum_i (\log X_i)^2 - (\sum_i \log X_i)^2/n} = -1.003$$

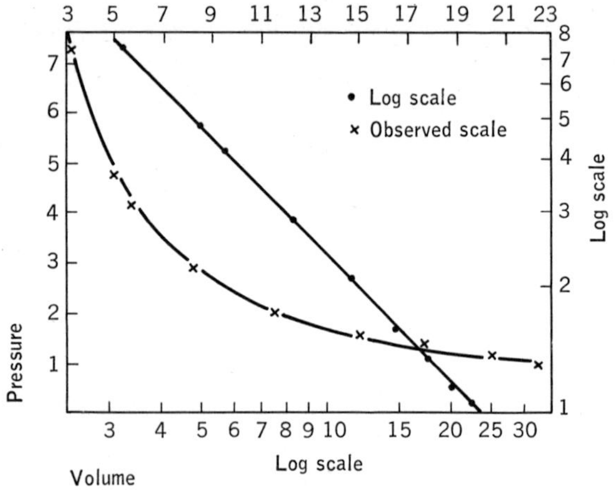

FIG. 16.3 Observed points plotted on fixed-interval and log scales

The regression equation is

$$\log \hat{Y} = 0.374 - 1.003 \log x = 1.369 - 1.003 \log X$$

TABLE 16.2

PRESSURE OF OXYGEN GAS KEPT AT 25°C WHEN MADE TO OCCUPY VARIOUS VOLUMES

Volume, liters X	Pressure, atm Y	Common logarithm	
		Volume X	Pressure Y
22.40	1.06	1.350	0.025
20.35	1.17	1.309	0.068
17.49	1.36	1.243	0.134
14.95	1.59	1.175	0.201
11.50	2.07	1.061	0.316
8.27	2.88	0.918	0.459
5.71	4.18	0.757	0.621
5.00	4.77	0.699	0.679
3.25	7.34	0.512	0.866

Exercise 16.3.1 Brockington et al. (16.2) collected the data shown in the accompanying table on the relation between moisture content and interstitial relative humidity for whole-kernel corn.

Sample no.	Moisture content		Equilibrium relative humidity, %
	Brown-Duval	Two-stage oven	
1	7.0	9.4	40.0
2	7.5	9.9	40.0
3	11.6	12.9	59.0
4	11.8	12.6	63.5
5	12.9	14.1	71.5
6	13.2	14.7	71.0
7	14.0	15.2	76.5
8	14.2	14.6	75.5
9	14.6	15.2	79.0
10	14.8	15.8	79.0
11	15.7	15.8	82.0
12	17.3	17.2	85.5
13	17.4	17.0	85.0
14	17.8	18.2	87.5
15	18.0	18.5	86.5
16	18.8	18.2	88.0
17	18.9	19.4	90.0
18	20.0	20.3	90.5
19	20.7	19.9	88.5
20	22.4	19.5	89.5
21	22.5	19.8	91.0
22	26.8	22.6	92.0

Plot moisture content (either measurement) against equilibrium on ordinary scale and log-scale graph paper. What seems to be an appropriate choice of scales to give a straight line? Compute the linear regression of equilibrium relative humidity on log of moisture content.

Exercise 16.3.2 In a study of optimum plot size and shape, Weber and Horner (16.7) considered length of plot and variance of plot means per basic unit for yield in grams and protein percentage (among other plot shapes and characters). They obtained the data shown in the accompanying table.

Shape	Number of units	Variance	
		Yield, grams	Protein, %
8 × 1	1	949	.116
16 × 1	2	669	.080
24 × 1	3	540	.053
32 × 1	4	477	.048

Plot the two sets of variances against number of units after transforming all three variables to log scales. Do the resulting data appear to be reasonably linearly related?

16.4 The second-degree polynomial.

The equation for this curve is $Y = a + bX + cX^2$, where a, b, and c are real numbers. For computational convenience, we use the mathematical description of an observation given as Eq. (16.1). Clearly, linearity applies to the parameters to be estimated and not to the observable ones.

$$Y_i = \mu + \alpha(X_i - \bar{x}) + \beta(X_i^2 - \overline{x^2}) + \epsilon_i \qquad (16.1)$$

In this equation α and β are partial regression coefficients, as discussed in Chap. 14, but cannot be interpreted in other than a two-dimensional space. The problem is to estimate μ, α, β, and σ^2, where σ^2 is the variance of the ϵ's. The procedure is to find estimates $\hat{\mu}$, $\hat{\alpha}$, and $\hat{\beta}$ such that Eq. (16.2) holds.

$$\Sigma(Y - \hat{Y})^2 = \Sigma[Y_i - \hat{\mu} - \hat{\alpha}x_i - \hat{\beta}(X_i^2 - \overline{x^2})]^2 = \min \qquad (16.2)$$

The method is precisely that of Sec. 14.5 with $X = X_1$ and $X^2 = X_2$. The estimate of μ is $\hat{\mu} = \bar{y}$. For estimates of α and β, designated by a and b, normal equations are given as Eqs. (16.3).

$$a\Sigma(X - \bar{x})^2 + b\Sigma(X - \bar{x})(X^2 - \overline{x^2}) = \Sigma(X - \bar{x})(Y - \bar{y})$$
$$a\Sigma(X - \bar{x})(X^2 - \overline{x^2}) + b\Sigma(X^2 - \overline{x^2})^2 = \Sigma(X^2 - \overline{x^2})(Y - \bar{y}) \qquad (16.3)$$

These correspond to Eqs. (14.11). Their solution is given in Eqs. (16.4).

$$a = \frac{[\Sigma(X - \bar{x})(Y - \bar{y})][\Sigma(X^2 - \overline{x^2})^2] - [\Sigma(X^2 - \overline{x^2})(Y - \bar{y})][\Sigma(X - \bar{x})(X^2 - \overline{x^2})]}{[\Sigma(X - \bar{x})^2][\Sigma(X^2 - \overline{x^2})^2] - [\Sigma(X - \bar{x})(X^2 - \overline{x^2})]^2}$$

$$b = \frac{[\Sigma(X - \bar{x})^2][\Sigma(X^2 - \overline{x^2})(Y - \bar{y})] - [\Sigma(X - \bar{x})(X^2 - \overline{x^2})][\Sigma(X - \bar{x})(Y - \bar{y})]}{[\Sigma(X - \bar{x})^2][\Sigma(X^2 - \overline{x^2})^2] - [\Sigma(X - \bar{x})(X^2 - \overline{x^2})]^2} \qquad (16.4)$$

In Eqs. (16.3) and (16.4), $\overline{x^2}$ is the mean of the X^2's. Computation of $\Sigma(X - \bar{x})^2$, $\Sigma(Y - \bar{y})^2$, and $\Sigma(X - \bar{x})(Y - \bar{y})$ is as usual. We compute the additional necessary quantities by Eqs. (16.5).

$$\Sigma(X - \bar{x})(X^2 - \overline{x^2}) = \Sigma X^3 - (\Sigma X)(\Sigma X^2)/n$$
$$\Sigma(X^2 - \overline{x^2})^2 = \Sigma X^4 - (\Sigma X^2)^2/n \qquad (16.5)$$
$$\Sigma(X^2 - \overline{x^2})(Y - \bar{y}) = \Sigma X^2 Y - (\Sigma X^2)(\Sigma Y)/n$$

The sum of squares attributable to regression is given by Eq. (16.6). The residual sum of squares is found by subtracting this from Σy^2.

$$\text{Regression SS} = a\Sigma xy + b\Sigma(X^2 - \overline{x^2})y \qquad (16.6)$$

16.5 The second-degree polynomial, an example.

The data in Table 16.3 will be used for illustrative purposes. The results of the application

TABLE 16.3
YIELD Y, IN POUNDS PER PLOT, AND TENDEROMETER READING X, OF ALASKA PEAS GROWN AT MADISON, WIS., 1953

Y {Yield, pounds per plot: 24.0 22.0 26.5 22.0 25.0 37.5 36.0 39.5 32.0 26.5 55.5 49.5 56.0 55.5

X {Tenderometer reading: 76.2 76.8 77.3 79.2 80.0 87.8 93.2 93.5 94.3 96.8 97.5 99.5 104.2 106.3

Y {Yield, pounds per plot: 58.0 61.5 69.0 71.5 73.0 76.5 78.5 74.0 71.5 77.0 85.5

X {Tenderometer reading: 106.7 119.0 119.7 119.8 119.8 123.5 141.0 142.3 145.5 149.0 150.0

$\Sigma X = 2,698.9$ $\Sigma X^2 = 305,148.35$ $\Sigma X^3 = 36,045,287.22$ $\Sigma X^4 = 4,429,289,685.23$
$\Sigma Y = 1,303.5$ $\Sigma Y^2 = 78,797.25$ $\Sigma XY = 152,129.55$ $\Sigma X^2 Y = 18,424,791.12$

$\Sigma x^2 = 13,785.90$ $\Sigma x^3 = 3,102,691.95$ $\Sigma x^4 = 704,669,064.92$
$\Sigma y^2 = 10,832.76$ $\Sigma xy = 11,408.90$ $\Sigma x^2 y = 2,514,356.15$

of Eqs. (16.5) are given in Table 16.3. From $\hat{\mu} = \bar{y}$ and Eqs. (16.3), we find

$$\hat{\mu} = \bar{y} = 52.14$$
$$\hat{\alpha} = a = 2.71328$$
$$\hat{\beta} = b = -0.00838$$

The regression equation is

$$\hat{Y} = 52.14 + 2.713x - 0.00838(X^2 - \overline{x^2})$$
$$= -138.48 + 2.713X - 0.00838X^2$$

The sum of squares attributable to regression is, by Eq. (16.6),

$$2.71328(11,408.90) + (-0.00838)(2,514,356.15) = 9,885.24 \text{ with 2 } df$$

The analysis of variance presentation is:

Source	df	SS	MS	F
Regression	2	9,885.24	4,942.62	115**
Residual	22	947.52	43.07	
Total	24	10,832.76		

If it is desired to determine whether the reduction due to fitting a second-degree polynomial is significantly larger than that attributable to linear regression, we compute the additional reduction attributable to inclusion of X^2 in the regression equation. To do this, we compute the linear regression equation and the reduction attributable to linear regression. For this,

$$\bar{y} = 52.14 \quad \text{and} \quad b = 0.82758$$

$$\text{Regression SS} = 0.82758(11,408.90) = 9,441.78 \text{ with 1 } df$$

The regression equation is

$$\hat{Y} = 52.14 + 0.828x = -37.20 + 0.828X$$

Next, the sum of squares attributable to fitting X^2 is

$$SS(X, X^2) - SS(X) = 9{,}885.24 - 9{,}441.78 = 443.46 \text{ with 1 } df$$

The test criterion for the null hypothesis that X^2 contributes nothing to variation in Y, or that $\beta = 0$, is $F = 443.46/43.07 = 10.30^{**}$ with 1 and 22 degrees of freedom. Figure 16.4 shows the two equations.

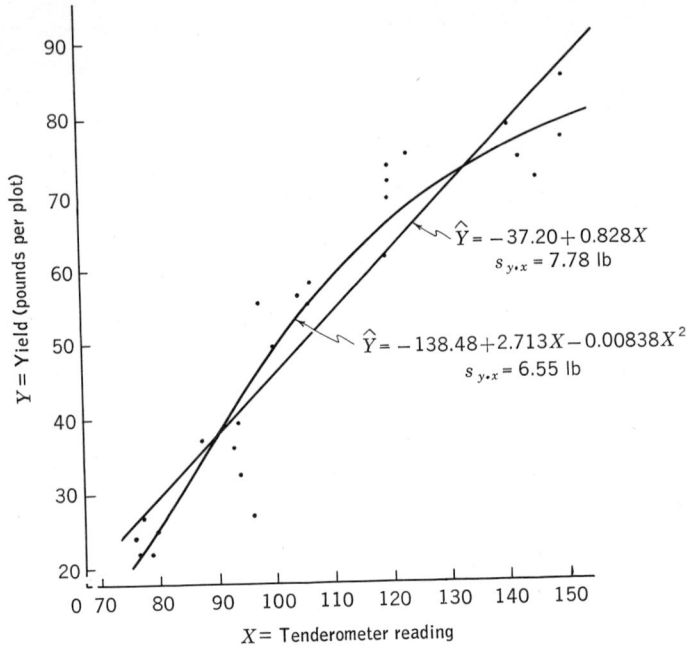

FIG. 16.4 Relationship between yield and tenderometer reading for data in Table 16.3

The standard error of estimate is $\sqrt{43.07} = 6.56$ lb. The multiple correlation coefficient is $R = \sqrt{9{,}885.24/10{,}832.76} = 0.955$ and is based on $n - 3 = 22$ degrees of freedom. Other quantities which may be desired can be found by the methods of Chap. 14.

Exercise 16.5.1 To Table 9.5 add the pair of observations (1,949, 1,796). Compute the quadratic regression of number of horses on year. Is the additional reduction due to including X^2 in the regression equation significant?

16.6 Higher-degree polynomials. When the X's are equally spaced and each Y is based on the same number of observations, it is possible to use orthogonal polynomials as discussed in Sec. 11.8; otherwise the methods of Chap. 14 must be used. For the third-degree polynomial, X, X^2, and X^3 correspond to X_1, X_2, and X_3; the problem becomes one of partial regression. However, the coefficients of the powers of X cannot be literally interpreted as partial regression coefficients since the polynomial remains in a two-dimensional space regardless of the degree of the polynomial. Since powers of X are involved, it is well to code X to avoid large numbers as much as possible.

NONLINEAR REGRESSION

16.7 Polynomials and covariance. In the analysis of experimental data, the regression of Y on X for residuals may be nonlinear. Consequently, it may be desired to use the covariate in a polynomial of degree greater than one. The methods of Sec. 15.12 apply in the case of a second-degree polynomial. The more general methods of Chap. 15 on multiple regression and covariance are easily extended for other polynomials.

16.8 Orthogonal polynomials. When increments between successive levels of X are equal and the Y values have a common variance, tables of orthogonal polynomial coefficients can be used in computations leading to tests of hypotheses about the goodness of fit of polynomials of various degrees. In the most common situation, the Y values are treatment means based on equal numbers of observations. Regression equations may also be found, though one is not likely to use this procedure except in the case of linear regression. The procedure, including decoding, is illustrated in Sec. 11.8 for linear regression. Since computation of the other regression coefficients requires knowledge of the polynomials themselves, it cannot be done with only a table of coefficients. In particular, the coefficient of X changes as we proceed from linear to quadratic regression and it is necessary to know the polynomials in order to make the change. However, the use of tabled coefficients greatly reduces the work of computing sums of squares and testing goodness of fit.

Orthogonal polynomials, but not the table of coefficients, can be used in situations where the X's are not equally spaced and/or observations have unequal but known variances. The successive polynomials are independent of one another and enable us to compute *additional sums of squares* attributable to the various powers of X. This may be done directly rather than by the procedure of Sec. 16.5, for example. If such polynomials were available, then considerable computational effort might be spared in some cases, especially if there were relatively few pairs of observations. For example, we might desire to use them in cases such as the following:

1. Observations made in time where it is desired to have more observations during a certain stage of some process. Thus, in measuring the percentage of cellulose digested by cattle, we might wish to make observations at 6-hr intervals during the first 24 hr but less frequently thereafter.

2. Observations made in space across some center such as a pollen source or irrigation ditch in such a way that there is one plot at the center and one plot on each side for each distance. Thus, there may be only half as many observations in the mean for the central treatment as there are in the mean for any specific distance from the center.

Fisher (16.3; 16.4) presents the method for equally spaced X's. Anderson and Houseman (16.1) and Pearson and Hartley (16.5) have prepared extensive tables of coefficients. Other authors have given methods for unequally spaced X's, a case which includes Y means based on unequal numbers. Robson (16.6) uses what is essentially Fisher's argument to obtain a recursion formula for computing coefficients. (A recursion formula is one which is applied over and over again, each new application making use of the previous application.) Robson proceeds as follows:

Any least-squares regression equation may be expressed in the form

$$\hat{Y}_i = b_0 f_0(X_i) + b_1 f_1(X_i) + \cdots + b_r f_r(X_i) \qquad (16.7)$$

$i = 1, \ldots, n > r$, where $\hat{Y}_i$ is the estimated population mean of the Y's at $X = X_i$. The polynomial $f_j(X_i)$ is of degree j and will provide the orthogonal polynomial coefficients, one for each X, required to determine the regression coefficients. The coefficients are given by Eq. (16.8).

$$b_j = \sum_i Y_i f_j(X_i) \tag{16.8}$$

Each regression coefficient represents a contrast or comparison as defined in Sec. 11.6. These contrasts are orthogonal and give additional sums of squares attributable to including X to the jth power in the regression equation. Thus b_0 is the zero degree or mean (not main) effect, b_1 is the first degree or linear effect, and so on. Finally

$$\sum_i Y_i^2 = b_0^2 + \cdots + b_{n-1}^2$$

if we carry r out to $n - 1$.

The recursion formula is given by Robson as Eq. (16.9) with c_h defined by Eq. (16.10).

$$f_h(X_i) = \frac{1}{c_h} \left[X_i^h - \sum_{j=0}^{h-1} f_j(X_i) \sum_{g=1}^{n} X_g^h f_j(X_g) \right] \tag{16.9}$$

$$c_h^2 = \sum_i \left[X_i^h - \sum_{j=0}^{h-1} f_j(X_i) \sum_{g=1}^{n} X_g^h f_j(X_g) \right]^2 \tag{16.10}$$

Notice that c_h^2 is the sum of the squares of quantities like the one, namely the ith, in the bracket of Eq. (16.9).

We shall now illustrate the use of the recursion formula. Suppose we have means for the percentage of digestion of cellulose at 6, 12, 18, 24, 36, 48, and 72 hr. Application of Eqs. (16.9) and (16.10) gives

$$f_0(X_i) = \frac{1}{c_0}(X_i^0) = \frac{1}{c_0}$$

$$c_0^2 = X_1^0 + \cdots + X_n^0 = n$$

(A quantity raised to the zero power equals 1.) Now

$$f_0(X_1) = \frac{1}{\sqrt{n}} = f_0(X_2) = \cdots = f_0(X_n)$$

From Eq. (16.8),

$$b_0 = \Sigma Y_i \frac{1}{\sqrt{n}}$$

and the reduction attributable to this polynomial is

$$b_0^2 = \frac{(\Sigma Y)^2}{n}$$

This is the correction term as we might have expected. The remaining $n - 1$ regression coefficients are concerned with the partitioning of the $n - 1$ degrees of freedom for treatment means.

Next,
$$f_1(X_i) = \frac{1}{c_1}\left[X_i - f_0(X_i)\sum_{g=1}^{n} X_g f_0(X_g)\right]$$
$$= \frac{1}{c_1}\left(X_i - \frac{1}{\sqrt{n}}\sum_{g=1}^{n} X_g \frac{1}{\sqrt{n}}\right)$$
$$= \frac{1}{c_1}(X_i - \bar{x})$$
$$c_1^2 = \Sigma(X_i - \bar{x})^2$$

Now $\bar{x} = 216/7 = 30.86$ and $\Sigma(X_i - \bar{x})^2 = 3{,}198.86$; $f_1(X_1) = (6 - 30.86)/\sqrt{3{,}198.86}, \ldots, f_1(X_n) = (72 - 30.86)/\sqrt{3{,}198.86}$. Finally

$$b_1 = \Sigma Y_i \frac{(X_i - \bar{x})}{\sqrt{\Sigma(X_i - \bar{x})^2}}$$

and the reduction attributable to linear regression is

$$b_1^2 = \frac{[\Sigma Y_i(X_i - \bar{x})]^2}{\Sigma(X_i - \bar{x})^2}$$

Again, this result was expected. The quantity in the brackets of the numerator is usually written in the form $\Sigma(Y_i - \bar{y})(X_i - \bar{x})$ but the two forms can be shown to be equivalent. Note that for $r = 2$, Eq. (16.7) becomes

$$\hat{Y} = \frac{(\Sigma Y_i)}{\sqrt{n}}\frac{1}{\sqrt{n}} + \frac{\Sigma Y_i(X_i - \bar{x})}{\sqrt{\Sigma(X_i - \bar{x})^2}}\frac{(X - \bar{x})}{\sqrt{\Sigma(X_i - \bar{x})^2}} = \bar{y} + \frac{\Sigma xy}{\Sigma x^2}(X - \bar{x})$$

For the second-degree equation,

$$f_2(X_i) = \frac{1}{c_2}\left[X_i^2 - \frac{1}{\sqrt{n}}\left(\sum_{g=1}^{n} X_g^2 \frac{1}{\sqrt{n}}\right) - \frac{X_i - \bar{x}}{\sqrt{\Sigma_i(X_i - \bar{x})^2}}\sum_{g=1}^{n} X_g^2 \frac{X_g - \bar{x}}{\sqrt{\Sigma_i(X_i - \bar{x})^2}}\right]$$
$$= \frac{1}{c_2}\left[X_i^2 - \frac{1}{n}\sum_g X_g^2 - \frac{(X_i - \bar{x})\sum_g X_g^2(X_g - \bar{x})}{\Sigma_i(X_i - \bar{x})^2}\right]$$

Again, we must find c_2^2 by Eq. (16.10). Notice that $f_2(X_i)$ is a polynomial of degree 2; it has an X_i^2, an X_i as an $X_i - \bar{x}$ times a constant, and a constant term, namely, $-\Sigma X_g^2/n$. The fact that there is an X_i shows that the complete quadratic equation will have a coefficient of X that differs from the coefficient of X in the linear equation; the coefficient of X_i in $f_2(X_i)$ supplies the adjustment.

Notice that we have computed the orthogonal functions for the general case. Then, for our example, we have used the X's of the experiment to get the coefficients of the Y's for the linear function of the Y's which gives us the additional reduction attributable to the highest power of X that has been introduced.

Exercise 16.8.1 A. van Tienhoven, Cornell University, conducted a 5 × 5 factorial experiment to study the effects of thyroid hormone (TH) and thyroid stimulating hormone (TSH) on follicular epithelium heights (among other responses) in chicks. TH is measured in γ units, TSH in Junkmann-Schoeller units, and the response in micrometer units. Treatment totals are shown in the accompanying table. Each total is of five observations.

		TSH				
		.00	.03	.09	.27	.81
	.00	3.42	6.21	11.21	14.40	19.40
	.04	5.64	5.85	9.16	18.30	19.65
TH	.16	5.13	8.39	12.74	15.20	15.07
	.64	5.37	5.24	9.14	17.66	16.30
	2.56	4.54	6.49	8.37	14.23	16.90

SOURCE: Unpublished data used with permission of A. van Tienhoven, Cornell University, Ithaca, N.Y.

The response as measured by epithelium heights is supposed to follow a logarithmic curve. The first levels of TH and TSH are not in equally spaced log sequences with the other levels. However, information was wanted for this particular treatment.

A preliminary analysis follows.

Ignore treatments TSH = .00 and TH = .00. Plot the response to TH, for each level of TSH, on log paper with treatment on the log scale. Does the response appear to be linear on this scale? Repeat for the response to TSH.

Use orthogonal polynomials to compute sums of squares for the over-all linear, quadratic, and cubic responses to TSH for the four nonzero levels. Choose the log scale for treatments so that the table of coefficients may be used directly. Test each response for significance.

Consider how you would test the homogeneity of these various responses over levels of TH. Test the homogeneity of the cubic responses since they measure deviations from quadratic responses and might be considered as candidates for an error term.

Compute sums of squares for over-all linear, quadratic, and cubic responses to TH.

How do you find coefficients for measuring TSH(linear) times TH(linear), TSH(linear) times TH(quadratic), and TSH(quadratic) times TH(linear)? Find and test these components (see Chap. 11).

PRELIMINARY ANALYSIS

Source	df	SS
Total	123	1,378.85
TSH	4	540.33
TH	4	48.09
TSH × TH	16	154.67
Error	99†	635.76

† There was one missing observation

Exercise 16.8.2 Use Eq. (16.9) to compute the cubic orthogonal polynomial.

Exercise 16.8.3 From the results in the text and of Exercise 16.8.2, compute your own table of orthogonal polynomial coefficients for four and five equally spaced and equally replicated treatments. Comment.

References

16.1 Anderson, R. L., and E. E. Houseman: "Tables of Orthogonal Polynomial Values Extended to n = 104," *Iowa Agr. Expt. Sta. Research Bull.* no. 297, 1942.

16.2 Brockington, S. F., H. C. Dorin, and H. K. Howerton: "Hygroscopic equilibria of whole kernel corn," *Cereal Chem.*, **26**: 166–173 (1949).

16.3 Fisher, R. A.: "The influence of rainfall on the yield of wheat at Rothamsted," *Phil. Trans. Roy. Soc.*, **B.213**: 89–142 (1925).

16.4 Fisher, R. A.: *Statistical Methods for Research Workers*, 11th ed., revised, Hafner Publishing Company, New York, 1950.

16.5 Pearson, E. S., and H. O. Hartley (eds): *Biometrika Tables for Statisticians*, vol. 1, Cambridge University Press, Cambridge, 1954.

16.6 Robson, D. S.: "A simple method for constructing orthogonal polynomials when the independent variable is unequally spaced," *Biometrics*, **15**: 187–191 (1959).

16.7 Weber, C. R., and T. W. Horner: "Estimation of cost and optimum plot size and shape for measuring yield and chemical characters in soybeans," *Agron. J.*, **49**: 444–449 (1957).

Chapter 17

SOME USES OF CHI-SQUARE

17.1 Introduction. The chi-square test criterion is most commonly associated with enumeration data. However, the chi-square distribution is a continuous distribution based upon an underlying normal distribution. At this point, we introduce a chapter on chi-square to stress, to some degree, its true nature by associating it with data from continuous distributions in several useful situations. We then proceed in two subsequent chapters to illustrate its use with enumeration data.

Chi-square is defined in Sec. 3.8 as the sum of squares of independent, normally distributed variables with zero mean and unit variance, as illustrated in Eqs. (3.9) and (3.10). In this chapter, we show how to compute a confidence interval for σ^2 using the χ^2 distribution. This is an exact procedure. The test criterion χ^2 is also used where it is an obvious approximation, for example, see Sec. 10.5 on testing the homogeneity of correlation coefficients. Two similar uses, tests of the homogeneity of variances and of the goodness of fit of observed continuous data to theoretical distributions, are discussed in this chapter.

17.2 Confidence interval for σ^2. Consider the statement

$$P(\chi_0^2 \leq \chi^2 \leq \chi_1^2) \leq .95$$

This is a statement about the random variable χ^2. An obvious choice of χ_0^2 and χ_1^2 would be the values $\chi^2_{.975}$ and $\chi^2_{.025}$, values such that

$$P(\chi^2 \leq \chi^2_{.975}) = .025 \quad \text{and} \quad P(\chi^2 \geq \chi^2_{.025}) = .025$$

These two values are convenient to use when computing a confidence interval for σ^2. However, they do not give the shortest possible confidence interval.

The customary procedure for computing a 95% confidence interval for σ^2 starts by combining the above pair of probability statements. We obtain

$$P(\chi^2_{.975} \leq \chi^2 \leq \chi^2_{.025}) = .95$$

where $\chi^2 = (n-1)s^2/\sigma^2$ by definition. (For a 99% confidence interval, use $\chi^2_{.995}$ and $\chi^2_{.005}$; and so on.) This formula leads to Eq. (17.1).

$$P\left[\frac{(n-1)s^2}{\chi^2_{.025}} \leq \sigma^2 \leq \frac{(n-1)s^2}{\chi^2_{.975}}\right] = .95 \qquad (17.1)$$

SOME USES OF CHI-SQUARE

This equation is still an equation about the random variable χ^2 or about s^2, though it is now made to sound like an equation about σ^2. From a particular sample, we now compute $(n-1)s^2/\chi^2_{.025}$ and $(n-1)s^2/\chi^2_{.975}$ as the end points of the 95% confidence interval for σ^2.

For example, Table 4.3, sample 1, has SS $= (n-1)s^2 = 2{,}376.40$ for $n = 10$. (We know that $\sigma^2 = 144$.) For a confidence interval for σ^2, compute

$$\frac{2{,}376.40}{\chi^2_{.025}} = \frac{2{,}376.40}{19.02} = 124.94$$

and

$$\frac{2{,}376.40}{\chi^2_{.975}} = \frac{2{,}376.40}{2.70} = 880.15$$

We now say that σ^2 lies between 124.94 and 880.15 unless the sample was an unusual one. In this case, we know that σ^2 does lie in the confidence interval because we sampled a known population.

Exercise 17.2.1 Compute a 90% confidence interval for σ^2 for the second sample in Table 4.3. A 95% confidence interval for the third sample. A 99% confidence interval for the fourth sample.

Exercise 17.2.2 Compute 95% confidence intervals for σ^2 for each soil type sampled in Table 5.6. Do the confidence intervals overlap? Compare this result with the result of the F test of $H_0: \sigma_1^2 = \sigma_2^2$ versus $H_1: \sigma_1^2 \neq \sigma_2^2$. Comment.

17.3 Homogeneity of variance. In a study including the inheritance of seed size in flax, Myers (17.7) obtained the variances among weights per 50 seeds from individual plants for parents and the F_1 generation. These are given in Table 17.1.

TABLE 17.1

TEST OF THE HOMOGENEITY OF VARIANCES

Class	df	Σx^2	s^2	$\log s^2$	$(n-1)\log s^2$	$1/(n-1)$
Redwing	81	4,744.17	58.57	1.76768	143.18208	.01235
Ottawa 770B	44	3,380.96	76.84	1.88559	82.96596	.02273
F_1	13	1,035.71	79.67	1.90129	24.71677	.07692
Totals	138	9,160.84			250.86481	.11200
Pooling			66.38	1.82204	251.44152	

$$\chi^2 = 2.3026\{[\Sigma(n_i - 1)]\log \bar{s^2} - \Sigma(n_i - 1)\log s_i^2\}$$
$$= 2.3026(251.44152 - 250.86481) = 1.3279 \text{ with 2 } df$$

$$\text{Correction factor} = 1 + \frac{1}{3(k-1)}\left[\Sigma\frac{1}{n_i - 1} - \frac{1}{\Sigma(n_i - 1)}\right]$$

$$= 1 + \frac{1}{3(2)}\left[.11200 - \frac{1}{138}\right] = 1.01746$$

$$\text{Corrected } \chi^2 = \frac{1.3279}{1.01746} = 1.305, \text{ not significant}$$

TABLE 17.2
Observed and Expected Values of Yield, in Grams, of 229 Spaced Plants of Richland Soybeans

Yield	3	8	13	18	23	28	33	38	43	48	53	58	63	68
Observed frequency	7	5	7	18	32	41	37	25	22	19	6	6	3	1
End point	5.5	10.5	15.5	20.5	25.5	30.5	35.5	40.5	45.5	50.5	55.5	60.5	65.5	
Deviation from mean	−26.43	−21.43	−16.43	−11.43	−6.43	−1.43	3.57	8.57	13.57	18.57	23.57	28.57	33.57	
Standard deviations from mean	−2.065	−1.674	−1.284	−0.893	−0.502	−0.112	0.279	0.670	1.060	1.451	1.841	2.232	2.623	
Probability	.0194	.0277	.0525	.0863	.1219	.1477	.1545	.1386	.1068	.0712	.0405	.0201	.0084	.0044
Expected frequency	4.4	6.3	12.0	19.8	27.9	33.8	35.4	31.7	24.5	16.3	9.3	4.6	1.9	1.0
Contribution to χ^2	1.54	0.27	2.08	0.16	0.60	1.53	0.07	1.42	0.26	0.45	1.17	0.43	0.64	0.00

$\bar{x} = 31.93 \qquad s = 12.80 \qquad \chi^2 = 10.62 \text{ with } 14 - 3 = 11 \, df$

SOME USES OF CHI-SQUARE

Let us suppose it is desired to test the homogeneity of the variances of Table 17.1. The test procedure, an approximate one due to Bartlett (17.1; 17.2), is carried out in Table 17.1. If natural logarithms (base e) are used, the multiplier 2.3026 is not required; this is used only with common logarithms (base 10).

Since the correction factor is always greater than one, its use decreases the crude χ^2. Thus, we generally compute the corrected χ^2 only when the crude χ^2 is significant.

Independent χ^2 values are additive, that is, the distribution of such a sum is also distributed as χ^2 with degrees of freedom equal to the sum of those for the individuals. This is true only for the crude χ^2's and not for the corrected ones. Thus, if variances were available for several years, it might be informative to make within- and among-year comparisons. Also, if variances were available for several segregating generations, one might compare among parents and F_1 (as here), among segregating generations, and between the two groups. The total χ^2 can be used as an arithmetic check, at least.

Exercise 17.3.1 Variances for the data of Table 7.1 vary from 1.28 to 33.64. While this is quite a range, each has only 4 degrees of freedom. Test the homogeneity of these variances. Can you see how to reduce the computations when all variances are based on the same number of degrees of freedom?

Exercise 17.3.2 Check the homogeneity of the variances in Exercise 7.3.1.

Exercise 17.3.3 The variances in Exercise 7.9.1 are based on different numbers of degrees of freedom. Check their homogeneity.

17.4 Goodness of fit for continuous distributions. It is often desired to know whether or not a given set of data approximates a given distribution such as the normal or chi-square. As an example of this, consider the data of Table 17.2, earlier presented in Table 2.5. Let us test the observed distribution against the normal distribution.

To compare an observed distribution with a normal distribution, expected cell frequencies are required. The yield value of 3 grams includes yields to 5.5 grams, of 8 grams includes yields greater than 5.5 and up to 10.5 grams, etc. To compute expected frequencies, the probabilities associated with each interval are necessary. These are found from Table A.4. Values for μ and σ must be estimated from our data. We find $\bar{x} = 31.93$ and $s = 12.80$ and consider them to be μ and σ. Now,

$$P(X < 5.5) = P\left(z < \frac{5.5 - 31.93}{12.80}\right)$$
$$= P(z < -2.065) = .0194$$

$$P(5.5 < X < 10.5) = P\left(\frac{5.5 - 31.93}{12.80} < z < \frac{10.5 - 31.93}{12.80}\right)$$
$$= P(-2.065 < z < -1.674) = .0471 - .0194$$
$$= .0277$$

and so on. Each probability times the total frequency 229 gives an expected frequency. The probability associated with the last cell is the probability of a value greater than 65.5 and is not the probability of a value greater than 65.5 but not more than 70.5. The computations are shown in Table 17.2. The sum of the probabilities should differ from one by rounding errors only. The sum of the expected frequencies should differ from 229 by rounding errors only.

Compute the value of the test criterion defined by Eq. (17.2).

$$\chi^2 = \Sigma \frac{(\text{observed} - \text{expected})^2}{\text{expected}} \qquad (17.2)$$

$= 10.62$ with $(14 - 1 - 2) = 11$ df, not significant.

The number of degrees of freedom is the number of cells decreased by one and the number of parameters estimated. There are two parameters in the normal distribution, namely, μ and σ^2. Notice that the number of degrees of freedom can never exceed $k - 1$ where k is the number of cells. This criterion, the right side of Eq. (17.2), is distributed approximately as χ^2 provided expected frequencies are large; some writers have suggested five to ten as a minimum. Such writers would recommend that we pool our first two cells as a new first and our last three or four as a new last. However, since the tails of a distribution often offer the best source of evidence for distinguishing among hypothesized distributions, the χ^2 approximation is improved at the expense of the power of the test. Cochran (17.3; 17.4; 17.5) has shown that there is little disturbance to the 5% level when a single expectation is as low as .5 and two expectations may be as low as 1 for fewer degrees of freedom than for our example. The 1% level shows a somewhat greater disturbance than the 5%.

Exercise 17.4.1 Fit a normal distribution to the table of butterfat data given as Table 4.1 assuming $\mu = 40$ and $\sigma = 12$ lb. Test the goodness of fit.

Exercise 17.4.2 Fit a normal distribution to the 500 means of Table 4.4, assuming no knowledge of the population mean and variance. Test the goodness of fit.

17.5 Combining probabilities from tests of significance.

Fisher (17.6) has shown that $-2 \log P$, where logs are to the base e and P is the probability of obtaining a value of the test criterion as extreme or more extreme than that obtained in a particular test, is distributed as χ^2 with two degrees of freedom. Such values may be added. In practice, it is customary to use logs to the base 10 and multiply the result by 2.3026 to give logs to the base e. Multiplication is conveniently done after addition rather than for the separate values.

This relation between P and χ^2 may be used when it is desired to pool information available from data which are not suitable for pooling. Such data must be from independent trials.

For example, we may have information on the difference in response to a check or standard treatment and a particular experimental treatment from several quite different experiments. These treatments may be the only ones that are common to the experiments, which may include different experimental designs. Each experiment may show the probability of a larger value of the test criterion, not necessarily the same criterion for all tests, to be between .15 and .05 for the particular comparison. Probabilities are computed on the assumption that the null hypothesis is true. No one value may be significant yet each may suggest the possibility of a real difference. The test would be applied to the combined probabilities, combination being carried out by adding the values of $-2 \log P$.

The use of this pooling procedure requires that the tables of the original test criteria be reasonably complete with respect to probability levels; interpolation must then be used. The sum of the $-2 \log P$ values is then referred to the χ^2 table, Table A.5, where the value of P for the pooled information is obtained. This pooled χ^2 has $2k$ degrees of freedom when k probabilities are being combined.

Wallis (17.8) suggests that this test is rarely or never ideal but is likely to be highly satisfactory in practice.

References

17.1 Bartlett, M. S.: "Properties of sufficiency and statistical tests," *Proc. Roy. Soc.*, **A160:** 268–282 (1937).

17.2 Bartlett, M. S.: "Some examples of statistical methods of research in agriculture and applied biology," *J. Roy. Stat. Soc. Suppl.*, **4:** 137–183 (1937).

17.3 Cochran, W. G.: "The χ^2 correction for continuity," *Iowa State Coll. J. Sci.*, **16:** 421–436 (1942).

17.4 Cochran, W. G.: "The χ^2 test of goodness of fit," *Ann. Math. Stat.*, **23:** 315–345 (1952).

17.5 Cochran, W. G.: "Some methods for strengthening the common χ^2 tests," *Biometrics*, **10:** 417–451 (1954).

17.6 Fisher, R. A.: *Statistical Methods for Research Workers*, 11th ed., revised, Hafner Publishing Company, New York, 1950.

17.7 Myers, W. M.: "A correlated study of the inheritance of seed size and botanical characters in the flax cross, Redwing × Ottawa 770B," *J. Am. Soc. Agron.*, **28:** 623–635 (1936).

17.8 Wallis, W. A.: "Compounding probabilities from independent significance tests," *Econometrica*, **10:** 229–248 (1942).

Chapter 18

ENUMERATION DATA I: ONE-WAY CLASSIFICATIONS

18.1 Introduction. This chapter is concerned with *enumeration data* classified according to a single criterion.

Enumeration data generally involve a discrete variable, that is, a qualitative rather than a quantitative characteristic. Thus, they consist of the numbers of individuals falling into well-defined classes. For example, a population is sampled and the numbers of males and females in the sample are observed, or the numbers of affirmatives, negatives, and maybes in response to a question on a questionnaire, and so on.

18.2 The χ^2 test criterion. In Sec. 3.8, Eq. (3.9), the statistic chi-square with n degrees of freedom is defined as the sum of squares of n independent, normally distributed variates with zero means and unit variances. In other words

$$\chi^2 = \sum_i \frac{(X_i - \mu_i)^2}{\sigma_i^2} \qquad (18.1)$$

where the X_i are independent.

Except for Chap. 17, χ^2 has been mentioned infrequently and only when it was involved in other distributions; for example, F is a ratio of two χ^2's divided by their degrees of freedom. In this chapter, test criteria called χ^2 are frequently used. However, we are now dealing with discrete data so that the test criteria are not χ^2 quantities as defined in Eq. (18.1), even when the null hypothesis is true. This means the test criteria can be distributed only approximately as χ^2 quantities.

The χ^2 distribution, when associated with discrete data, is usually in conjunction with a test of *goodness of fit*. The test criterion is defined by Eq. (18.2).

$$\chi^2 = \sum \frac{(\text{observed} - \text{expected})^2}{\text{expected}} \qquad (18.2)$$

The sum is taken over all cells in the classification system. *Observed* refers to the numbers observed in the cells; *expected* refers to the average numbers when the hypothesis is true, that is, to the theoretical values. The sum of such deviations, namely, values of (observed − expected), will equal zero within rounding errors. The number of degrees of freedom involved will be discussed for several situations as they arise.

18.3 Two-cell tables, confidence limits for a proportion or percentage.

Many sampling situations allow only two possible outcomes, for example, the numbers of Yeses and Noes in response to a question, or the numbers of individuals showing the presence or absence of a qualitative characteristic. An estimate of a population proportion or a test of hypothesis about a proportion is often required. Consider the following example. In studying *Dentaria* in an area near Ithaca, N.Y., a class in taxonomy and ecology observed the numbers of plants which did or did not flower. This was done for a number of samples. In one sample, there were 42 flowering and 337 nonflowering plants. Let us set a confidence interval on the proportion of flowering plants in the population.

Assume that sampling has been random and from a stable population. Since there are only two possible outcomes associated with an individual, we have a so-called *binomial population* (discussed in some detail in Chap. 20). The parameter to be estimated by a confidence interval is the proportion of flowering plants in the population of plants or the probability that a randomly selected plant will be a flowering plant. The parameter is generally denoted by p and its estimate, the observed proportion by $\hat{p}$. The mean of all possible $\hat{p}$ values is p, that is, $\hat{p}$ is an unbiased estimator of p, the population mean. The variance of the statistic $\hat{p}$ is $p(1-p)/n$, where n is the total number of observations in the sample; it is estimated by $\hat{p}(1-\hat{p})/n$ when necessary. Note that the variance is computed from the mean. Often, $1-p$ and $1-\hat{p}$ are written as q and $\hat{q}$, respectively.

We now proceed to use the normal distribution as an approximation to the binomial distribution, knowing that a basic assumption is false.

1. *The normal approximation.* As an approximation, we say p is normally distributed with mean p and variance $p(1-p)/n$, estimated by $\hat{p}(1-\hat{p})/n$. Hence the 95% confidence interval for p is given by Eq. (18.3).

$$\hat{p} \pm z_{.05}(\text{normal})\sqrt{\frac{\hat{p}(1-\hat{p})}{n}} \tag{18.3}$$

$$= \frac{42}{379} \pm 1.96\sqrt{\frac{42}{379}\frac{337}{379}\frac{1}{379}} = .111 \pm .032 = (.079, .143)$$

This approximation is not the only possible one based on the normal distribution but it is convenient and should differ little from any other based on reasonable sample size. Suggested lower limits for sample sizes are given in Table 18.1 and are due to Cochran (18.2).

2. *The exact distribution.* When it is decided that an approximation is inadequate, tables of the binomial distribution are required. Many such tables are available, for example, *Tables of the Cumulative Binomial Probability Distribution* by the Computation Laboratory, Harvard University (18.7), contain sums of the probabilities of k terms for $k = 1, 2, \ldots, n+1$ for $n = 1, 2, \ldots, 1{,}000$ observations in the sample and for $p = 0.01$ to 0.50. Confidence intervals computed from these tables will rarely be symmetric about $\hat{p}$ since they associate about half of the probability of a Type I error with each tail of the distribution.

Essentially we choose two values of p, say p_1 and p_2, with the property that they are the lowest and highest values of p that can be hypothesized and found acceptable on the basis of the observed data. In practice, it will usually be impossible to find probabilities exactly equal to half the probability of a Type I error; in some cases, it may be necessary to put the full probability in one tail, for example, if all or very nearly all observations fall in one class, then one confidence limit will be .00 or 1.00.

TABLE 18.1

BINOMIAL SAMPLE SIZE FOR NORMAL APPROXIMATION TO APPLY

$\hat{p}$	$n\hat{p}$ = number observed in *smaller* class	$n =$ sample size
0.5	15	30
0.4	20	50
0.3	24	80
0.2	40	200
0.1	60	600
0.05	70	1,400

SOURCE: Reprinted with permission from W. G. Cochran, "Sampling Techniques," Table 3.3, page 41, John Wiley & Sons, Inc., New York, 1953.

Tables A.14 give 95 and 99% confidence intervals for the binomial distribution. These are obtained from more extensive tables by Mainland et al. (18.3). To illustrate their use, a coin was tossed 30 times and 13 heads and 17 tails observed. From the tables, the 95 and 99% confidence intervals for the probability p, of obtaining a head in a single toss, are given as $.2546 < p < .6256$ and $.2107 < p < .6772$, respectively. The confidence limits for the probability of a tail may be found in the table or by subtracting each of the above limits from 1.00.

3. *Clopper and Pearson charts*. Alternative to computing a confidence interval, one may use a chart entitled *Confidence Belts for* p, as given in Clopper and Pearson (18.1), or a similar chart. Table A.15 is one such chart. For the *Dentaria* data, compute $\hat{p} = 42/379 = .111$ and locate it on the $\hat{p} = X/n$ scale; vertically above this axis find two lines marked for sample size near 379 (400 is the closest value); now move horizontally over to the p scale where the confidence interval values are found. For the 95% confidence interval, we obtain (.07,.15) without attempting to do any real interpolation for the observed sample size of 379.

Such charts as in Table A.15 are computed by an approximate method of interpolation, which gives a more accurate result than the normal approximation. Confidence intervals will not be symmetric about $\hat{p}$.

4. *Binomial probability paper*. Another procedure makes use of binomial probability paper, designed by Mosteller and Tukey (18.4). This paper and the results obtained from it are based on transforming the data to a scale upon which the resulting numbers are approximately normally distributed with nearly constant variance; the implication is that the mean and variance are nearly independent.

ENUMERATION DATA I: ONE-WAY CLASSIFICATIONS

Table A.16 is binomial paper (see also Fig. 18.1, where an example is worked). For the *Dentaria* data, plot the point (42,337) or (337,42). Because of the dimensions of the paper, we plot the latter. A line through this point to the origin cuts the quarter circle at the point whose coordinates are the two percentages of the problem. We read .89 and .11 as proportions. Obtaining the confidence interval involves a triangle whose long side or hypotenuse faces away from the origin and with coordinates (337,43), (337,42), and (338,42). Notice that one number of the observed pair is increased by one to give the first coordinate, then the other to give the third; only one number is increased to give each new coordinate. Our triangle cannot be distinguished from a point. About this point, draw a circle of radius 1.96 Full Scale standard deviations. These are given on the paper. Tangents from the origin to this circle will cut the quarter circle at points whose ordinates are the confidence limits. We read .08 and .14. If we had wished a confidence interval for the proportion of nonflowering plants, we would have read the two abscissas. If the point had been distinguishable as a triangle, we would have measured two standard deviations upward from the upper acute angle and downward from the lower acute angle before drawing tangents through the quarter circle to the origin.

All our methods are seen to give virtual agreement for this problem. This depends on both the sample size and the p value. Change to percentages involves multiplication by 100.

Exercise 18.3.1 Two other samples of *Dentaria* from the same area yielded 6 flowering to 20 nonflowering and 29 flowering to 485 nonflowering plants. Obtain 95% confidence intervals in each case for the proportion of flowering plants in the population. Comment on the validity of each procedure.

Exercise 18.3.2 From a 30 × 10 m transect through the *Dentaria* area, all plants were enumerated and consisted of 296 flowering and 987 nonflowering plants. Assuming that these plants are a random sample from the *Dentaria* population, obtain a 95% confidence interval for the proportion of flowering plants in the population. Use binomial probability paper for this purpose. [You will have to change scales by plotting (296,987) as (29.6,98.7) and using the Tenth Scale of standard errors.]

18.4 Two-cell tables, tests of hypotheses. A confidence interval procedure includes a test of hypothesis against two-sided alternatives. However, where it is desired to test a hypothesis, it may be more convenient to compute a test criterion than to compute a confidence interval.

Consider a certain F_1 generation of *Drosophila melanogaster* with 35 males and 46 females. It is required to test the hypothesis of a 1:1 sex ratio.

1. *The normal approximation.* Using the normal approximation, we test the hypothesis $\mu = .5$, the 1:1 ratio, using variance $p(1-p)/n = (.5)(.5)/81$ where $81 = 35 + 46$. Compute

$$z = \frac{35/81 - .5}{\sqrt{(.5)(.5)/81}} = 1.22$$

From Table A.4, a z value of 1.96 is required for significance at the 5% level. Also from Table A.4, $P(|z| \geq 1.22) = 2(.1112) = .2224$. There is no evidence to deny the null hypothesis.

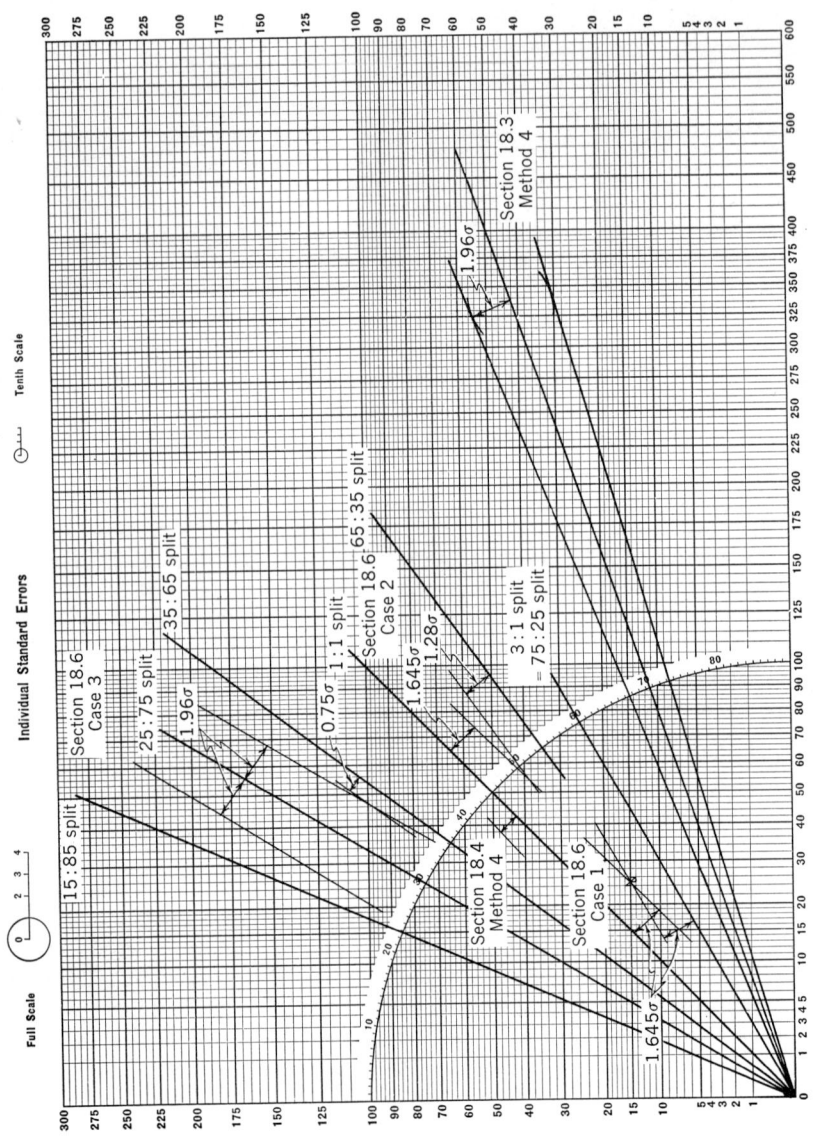

FIG. 18.1 Some examples of the use of binomial probability paper

A test against one-sided alternatives can also be made. For example, the 5% significance point is at $z = 1.645$ if we are looking for a p that is larger than .5; it is at $z = -1.645$ if we are looking for a p that is less than .5. For other significance levels, see Sec. 3.5, Cases 1.

2. *The χ^2 criterion.* Applying Eq. (18.2) to the *Drosophila* example, we find

$$\chi^2 = \Sigma \frac{(\text{observed} - \text{expected})^2}{\text{expected}}$$

$$= \frac{(35 - 40.5)^2}{40.5} + \frac{(46 - 40.5)^2}{40.5} = 1.49 \text{ with 1 } df$$

From Table A.5, we find $.25 > P$ (larger χ^2 by chance if ratio is $1:1$) $> .10$. Since a single degree of freedom is involved, we have the square of a single normal deviate. The single normal deviate was just computed as $z = 1.22$ and $P(|z| \geq 1.22) = .2224$. Now $z^2 = 1.49 = \chi^2$. This relation between z and χ^2 holds only for a single degree of freedom. The equivalence of these tests is easily shown by elementary algebra. The χ^2 test is against two-sided alternatives. Interpretation of the sample evidence is unchanged.

For χ^2 with a single degree of freedom, there is only one independent normal deviate. For the two-cell table, the two deviations are of the same size but opposite sign. This has led to the following short-cut formula for two-cell tables:

$$\chi^2 = (X - \mu)^2 \frac{n_1 + n_2}{n_1 n_2} \qquad (18.4)$$

where X is one of the observed cell numbers n_1 or n_2 and μ is the expected value or mean for that cell, namely, $(n_1 + n_2)p$ or $(n_1 + n_2)(1 - p)$. Equation (18.4) does not give a result that is identical with that of Eq. (18.2). We must replace $(n_1 + n_2)/n_1 n_2$ by $1/(n_1 + n_2)p(1 - p)$ to get an identical result.

One may also use

$$\chi^2 = \frac{(r_2 n_1 - r_1 n_2)^2}{r_1 r_2 (n_1 + n_2)} \qquad (18.5)$$

for a ratio $r_1 : r_2$ with observed numbers n_1 and n_2. If the ratio is expressed as $r:1$, with $r \geq 1$, then

$$\chi^2 = \frac{(n_1 - r n_2)^2}{r(n_1 + n_2)} \qquad (18.6)$$

where n_2 is the observed number in the cell with the smaller expected value. Equations (18.5) and (18.6) give the same result as Eq. (18.2).

To improve the approximation to the χ^2 distribution and thus be able to obtain a more exact probability value from the χ^2 table, Yates (18.9) has proposed a *correction for continuity*, applicable when the criterion has a single degree of freedom. This correction is intended to make the actual distribution of the criterion, as calculated from discrete data, more nearly like the χ^2 distribution based on normal deviates. The approximation calls for the absolute value of each deviation to be decreased by $\frac{1}{2}$. Thus,

$$\text{Adjusted } \chi^2 = \Sigma \frac{(|\text{observed} - \text{expected}| - .5)^2}{\text{expected}} \tag{18.7}$$

$$= \frac{(|35 - 40.5| - .5)^2}{40.5} + \frac{(|46 - 40.5| - .5)^2}{40.5}$$

$$= 1.23, \text{ for the } Drosophila \text{ data}$$

Adjustment results in a lower chi-square. Consequently, in testing hypotheses, it is worthwhile only when the unadjusted χ^2 is larger than the tabulated χ^2 at the desired probability level.

3. *The exact distribution.* When the sample size is small, it is advisable to use the exact distribution, namely, the binomial. "Small" may be defined as a number less than the appropriate number in Table 18.1. Tables of the binomial distribution can be used for testing hypotheses about p without computing confidence limits. In view of the work involved, computation of such limits or the use of existing tables of confidence limits may be the more profitable procedure. Use Tables A.14 or Mainland's (18.3) tables for two-sided alternatives. If the alternatives are one-sided, obtain the confidence interval for double the probability of a Type I error acceptable for test purposes or put all the probability in the appropriate tail.

4. *Binomial probability paper.* Binomial probability paper may also be used to test the null hypothesis. For the *Drosophila* data, plot the triangle (35,46), (35,47), (36,46). This triangle is distinguishable from a point. The hypothesized probability is obtained as a *split*, here 1:1 or 50:50 and so on, which is plotted as a line through the origin and any pair of values which describe the split (see Fig. 18.1). The shortest distance from the triangle to the split, in this case from (36,46) to the split, is compared with the standard error scale and is seen to be a little more than one standard deviation (compare $z = 1.22$).

Exercise 18.4.1 Woodward and Rasmussen (18.8) studied hood and awn development in barley. They conclude that this development is determined by two gene pairs. From an F_2 generation, 5 of 9 genotypes should breed true in the F_3 generation. Of these 5, 4 should give 3:1 segregation. The F_3 results for these 4 genotypes were:

2,376 hoods to 814 short awns
1,927 hoods to 642 long awns
1,685 long awns to 636 short awns
623 long awns to 195 short awns

Test $H_0: p = \frac{3}{4}$ for the appropriate character in each of the four cases. Do not use the same test in each case but comment on the appropriateness of the test you use. (One significant value was attributed to failure to classify awn types correctly in brachytic plants.)

18.5 Tests of hypotheses for a limited set of alternatives. When data have been obtained and a hypothesis tested, the value of the test criterion may lead us to accept the hypothesis. [Acceptance of the null hypothesis means that it and chance offer a not unreasonable explanation of the existing data.] However, chance and any value of p within the confidence interval will not lead us to deny the null hypothesis. Thus, acceptance of the null hypothesis, precisely as stated, would be a very strong statement. On the other

ENUMERATION DATA I: ONE-WAY CLASSIFICATIONS   359

hand, the type of alternative hypothesis discussed so far has been a set of alternatives. Hence acceptance of the alternative hypothesis is a general but somewhat indefinite statement. It is strong with respect to the null hypothesis only, since it rejects that hypothesis.

In certain genetic problems, there may be a limited number of possible null hypotheses to choose among and no one of them may be an obvious choice for the null hypothesis. For example, the experimenter may have to choose between the ratios 1:1 and 3:1. Which ratio should be tested as the null hypothesis, and how do we test against a single alternative as opposed to a set of alternatives? If both hypotheses are tested, one may be led to the conclusion that either hypothesis is satisfactory. In this case, it is clear that the sample size has been insufficient to distinguish between ratios at the chosen confidence level. When more than two ratios are possible, the results of testing each ratio as a null hypothesis may be more confusing.

Problems such as those of the previous paragraph are a special class of problems and require a special solution. Thus, in the analysis of variance, an F test is a correct general solution. However, if the experimenter wishes to test differences between all possible pairs of means, procedures such as Duncan's and Tukey's are advised; if treatments are a factorial set, then a test of main effects and interactions is advised. Here also, where there is a finite number of possible ratios, an alternative to the methods so far presented in this chapter is advised.

A partial solution to the problem of choosing between two or among more binomial ratios is given in Tables A.17, constructed by Prasert NaNagara (18.5). These tables give regions of acceptance for the various ratios, together with the probabilities of making a wrong decision according to which hypothesis is true. This solution begins with the premise that it is not always desirable to fix the probability of a Type I error in advance at one of the customary levels and to disregard the probability of a Type II error. This premise is particularly valid when no hypothesis is obviously the null hypothesis. Here, the probabilities of the possible errors are best fixed in advance and the sample size is then chosen so that the probabilities are not exceeded.

The method of solution, called *minimax*, is to minimize the maximum probability of an error of any kind. For example, suppose an individual falls into one of two distinct classes and we have only two hypotheses about the ratio involved, namely, that it is 1:1 or 3:1. Suppose the sample size is 20. If we observe 0:20, then we must logically accept the 1:1 hypothesis. The same is true if we observe 1:19, 2:18, ..., 10:10. Let us now jump to 15:5; here we must logically accept the 3:1 hypothesis. The same is true if we observe 16:4, ..., 20:0. The most difficult decisions will be for the observed values 11:9, ..., 14:6.

Let us now consider each of these possible cases, associate a rule for deciding between the two ratios in each case, consider the performance of each rule, and finally decide which rule is best.

Proposed rule 1. If we observe 0, 1, ..., 11 in the first class (the first 1 of 1:1 or the 3 of 3:1), accept the 1:1 hypothesis and reject the 3:1 hypothesis; if we observe 12, ..., 20 in the first class, reject the 1:1 hypothesis and accept the 3:1 hypothesis.

Performance of proposed rule 1. To judge this rule, either solely on its own merit or relative to other rules, we have to know what it will do for us if the true ratio is 1:1 and what it will do if the true ratio is 3:1.

Performance when the true ratio is 1:1. With this rule and a true ratio of 1:1, we make a wrong decision if we observe 12, ..., 20 in the first class, because the rule requires us to accept the 3:1 hypothesis. Let us compute the probability of making a wrong decision. To do this, add the probabilities associated with 12, ..., 20 in the first class when the true ratio is 1:1. This requires a table of the binomial distribution for $n = 20$ and $p = .5$. The probability is 0.2517.

Performance when the true ratio is 3:1. With the same rule but a true ratio of 3:1, we make a wrong decision if we observe 0, ..., 11 in the first class, because the rule requires us to accept the 1:1 hypothesis. The probability of a wrong decision is found by adding the probabilities associated with 0, ..., 11 in the first class when 3:1 is the true ratio. For these probabilities, we need a binomial probability table for $n = 20$ and $p = .75$ (or .25). The total probability is 0.0409.

If we now examine the performance of the proposed rule, we see that it is not too satisfactory if the true ratio is 1:1 but is satisfactory if the true ratio is 3:1. We would certainly like to have a better balance between the two probabilities. Hence, let us consider other rules.

Proposed rule 2. If we observe 0, ..., 12 in the first class, accept the 1:1 hypothesis and reject the 3:1 hypothesis; if we observe 13, ..., 20 in the first class, reject the 1:1 hypothesis and accept the 3:1 hypothesis.

Performance of proposed rule 2 if 1:1 *is the true ratio*. Now we make a wrong decision if we observe 13, ..., 20 in the first class, because the rule requires us to accept the 3:1 hypothesis. Again we refer to the binomial probability table for $n = 20$ and $p = .5$ and find the sum of the probabilities associated with 13, ..., 20 in the first class. This is 0.1316.

Performance of proposed rule 2 if 3:1 *is the true ratio*. Now we make a wrong decision if we observe 0, ..., 12 in the first class. We refer to the binomial probability table for $n = 20$ and $p = .75$ (or $p = .25$) and find the sum of the probabilities associated with 0, ..., 12 in the first class. The sum is 0.1018.

This rule has better performance than the previous one if we are trying to have the probabilities of a wrong decision about equal, regardless of which hypothesis is true. We may consider that both probabilities are too high.

Other proposed rules. The other rules to be considered are: (3) to accept the 1:1 hypothesis and reject the 3:1 hypothesis if we observe 0, ..., 13 in the first class; to reject the 1:1 hypothesis and accept the 3:1 hypothesis if we observe 14, ..., 20 in the first class; and (4) to accept the 1:1 hypothesis and reject the 3:1 hypothesis if we observe 0, ..., 14 in the first class; to reject the 1:1 hypothesis and accept the 3:1 hypothesis if we observe 15, ..., 20 in the first class.

The probabilities of making wrong decisions are given in Table 18.2 for three of the four rules just discussed. The numbers under the heading "Acceptance regions for ratios" refer to the numbers observed in each class.

Deciding among possible rules. We must now decide which rule is the best. Our criterion of "best" is that the worst possible situation into which we can get ourselves should be kept within reasonable bounds. In Table 18.2, we observe that rule 1 can lead us to wrong decisions about 25% of the time; rule 2 about 13% of the time; rule 3 about 21% of the time. These values, underlined in Table 18.2, are from the group which contains the worst possible situation. The smallest of these values, the minimum of the maximum probabilities of making a wrong decision, is 0.1316. This is the best we can do and the best rule is the corresponding rule, rule 2.

TABLE 18.2

PROBABILITIES OF MAKING WRONG DECISIONS FOR RATIOS 1:1 AND 3:1, $n = 20$

Rule	Acceptance regions for ratios		Probability of making a wrong decision when the true ratio is:	
	1:1	3:1	1:1	3:1
1	0–11	12–20	.2517	.0409
2	0–12	13–20	.1316	.1018
3	0–13	14–20	.0577	.2142

In review and with Table 18.2 in mind: with a sample size of 20, suppose the 1:1 ratio is accepted when there are less than 12 in the potentially larger group and the 3:1 ratio is accepted when the number is more than 11. Then the wrong decision is made with probability .0409 if 3:1 is the true ratio and with probability .2517 if 1:1 is the true ratio. This is essentially what happens if 3:1 is made the ratio for the null hypothesis, if the alternative hypothesis is appropriately one-sided, and if the Type I error is fixed at 5%; 4.09% is as close as we can get to 5%. On the other hand, if the acceptance region for the 1:1 ratio calls for less than 14 in the potentially larger class and for the 3:1 ratio for more than 13 in the class, the wrong decision is made with probability .0577 if the true ratio is 1:1 and with probability .2142 if the true ratio is 3:1. This is roughly equivalent to testing the null hypothesis of a 1:1 ratio against appropriately one-sided alternatives and with a fixed Type I error of .05; 5.77% is as close as we can get to 5%. With the division point between 12 and 13, the probabilities of a wrong decision are more nearly equal regardless of which hypothesis is true. For the three situations, the maximum probabilities of error are .2517, .1316, and .2142. In choosing 0–12 for the first class, we choose the solution for which the maximum probability of a wrong decision is a minimum, namely, .1316.

To visualize the situation which exists when the acceptance region for the 1:1 ratio is 0–13, refer to Fig. 18.2. Suppose the true ratio is 1:1. If we obtain more than 13 individuals in the first class, we wrongly accept the 3:1 ratio. The probability associated with this error is the sum of the probabilities of observing exactly 14, exactly 15, ..., exactly 20, namely, $P = .0577$, computed for the 1:1 ratio. This is the probability of a Type I error if the null hypothesis is that the ratio is 1:1. Now suppose the true ratio is 3:1. If we obtain fewer than 14 individuals in the first class, we wrongly accept the ratio 1:1. The probability of this kind of error is the sum of the probabilities of observing exactly 13, exactly 12, ..., exactly zero, namely, $p = .2142$, computed for the 3:1 ratio. This is the probability of a Type II error for the null hypothesis that the ratio is 1:1. Probabilities in Fig. 18.2 are shown as solid columns rather than vertical lines for reasons of clarity.

Tables A.17 contain only minimax solutions. To illustrate their use, consider Table A.17A for sample size 44. Accept the 1:1 hypothesis if 0 to 27 observations fall in the potentially larger group; if this group contains 28 to 44 observations, accept the 3:1 hypothesis. At worst, one would be in error 4.81% of the time on the average; this would be the case if data were always from a distribution with a 1:1 ratio. If one were never presented with data from other than a distribution with a 3:1 ratio, then one would be wrong 3.18% of the time on the average.

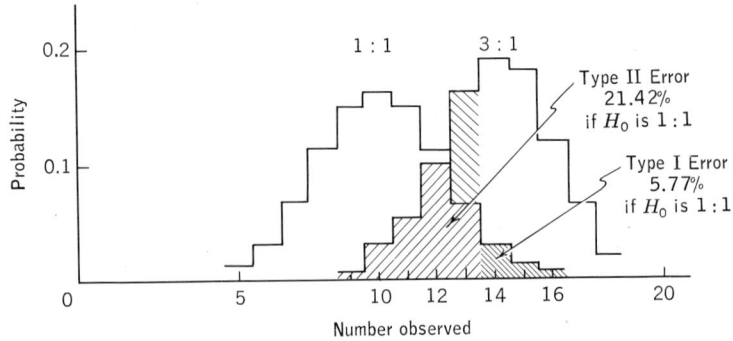

FIG. 18.2 Binomial probability distributions for ratios 1:1 and 3:1 and $n = 20$

Since the binomial distribution is used in obtaining the tabled probabilities, it is virtually impossible to find the exact probabilities of .05 and .01 in these tables. The remaining tables of this group give rules for acceptance of various alternative hypotheses for given sample sizes, and the probabilities of making wrong decisions.

Exercise 18.5.1 In a genetic problem, it is required to have sufficient data to distinguish between test cross ratios 3:1 and 7:1. How many offspring should be observed if the experimenter will be satisfied with a 10% error rate, regardless of which ratio is the true one?

Exercise 18.5.2 An experimenter has 727 F_2 plants to classify and knows that he will have to distinguish between the ratios 9:7, 13:3, and 15:1. What should his rule be to distinguish among these ratios if he wishes to have maximum protection against the worst possible situation with which nature may present him?

18.6 Sample size. In the general case, we wish to test the null hypothesis that the unknown parameter p, of a binomial population, is a specified value. In testing, we require that we should be able to detect some alternative with reasonable frequency. In the previous section, we discussed the minimax procedure for choosing acceptance and rejection regions when we are confronted with a limited set of possible p values. The associated tables, Tables

ENUMERATION DATA I: ONE-WAY CLASSIFICATIONS

A.17, may be used when it is desired to see what sort of protection is available for the rules given there; they may also be used to some extent in choosing sample size.

Let us consider the use of binomial probability paper for choosing sample size for a number of cases.

Case 1. Suppose that only two ratios are considered to be possible. It is desired to know the sample size required to distinguish between the two ratios.

For example, let the ratios 1:1 and 3:1 be the only possible ratios. Further, no matter which ratio is the true value, the seriousness of a wrong decision is about the same. In other words, no matter which ratio is true, we wish to have about the same probability of making a wrong decision. Let us set this probability at .05.

The procedure we present is based on the use of binomial probability paper. First, draw the 1:1 and 3:1 splits on the paper (see Fig. 18.1). On the side of each split nearest the other split, draw a parallel line 1.645 Full Scale units (about $1.645 \times 5 = 8.2$ mm) from the former split. The value 1.645 is such that $P(z \geq 1.645) = .05$ and is chosen because we are looking for an essentially one-sided test procedure regardless of which hypothesis is true. These two lines, one parallel to each split, intersect approximately at the point (25,15). This point is at the lower right corner of the right-angled triangle with right angle at (24,15), and at the upper left corner for the one with right angle at (25,14). In other words, the triangle with right angle at (25,15) has its other angles outside both the one-sided 95% confidence limits. Hence we require $25 + 15 - 1 = 39$ observations in our sample. A sample of size 39 will permit us to make a decision in every case; a larger sample size will give us greater protection than desired. For the larger sample size, we will have to alter our probabilities in order to obtain a rule for deciding between alternatives that will cover all possible sample results.

If we can compare this result with that obtained from Tables A.17, we find a slight difference. The appropriate table says that 44 is the required number to guarantee the desired protection but that it actually does slightly better than required. The table also shows that 40 observations will give almost the desired protection. We must recall that binomial probability paper relies on a transformation which gives a nearly normally distributed variable and that it can give us only an approximate sample size. This accounts for the discrepancy.

Case 2. Suppose we have only one candidate for the null hypothesis but wish to have reasonable assurance that we can detect an alternative value of a specified size.

For example, suppose we are doing public opinion sampling using a question which requires a Yes or No for an answer. We wish to test $H_0: p = .5$ but want to detect $p = .65$ if it is in a specified direction. We decide to test H_0 at the 5% level using a one-sided test and to detect $p = .65$, if it is the true ratio, 90% of the time.

Draw the 1:1 and 65:35 splits (see Fig. 18.1). Draw a line parallel to the 1:1 split and 1.645 Full Scale standard deviations from it. This line will be between the two splits and represents a one-tailed test of $H_0: p = .5$ at the 5% level. Also between the two splits, draw a line parallel to the 65:35 one and

1.28 Full Scale (about 1.28 × 5 = 6.4 mm) standard deviations from it. This line corresponds to a one-tailed test of $H_0: p = .65$ at the 10% level. We read the point (52,38) at the intersection of the parallel lines. The appropriate sample size should be approximately 90. In some cases, such as this, it becomes very difficult to read the scale and the paper exactly. If more accuracy is desired use the fact that $\sigma = 5.080$ mm on the Full Scale.

Case 3. Suppose we have only one candidate for the null hypothesis but wish to have a reasonable chance of detecting an alternative if it differs by as much as 10 percentage points.

For example, suppose $H_0: p = .25$. We wish to test this null hypothesis using a two-tailed test at the 5% level of significance. We wish to detect a proportion of 15 or 35%, if such is the true proportion, about 75% of the time.

Draw the 25:75 split. In addition, draw the 35:65 and 15:85 splits (see Fig. 18.1). On each side of the 25:75 split, draw a line parallel to the split and 1.96 Full Scale standard deviations from it. These correspond to the two-sided test procedure. If you simply look at the sums of the coordinates of the points where the parallel lines intersect the 35:65 and 15:85 splits, you will realize that there is no test that will guarantee us precisely some stated probability of detection for each alternative. (The points referred to are such that they give us 50% detection.) We will be cautious and ask for 75% protection in the less favorable case, namely, when $p = .35$. Now draw a line parallel to the 35:65 split and .67 units from it. $[P(z \geq .67) = .75$ for the normal distribution.] The two lines intersect at approximately (46,98). Hence the required sample size is approximately 144.

Exercise 18.6.1 Use binomial probability paper to determine the sample size necessary for distinguishing between the following ratios: 3:1 and 7:1; 9:7 and 15:1; 27:37 and 3:1. Assume that we are prepared to reject any ratio falsely about 5% of the time but not more frequently.

Exercise 18.6.2 For a sampling problem with discrete data, it has been decided to test $H_0: p = .40$. A one-tailed test at the 1% level is to be used and is to be one-sided. It is required to detect $H_1: p = .65$, if it is the true hypothesis, about 80% of the time. What is the necessary sample size?

Exercise 18.6.3 Suppose we must distinguish between the ratios 1:1, 3:1, and 7:1. We would like to set the probability of rejecting each hypothesis, when true, at about .05 but will obviously have to arrange things so that this is the maximum probability of a wrong decision. Can you construct a procedure for finding the necessary sample size, using binomial probability paper? Compare your result with that obtained from Table A.17.

18.7 One-way tables with n cells. Robertson (18.6) gives data which include the F_2 progeny of a barley cross from F_1 normal green plants. The observed characters are non-two-row versus two-row and normal or green versus chlorina plant color. These data are given in Table 18.3. It is desired to test the hypothesis of a 9:3:3:1 ratio, normal dihybrid segregation. Calculate χ^2 as

$$\chi^2 = \Sigma \frac{(\text{observed} - \text{expected})^2}{\text{expected}}$$
$$= 54.36^{**} \text{ with } 3 \, df$$

ENUMERATION DATA I: ONE-WAY CLASSIFICATIONS

The degrees of freedom are one less than the number of cells. The evidence is against the theoretical ratio of $9:3:3:1$. The geneticist would probably conclude that linkage caused such a poor fit. (These data will be discussed further in Chap. 19.)

TABLE 18.3
OBSERVED AND EXPECTED VALUES FOR AN F_2 PROGENY

	Green		Chlorina		Totals
	Non-2-row	2-row	Non-2-row	2-row	
Observed	1,178	291	273	156	1,898
Expected	1,067.6	355.9	355.9	118.6	1,898
(Diff)^2/expected	11.416	11.835	19.310	11.794	54.36

What was our alternative hypothesis? Simply that the ratio was other than $9:3:3:1$. When an alternative hypothesis is specified, we would like a more efficient method of making a decision among the possible alternatives. Although a multinomial distribution is indicated as the exact distribution, it is difficult to define regions of acceptance and rejection. Table A.17 is available for choosing among a few sets of three-outcome ratios.

Exercise 18.7.1 Woodward and Rasmussen (18.8) hypothesized a 9 hooded to 3 long-awned to 4 short-awned types in the F_2 generation. The observed data were $348:115:157$. Test their hypothesis. How many degrees of freedom are there for the χ^2 criterion?

References

18.1 Clopper, C. J., and E. S. Pearson: "The use of confidence or fiducial limits illustrated in the case of the binomial," *Biometrika*, **26:** 404 (1934).
18.2 Cochran, W. G.: *Sampling Techniques*, John Wiley & Sons, Inc. New York, 1953.
18.3 Mainland, D., L. Herrera, and M. I. Sutcliffe: *Tables for Use with Binomial Samples*, published by the authors, New York, 1956.
18.4 Mosteller, F., and J. W. Tukey: "The uses and usefulness of binomial probability paper," *J. Am. Stat. Assoc.*, **44:** 174–212 (1949).
18.5 NaNagara, P.: "Testing Mendelian ratios," M.S. Thesis, Cornell Univ., Ithaca, N.Y., 1953.
18.6 Robertson, D. W.: "Maternal inheritance in barley," *Genetics*, **22:** 104–113 (1937).
18.7 *Tables of the Cumulative Binomial Probability Distribution*, Ann. Computation Lab. of Harvard Univ., vol. 35, Harvard University Press, Cambridge, Mass. (1955).
18.8 Woodward, R. W., and D. C. Rasmussen: "Hood and awn development in barley determined by two gene pairs," *Agron. J.*, **49:** 92–94 (1957).
18.9 Yates, F.: "Contingency tables involving small numbers and the χ^2 test," *J. Roy. Stat. Soc. Suppl.*, **1:** 217–235 (1934).

Chapter 19

ENUMERATION DATA II: CONTINGENCY TABLES

19.1 Introduction. Enumeration data are often classified according to several variables. For example, a person may be classified as a smoker or a nonsmoker and as one with or without coronary disease; a fruit fly may be classified as male or female and according to parent mating, and so on. There are two variables of classification in each case.

This chapter deals with the processing of enumeration data classified according to more than one variable, that is, with data in *contingency tables*.

19.2 Independence in $r \times c$ tables. Federer (19.3) observed numbers of plants emerging from 600 seeds for each of three treatments and many strains. Part of the results are shown in Table 19.1. This is an $r \times c$ contingency table with $r = 4$ rows and $c = 3$ columns. The two variables of classification are strain, with four categories, and treatment, with three categories. It is desired to test the hypothesis of the independence of the two variables of classification.

The hypothesis of *independence* implies that the ratio of numbers of plants emerging from treatment to treatment is the same, within random sampling errors, for each strain. If there is not independence, there is said to be *interaction* and the test is often called a test of interaction. In the case of interaction, the ratio from treatment to treatment is dependent on the strain, that is, the variables are not independent.

The test of independence is a symmetric test. Thus, if the categories of the strain variable have a meaning, say according to pollen parent, then the ratio from strain to strain may be tested for constancy over the three treatments. This is the same test as that in the previous paragraph. In general, one's point of view will depend upon the particular set of data. You may be interested only in the dependence or independence of the variables or you may be interested in the constancy of the set of ratios associated with one variable of classification as you go from category to category of the other variable.

No assumptions need be made about the true ratios for either variable, either within any category of the other variable or over all categories.

The test criterion is given by Eq. (19.1).

$$\chi^2 = \Sigma \frac{(\text{observed} - \text{expected})^2}{\text{expected}}, \qquad (r-1)(c-1) \ df \qquad (19.1)$$

Although called χ^2, it is clear that it can be distributed only approximately as χ^2. The approximation is surprisingly good even for small samples as long as there is more than one degree of freedom.

Table 19.1
Calculation of χ^2 for an $r \times c$ Table

Strain		Treatment			Totals
		Threshed, untreated	Unthreshed, untreated	Unthreshed, treated	
42473	Observed	335	60	18	413
	Expected	336.98	58.05	17.97	413
	Deviation	−1.98	+1.95	+.03	0.0
	Chi-square	.0116	.0655	.0001	
42449	Observed	300	57	19	376
	Expected	306.79	52.85	16.36	376
	Deviation	−6.79	+4.15	+2.64	0.0
	Chi-square	.1503	.3259	.4260	
42450	Observed	294	51	16	361
	Expected	294.55	50.74	15.71	361
	Deviation	−.55	+.26	+.29	0.0
	Chi-square	.0010	.0013	.0054	
42448	Observed	290	42	12	344
	Expected	280.68	48.35	14.97	344
	Deviation	+9.32	−6.35	−2.97	0.0
	Chi-square	.3095	.8340	.5892	
Totals	Observed	1,219	210	65	1,494
	Expected	1,219	209.99	65.01	1,494
	Deviation	0.0	+.01	−.01	0.0

$\chi^2 = 2.72$ with 6 df, ns

Expected values are computed under the assumption that the null hypothesis is true. Thus, when no ratios have been assumed, the proportion of emergent seeds in the threshed, untreated category is 1,219/1,494, obtained from column totals. Likewise, the proportion of emergent seeds in strain 42473 is given by 413/1,494, based on row totals. (Note that these are not the proportions of seeds emerging. Such proportions would involve multiples of 600.) If the variables are independent, the number in the upper left cell should be (1,219/1,494)(413/1,494) 1,494 = 336.98 seeds. If the variables are dependent, this procedure will not give appropriate expected numbers.

Denote the number of observations in the i,jth cell by n_{ij}. Then $n_{i.}$, $n_{.j}$, and $n_{..}$ represent the totals in the ith row, the jth column, and the grand total. Let E_{ij} be the expected number in the i,jth cell. Then E_{ij} is determined by Eq. (19.2).

$$E_{ij} = \frac{n_{i.}n_{.j}}{n_{..}} \qquad (19.2)$$

For an example, see Table 19.1. Note that the sum of the deviations in each row and column is zero. Consider the restrictions on assigning random deviations to such a table: there is one restriction per row, namely, that the sum of the deviations in any row be zero. Now we need only $c - 1$ further restrictions, one for each of the $c - 1$ columns, since these and the over-all restriction imply the final column restriction. Hence, such a table has $rc - (r + c - 1) = (r - 1)(c - 1)$ degrees of freedom.

Notice that each contribution to χ^2 is measured relative to the expected value E_{ij}, that is, the contribution from the i,jth cell is $(n_{ij} - E_{ij})^2/E_{ij}$. Thus a large deviation where E_{ij} is large and a small one where E_{ij} is small will contribute about equally to χ^2.

In the example, $\chi^2 = 2.72$ with $3(2) = 6$ df and is not significant. There is no reason to believe that the ratios from treatment to treatment differ with the strain. Had there been a significant χ^2, the table of contributions would supply information on possible causes and would suggest specific alternatives. While a test of a hypothesis suggested by the data is not valid, the investigator would very likely use such a test when a valid procedure had led to the conclusion that the original null hypothesis was false. However, the tabulated probability levels could not be considered truly applicable.

A common procedure for locating the cause of significance is to eliminate the rows or columns which appear to be the cause and run a new test of significance on the reduced data. The remarks of the previous paragraph about probability levels still apply.

Alternate calculations. An alternative computation which also requires the expected values is given as Eq. (19.3).

$$\chi^2 = \Sigma \frac{n_{ij}^2}{E_{ij}} - n.. \tag{19.3}$$

A much shorter alternative is given by Skory (19.17). The three steps follow, the third being Eq. (19.4) and giving χ^2.

1. Compute $\sum_i n_{ij}^2/n_i.$ for each column. Denote these by A_j.

2. Compute $\sum_j A_j/n._j$. Denote this by B.

3. Then $\chi^2 = (B - 1)n... \tag{19.4}$

Each A_j and B is obtained in a single operation on most automatic desk calculators. The computations can be done for rows rather than columns. In practice, keep the number of A_j's as small as possible.

Applying the above procedure to our example, we obtain

$$A_1 = \frac{335^2}{413} + \frac{300^2}{376} + \frac{294^2}{361} + \frac{290^2}{344} = 995.00458$$

$$A_2 = \frac{60^2}{413} + \frac{57^2}{376} + \frac{51^2}{361} + \frac{42^2}{344} = 29.69056$$

$$A_3 = \frac{18^2}{413} + \frac{19^2}{376} + \frac{16^2}{361} + \frac{12^2}{344} = 2.87235$$

$$B = \frac{995.00458}{1219} + \frac{29.69056}{210} + \frac{2.87235}{65} = 1.00182$$

$$\chi^2 = (.00182)1494 = 2.72 \text{ on } 6 \ df$$

ENUMERATION DATA II: CONTINGENCY TABLES

This is probably the simplest computational procedure but suggests no alternative to the null hypothesis when this is declared false.

A large chi-square indicates lack of independence of the variables of classification but gives little information about the degree of dependence. A measure of dependence is given by Eq. (19.5).

$$\frac{\chi^2}{n(t-1)} \tag{19.5}$$

t is the smaller of r and c, and n is the total number of observations in the table. This criterion lies between 0 and 1. Its distribution is obviously related to χ^2 by a simple change of variable.

There are other coefficients of contingency.

Exercise 19.2.1 In a study of the inheritance of combining ability in maize, Green (19.4) obtained three frequency distributions. The evidence favored a common variance. It then became of interest to see if the distributions had the same form regardless of the location of the distributions. The frequency distributions used to compare the forms are shown in the accompanying table.

F_2 sample	Class centers: Standard errors above or below the mean							Total
	−3	−2	−1	0	1	2	3	
High × high	1	5	28	19	22	8	—	83
High × low	2	5	19	28	23	6	—	83
Low × low	3	9	18	24	18	9	2	83

Test the hypothesis of homogeneity of form. (Recall that much information concerning the form of a distribution is associated with the tails, as was shown in Sec. 17.4. Hence, pool only the +2 and +3 classes.) Use Eqs. (19.1), (19.3), and (19.4) and compare them for ease of use.

Exercise 19.2.2 In a study of the inheritance of coumarin in sweet clover, Rinke (19.16) observed the data in the accompanying table.

Classes for growth habit	Coumarin classes in 1/100 of 1%			Total
	0–14	15–29	Above 30	
Standard	8	24	7	39
Intermediate standard	11	28	17	56
Intermediate alpha	9	31	11	51
Alpha	1	14	11	26
Total	29	97	46	172

What is the evidence for the independence of growth habit and coumarin content?

Exercise 19.2.3 Compute the coefficient of contingency for the data in Exercise 19.2.2.

Exercise 19.2.4 Would coefficients of contingency be meaningful for the data of Tables 19.1 and 19.2?

19.3 Independence in $r \times 2$ tables. The $r \times 2$ table is a special case of the $r \times c$ table. As an example, consider some data of Di Raimondo (19.2) on mice untreated with penicillin, presented in Table 19.2. The mice were

TABLE 19.2

CHI-SQUARE IN A $k \times 2$ TABLE

Inoculum	Alive = A	Dead	Total	$\hat{p}$	$\hat{p}A$
NA	10	30	40	.250	2.500
FA	9	31	40	.225	2.025
Paba	9	41	50	.180	1.620
B_6	13	27	40	.325	4.225
Totals	41	129	170		10.370
				$\hat{\hat{p}} = 41/170 = .241176$	9.888

$$\chi^2 = \frac{10.370 - 9.888}{.241176(1 - .241176)} = 2.63 \text{ with 3 } df, ns$$

injected with bacterial inoculum (*Staphylococcus aureus*) cultured in a broth enriched with the vitamins niacinamide (NA), folic acid (FA), p-aminobenzoic acid (Paba), and B_6 as pyridoxin, each in excess of 10 micrograms per milliliter.

The computations can be carried out as required for Eq. (19.1) where expected values are computed. Also, Eqs. (19.3) or (19.4) may be used. In addition, an alternative procedure not available for the general $r \times c$ table is often used; see Table 19.2 and the next paragraph.

The procedure used in Table 19.2 requires an estimate of the probability associated with one category for each row of the table. Estimates for each row are given by Eq. (19.6).

$$\hat{p}_i = \frac{n_{ij}}{n_{i.}}, \quad i = 1, \ldots, r \quad \text{and} \quad \hat{\hat{p}} = \frac{n_{.j}}{n_{..}}, \quad j = 1 \text{ or } 2 \quad (19.6)$$

For our example, $\hat{p}_1 = 10/40 = .250, \ldots, \hat{p}_4 = 13/40 = .325$, and $\hat{\hat{p}} = 41/170 = .241176$. Compute χ^2 by Eq. (19.7). (The subscript 1 may be replaced by the subscript 2.)

$$\chi^2 = \frac{\Sigma \hat{p}_i n_{i1} - \hat{\hat{p}} n_{.1}}{\hat{\hat{p}}(1 - \hat{\hat{p}})} \quad (19.7)$$

$$= \frac{10.370 - 9.888}{(.241176)(.758824)} = 2.63 \text{ with 3 } df, ns$$

On the basis of this sample, we conclude that the ratio of live to dead mice does not vary by more than chance, from inoculum to inoculum. If there are

ENUMERATION DATA II: CONTINGENCY TABLES

real differences in the population ratios, then our sample has not been large enough to detect these differences.

An alternate computational procedure can be obtained by rearranging Eq. (19.7) to give Eq. (19.8).

$$\chi^2 = \frac{\sum_i (n_{ij}^2/n_{i.}) - n_{.j}^2/n_{..}}{n_{.1}n_{.2}/n_{..}^2}, \qquad j = 1 \text{ or } 2 \tag{19.8}$$

This form has the advantage that each of the two terms in the numerator can be obtained without the necessity of recording any intermediate terms. In particular, $\sum_i n_{ij}^2/n_{i.}$ can be accumulated continuously on most desk calculators.

Exercise 19.3.1 From Exercise 19.2.1, obtain the data for the high × high and low × low F_2 samples. Reduce these data to a 2 × 6 table and, for computing experience, apply the methods of this section to determine χ^2.

19.4 Independence in 2 × 2 tables. The data of Table 18.3 fall naturally into a 2 × 2 table. Such situations are common. The usual question asked of such data is "Are the data homogeneous?" The question is answered by computing the interaction χ^2.

Interaction χ^2. Any one of the formulas used to compute interaction for an $r \times c$ or $r \times 2$ contingency table is applicable to a 2 × 2 table. However, probably the most common formula is Eq. (19.9).

$$\chi^2 = \frac{(n_{11}n_{22} - n_{12}n_{21})^2 \, n_{..}}{n_{1.}n_{2.}n_{.1}n_{.2}} \tag{19.9}$$

for the table

n_{11}	n_{12}	$n_{1.}$
n_{21}	n_{22}	$n_{2.}$
$n_{.1}$	$n_{.2}$	$n_{..}$

Di Raimondo (19.2) used two control broths in the experiment mentioned in Sec. 19.3. They were (1) a simple broth referred to as standard and (2) one with penicillin in amount 0.15 U per milliliter. Table 19.3 contains the pooled data from the experiments with untreated mice.

Since $\chi^2 = .0913$, the sample provides no evidence that the responses to the standard and penicillin broths differ. The data appear to be homogeneous.

Since a single degree of freedom is involved in χ^2 for 2 × 2 tables, a correction for continuity is appropriate. The *adjusted* χ^2 is given by Eq. (19.10).

$$\text{Adjusted } \chi^2 = \frac{(|n_{11}n_{22} - n_{12}n_{21}| - n_{..}/2)^2 n_{..}}{n_{1.}n_{2.}n_{.1}n_{.2}} \tag{19.10}$$

For Table 19.3, adjusted $\chi^2 = .0032$.

Normal approximation. Equation (19.9) is equivalent to comparing two observed probabilities using the normal approximation given by Eq. (19.11).

$$z = \chi = \frac{\hat{p}_1 - \hat{p}_2}{\sqrt{\hat{p}(1-\hat{p})[(1/n_1.) + (1/n_2.)]}} \quad (19.11)$$

Squaring Eq. (19.11) gives

$$z^2 = \chi^2 = \frac{(\hat{p}_1 - \hat{p}_2)^2}{\hat{p}(1-\hat{p})[(1/n_1.) + (1/n_2.)]}$$

$$= \frac{(8/20 - 48/110)^2}{56/130 \; 74/130 (1/20 + 1/110)} = .0913, \text{ as before}$$

TABLE 19.3

CHI-SQUARE FOR A 2 × 2 TABLE

Treatment	Alive	Dead	Total
Standard	8	12	20
Penicillin	48	62	110
Total	56	74	130

$$\text{Unadjusted } \chi^2 = \frac{[8(62) - 12(48)]^2 130}{56(74)20(110)} = .0913$$

$$\text{Adjusted } \chi^2 = \frac{(|8(62) - 12(48)| - 130/2)^2 130}{56(74)20(110)} = .0032$$

Binomial paper. Two observed probabilities can be compared on binomial probability paper, Table A.16 (reference 19.13). To do so:

1. Plot each paired count and the total paired count.
2. Sum the middle distances, that is, from the center of each hypotenuse, from the two paired counts to the line through the total paired count.
3. Compare this sum with $1.96 \sqrt{2} = 2.77$ or $2.57 \sqrt{2} = 3.64$ Full Scale units for 5 and 1% levels, respectively.

There is an interesting difference between 2 × 2 tables such as 19.3 and 18.3. In Table 18.3, the size of the single experiment is fixed and, within this limitation, the four border totals are random. Here it is quite correct to compare either the two row-number probabilities or the two color probabilities. In Table 19.3, two experiments are involved, each with a fixed number of animals. Each experiment supplies an estimate of the probability of an animal living or dying. These probabilities can be validly compared. Now the experiments are tabulated side by side as a single experiment to compare two treatments, the original experiments, but the only randomness in border totals is that of columns. For this reason, it is not valid to compare the ratios 8/56 and 12/74. On the other hand, the numerical results will be the same.

Exercise 19.4.1 Obtain interaction χ^2 for the data in Table 19.3 by using binomial probability paper. Compare your result with that given in the text.

Exercise 19.4.2 C. A. Perrotta, University of Wisconsin, observed the frequency of visitation of mice to traps (small pieces of board) which were previously urinated on by mice and to others which were clean. The following results were obtained in 1954.

	Urinated	Clean
Visited	17	3
Not visited	9	23

Test the null hypothesis that frequency of visitation is not affected by treatment. For this, compute both adjusted and unadjusted interaction χ^2. Compare the result with that read from binomial probability paper.

Exercise 19.4.3 Smith and Nielsen (19.18) have studied clonal isolations of Kentucky bluegrass from a variety of pastures. Their observations include the data shown in the accompanying table. Is there evidence of heterogeneity in response to mildew?

No.	Character of pasture	Mildewed	Not mildewed	Total
1	Good lowland, moderately grazed	107	289	396
5	Good lowland, moderately grazed	291	81	372

Exercise 19.4.4 Smith and Nielsen (19.18) observed the same fields for rust and obtained the data shown below. Is there evidence of heterogeneity in response to rust?

Field	Rust	No rust	Total
1	372	24	396
5	330	48	378

19.5 Homogeneity of two-cell samples.

The two-cell table is probably the most common tabular presentation for discrete data. In other words, most discrete data are concerned with the presence or absence of a character, that is, with a *dichotomy*. Often many samples with similar information are available and it is desired to pool them to obtain a better estimate of the population proportion. Pooling is appropriate when the samples are homogeneous and it then becomes important to test the hypothesis of homogeneity.

Consider the following example. Smith (19.19) observed annual versus biennial growth habit and its inheritance in sweet clover. He examined 38 segregating progenies, of which the results from the first six appear in Table 19.4. Applying the test of homogeneity given by Eq. (19.8), we obtain

$$\chi^2 = \frac{\Sigma(n_{i2}^2/n_{i.}) - n_{.2}^2/n_{..}}{n_{.1}n_{.2}/n_{..}^2} = \frac{8.711885 - 8.404545}{.157252} = 1.95 \text{ with } 5 \text{ } df \quad ns$$

Table 19.4
Segregating Progenies of Sweet Clover

Culture	Observed values			Expected (3:1) values		
	Annual	Biennial	Total	Annual	Biennial	χ^2 (1 df)
4–3	18	6	24	18.00	6.00	0.0000
4–11	33	7	40	30.00	10.00	1.2000
4–14	38	12	50	37.50	12.50	0.0267
4–15	19	5	24	18.00	6.00	0.2222
4–16	39	7	46	34.50	11.50	2.3478
4–21	30	6	36	27.00	9.00	1.3333
Totals	177	43	220	165.00	55.00	5.1300 (6 df)

There is no evidence from which to conclude that the cultures differ in the proportion of annual to biennial habit. One would estimate the common proportion of biennial habit plants to be $\hat{p} = 43/220 = .195$ for the segregating progenies of these six cultures.

In some cases, there will be sound biological or other reasons to hypothesize the common ratio. In this particular case, there is the null hypothesis of a 3:1 ratio of annual to biennial habit. When the population proportion is known or hypothesized, Eq. (19.8) is modified to give Eq. (19.12).

$$\chi^2 = \frac{\sum_i (n_{ij}^2/n_{i.}) - n_{.j}^2/n_{..}}{p(1-p)}, \quad j = 1 \text{ or } 2 \qquad (19.12)$$

Here, p is the population proportion, and $p = \frac{1}{4}$ or $\frac{3}{4}$. For our example,

$$\chi^2 = \frac{8.711885 - 8.404545}{.25(.75)} = 1.64 \text{ with } 5 \text{ df}$$

Equations (19.8) and (19.12) have the same numerator. Equation (19.12) gives a lower χ^2 than Eq. (19.8) whenever the population proportion is nearer .5 than is the observed proportion.

Exercise 19.5.1 Mendel (19.12), in his classic genetic study, observed plant to plant variation. In a series of experiments on form of seed, the first ten plants gave the results shown below:

Plant	1	2	3	4	5	6	7	8	9	10
Round seed	45	27	24	19	32	26	88	22	28	25
Angular seed	12	8	7	10	11	6	24	10	6	7

Test the null hypothesis H_0: 3 round:1 angular, using totals. Test the homogeneity of the 3:1 ratio for the ten plants, using the assumption that the 3:1 ratio is the true ratio. What is the value of the homogeneity χ^2 when the 3:1 ratio is not assumed?

Exercise 19.5.2 Mendel (19.12) also observed plant to plant variation in an experiment on the color of the seed albumen. The results for the first ten plants follow:

Plant	1	2	3	4	5	6	7	8	9	10
Yellow albumen	25	32	14	70	24	20	32	44	50	44
Green albumen	11	7	5	27	13	6	13	9	14	18

Repeat Exercise 19.5.1 for this set of data.

19.6 Additivity of χ^2. The information in the data of tables such as 19.4 is not completely extracted by a homogeneity χ^2 when the investigator has a hypothesis about the population proportion. For example, a χ^2 can be computed to test the null hypothesis for each culture (row). For Table 19.4, these are shown in the last column. No individual χ^2 is significant.

Independent χ^2's can be added. We obtain a total χ^2 based on 6 degrees of freedom. The value is 5.1300 and is not significant. (Only unadjusted χ^2 values may be added.)

Also, we may compute a χ^2 based on the totals 177 and 43. Here $\chi^2 = 3.4909$ with 1 degree of freedom and is not significant though it is approaching the 5% point; $\chi^2_{.05} = 3.841$ for 1 degree of freedom.

Finally, the difference between the sum of the χ^2's and the over-all χ^2 measures interaction. Thus, $5.1300 - 3.4909 = 1.6391$ with 5 degrees of freedom as compared with 1.64 by direct computation. This is far from significant; about two-thirds of the sum of the χ^2's is associated with a test of the ratio of the observation totals.

Now consider the χ^2's computed for this example:

1. *Individual χ^2's.* Each gives information about a particular sample. When each sample is small, it will be difficult to detect any but large departures from the null hypothesis. Also, if there are many samples and the null hypothesis is true, we expect the occasional one, about one in twenty, to show significance falsely.

2. *The sum of the individual χ^2's.* Here we have a pooling of the information in the samples. For example, suppose the true population was such that χ^2 was 2.000, on the average; we find $.20 > P > .10$. The individual χ^2's would be distributed about this value, some being larger and others smaller; some χ^2's would be significant, others not. This makes the information somewhat difficult to evaluate unless we sum the χ^2's. If we have 20 samples, then the sum should be about $20(2.000) = 40.000$ with 20 degrees of freedom; this is highly significant.

One obvious difficulty arises in the interpretation of a sum of χ^2's. Suppose the population ratios differ from sample to sample, from that of the null hypothesis. This tends to make all individual χ^2's too large. A significant sum may be attributable to more than one alternative hypothesis, that is, to heterogeneous samples.

3. *χ^2 on totals.* Here we pool the samples and compute χ^2 with 1 degree of freedom. If the samples are homogeneous but the null hypothesis is false, then

we have a larger sample and are able to detect a smaller departure from the null hypothesis. For example, suppose we sample from a population in which the hypothesis is 3:1 and observe 17:3 individuals. Here, $\chi^2 = 1.07$ and is not significant. However, if we have four times as many individuals and observe exactly the same ratio, namely, if we observe $(17 \times 4):(3 \times 4) = 68:12$, then $\chi^2 = 4.27$ and is significant.

If the samples are heterogeneous, some with higher and others with lower ratios than that hypothesized, then pooling the samples can lead to a low χ^2 and acceptance of the null hypothesis. For example, if 3:1 is the ratio of the null hypothesis, then a sample of 67:13 gives $\chi^2 = 3.27$, near the tabulated 5% point. Also, a sample of 53:27 has $\chi^2 = 3.27$. This sample departs from the null hypothesis in the other direction. Pooling the samples, we have an observed ratio of 120:40, which fits the null hypothesis perfectly.

4. *Homogeneity χ^2*. It is now apparent just how important homogeneity is; it is a measure of the likeness or unlikeness of the samples. It is independent of the ratio of the null hypothesis in that the individual samples can depart from this ratio without increasing heterogeneity χ^2, provided they depart in the same direction and to about the same extent.

In summary, χ^2 on totals and homogeneity χ^2 are a reasonably adequate analysis of the data. If homogeneity χ^2 is significant, then it may be wise to consider the individual χ^2's. The sum of the individual χ^2's may be difficult to interpret by itself.

Exercise 19.6.1 For the data of Exercise 19.5.1, compute χ^2 for each plant to test H_0:3:1 ratio. Sum these χ^2's and show that the sum is equal to the sum of the two χ^2's computed for Exercise 19.5.1. Note also that the degrees of freedom are additive.

Exercise 19.6.2 Repeat Exercise 19.6.1 for the data in Exercise 19.5.2.

Exercise 19.6.3 Discuss the results obtained in Exercises 19.5.1 and 19.6.1. In 19.5.2 and 19.6.2.

19.7 More on the additivity of χ^2. The data in Table 19.2 are homogeneous; $\chi^2(3\ df) = 2.63$. The data in Table 19.3 are also homogeneous; $\chi^2(1\ df) = .0913$. Hence, it seems appropriate to pool the data in each table and see if the treatment and control data are homogeneous. Pooling the data gives Table 19.5.

TABLE 19.5

POOLED DATA FROM TABLES 19.2 AND 19.3

Treatment	Alive	Dead	Total
Vitamins	41	129	170
Controls	56	74	130
Total	97	203	300

$$\chi^2 = \frac{[129(56) - 41(74)]^2 300}{97(203)170(130)} = 12.10^{**} \text{ with } 1\ df$$

The evidence in Table 19.5 is against a common probability of death for untreated mice injected with *Staphylococcus aureus* cultured under the two sets of conditions.

The handling of the Di Raimondo data up to this point must suggest additivity of χ^2. Thus we ask the question, "Does χ^2 (within vitamins) $+ \chi^2$ (within controls) $+ \chi^2$ (vitamins versus controls) $= \chi^2$ (vitamins and controls)?" The answer is No. In this case, the sum of the χ^2's is $2.63 + .09 + 12.10 = 14.82$ with $3 + 1 + 1 = 5$ df; χ^2 (vitamins and controls) $= 14.41$ with 5 degrees of freedom (see Table 19.6). The difference between the two χ^2's, each with 5 degrees of freedom, is small but is not due to rounding errors.

TABLE 19.6

DATA FROM TABLES 19.2 AND 19.3

Treatment	Alive	Dead	Total
Standard	8	12	20
Penicillin	48	62	110
NA	10	30	40
FA	9	31	40
Paba	9	41	50
B_6	13	27	40
Total	97	203	300

$\chi^2 = 14.41**$ with 5 df

If the three comparisons, within vitamins, within controls, between vitamins and controls, had been in an analysis of variance, then they would clearly have been independent and the sum of their sums of squares would have been the same as that for among treatments, including both vitamins and controls. The same is not true of discrete data and χ^2 values.

Probably few people would object to the three nonindependent comparisons made, if the investigator regarded these as meaningful. However, independent comparisons can be made and will now be illustrated. Situations will be pointed out where such comparisons are meaningful and, because of their independence, are to be preferred. Our illustration will be seen to be somewhat artificial.

Consider the data of Table 19.6, for which independent comparisons will now be made. Independent comparisons and their computations are due to Irwin (19.5), Lancaster (19.8, 19.9), and Kimball (19.7). Equation (19.1) was applied to give $\chi^2 = 14.41**$ with 5 degrees of freedom. This value is to be partitioned into independent χ^2's with single degrees of freedom. We begin by applying Eq. (19.13). This equation is very similar to Eq. (19.9) but if you examine it carefully, you will find it differs in both numerator and denominator.

The remaining equations are obvious modifications of Eq. (19.13). Their application yields:

$$\chi_1^2 = \frac{n_{..}^2(n_{11}n_{22} - n_{12}n_{21})^2}{n_{.1}n_{.2}n_{1.}(n_{1.} + n_{2.})} \tag{19.13}$$

$$= \frac{300^2[8(62) - 12(48)]^2}{97(203)20(110)130} = .1023, \text{ ns}$$

$$\chi_2^2 = \frac{n_{..}^2[(n_{11} + n_{21})n_{32} - (n_{12} + n_{22})n_{31}]^2}{n_{.1}n_{.2}n_{3.}(n_{1.} + n_{2.})(n_{1.} + n_{2.} + n_{3.})} \tag{19.14}$$

$$= \frac{300^2[56(30) - 74(10)]^2}{97(203)40(130)170} = 4.5686*$$

$$\chi_3^2 = \frac{n_{..}^2[(n_{11} + n_{21} + n_{31})n_{42} - (n_{12} + n_{22} + n_{32})n_{41}]^2}{n_{.1}n_{.2}n_{4.}(n_{1.} + n_{2.} + n_{3.})(n_{1.} + n_{2.} + n_{3.} + n_{4.})} \tag{19.15}$$

$$= \frac{300^2[66(31) - 104(9)]^2}{97(203)40(170)210} = 3.9436*$$

$$\chi_4^2 = \frac{n_{..}^2[(n_{11} + n_{21} + n_{31} + n_{41})n_{52} - (n_{12} + n_{22} + n_{32} + n_{42})n_{51}]^2}{n_{.1}n_{.2}n_{5.}(n_{1.} + n_{2.} + n_{3.} + n_{4.})(n_{1.} + \cdots + n_{5.})} \tag{19.16}$$

$$= \frac{300^2[75(41) - 135(9)]^2}{97(203)50(210)260} = 5.7921*$$

$$\chi_5^2 = \frac{n_{..}^2[(n_{11} + \cdots + n_{51})n_{62} - (n_{12} + \cdots + n_{52})n_{61}]^2}{n_{.1}n_{.2}n_{6.}(n_{1.} + \cdots + n_{5.})n_{..}} \tag{19.17}$$

$$= \frac{300^2[84(27) - 176(13)]^2}{97(203)40(260)300} = .0006$$

The sum of these χ^2's is 14.41 with 5 degrees of freedom, the same as for the over-all table.

The general formulas for independent χ^2's in any $r \times 2$ table are apparent from the particular cases shown.

In reviewing this illustration, χ_1^2 is meaningful in that it compares two standards. Also, χ_2^2 is meaningful in that we compare the average of the responses to two standards, for which there is no evidence of a different response with the response to a treatment. Since significance is found at this stage, there is little meaning in averaging the responses of three treatments, one of which is apparently different from the other two, with the response to a fourth treatment. For this reason, we have referred to our illustration as artificial.

The use of independent χ^2's as computed here seems to have real applicability in the case of two treatments and a standard or two standards and a treatment. If the two treatments or two checks appear to differ, the value in averaging them and comparing them with the remaining entries is doubtful. Having found significance, one might well proceed to make those nonindependent tests which seemed most profitable.

Exercise 19.7.1 In what way do Eqs. (19.9) and (19.13) differ?

Exercise 19.7.2 To the mildew data of Exercise 19.4.3, add the following:

No.	Character of pasture	Mildewed	Not mildewed	Total
9	Good upland, moderately grazed	280	144	424

To the rust data of Exercise 19.4.4, add:

Field	Rust	No rust	Total
9	371	54	425

If you feel that the methods of this section apply to either 3×2 contingency table, proceed with the application. In either case, justify your position.

19.8 Exact probabilities in 2×2 tables. Occasionally it is possible to obtain only limited amounts of data, for example if, to obtain the data, it is necessary to destroy expensive or hard to obtain experimental units. When the numbers in a 2×2 table are very small, it may be best to compute exact probabilities rather than to rely upon an approximation. Consider the following data:

	Have	Have not	Total
Standard	5	2	7
Treatment	3	3	6
Total	8	5	13

When it is required to test homogeneity, we compute the probability of obtaining the observed distribution or a more extreme one, the more extreme ones being:

6	1	7
2	4	6
8	5	13

and

7	0	7
1	5	6
8	5	13

Thus, we require the sum of the probabilities associated with the three distributions given. Marginal totals are the same for all three tables. The sum of the probabilities will be used in judging significance.

The probability associated with the distribution

n_{11}	n_{12}	$n_{1\cdot}$
n_{21}	n_{22}	$n_{2\cdot}$
$n_{\cdot 1}$	$n_{\cdot 2}$	$n_{\cdot\cdot}$

is

$$P = \frac{n_{1\cdot}!\, n_{2\cdot}!\, n_{\cdot 1}!\, n_{\cdot 2}!}{n_{11}!\, n_{12}!\, n_{21}!\, n_{22}!\, n_{\cdot\cdot}!} \qquad (19.18)$$

where $n_{ij}!$ is defined by Eq. (19.19).

$$n! = n(n-1) \cdots 1 \text{ and } 0! = 1 \qquad (19.19)$$

Read $n!$ as n factorial.

The probabilities for our three tables are

$$P = \frac{7!6!8!5!}{5!2!3!3!13!} = .3263$$

$$P = \frac{7!6!8!5!}{6!1!2!4!13!} = .0816$$

$$P = \frac{7!6!8!5!}{7!0!1!5!13!} = .0047$$

The sum of the probabilities is .4126. It is clear that computation of the first or second probability alone was sufficient to answer the question of significance. In practice, one uses this approach by computing the largest individual probability first, and so on. Notice that the probabilities computed concern more extreme events in one direction. Presumably, this is the sort of test we require in a comparison of a standard and a treatment, namely, a one-sided test. If the alternatives are two-sided, the 5 and 1% significance levels call for probabilities of .025 and .005.

While the computation of probabilities is relatively simple, Mainland (19.10, 19.11) has prepared convenient tables which eliminate the necessity. The first of these references includes exact probabilities for each term for equal and unequal size samples of up to 20 observations and a tabulation of those 2×2 tables which show significance together with the exact level of significance. The second reference includes minimum contrasts required for significance for a selection of sample sizes up to 500. Significance tables are also given by Pearson and Hartley (19.15).

It is to be noted that the 2×2 table of this section is a third type. In the first type, only the grand total was fixed; a certain progeny was observed and classified by a row and a color variable. In the second type, two totals were fixed; two samples were given different treatments with no restrictions, other than implied in a fixed total, on the outcome. In this third type, all marginal totals are considered fixed at their observed values. Probabilities are then computed for these fixed totals.

The present 2×2 table involves an experiment self-contained in a manner not previously considered. The 13 observations of our example might have been made on 13 animals. At the initiation of the experiment, the animals were assigned at random, 7 to a standard treatment and 6 to a treatment. At the end of the experiment, 8 were observed to "have" a certain character, for example, showed a nutritional deficiency, were dead, etc. The data are now analyzed as follows: Given 7 on standard, 6 on treatment, and 8 showing "have" at the end of the experiment, what is the probability associated with the 5 to 3 partition and more extreme partitions of 8 "haves"? Thus all marginal totals are seen to be fixed when probabilities are computed. Such probabilities are termed *conditional probabilities*.

Exercise 19.8.1 For the data in Sec. 19.8, set down the remaining possible tables for fixed marginal totals. Compute the exact probability for each remaining table and show that the sum of the probabilities for all possible tables is unity. Are the probabilities in the set symmetrical? With this in mind, compare the choice of rejection region used for the exact binomial with that proposed for the exact test of Sec. 19.8, when a two-tailed test is required.

Exercise 19.8.2 Test the hypothesis of homogeneity for the data of Exercise 19.4.2. Use the exact procedure of Sec. 19.8 and assume that the alternatives are two-sided. Compare the result with that previously obtained.

19.9 Two trials on the same subjects.

Suppose we have a group of subjects who are prepared to assist us in evaluating two treatments. While we can assign half the subjects to each group, we can also arrange to have each subject receive both treatments but at different times. For example, we might wish to compare headache remedies or seasickness preventives. If we can duplicate the necessary conditions for each subject, then we have twice as many subjects on each treatment as when we split the group. Also, the comparison of the two treatments should be subject to the minimum variation since we try both treatments on every individual.

The results of a trial such as has just been discussed may be presented in a 2 × 2 table when the trial results in enumeration data. Thus, we may have:

Remedy		A	
		Helped	Not helped
B	Helped	n_{11}	n_{12}
	Not helped	n_{21}	n_{22}

The comparison that suggests itself is that of the proportion helped by A to the proportion helped by B, namely, $(n_{11} + n_{21})/n_{..} = n_{\cdot 1}/n_{..}$ to $(n_{11} + n_{12})/n_{..} = n_{1\cdot}/n_{...}$. Both of these ratios include n_{11} in the numerator so that they are correlated. This creates a difficulty in constructing a test criterion based on the two ratios. Since these n_{11} individuals were helped by both treatments, it has been suggested that we consider only those individuals who were helped by one or the other but not by both. That is, the proposed test criterion is based on comparing the cells with n_{12} and n_{21} observations to see whether or not these two numbers depart significantly from a 1:1 ratio. Equation (19.1) is the appropriate test criterion with the expected value equal to $(n_{12} + n_{21})/2$ for both cells, as seen also in Eqs. (18.5) and (18.6). This test criterion reduces to Eq. (19.20) when $p = .5$, as here.

$$\chi^2 = \frac{(n_{12} - n_{21})^2}{n_{12} + n_{21}} \text{ with 1 } df \qquad (19.20)$$

19.10 Linear regression, $r \times 2$ tables.

Cochran (19.1) gives a method for determining linear regression in $r \times 2$ tables, where rows fall in a natural order. Scores, z_i, are assigned to rows, in an attempt to convert them to values on a continuous scale. The regression coefficient b of the estimate of the row probability $\hat{p}_i$ on z_i is a weighted regression given by Eq. (19.21); $\hat{p}_i$ is given weight $n_i./\hat{\bar{p}}(1 - \hat{\bar{p}})$ where $\hat{\bar{p}} = \sum_i n_{i2}/\sum_i n_i.$

$$b = \frac{\sum_i n_i.(\hat{p}_i - \hat{\bar{p}}_i)(z_i - \bar{z}_w)}{\sum_i n_i.(z_i - \bar{z}_w)^2} \quad \text{(definition formula)} \quad (19.21)$$

$\bar{z}_w$ is the weighted mean of z_i defined by $\bar{z}_w = \sum_i n_i.z_i/\sum_i n_i.$. The criterion for testing b is given by Eq. (19.22), though distributed only approximately as χ^2, and has one degree of freedom.

$$\chi^2 = \frac{[\sum_i n_i.(\hat{p}_i - \hat{\bar{p}}_i)(z_i - \bar{z}_w)]^2}{\hat{\bar{p}}(1 - \hat{\bar{p}})\sum_i n_i.(z_i - z_w)^2} \quad (19.22)$$

An experiment performed with white pine by R. F. Patton of the University of Wisconsin compared the effect of age of parent tree, from which scions were taken, upon susceptibility of grafts to blister rust, as shown in Table 19.7. All grafts were made on four-year-old transplants.

TABLE 19.7

AGE OF PARENT TREE AND REACTION OF GRAFTS TO BLISTER RUST

Age of parent tree, years	Score z_i	Healthy n_{i1}	Diseased n_{i2}	Total $n_i.$	$\hat{p}_i = n_{i2}/n_i.$ %	$n_i.z_i$
4	1	7	14	21	67	21
10	2	6	11	17	65	34
20	4	11	5	16	31	64
40 and greater	8	15	8	23	35	184
Total		39 $= n._1$	38 $= n._2$	77 $= n..$	0.4935 $= \hat{\bar{p}}$	303

SOURCE: Data used with permission of R. F. Patton, University of Wisconsin, Madison, Wis.

Compute χ^2 by Eq. (19.23), a working formula for Eq. (19.22).

$$\chi^2 = \frac{[\sum_i n_{i2}z_i - n._2(\Sigma n_i.z_i/n..)]^2}{[\sum_i n_i.z_i^2 - (\sum_i n_i.z_i)^2/n..](\hat{\bar{p}})(1 - \hat{\bar{p}})} \quad (19.23)$$

$$= \frac{[14(1) + \cdots + 8(8) - (38)(303)/77]^2}{[21(1) + \cdots + 184(8) - 303^2/77](.4935)(.5065)}$$

$$= \frac{29.53^2}{156.17} = 5.58^* \text{ with } 1 \text{ } df.$$

The total χ^2, calculated by Eq. (19.8), can be partitioned as below.

	df	χ^2
Regression of $\hat{p}_i$ on z_i	1	5.58*
Deviations from regression	2	2.58
Total	3	8.16*

19.11 Sample size in 2×2 tables. The problem of sample size in two-cell tables is discussed in Secs. 18.5 and 18.6. We now turn our attention to the problem of the sample size necessary to detect differences in 2×2 tables.

Consider the data in Table 19.5. The treatment "vitamins" consists of a homogeneous set of treatments; the treatment "controls" consists of a homogeneous pair. It is conceivable that an investigator might wish to experiment further with one of the treatments and one of the standards, or might wish to continue with a different but similar experiment where the only related information was that available from this experiment. Since the controls give a higher proportion of "alives," it is reasonable to assume this will continue to be the case and to test only for a one-sided set of alternatives.

Paulson and Wallis (19.14) give Eq. (19.24) for determining sample size.

$$n = 1{,}641.6 \left(\frac{z_\alpha + z_\beta}{\arcsin \sqrt{p_S} - \arcsin \sqrt{p_E}} \right)^2 \qquad (19.24)$$

In this equation, angles (arcsin means "the *angle* whose sine is") are in degrees as in Table A.10, n is the number of observations in each sample, z_α is the normal deviate such that $P(z \geq z_\alpha) = \alpha$, z_β is the normal deviate such that $P(z \geq z_\beta) = \beta$, p_S is the population probability or proportion associated with the standard or control treatment, and p_E is that for the experimental treatment. The quantities z_α and z_β are obtained from Table A.4. Since p_S and p_E are never available, they must be estimated; this applies particularly to p_S while p_E may be chosen quite arbitrarily. Reasonably good estimates should not seriously affect the result unless at least one of the p's is very small or very large. The test procedure implied by Eq. (19.24) is χ^2 against one-sided alternatives and adjusted for continuity; the null hypothesis is that of homogeneity. If the null hypothesis is true, then it will be rejected with probability α; if the alternative that p_S is as large as stated is true, then it will be rejected with probability β. If the true p_S is closer to p_E than stated, then we fail to detect this fact more than $100\beta\%$ of the time; if the true p_S is farther from p_E than stated, then we fail to detect this fact less than $100\beta\%$ of the time, that is, we detect it with probability greater than $1 - \beta$. (If radians rather than degrees are used for angles, replace 1,641.6 by 0.5.)

We shall now illustrate the use of Eq. (19.24). Suppose we plan to conduct an experiment much like that summarized in Table 19.5 but with only one standard or control and one treatment. If the null hypothesis of homogeneity is true, we are prepared to reject it only 1% of the time, hence $\alpha = .01$. We do not know p_S but can estimate it from Table 19.5 as $\hat{p}_S = 56/130 = .43$. We wish to detect p_E, when it is about the size observed in Table 19.5, namely, $p_E = 41/170 = .24$, with probability .75. In other words, $1 - \beta = .75$ and $\beta = .25$.

Using Eq. (19.24), we obtain

$$n = 1{,}641.6 \left(\frac{2.327 + 0.675}{40.98 - 29.33}\right)^2 = 109 \text{ mice per treatment}$$

If only one method is involved in the experiment, in other words, if we must decide which of two theoretical ratios applies to a single sample, then Eq. (19.24) may be used but gives double the required sample size.

Reference (19.14) gives a nomograph based on Eq. (19.24) which may be used to determine sample size. A straightedge is the only additional equipment required. Kermack and McKendrick (19.6) also give a solution to this problem, including a table of values of n.

Exercise 19.11.1 Suppose a standard procedure gives a response of about 65% kill in a binomial situation, but that the response is sufficiently variable that the standard is usually included in any experiment. It is decided to consider a proposed treatment for which it is claimed that 80% kill will result. A one-sided z (or χ^2) is to be the test criterion with a 5% significance level; it is desired to detect an 80% kill in the alternative with a probability of .90. What is the necessary sample size for each treatment?

19.12 n-way classification. Binomial data in tables of more than two dimensions present problems of statistics and interpretation. In Sec. 20.5, we discuss the use of a transformation for some such problems. For the present, we consider only the $2 \times 2 \times 2$ table.

In Fig. 19.1, the letters might be numbers of germinating seeds which damp off and which do not damp off (the response), according to variety and treatment. We may hypothesize that whatever treatment times response interaction there is, it differs from variety to variety by no more than random variation. In testing this hypothesis, we test a *three-factor* or *second-order interaction*.

Let $p_1, \ldots, p_8$ be the true relative frequencies or probabilities, not generally known, of an observation falling in cells with entries $a, \ldots, h$, respectively. If there is no treatment times response interaction for variety A, then $p_1/p_2 = p_3/p_4$ and $p_1 p_4 = p_2 p_3$. Similarly for variety B, $p_5/p_6 = p_7/p_8$ and $p_5 p_8 = p_6 p_7$. If there is interaction, the equalities do not hold. In any case, the ratios $p_1 p_4 / p_2 p_3$ and $p_5 p_8 / p_6 p_7$ are directly related to interaction in each table.

The null hypothesis for a three-factor interaction is that the treatment times response interaction, whether present or absent, is equal for the two varieties. This is equivalent to the hypothesis $p_1 p_4 / p_2 p_3 = p_5 p_8 / p_6 p_7$, or

$$H_0: p_1 p_4 p_6 p_7 = p_2 p_3 p_5 p_8 \tag{19.25}$$

This hypothesis involves a single degree of freedom and, hence, a single deviation. This can be obtained by solving the cubic Eq. (19.26).

$$(a + x)(d + x)(f + x)(g + x) = (b - x)(c - x)(e - x)(h - x) \tag{19.26}$$

ENUMERATION DATA II: CONTINGENCY TABLES

The test criterion for H_0 is given by Eq. (19.27).

$$\chi^2 = x^2 \sum_i \frac{1}{E_i} \qquad (19.27)$$

E_i is the expected value for the ith cell.

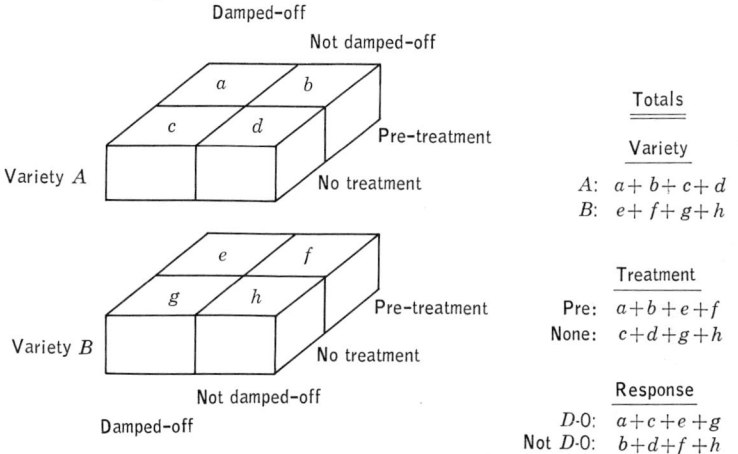

FIG. 19.1 Representation of a $2 \times 2 \times 2$ table

Example. Galton obtained the data of Table 19.8 from 78 families. Offspring were classified as light-eyed or not, as to whether or not they had a light-eyed parent, and as to whether or not they had a light-eyed grandparent.

TABLE 19.8

NUMBERS OF CHILDREN ACCORDING TO A LIGHT-EYED CHARACTERISTIC IN CHILD, PARENT, AND GRANDPARENT

Grandparent		Light		Not	
Parent		Light	Not	Light	Not
Child	Light	1,928	552	596	508
	Not	303	395	225	501

The cubic equation is

$$(1{,}928 + x)(395 + x)(508 + x)(225 + x)$$
$$= (552 - x)(303 - x)(596 - x)(501 - x)$$

which becomes

$$5{,}008x^3 + 1{,}174{,}832x^2 + 1{,}262{,}521{,}792x + 37{,}104{,}335{,}424 = 0$$

The solution of this equation may be obtained by trial and error. We recommend multiples of 10 as initial trial values. In this case, they must obviously be negative multiples. Once two values are obtained such that the left side of the equation changes sign, the location of the solution is reasonably well established. The final solution should involve very few additional trial values. For our equation, the solution is $x = -30.1$. From Eq. (19.27),

$$\chi^2 = (-30.1)^2 \left(\frac{1}{1,928 - 30.1} + \cdots + \frac{1}{501 + 30.1} \right) = 16.93 \text{ with } 1 \ df$$

Since this χ^2 has a single degree of freedom, the correction for continuity is applicable.

$$\text{Corrected } \chi^2 = (30.1 - .5)^2 \left(\frac{1}{1,928 - 30.1} + \cdots + \frac{1}{501 + 30.1} \right) = 11.78$$

The data leave little doubt about a second-order interaction. Alternately, the two first-order interactions are not homogeneous. The test is symmetric in that a comparison of any pair of first-order interactions, that is, the two variety times response, the two variety times treatment, or the two response times treatment interactions, lead to the same null hypothesis, namely, Eq. (19.25).

References

19.1 Cochran, W. G.: "Some methods for strengthening the common χ^2 tests," *Biometrics*, **10**: 417–451 (1954).

19.2 Di Raimondo, F.: "In vitro and in vivo antagonism between vitamins and antibiotics," *International Review of Vitamin Research*, **23**: 1–12 (1951).

19.3 Federer, W. T.: "Variability of certain seed, seedling, and young-plant characters of guayule," *U.S. Dept. Agr. Tech. Bull.* 919, 1946.

19.4 Green, J. M.: "Inheritance of combining ability in maize hybrids," *J. Am. Soc. Agron.*, **40**: 58–63 (1948).

19.5 Irwin, J. O.: "A note on the subdivision of χ^2 into components," *Biometrika*, **36**: 130–134 (1949).

19.6 Kermack, W. O., and A. G. McKendrick: "The design and interpretation of experiments based on a four-fold table: The statistical assessment of the effects of treatment," *Proc. Roy. Soc. Edinburgh*, **60**: 362–375 (1940).

19.7 Kimball, A. W.: "Short-cut formulas for the exact partition of χ^2 in contingency tables," *Biometrics*, **10**: 452–458 (1954).

19.8 Lancaster, H. O.: "The derivation and partition of χ^2 in certain discrete distributions," *Biometrika*, **36**: 117–129 (1949).

19.9 Lancaster H. O.: "The exact partition of χ^2 and its application to the problem of pooling of small expectations," *Biometrika*, **37**: 267–270 (1950).

19.10 Mainland, D.: "Statistical methods in medical research. I. Qualitative Statistics (Enumeration Data)." *Can. J. Research, sec. E, Medical Sciences*, **26**: 1–166 (1948).

19.11 Mainland, D., L. Herrera, and M. I. Sutcliffe: *Tables for Use with Binomial Samples*, published by the authors, New York, 1956.

19.12 Mendel, G.: *Versuche über Pflanzen Hybriden*, 1866, English translation from the Harvard University Press, Cambridge, Mass., 1948.

19.13 Mosteller, F., and J. W. Tukey: "The uses and usefulness of binomial probability paper," *J. Am. Stat. Assoc.*, **44**: 174–212 (1949).

19.14 Paulson, E., and W. A. Wallis: Chap. 7 of *Techniques of Statistical Analysis*, edited by C. Eisenhart, M. W. Hastay, and W. A. Wallis, McGraw-Hill Book Co., Inc., New York, 1947.

19.15 Pearson, E. S., and H. O. Hartley (eds): *Biometrika Tables for Statisticians*, vol. 1, Cambridge University Press, Cambridge, 1954.

19.16 Rinke, E. H.: "Inheritance of coumarin in sweet clover," *J. Am. Soc. Agron.*, **37**: 635–642 (1945).

19.17 Skory, J.: "Automatic machine method of calculating contingency χ^2," *Biometrics*, **8**: 380–382 (1952).

19.18 Smith, D. C., and E. L. Nielsen: "Comparisons of clonal isolations of *Poa pratensis L.*, etc." *Agron. J.*, **43**: 214–218 (1951).

19.19 Smith, H. B.: "Annual versus biennial growth habit and its inheritance in *Melilotus alba*," *Am. J. Botany*, **14**: 129–146 (1927).

Chapter 20

SOME DISCRETE DISTRIBUTIONS

20.1 Introduction. Chapters 18 and 19 dealt with enumeration data. Enumeration data arise from discrete distributions. However, in discussing enumeration data, we rarely commented on the type of discrete distribution unless we stated specifically that we were assuming an underlying binomial or multinomial distribution and wished to test a null hypothesis concerning population ratios.

In this chapter, a brief discussion of several useful discrete distributions is given. The discussion includes some of the uses found for the distributions and some of the tests associated with them.

20.2 The hypergeometric distribution. Suppose that we are presented with a fish tank containing two angelfish and three guppies. We have a dip net with which to catch these fish and they are equally easy (or difficult) to catch. Hence, the probability that the first fish caught will be an angelfish is $2/5$ and the probability that it will be a guppy is $3/5$. Now if we catch an angelfish, then the probabilities associated with the second fish caught are $1/4$ and $3/4$; if we catch a guppy, then they are $2/4$ and $2/4$. In either case, they have changed because sampling has been *without replacement*. This is quite different from a binomial problem where the probability associated with a specified event is constant from trial to trial. Here, the occurrence of one event does not affect the occurrence of any other. In tossing coins, the binomial distribution applies and we state this independence-of-events property by saying that a coin has no memory.

The fish in a tank problem is much simpler than any practical problem is likely to be, but it serves to introduce the *hypergeometric distribution*. A more general hypergeometric problem may be stated as follows: Given N elements consisting of N_1 with property A and $N - N_1$ with property not-A, what is the probability that a sample of n elements, drawn without replacement, will consist of n_1 with property A and $n - n_1$ with property not-A? The probability associated with this event is given by Eq. (20.1).

$$P(n_1) = \frac{\binom{N_1}{n_1}\binom{N - N_1}{n - n_1}}{\binom{N}{n}} \quad (20.1)$$

where

$$\binom{N_1}{n_1} = \frac{N_1!}{n_1!(N_1 - n_1)!} \quad (20.2)$$

SOME DISCRETE DISTRIBUTIONS

for $N_1!$ defined by Eq. (19.19). The system of probabilities is called the *hypergeometric distribution*. Notice that we must say sampling is without replacement, that is, once a sample element is drawn, it is permanently removed from the population. If sampling were with replacement, its return to the population would result in independent probabilities being associated with each draw, and we would have a binomial population.

Two common statistical problems involve the hypergeometric distribution. They are (1) quality inspection and (2) tag-recapture or mark-recapture problems.

Quality inspection. The problem here is to estimate the proportion of defectives. A defective may be a raisin cookie without a raisin, a punctured dry-yeast container which should be airtight, or a bridge hand without a face card. Here, the population size N is known and the number of defectives N_1 is to be estimated. If a sample of n observations contains X defectives, then N_1 may be estimated by using Eq. (20.3).

$$\hat{N}_1 = \frac{X}{n} N \tag{20.3}$$

The variance of the estimate, $\hat{N}_1$, is estimated by using Eq. (20.4).

$$s_{\hat{N}_1}^2 = \frac{N(N-n)}{n} \frac{X}{n}\left(1 - \frac{X}{n}\right)\frac{n}{n-1} \tag{20.4}$$

Tag-recapture problems. Here we have a problem that is fairly common where wildlife populations are concerned. A known number of birds, fish, or animals are captured, tagged, or otherwise marked and returned to the population. A second sample is captured and the number of recaptured individuals is observed. Here the problem is one of estimating the size of the population. An estimate is given by Eq. (20.5).

$$\hat{N} = \frac{N_1 n}{X} \tag{20.5}$$

Here N_1 is the number of tagged individuals in the population and X is the number of tagged individuals in the sample of size n.

Hypergeometric distributions may be further generalized to include more than two classes of elements.

Exercise 20.2.1 If you have any facility with algebra, you will find it easy to rearrange Eq. (19.18) and show that it is a probability from a hypergeometric distribution.

20.3 The binomial distribution. Sampling a hypergeometric distribution is like drawing beans from a bag without replacement, whereas sampling a binomial distribution is like drawing beans from a bag *with replacement*. In other words, the probability associated with an event, such as drawing a specially marked bean, is constant from trial to trial. The events are said to be *mutually independent* or simply *independent*.

Sections of Chaps. 18 and 19 dealt with events for which there were only two possible outcomes. In these sections, we assumed that we were dealing with a binomial distribution and then proceeded to test some null hypothesis about a parameter. In this chapter, we shall be more concerned with the assumption of a binomial distribution and shall make it a part of our hypothesis and subject to testing.

In a single *binomial event* there are two possible outcomes, only one of which can occur. Mathematicians refer to these as a "success" and a "failure." For us, they will be red versus white, awned versus awnless, alive versus dead, etc. The usual sampling procedure consists of observing a number of events and recording the numbers of successes and failures. Thus, the probability distribution usually required is the distribution of the number of successes. This is given by Eq. (20.6).

$$P(X = n_1|n) = \binom{n}{n_1} p^{n_1}(1-p)^{n-n_1} \qquad (20.6)$$

Equation (20.6) is read as "the probability that the number of successes X will equal n_1 in n random trials equals . . . ." Here $\binom{n}{n_1}$ is defined by Eq. (20.2) and p is the probability of a success. The mean or expected number of successes is np and the variance of the number of successes is $np(1-p)$.

For example, what is the probability that a family of six children will consist of six girls? We assume that the binomial distribution holds and that, prior to the birth of any child, the probability of a girl is $\frac{1}{2}$. In assuming a binomial distribution, we assume independence of events; in other words, the probability remains constant from event to event. The required probability is

$$\binom{6}{6}\left(\frac{1}{2}\right)^6\left(\frac{1}{2}\right)^0 = \frac{6!}{6!0!}\frac{1}{64}1 = .015625$$

Apparently, between one and two out of 100 families of six children will be all girls. Also, the probability of a family of six children being all of the same sex is $2(.015625) = .03125$. If p were other than .5, we would have to compute two probabilities (they would be different) and add them.

The binomial distribution may be generalized to give the multinomial distribution where there are more than two possible outcomes for each trial. Of the possible outcomes, only one will occur at each trial.

20.4 Fitting a binomial distribution. We have just seen that a family of six girls constitutes an unusual event. However, if we observe a great many families of six children, we will almost certainly observe families of six girls. It is of interest to know whether they occur more or less frequently than they ought if their probabilities are binomial. In other words, do we have a binomial distribution with the required independence of events and a constant probability from event to event?

To determine whether we have a binomial distribution, we must observe many families of size six. The results of our observations will be recorded as six girls and no boys, five girls and one boy, . . . , no girls and six boys—seven possible outcome groups in all. Since each outcome group supplies information relative to the true distribution, we must avoid pooling groups if at all possible. This common problem may involve some genetic character for animals in the same litter size, or the response to some technique where the same number of plants or animals is used from trial to trial. The answer may concern the effect of environment on a genetic character, the ability to get a constant response (within binomial sampling variation) to a technique, or simply the ability to determine if the analysis of a set of data should be based on the assumption of a binomial distribution.

To illustrate the procedure of fitting a binomial distribution, consider the data of Table 20.1. They were obtained as follows: each student in a class of

TABLE 20.1

FITTING A BINOMIAL DISTRIBUTION

Data		Probability	Computations				
X	Frequency, f	$P(X = n_1\|5)$	Coefficient	$.4^X$	$.6^{5-X}$	P	Expected frequency
0	13	$\binom{5}{0}.4^0\ .6^5$	1	1	.07776	.07776	6.2208
1	18	$\binom{5}{1}.4\ .6^4$	5	.4	.1296	.25920	20.7360
2	20	$\binom{5}{2}.4^2\ .6^3$	10	.16	.216	.34560	27.6480
3	18	$\binom{5}{3}.4^3\ .6^2$	10	.064	.36	.23040	18.4320
4	6	$\binom{5}{4}.4^4\ .6$	5	.0256	.6	.07680	6.1440
5	5	$\binom{5}{5}.4^5\ .6^0$	1	.01024	1	.01024	.8192
Totals	80	1.00				1.00	80.0000

All data: $p = .4$,
$\hat{p} = \Sigma Xf(X)/5(80) = 161/400 = .4025$
$\sigma_{\hat{p}}^2 = p(1-p)/n = .4(.6)/400 = .0006$,
$\hat{\sigma}_{\hat{p}}^2 = \hat{p}(1-\hat{p})/n = .4025(.5975)/400 = .0060$

Set of five samples: $p = .4$,
$\hat{p} = $ (80 values ranging from 0 to 1),
$\sigma_{\hat{p}}^2 = .4(.6)/5 = .048$
$\mu_X = np = 5(.4) = 2$, $\sigma_X^2 = np(1-p) = 5(.4)(.6) = 1.2$
$\bar{x} = \Sigma Xf(X)/\Sigma f(X) = 161/80 = 2.0125$

80 was assigned the problem of drawing 10 random samples of 10 observations from Table 4.1. The 10 samples were randomly paired and differences computed. For the 5 samples of 10 random differences, $t = \bar{d}/s_{\bar{d}}$ was computed. The results should follow the t distribution since the true mean of the differences is zero and the parent population is normal. However, computing machines and statistics were new to all students and this introduced the possibility of error.

Since we first wish to illustrate the binomial distribution *for known p*, let us arbitrarily choose $t_{.40}(9\ df) = .883$ and observe the number of t values greater than .883 for each of the 80 sets of five samples. Denote this number by X. We now have to see how well our data fit a binomial distribution with $p = .4$. From Eq. (20.6),

$$P(X = n_1|5) = \binom{5}{n_1} .4^{n_1} .6^{5-n_1} \text{ for } n_1 = 0, 1, \ldots, 5$$

Each probability when multiplied by 80 gives the expected frequency. These frequencies and the necessary computations are shown in Table 20.1. Note that the observed frequencies are not necessary for the computations. Additional population and sample information is given at the foot of the table.

While the coefficients are easy to obtain from Eq. (20.2), they may also be found from Pascal's triangle. Construction of the triangle is simple when you observe that each number is the sum of the two numbers in the preceding line which are immediately above it to the right and left; the first and last numbers are always ones.

Number observed = n *Binomial coefficients*

```
1                    1   1
2                  1   2   1
3                1   3   3   1
4              1   4   6   4   1
5            1   5  10  10   5   1
.
.
.
```

We are now ready to test the null hypothesis that we have a binomial distribution with $p = .4$. This is a goodness-of-fit test as described in Chap. 17; the test criterion is χ^2. The computations are carried out in Table 20.2 and

TABLE 20.2
TEST OF THE BINOMIAL DISTRIBUTION WITH $p = .4$

X	Observed frequency	Expected frequency	Deviation	$\dfrac{(\text{Deviation})^2}{\text{Expected frequency}}$
0	13	6.2208	6.7792	7.388
1	18	20.7360	−2.7360	.361
2	20	27.6480	−7.6480	2.116
3	18	18.4320	−.4320	.010
4	6	6.1440	−.1440	.003
5	5	.8192	4.1808	21.337
Totals	80	80	0.0	31.215

give $\chi^2 = 31.215^{**}$ with 5 degrees of freedom. The null hypothesis of a binomial distribution with $p = .4$ is not supported. Since we did not have to estimate p, we lose only one degree of freedom which is associated with the restriction that the sum of the deviations must be zero.

SOME DISCRETE DISTRIBUTIONS

When the data do not support the null hypothesis, we may look for the cause in the value chosen for p, in the nature of the distribution apart from the value of p, or in both. The observed proportion does not differ appreciably from the hypothetical one, so we conclude that the data are not binomially distributed. The cells with $X = 0$ and $X = 5$ are the only cells with positive deviations; also, they are the biggest contributors to χ^2. It seems advisable to check the computational procedures of those students obtaining consistently high or consistently low values of t.

Let us now compare the procedure of fitting the binomial distribution with the summary of procedures, Sec. 19.6, used for the data of Table 19.4. These procedures, applied to the present data, would call for an 80×2 table. Individual χ^2's would be of little use until we found out whether or not the results of the 80 trials were homogeneous. The sum of the individual χ^2's would normally be partitioned to test a hypothesis about p and one of homogeneity. The χ^2 value for totals would be a test of $H_0: p = .4$ and we would accept the null hypothesis ($\hat{p} = .4025$). Homogeneity χ^2 tests the hypothesis of a constant p from trial to trial. We hope it would show significance. For this example, the two procedures might well lead to the same conclusions.

Can the procedures lead to different conclusions? The answer is Yes. Suppose that each of our 80 trials gave three values less than .883 and two greater. The observed ratio is the hypothesized one and the data are homogeneous. In other words, we could have less than binomial variation and not be able to determine so by the methods of Chap. 19. The methods of this chapter are designed expressly for detecting departures from the binomial distribution.

One other problem of fitting a binomial distribution arises. This is the problem of fitting a binomial distribution when there is not a known or hypothesized value of p. For an unknown p, testing the fit of a binomial distribution tests the binomial nature of the variation, including the constancy of p from trial to trial. It is necessary to estimate p, and this accounts for a degree of freedom. Thus, for the data in Table 20.1, we would estimate p as $\hat{p} = .4025$, complete the fitting process of Table 20.1 using .4025 as the value of p, and proceed to the test procedure of Table 20.2. The resulting χ^2 has $6 - 2 = 4$ degrees of freedom rather than 5.

Exercise 20.4.1 Complete Pascal's triangle to $n = 10$. How many coefficients are there for $n = 7$? How many are there in general?

Exercise 20.4.2 For the data in Table 20.1, test the null hypothesis that $p = .4$, assuming a binomial distribution.

Exercise 20.4.3 Compute the theoretical mean and variance of the number of successes in five trials, using the columns headed X and Expected frequency. Observe that they equal $5(.4)$ and $5(.4)(.6)$ as computed by formula. (See Sec. 2.18; it is hardly necessary to code.)

Exercise 20.4.4 For each of the 80 sets of five samples, the number of t values less than 1.833 was also observed. The data follow:

No. of t's less than 1.833:	5	4	3	2	1	0
No. of samples:	51	18	6	2	2	1

What is the theoretical value of p for this binomial distribution? Fit a binomial distribution for the theoretical value of p. In testing the goodness of fit of your data, will you want to pool any of the results? Test the goodness of fit, being sure to state the number of degrees of freedom for the test criterion.

Exercise 20.4.5 The same 80 sets of five t values were observed with respect to another tabulated value of t. While the results are available and given below, the value of t is no longer available.

No. of t's less than unstated value:	5	4	3	2	1	0
No. of samples:	23	30	16	5	4	2

Fit a binomial distribution to these data and test the goodness of fit. How many degrees of freedom does your test criterion have? Estimate the binomial p, using all the data, by means of a 95% confidence interval. In the light of the goodness of fit test, do you believe your interval is too wide, too narrow, just right?

20.5 Transformation for the binomial distribution. In Sec. 8.15, the arcsin $\sqrt{X}$ was recommended for binomial data. Table A.10 is used for making the transformation and the resulting angles are given in degrees. The variance of an observation is approximately $821/n$.

When an investigator is confident that the variation in his data is purely binomial, this transformation and the theoretical variance may be very useful. For example, many $r \times c \times 2$ *contingency tables* may be presented as $r \times c$ tables of proportions. If these proportions are transformed according to the arcsin transformation, standard analyses of variance procedures may be used for obtaining main effect and interaction sums of squares. Each sum of squares may then be tested by χ^2. For example, suppose the proportions are based on unequal numbers of observations. The proportions are transformed to give a two-way table with only one observation per cell, this being the single transformed proportion. Thus it is not possible to compute a within-cells variance. In addition, we may wish to test for interaction. The procedures of Chap. 13 apply for computing main effects and interaction but there is no procedure for obtaining an error term. We must use the theoretical variance.

To test any null hypothesis, the procedure is as follows. Compute the appropriate sum of squares using the transformed proportions. Such sums of squares are computed by standard analysis of variance procedures. *If the denominators of the proportions are the same*, then the computations are such that all sums of squares are on a per-observation basis where per-observation refers to the common denominator. The variance to be used for each sum of squares is, then, $821/n$ where n is the common denominator. *If the denominators are disproportionate*, then the computations are carried out as in Chap. 13, with the weights based on these denominators. Now, per-observation refers to the single binomial trial and the variance to be used for each sum of squares is $821/n$ where $n = 1$.

Since χ^2 equals a sum of squares divided by σ^2, mean squares are not computed. Instead, each sum of squares to be tested is divided by $821/n$ to give a χ^2 based on the number of degrees of freedom associated with the sum of squares in the numerator. The resulting χ^2's are referred to Table A.5 for judging significance. In the case of unequal denominators, values of χ^2 will not be additive.

SOME DISCRETE DISTRIBUTIONS

Exercise 20.5.1 Review Exercise 8.15.2 in the light of your new knowledge of the arcsin transformation. Was the error term you used at that time a reasonable error? Are the χ^2's additive in this exercise? Why?

20.6 The Poisson distribution. This discrete distribution is sometimes related to the binomial distribution with small p and large n. However, it is a distribution in its own right and random sampling of organisms in some medium, insect counts in field plots, noxious weed seeds in seed samples, numbers of various types of radiation particles emitted, may yield data which follow a Poisson distribution.

Probabilities for a Poisson distribution are given by Eq. (20.7).

$$P(X = k) = \frac{e^{-\mu}\mu^k}{k!} \qquad (20.7)$$

This is read as "the probability that the random variable X takes the value k is equal to . . . ." The value of k may be 0, 1, 2, . . . ; there is no stopping point. The mean of the distribution is μ; the variance is also μ. It is customary to make a number of observations and the mean of these, $\bar{x}$, provides an estimate of both μ and σ^2.

An exact procedure for obtaining a confidence interval for the mean of a Poisson distribution (including a table) is given by Fisher and Yates (20.3). Blischke also discusses the problem (20.1).

To illustrate *fitting a Poisson distribution*, we have chosen a sample giving the actual distribution of yeast cells over 400 squares of a haemacytometer. These data were obtained by Student (20.6) and are presented in Table 20.3.

TABLE 20.3

FITTING A POISSON DISTRIBUTION

X	Observed frequency	Probability, Eq. (20.7)	Computation†	Probability	Expected frequency
0	213	$P_0 = e^{-\mu}$	.5054	.5054	202.16
1	128	$P_1 = \mu P_0$	.6825(.5054)	.3449	137.96
2	37	$P_2 = \dfrac{\mu}{2} P_1$	.34125 P_1	.1177	47.08
3	18	$P_3 = \dfrac{\mu}{3} P_2$	.2275 P_2	.0268	10.72
4	3	$P_4 = \dfrac{\mu}{4} P_3$	.170625 P_3	.0046	1.84
5	1	$P_5 = \dfrac{\mu}{5} P_4$	.1365 P_4	.0006	.24
>5	0	$1 - \sum_{i=0}^{5} P_i$	1−1.0000	.0000	.00
$\bar{x} = 273/400 = 0.6825$				1.0000	400.00

† $\log P_0 = -\mu \log_{10} e = -.6825(.434295) = -.296406 = \bar{1}.703594$; $P_0 = .5054$

Since the parameter μ is unknown, it will be estimated from the data. Computations proceed as in Table 20.3. We have chosen to rewrite Eq. (20.7) as a recursion formula, Eq. (20.8), and to make use of this formula in obtaining probabilities.

$$P(X = k) = \frac{\mu}{k} P(X = k - 1) \qquad (20.8)$$

The first step is to find P_0 as indicated at the foot of the table. This requires that we find the antilog of $\bar{1}.703594$, the only time the log table is required. If a table of e^μ is available, even this single use of a log table becomes unnecessary ($e^{-\mu} = 1/e^\mu$). From here on, two computing techniques for desk calculators are available. They are:

1. Compute $\mu/2, \mu/3, \mu/4, \ldots$. Carry out the first multiplication and record P_1. Use the transfer slide (if available) so that P_1 becomes the next multiplicand. Multiply by $\mu/2$. Continue to use transfer slide with $\mu/3$, $\mu/4, \ldots$ as successive multipliers.

2. (Alternative to 1.) Put μ in the calculator as a constant multiplicand (where possible). Multiply by P_0 and record P_1. Use P_1 as next multiplier, followed by division by 2 to give P_2. Use P_2 as next multiplier and 3 as next divisor to give P_3. Continue similarly.

This method of fitting a Poisson distribution has the advantage that it requires a single use of a logarithmic or exponential table. After the first probability is determined, succeeding ones are taken directly from the computing machine without the necessity of finding antilogs. Also, operator errors are minimized where the computing machine has either a transfer slide or a multiplicand lock. Finally, rounding errors seem less likely to pile up as the result of introducing new rounding errors through use of many different logarithms.

The final step in fitting is to multiply each probability by the total frequency; in this case, 400.

To test the goodness of fit, χ^2 is an appropriate test criterion. The method is shown in Table 20.4; expected frequencies are from Table 20.3. For this test, we have five classes, the last two observed being pooled because one of the expected frequencies is less than one. In addition, it was necessary to estimate

TABLE 20.4
TESTING THE GOODNESS OF FIT TO A POISSON DISTRIBUTION

X	Observed frequency	Expected frequency	Deviation	$\dfrac{(O - E)^2}{E}$
0	213	202.16	10.84	.581
1	128	137.96	−9.96	.719
2	37	47.08	−10.08	2.158
3	18	10.72	7.28	4.944
4	3 ⎫ 4	1.84 ⎫ 2.08	1.92	1.772
5	1 ⎭	.24 ⎭		
	400	400.00	0.00	10.174

SOME DISCRETE DISTRIBUTIONS

μ. Hence, the degrees of freedom are $5 - 1 - 1 = 3$ df. There is evidence that these data do not fit a Poisson distribution. If there is sound reason to believe that such data arise with Poisson probabilities, then we have an unusual sample or the mean of the distribution is not stable.

Fisher (20.2) has proposed an alternative measure to χ^2 for testing goodness of fit to Poisson distributions. This test is intended more particularly for observed distributions when expected frequencies are small. Rao and Chakravarti (20.4) have considered the problem further.

Exercise 20.6.1 Student also observed the accompanying distributions. Fit a Poisson distribution to one or more of these samples.

Sample	X:	0	1	2	3	4	5	6	7	8	9	10	11	12
2		103	143	98	42	8	4	2						
3		75	103	121	54	30	13	2	1	0	1			
4		0	20	43	53	86	70	54	37	18	10	5	2	2

Test the goodness of fit.

20.7 Other tests with Poisson distributions. The fact that the mean and variance of a Poisson distribution are equal suggests that their ratio should provide a test of significance. The ratio usually used is given in Eq. (20.9).

$$\chi^2_{n-1} = \frac{\Sigma(X_i - \bar{x})^2}{\bar{x}} \tag{20.9}$$

Notice that it is the ratio of the sum of squares, rather than the variance, to the mean. It is the equation $\chi^2 = \Sigma[(O - E)^2/E]$ expressed in different terms.

Let us apply this test criterion to sample 2, Exercise 20.6.1. We obtain:

$$\chi^2_{399} = \frac{513.40}{1.3225} = 388.20$$

Notice that the degrees of freedom are $400 - 1 = 399$ df. The X values are $0, 1, \ldots, 6$. Since Table A.5 does not give χ^2 values for 399 degrees of freedom, we rely on the fact that $\sqrt{2\chi^2} - \sqrt{2n - 1}$ is approximately normally distributed with zero mean and unit variance. Hence, we can use Table A.4. However, a glance at Table A.5 shows that the median ($P = .5$) χ^2 value is approximately equal to the degrees of freedom. There seems little use in completing the computations.

The null hypothesis for the test criterion given by Eq. (20.9) is essentially one of homogeneity. In other words, we are testing for a stable μ.

This test criterion may be used in experimental situations such as that of Student or in more general situations. Thus, we may have t treatments and make n observations on each treatment. The experimental design may be a completely random one with equal replications, a randomized complete-block design, or a Latin square. The sum of Poisson variables is also a Poisson variable. Hence, if the null hypothesis of no treatment differences is valid, the treatment totals will follow a Poisson distribution even though there are real block differences. Equation (20.9) with $t-1$ degrees of freedom is appropriate for testing the null hypothesis of no differences among treatment means.

A somewhat special situation arises when extremely small binomial probabilities are acting, for example, when numbers of mutants are being observed. If the samples are large and the numbers are small, we may assume a Poisson distribution. For example, two lines of corn were observed and the following data resulted:

	Nonmutants	Mutants	Total (approx.)
A	5×10^5	10	5×10^5
B	6×10^5	4	6×10^5

We would like to test the null hypothesis that the probability of a mutation is the same for each line. A test may be based on determining whether the 10:4 split is improbable in sampling a population where the true proportions are $5 \times 10^5 : 6 \times 10^5$ and we stop after 14 mutants have been observed. This problem involves a conditional Poisson distribution which can be related to a binomial distribution.

If we use the binomial distribution, the appropriate p value is $5(10^5)/[5(10^5) + 6(10^5)] = 5/11$, which lies between .45 and .46. We now use Eq. (20.6) and find

$$P(X \geq 10 | n = 14) = \sum_{n_1=10}^{14} \binom{14}{n_1} p^{n_1}(1-p)^{14-n_1}$$

This equals .0426 for $p = .45$ and .0500 for $p = .46$, as discussed in reference (18.7).

Since the probability of obtaining a 10:4 or more extreme split, under the null hypothesis, is small (very nearly .05), we reject the null hypothesis that the probability of mutation is the same for each line. Other test criteria for this example are discussed by Steel (20.5).

Exercise 20.7.1 Use Eq. (20.9) to test the null hypothesis of a stable μ for the data in Table 20.3. Repeat for any sample you may have used in completing Exercise 20.6.1. Compare the results obtained from the two criteria.

Exercise 20.7.2 Visualize a situation where the procedures of Secs. 20.6 and 20.7 would lead to different conclusions. Explain why this happens in terms of assumptions, null hypotheses, and alternative hypotheses.

Exercise 20.7.3 Apply the χ^2 test criterion to the data used in this section. Compare your result with that given in the text.

References

20.1 Blischke, W. R.: "A comparison of equal-tailed, shortest, and unbiased confidence intervals for the chi-square distribution," M.Sc. Thesis, Cornell University, Ithaca, N.Y., 1958.

20.2 Fisher, R. A.: "The significance of deviations from expectation in a Poisson series," *Biometrics*, **6:** 17–24 (1950).

20.3 Fisher, R. A., and F. Yates: *Statistical Tables for Biological, Agricultural and Medical Research*, 5th ed., Hafner Publishing Company, New York, 1957.

20.4 Rao, C. R., and I. M. Chakravarti: "Some small sample tests of significance for a Poisson distribution," *Biometrics*, **12:** 264–282 (1956).

20.5 Steel, R. G. D.: "A problem involving minuscule probabilities," *Mimeo Series* BU-81-M, *Biometrics Unit*, Cornell Univ., Ithaca, N.Y., 1957.

20.6 Student: "On the error of counting with a haemacytometer," *Biometrika*, **5:** 351–360 (1907).

Chapter 21

NONPARAMETRIC STATISTICS

21.1 Introduction. The techniques so far discussed, especially those involving continuous distributions, have stressed the underlying assumptions for which the techniques are valid. These techniques are for the estimation of parameters and for testing hypotheses concerning them. They are called *parametric* statistics. The assumptions generally specify the form of the distribution and Chaps. 1 to 16 are concerned largely with data where the underlying distribution is normal.

A considerable amount of collected data is such that the underlying distribution is not easily specified. To handle such data, we need *distribution-free* statistics, that is, we want procedures which are not dependent upon a specific parent distribution. If we do not specify the nature of the parent distribution, then we will not ordinarily deal with parameters. Thus, we have *nonparametric* statistics which compare distributions rather than parameters. The terms nonparametric and distribution-free refer to the same topic though occasionally both terms will not be wholly applicable. Most nonparametric statistics are intended to apply to a large class of distributions rather than to all possible distributions.

Nonparametric statistics have a number of advantages.

1. Since they may use ranks or the signs of differences, they are often, though not always, quick and easy to apply and to learn.

2. For the same reasons, they may reduce the work of collecting data. Thus, we may simply score a plant for virus infection or a plot for insect infestation, or we may rank cakes for flavor and texture. Even when there is a scale such that the resulting observations may be normally distributed, we may choose to use ranks if we plan to collect a great deal of data.

3. Sampling procedures may include several populations for which we prefer to assume very little. For example, in a variety trial consisting of one replicate at each of many locations, or in a summary of results over locations and years, there may be very different variances involved; if we observe ranks, these can be analyzed by nonparametric procedures.

4. Finally, probability statements are not qualified as severely as they are when parametric procedures are used, the qualifications being the set of assumptions.

Nonparametric procedures also have disadvantages, of which the greatest is probably the following. If the form of the parent population is known to be reasonably close to a distribution for which there is standard theory, or if the

NONPARAMETRIC STATISTICS

data can be transformed so that such is the case, then nonparametric procedures do not extract as much information from the data. If all an investigator's experiments result in data such that the null hypothesis is true, then nonparametric procedures are as good as any others since the investigator sets the error rate. However, if the null hypothesis is false, then the usual problem is to detect differences among means. Nonparametric procedures are not as good as classical procedures for this purpose, provided the assumptions about the parent distribution are valid. In addition, nonparametric procedures have not been generally developed for comparisons of single degrees of freedom.

21.2 The sign test. In this test, we consider medians rather than means. The median is the value such that half of the probability lies on each side. It is clear that the mean and median will be the same for symmetric distributions.

The sign test is based on the signs of the differences between paired values. This means it can also be used when the paired observations are simply ranked.

To illustrate the test procedure, consider the data of Exercise 5.6.1. These data are cooling constants of freshly killed mice and of the same mice reheated to body temperature. The differences, freshly killed minus reheated, are: $+92$, $+139$, -6, $+10$, $+81$, -11, $+45$, -25, -4, $+22$, $+2$, $+41$, $+13$, $+8$, $+33$, $+45$, -33, -45, and -12. There are 12 pluses and 7 minuses. These numbers serve to test the null hypothesis that each difference has a median of zero, in other words, that pluses and minuses occur with equal probability.

To test the null hypothesis, any one of the test criteria of Sec. 18.4 is applicable. Equation (21.1) gives a χ^2 adjusted for continuity appropriate to test $H_0: p = .5$, as is the case here.

$$\chi^2 = \frac{(|n_1 - n_2| - 1)^2}{n_1 + n_2} \tag{21.1}$$

The values n_1 and n_2 are the numbers of pluses and minuses. For our example, $\chi^2 = (12 - 7 - 1)^2/19 = 16/19 < 1$ and is clearly not significant.

This test has a number of advantages. It is easy to apply and, if we plan to use this test, we can simplify the collection of the data. We do not need homogeneity of variance in the usual sense since each difference may be from a different continuous distribution, provided all distributions have zero as median. Differences must, of course, be independent. In addition, the test is not very sensitive to gross recording errors. This may be of some importance when reviewing work of other years and other investigators.

The test has the disadvantage of throwing away a lot of information in the magnitude of the differences. Thus, it is impossible to detect a departure from the null hypothesis with fewer than six pairs of observations and one would hesitate to use it with less than 12 pairs. With 20 or more pairs of observations, it becomes more useful. When ties occur, as they do in practice, they may be assigned in equal numbers to the plus and minus categories or simply discarded along with the information they contain.

The sign test can be modified to handle a number of other situations. For example,

1. For a sample from a single population, we can test the null hypothesis that the median is a specified value. Observe the numbers of observations

that lie above or below the hypothesized value and use Eq. (21.1) as test criterion.

2. For paired observations, we may ask if treatment A gives a response that is k units better than that for B. This is the usual linear model without the customary restrictions. Observe the signs of differences $X_i - (Y_i + 10)$ and apply the sign test.

3. For paired observations, we may ask if treatment A gives a response that is $k\%$ better than that given by B. This is a nonadditive model that would call for a transformation if the procedures of the analysis of variance chapters were to be used. Observe the signs of differences $X_i - (1 + K)Y_i$ where K is the decimal fraction corresponding to $k\%$, and use the sign test.

Exercise 21.2.1 Apply the sign test to the data of Table 5.5, and Exercise 5.6.2. Do you draw the same conclusions as you drew in Chap. 5? Comment on the sample size relative to each test procedure.

21.3 Wilcoxon's signed rank test. This test (references 21.7, 21.8, and 21.9) is an improvement upon the sign test in the matter of detecting real differences with paired treatments. The improvement comes about through making some use of the magnitudes of the differences.

The steps in the procedure are:
1. Rank the differences between paired values from smallest to largest without regard to sign.
2. Assign to the ranks the signs of the original differences.
3. Compute the sum of the positive or negative ranks, whichever is smaller.
4. Compare the sum obtained at step 3 with the critical value.

Applying the above procedure to the mouse data of Sec. 21.2, we obtain:

Difference:	+2,	−4,	−6,	+8,	+10,	−11,	−12,	+13,	+22,	−25,
Signed rank:	+1,	−2,	−3,	+4,	+5,	−6,	−7,	+8,	+9,	−10,

Difference:	−33,	+33,	+41,	−45,	+45,	+45,	+81,	+92,	+139.
Signed rank:	−11½,	+11½,	+13,	−15,	+15,	+15,	+17,	+18,	+19.

The sum of the negative ranks is $T = 54.5$.

This value, disregarding sign, is referred to Table A.18 to judge significance. The critical value at the 5% level is 46 and we conclude that the evidence is not sufficient to deny the null hypothesis. Notice that small values of T are significant ones. For z, t, χ^2, and F, it is the large values which generally supply evidence against the null hypothesis.

Beyond the range of Table A.18, z and Table A.4 may be used to test significance. For $z = (T - \mu_T)/\sigma_T$, μ_T and σ_T are given by Eqs. 21.2 for n equal to the number of pairs.

$$\mu_T = \frac{n(n+1)}{4} \quad \text{and} \quad \sigma_T = \sqrt{\frac{n(n+\tfrac{1}{2})(n+1)}{12}} \quad (21.2)$$

Notice how ties are handled in the example. The average value is given to each rank. This is necessary when the tied ranks include both signs but not otherwise. In improving the sign test to give the ranked sign test, it has been necessary to introduce an assumption. The assumption is that each difference is from some symmetric distribution. It is not necessary that each difference have the same distribution. This test may also be used with a single sample where it is desired to test a null hypothesis about the median. Here the hypothesized value is subtracted from each observation and the resulting numbers are processed as in the preceding instructions.

Exercise 21.3.1 Apply Wilcoxon's signed rank test to the data of Table 5.5 and Exercise 5.6.2. Compare your results with those obtained previously. (See Exercise 21.2.1.)

21.4 Two tests for two-way classifications. Probably the most common experimental design is the randomized complete-block design with more than two treatments. Friedman (21.1) has proposed the following test for such designs:
1. Rank the treatments within each block from lowest to highest.
2. Obtain the sum of the ranks for each treatment.
3. Test the null hypothesis of no differences among population means for treatments, by Eq. (21.3).

$$\chi_r^2 = \frac{12}{bt(t+1)} \sum_i r_{i\cdot}^2 - 3b(t+1) \qquad (21.3)$$

with $t - 1$ degrees of freedom, where t is the number of treatments, b is the number of blocks, and $r_{i\cdot}$ is the sum of the ranks for the ith treatment. Note that 12 and 3 are constants, not dependent on the size of the experiment. Definition of $r_{i\cdot}$ implies that r_{ij} is the rank of the ith treatment in the jth block. This test criterion measures the homogeneity of the t sums and is distributed approximately as χ^2. The approximation is poorest for small values of t and b. Friedman has prepared tables of the exact distribution of χ_r^2 for some pairs of small values of t and b.

Let us now apply Friedman's procedure to the data of Table 8.2, presented again in Table 21.1. Treatment responses are given opposite block numbers, and ranks below. Tied treatments are given the average rank. The value of the test criterion is right at the 5% value for this example; the F value was beyond the 1% point.

An alternate procedure based on work by George W. Brown is given by Mood (21.4). This test assumes the distributions are continuous and identical except for location. (Notice that this requires homogeneity of variance.) We test whether the treatment contributions to the cell medians are all zero. The contributions are such that, if they are real, their median is zero. For the random or mixed model, no assumption about interaction is necessary; for the fixed model, zero interaction is assumed.

The test procedure for treatments follows.
1. Find the median of the observations in each block.
2. Replace each observation by a $+$ or a $-$, according to whether it is above or below the median of the observations in the block. (see Table 21.1.)
3. Record the numbers of plus and minus signs, by treatments, in a $2 \times t$ table.
4. Test the result as for an ordinary contingency table.

TABLE 21.1

OIL CONTENT OF REDWING FLAXSEED†

Block	Treatment					
	S	EB	FB	FB(1/100)	R	U
1	4.4(−) 2.5	3.3(−) 1	4.4(−) 2.5	6.8(+) 6	6.3(+) 4	6.4(+) 5
2	5.9(+) 4	1.9(−) 1	4.0(−) 2	6.6(+) 5	4.9(−) 3	7.3(+) 6
3	6.0(+) 4	4.9(−) 2	4.5(−) 1	7.0(+) 5	5.9(−) 3	7.7(+) 6
4	4.1(−) 2	7.1(+) 5.5	3.1(−) 1	6.4(−) 3	7.1(+) 5.5	6.7(+) 4
Rank totals	12.5	9.5	6.5	19	15.5	21

$$\chi_r^2 = \frac{12}{4(6)7}(12.5^2 + \cdots + 21^2) - 3(4)7 = 11.07 \text{ with } 5 \text{ df}; \quad \chi_{.05}^2(5 \text{ df}) = 11.1$$

† See Table 8.2.

We now apply the procedure to the data of Table 21.1. It will not generally be necessary to compute medians. For example, in block 1 the median is between 4.4 and 6.3; this is sufficient for assigning pluses and minuses. The signs given in Table 21.1 are for this test. The two-way contingency table is shown below. It is now necessary only to compute χ^2 and compare with tabulated values for $t - 1 = 5$ degrees of freedom.

Treatment	S	EB	FB	FB(1/100)	R	U
Above	2	1	0	3	2	4
Below	2	3	4	1	2	0

Exercise 21.4.1 Compute χ_r^2 for testing differences among treatment means, using the data of Exercises 8.3.1 and 8.3.2. Compare the results with those obtained by the standard analysis of variance procedure.

Exercise 21.4.2 Complete Brown's test for the data of Table 21.1 by finding the value of χ^2 and comparing with tabulated values. Compare the result with that obtained by Friedman's test.

Exercise 21.4.3 Use Brown's procedure and test differences among treatment means for the data of Exercises 8.3.1 and 8.3.2. Compare the results with those obtained by the analysis of variance and by Friedman's procedure.

21.5 Tests for the completely random design, two populations.
While the randomized complete-block design is probably the most used design,

NONPARAMETRIC STATISTICS

the completely random design is perhaps the simplest. Since it is used quite often and since tests for the resulting data are available, we shall consider its analysis.

We shall call the first test *Wilcoxon's two sample test*, though Wilcoxon (21.7) developed it for equal-sized samples. The test was extended to deal with unequal-sized samples by Mann and Whitney (21.3).

Wilcoxon's two sample test for unpaired observations is as follows where $n_1 \leq n_2$.
1. Rank the observations for both samples together from smallest to largest.
2. Add ranks for the smaller sample. Call this T.
3. Compute $T' = n_1(n_1 + n_2 + 1) - T$, the value you would get for the smaller sample if the observations had been ranked largest to smallest. (It is not the sum of the ranks for the other sample.)
4. Compare the smaller rank sum with tabulated values.

We now apply the test to the data of Table 5.2 on coefficients of digestibility for sheep (S) and steers (C). The observations are ordered and ranked as: 53.2 (S), 1; 53.6 (S), 2; 54.4 (S), 3; 56.2 (S), 4; 56.4 (S), 5; 57.8 (S), 6; 58.7 (C), 7; 59.2 (C), 8; 59.8 (C), 9; 61.9 (S), 10; 62.5 (C), 11; 63.1 (C), 12; 64.2 (C), 13. Tied observations, when they occur, are given the mean rank. The sums of the ranks for steers (C), the smaller sample, is $T = 60$. We also compute $T' = 6(6 + 7 + 1) - 60 = 24$.

The observed value of the test criterion is compared with the values in Table A.19, prepared by White (21.6). White also gives a table for $P = .001$. Again note that small values of this test criterion lead to rejection of the null hypothesis. Since the 5% value of the lesser rank sum is 27, we reject the null hypothesis. The analysis of variance gave a value of F which is just beyond the 1% level.

The difference between the two conclusions may result from one or more of several causes. First, if the assumptions underlying the analysis of variance are true, then we would expect it to be better able to detect real departures from the null hypothesis. Secondly, if the underlying assumptions are false, we may be detecting false assumptions rather than real differences. Tukey's test of additivity and the F test of homogeneity of variance (two-tailed F) may be used to examine the validity of the assumptions. In our example, the difference between conclusions seems trivial.

If the tables are inadequate, we may make use of the mean and standard deviation of T as given in Eqs. (21.4).

$$\mu_T = \frac{n_1(n_1 + n_2 + 1)}{2} \quad \text{and} \quad \sigma_T = \sqrt{\frac{n_1 n_2(n_1 + n_2 + 1)}{12}} \quad (21.4)$$

With these and T, we may compute the quantity $z = (T - \mu_T)/\sigma_T$ which is approximately normally distributed. Table A.4 may be used to judge significance.

A *median test* of the difference between population means is also available, as shown by Mood (21.4). The procedure follows.
1. Order the two samples as one from smallest to largest.
2. Find the median.
3. For each sample, observe the number of observations greater than the median.

4. Use these two numbers and the two sample sizes to complete a 2 × 2 contingency table.

5. Test significance by χ^2 with one degree of freedom if both sample sizes exceed 10; otherwise, Eq. (19.18) is appropriate, especially if the sum of the two sample sizes is small.

For the data used above, 58.7 is the median. There is 1 S greater than the median, and 5 C's. The 2 × 2 table is:

	S	C	
Above	1	5	6
n_i − above	5	2	7
	6	7	13

The total of the "n_i − above" row will equal $(n_1 + n_2 + 1)/2$ if $n_1 + n_2$ is odd, and $(n_1 + n_2)/2$ if $n_1 + n_2$ is even.

Exercise 21.5.1 Apply Wilcoxon's two-sample test to the data of Exercises 5.4.1 and 5.4.2. Compare your conclusions with those previously drawn.

Exercise 21.5.2 Complete the computations for the median test of the data discussed in the preceding section. Compare the results with those from the analysis of variance and Wilcoxon's test.

Exercise 21.5.3 Apply the median test to the data of Exercises 5.4.1 and 5.4.2. Compare your conclusions with those of the analysis of variance and Wilcoxon's test.

21.6 Tests for the completely random design, any number of populations. Kruskal and Wallis (21.2) have developed a test criterion based on ranks which is appropriate for the completely random design. As for the other rank tests, we assume that all populations sampled are continuous and identical, except possibly for location. The null hypothesis is that the populations all have the same location.

The procedure for applying the test follows.
1. Rank all observations together from smallest to largest.
2. Sum the ranks for each sample.
3. Compute the test criterion and compare with tabulated values.
The test criterion is given by Eq. (21.5).

$$H = \frac{12}{n(n+1)} \sum_i \frac{R_i^2}{n_i} - 3(n+1) \qquad (21.5)$$

Here n_i is the number of observations in the ith sample, $i = 1, \ldots, k$, $n = \Sigma n_i$, and R_i is the sum of the ranks for the ith sample. H is distributed as χ^2 with $k - 1$ degrees of freedom if the n_i are not too small. For $k = 2$, use Wilcoxon's test. For $k = 3$ and all combinations of the n_i's up to 5, 5, 5, a table of exact probabilities is given by Kruskal and Wallis (21.2). Ties are given the mean rank and, when this is necessary, a correction in H may be made. This correction will not ordinarily change the value of H appreciably.

It is given by Eq. (21.6).

$$\text{Divisor} = 1 - \frac{\Sigma T}{(n-1)n(n+1)} \qquad (21.6)$$

where $T = (t-1)t(t+1)$ for each group of ties and t is the number of tied observations in the group. This number is used as a divisor of H to give a corrected H.

We now apply the procedure to the data of Table 7.1. The observations and their ranks are (number or letter in parentheses refers to treatment): 9.1 (4), 1; 11.6 (13), 2; 11.8 (13), 3; 11.9 (4), 4; 14.2 (13), 5; 14.3 (13), 6; 14.4 (13), 7; 15.8 (4), 8; 16.9 (C), 9; 17.0 (4), 10; 17.3 (C), 11; 17.7 (5), 12; 18.6 (7), 13; 18.8 (7), 14; 19.1 (C), 15; 19.4 (4), 17; 19.4 (1), 17; 19.4 (C), 17; 20.5 (7), 19; 20.7 (7), 20; 20.8 (C), 21; 21.0 (7), 22; 24.3 (5), 23; 24.8 (5), 24; 25.2 (5), 25; 27.0 (1), 26; 27.9 (5), 27; 32.1 (1), 28; 32.6 (1), 29; 33.0 (1), 30. The sums of the ranks for each sample are: $R(1) = 130$, $R(5) = 111$, $R(4) = 40$, $R(7) = 88$, $R(13) = 23$, and $R(C) = 73$. Now

$$H = \frac{12}{30(31)} \frac{130^2 + \cdots + 73^2}{5} - 3(31) = 21.64 \text{ with } 6 - 1 = 5 \text{ df}$$

Since 21.64 is beyond the .005 probability, we reject the null hypothesis. The same conclusion was drawn when the analysis of variance was used.

The *median test* may also be applied to data from a completely random design. The procedure follows.

1. Rank all observations together from smallest to largest.
2. Find the median.
3. Find the number of observations above the median for each treatment.
4. Complete a $2 \times k$ table using the numbers obtained at step 3 and the differences between the n_i's and these numbers. (See the two-sample median test of Sec. 21.5.)
5. Compute χ^2 with $k - 1$ degrees of freedom for the contingency table obtained at step 4.

Exercise 21.6.1 Apply the H test to the data of Exercises 7.3.1 and 7.9.1. Compare the results with those obtained by the analysis of variance.

Exercise 21.6.2 Apply the correction for ties to the example worked in the preceding section. To H for the data of Exercise 7.3.1. Has the correction changed your conclusion appreciably?

Exercise 21.6.3 Apply the median test to the data used in the preceding section and to the data of Exercises 7.3.1 and 7.9.1. Compare your conclusions with those obtained by the analysis of variance and the H test.

Note. The χ^2 test applied to enumeration data may be considered to be a nonparametric test. It is partly for this reason that Secs. 20.4 and 20.6 were considered necessary.

21.7 Chebyshev's inequality.
This inequality states

$$P(|X - \mu| > k\sigma) \leq \frac{1}{k^2} \qquad (21.7)$$

Since X is simply a random variable, we may substitute $\bar{x}$ provided we replace

σ by $\sigma_{\bar{x}}$. Other substitutions are also possible. Since μ and σ are a part of the probability statement, it may be better described as distribution-free rather than nonparametric. The inequality is valid for any distribution with a finite variance.

Let us apply the inequality to the problem of determining whether the corn lines of Sec. 20.7 differ in mutation rate. The distribution may be binomial; in order to estimate a variance for the difference between the proportions, we will so assume. The standard deviation is

$$\hat{\sigma} = \sqrt{\frac{14}{11(10^5)} \frac{11(10^5) - 14}{11(10^5)} \left[\frac{1}{5(10^5)} + \frac{1}{6(10^5)} \right]} = \sqrt{\frac{14}{11(10^5)} \frac{11}{30(10^5)}} \text{ (approx)}$$

The null hypothesis is that μ, the difference between proportions, is zero. The observed value of X to be used in Eq. (21.7) is

$$X = \frac{10}{5(10^5)} - \frac{4}{6(10^5)} = \frac{40}{30(10^5)}$$

Now from the bracket part of Eq. (21.7), we have

$$\frac{40}{30(10^5)} > k \sqrt{\frac{14}{11(10^5)} \frac{11}{30(10^5)}}$$

If we solve for k, we get $k = 1.95$ and $1/k^2 = .26$. Hence, the probability of obtaining a difference in proportions larger than that obtained is less than .26.

If we had been prepared to assume a normal distribution, then k would have been a normal deviate and this two-tailed procedure would call for a corresponding probability of .05, a long way from .26. We see the importance of using a nonparametric procedure when it is impossible to put much reliance in any assumptions concerning an underlying distribution, and, at the same time, the importance of taking advantage of any reasonable assumptions. In addition, it appears that we will often have to rely upon an estimate of σ.

Chebyshev's inequality may also be used in determining sample size. Suppose we wish to estimate the mean number of colonies of a certain bacterium on a unit area of ham stored under normal conditions for a specified length of time. We wish our estimate to be within $\sigma/4$ of the true μ. How many samples are required if we wish to obtain this result with a probability of .90?

In terms of a probability statement we want

$$P\left(|\bar{x} - \mu| > \frac{\sigma}{4}\right) \leq .10$$

Relate this to Eq. (21.7), and we have

$$P\left(|\bar{x} - \mu| > \frac{\sqrt{n}}{4} \frac{\sigma}{\sqrt{n}}\right) \leq .10 = \frac{4^2}{n}$$

(Clearly $k = \sqrt{n}/4$ and $1/k^2 = 4^2/n$.) We must solve the right-hand equality for n. We obtain

$$n = \frac{4^2}{.10} = 160$$

The necessary sample size is 160. Here, we did not have to specify σ.

21.8 Spearman's coefficient of rank correlation.

The correlation coefficient r is applicable to the bivariate normal distribution, a distribution which is not too common. Several coefficients have been proposed that do not require the assumption of a bivariate normal distribution.

Spearman's coefficient of rank correlation applies to data in the form of ranks. The data may be collected as ranks or may be ranked after observation on some other scale. The procedure follows.

1. Rank the observations for each variable.
2. Obtain the differences in ranks for the paired observations. Let $d_i =$ the difference for the ith pair.
3. Estimate ρ by Eq. (21.8).
4. If the number of pairs is large, the estimate may be tested using the criterion given in Eq. (21.9).

Equations (21.8) and (21.9) follow.

$$r_s = 1 - \frac{6 \sum_i d_i^2}{(n-1)n(n+1)} \qquad (21.8)$$

where r_s is Spearman's rank correlation coefficient and n is the number of d's. The criterion

$$t = r_s \sqrt{\frac{n-2}{1-r_s^2}} \qquad (21.9)$$

is distributed as Student's t with $n - 2$ degrees of freedom.

We now apply the procedure to the data of Exercise 10.2.1 for the characters $T =$ tube length and $L =$ limb length. Ties will be given the mean rank.

T:	49,	44,	32,	42,	32,	53,	36,	39,	37,	45,	41,	48,	45,	39,	40,	34,	37,	35
Rank:	17,	13,	1.5,	12,	1.5,	18,	5,	8.5,	6.5,	14.5,	11,	16,	14.5,	8.5,	10,	3,	6.5,	4
L:	27,	24,	12,	22,	13,	29,.	14,	20,	16,	21,	22,	25,	23,	18,	20,	15,	20,	13
Rank:	17,	15,	1,	12.5,	2.5,	18,	4,	9,	6,	11,	12.5,	16,	14,	7,	9,	5,	9,	2.5
Difference:	0,	−2,	.5,	−.5,	−1,	0,	1,	−.5,	.5,	3.5,	−1.5,	0,	.5,	1.5,	1,	−2,	−2.5,	1.5

From Eq. (21.8),

$$r_s = 1 - \frac{6(37.50)}{17(18)19} = .9718$$

As a check, $\Sigma d_i = 0$. Also, r_s must lie between -1 and $+1$.

Obviously, this value of r_s is highly significant.

Another application is also available. In some instances, a set of objects may have a "true" order. For example, a set of paint chips may have an increasing amount of some color or a set of taste samples may have increasing amounts of a flavoring compound. If we ask a color or taste panelist to rank the set, we have a true standard with which to compare his rankings. Spearman's rank correlation is valid for rating his competence.

Exercise 21.8.1 Compute the value of t for the example in the preceding section.

Exercise 21.8.2 Compute values of r_s for T and N, and L and N as given in Exercise 10.2.1. Test these r_s's for significance. How do your estimates compare with those obtained when a bivariate normal distribution is assumed?

21.9 A corner test of association. Olmstead and Tukey (21.5) have developed the following nonparametric test for the association of two continuous variables. They have called it the "quadrant sum" test. It is computed as follows:
1. Plot the paired observations.
2. Draw the medians for each variable.
3. Beginning at the top, count down the number of observations (using the Y axis) which appear, until it is necessary to cross the vertical median. Record this number together with the sign of the quadrant.
4. Repeat as in step 3 from the right, using the horizontal median.
5. Repeat from the bottom and from the left.
6. Compute the quadrant sum and compare with tabulated values.

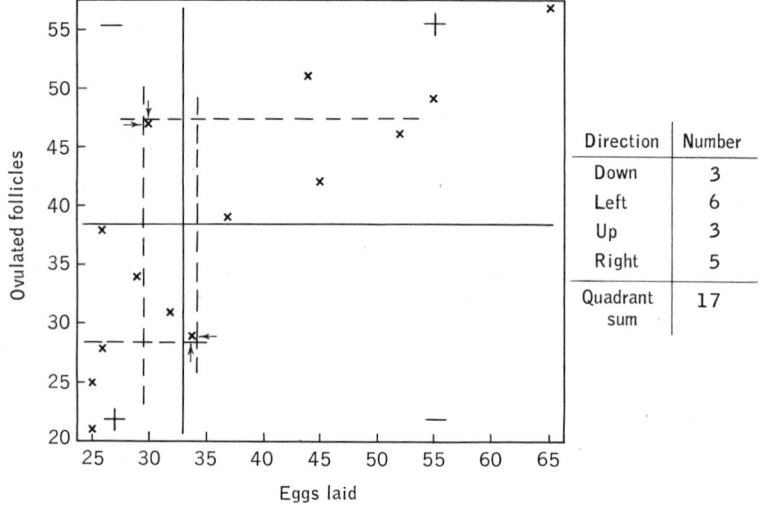

FIG. 21.1 The corner test of association

Table A.20 is appropriate for judging the significance of any quadrant sum.

An *even number of pairs* poses no problem in drawing medians and applying the test. For an *odd number of pairs*, each median passes through a point, presumably different. Call these points (X_m, Y) and (X, Y_m). For purposes of computing the quadrant sum, replace these two pairs by the single pair (X, Y). This leaves an even number of pairs.

We now apply the procedure to the data of Exercise 10.2.2. These data are plotted in Fig. 21.1 where the quadrant sum is also computed. We have used arrows in our figure to show each point which is the first one to be excluded as the result of having to cross a median. The quadrant sum is 17 and highly significant.

Ties occur under situations like the following. If we count down to the first observation that requires us to cross the vertical median, this point may have a Y value which is common to points on the other side. Thus for this

Y value, we have points which are favorable for inclusion in the quadrant sum and points which are not. Olmstead and Tukey (21.5) suggest that such tied groups be treated as if the number of points before crossing the median were

$$\frac{\text{Number favorable for inclusion}}{1 + \text{number unfavorable}}$$

It is now seen that the quadrant sum test is easy to apply and places special weight on extreme values of the variables.

References

21.1 Friedman, M.: "The use of ranks to avoid the assumption of normality implicit in the analysis of variance," *J. Am. Stat. Assoc.*, **32:** 675–701 (1937).

21.2 Kruskal, W. H., and W. A. Wallis: "Use of ranks in one-criterion variance analysis," *J. Am. Stat. Assoc.*, **47:** 583–621 (1952).

21.3 Mann, H. B., and D. R. Whitney: "On a test of whether one of two random variables is stochastically larger than the other," *Ann. Math. Stat.*, **18:** 50–60 (1947).

21.4 Mood, A. M.: *Introduction to the Theory of Statistics*, McGraw-Hill Book Company, Inc., New York, 1950.

21.5 Olmstead, P. S., and J. W. Tukey: "A corner test for association," *Ann. Math. Stat.*, **18:** 495–513 (1947).

21.6 White, C.: "The use of ranks in a test of significance for comparing two treatments," *Biometrics*, **8:** 33–41 (1952).

21.7 Wilcoxon, F.: "Individual comparisons by ranking methods," *Biometrics Bull.*, **1:** 80–83 (1945).

21.8 Wilcoxon, F.: "Probability tables for individual comparisons by ranking methods," *Biometrics Bull.*, **3:** 119–122 (1947).

21.9 Wilcoxon, F.: *Some Rapid Approximate Statistical Procedures*, American Cyanamid Company, Stamford, Conn., 1949.

Chapter 22

SAMPLING FINITE POPULATIONS

22.1 Introduction. The sampling discussed so far has been concerned with experiments in which the experimental units were not generally obtained by random procedures. Randomization was used to assign treatments to units and the populations were hypothetical ones, the units differing by random errors. We now turn our attention to populations which are no longer theoretical, but ones for which the experimental units can be enumerated and, consequently, randomly sampled. For example, silos are sampled for insecticide residues, soils for chemical analyses, plant populations for taxonomic purposes, fruit for quality, wheat fields for preharvest estimates of yield and quality, primitive races for many characteristics, people for opinions, and so on. In all these examples, populations of interest are finite populations.

A new problem arises in sampling finite populations. For example, if we want information from a sample of wholesale seed dealers in the state, we will be sampling a finite population. If our sample is large, say 25% of the wholesale seed dealers, we need to know if existing techniques are applicable or if it will be necessary to develop new ones.

In sampling *finite populations*, there are three fairly distinct ways in which selection may be made. These are:
1. Random sampling
2. Systematic sampling
3. Authoritative sampling

Random sampling will be our chief concern but first we shall briefly comment on systematic and authoritative sampling.

Systematic sampling is used when every kth individual in the population is included in the sample. Such a procedure is always very simple but clearly unsatisfactory if unrecognized trends or cycles are present in the population. Since populations must be listed before sampling, some relation between one or more of the characters being investigated and the order of listing may be introduced unconsciously. It is not generally safe to assume there is no such relation.

Systematic sampling can be conducted in such a way that an unbiased estimate of the sampling error may be obtained. This requires more than one systematic sample. For a single systematic sample, available formulas for estimating the variance of a mean require knowledge of the form of the population.

Authoritative sampling requires that some individual, who is well acquainted with the material to be sampled, draw the sample without regard to randomization. Such a procedure is completely dependent on the knowledge and skill of the sampler. It may produce good results in some instances but is rarely recommended.

Random sampling. Randomness may be introduced into the sampling procedure in a number of ways to give us various sample designs. Because of randomization, valid estimates of error are possible. Probability theory can be applied and objective conclusions drawn.

22.2 Organizing the survey. A considerable amount of effort must go into the planning and execution of a sample survey, quite apart from the actual sampling. We shall arbitrarily list five steps in the conduct of a survey.

1. Clarifying the objectives
2. Defining the sampling unit and population
3. Choosing the sample
4. Conducting the survey
5. Analyzing the data

These steps will now be discussed briefly.

1. *Clarifying the objectives.* This consists primarily of stating objectives as concisely as possible, each objective being stated as a hypothesis to be tested, a confidence interval to be computed, or a decision to be made.

With the primary objective in mind, we consider what data are to be collected. When there are several objectives in mind, we may have to modify our ideas as to what data are to be collected, in order to accomplish all objectives. We may even have to modify our objectives so that the survey will not become too complex and costly.

Usually, the survey group will have a fixed budget and will wish to maximize the amount of information per dollar spent. Alternatively, the objectives will include a statement of the amount of information desired, generally in the form of the size of a confidence interval, and we shall have to minimize the cost.

2. *Defining the sampling unit and population* has been done to some degree in step 1. For sampling, the population must be divided into distinct *sampling units* which, together, constitute the population. There may be a number of choices for sampling units. The final choice may be somewhat arbitrary but must be usable. If we are sampling people, we may choose the individual, the family, or the occupants of some specified living area as the sampling unit. Whatever the choice, it will be necessary to locate and identify the unit in the field.

For random sampling, we must be able to *list* all sampling units; we may even have to list all units. Existing lists, for example, lists of school children, farmers, etc., may need to be revised or new lists made, whichever seems more feasible and economical. Grids may be superimposed on maps of fields, forests, or other land areas when it is necessary to obtain crop samples or information about wildlife cover. Here we may need both ingenuity and arbitrariness, especially if we have an irregularly shaped area to sample.

While the sampling unit is being decided upon, it will be necessary to consider what is to be measured and what methods of measurement are to be used.

Are we measuring height, weight, or opinion? If so, how? Can a questionnaire be used to measure emotional stresses? If so, can interviewers be obtained from university students seeking part-time employment? Can these be trained in several days? Will a prospective engineer be as useful as a premed student?

3. *Choosing the sample.* Ways in which the sample may be drawn are termed *sample designs.* Sample designs will be discussed in later sections of this chapter.

The choice of sample size is related, in part, to available funds; if they are inadequate to permit a sample large enough to accomplish the stated objectives, the objectives should be revised or the survey delayed until adequate funds are available.

The sample design and size will give a fair idea of the extent and nature of the necessary tables and computations.

4. *Conducting the survey.* Probably, it will be necessary to train some of the personnel in order to have uniformity in locating and identifying the sampling units and in recording responses to questionnaires or other data. A time schedule will be necessary. Some scheme is generally required for early checking on the validity of the data recorded on the various forms. Provision must be made for making quick decisions when unforeseen events arise.

5. *Analyzing the data.* First, it will be necessary to edit the data for recording errors and invalidity. Finally, the survey should be reviewed for possible means of improving future surveys.

22.3 Probability sampling. Suppose our population is clearly defined and a listing of sampling units has been made. Now, we can also list all possible samples. We shall use the term *probability sampling* when

1. Each sampling unit has or is assigned a known probability of being in the sample.

2. There is random selection at some stage of the sampling procedure and it is directly related to the known probabilities. Random selection will involve a mechanical procedure for choosing the units to be included in the sample.

3. The method for computing any estimate of a mean is clearly stated and will lead to a single value of the estimate. This is a part of the analysis of the data. In estimating any mean, we shall use the selection probabilities assigned to the sampling units. These will provide weights, each of which will be some constant multiple of the reciprocal of the probability.

When these criteria are satisfied, a probability of selection can be assigned to each sample and to each estimate. Hence, we can construct a probability distribution of the estimates given by our sampling plan. In this way, we can evaluate the worth of our plan and compare it with other probability sampling plans. Evaluation consists of measuring the accuracy of any estimate by the size of its standard deviation.

When the probabilities assigned to each sampling unit are all equal, then the weights to be used in computing estimates of means are all equal. We do not need to think consciously of the weights because the sample is *self-weighting.* While such samples are easy to analyze, they lack certain advantages possessed by other probability sampling plans, advantages such as ease and low cost of administration per unit of information and the ability to obtain estimates for individual strata (see Sec. 22.5).

SAMPLING FINITE POPULATIONS

A probability sample does not assure us that all our estimates will be unbiased. We have already seen that, in random sampling from a normal population, we choose to use $s = \sqrt{\Sigma x^2/(n-1)}$ although it is a biased estimate of σ. Biased estimates are also used in sample surveys. They must, of course, be used with care since they may introduce distortion into probability statements. In particular, when biased estimates are averaged (not necessarily arithmetically), the effect on the average and its ultimate use may not be apparent. Several types of probability sampling will now be discussed.

22.4 Simple random sampling. For random sampling, the population is listed and the plan and sample size fixed. For *simple random sampling*, each possible sample has the same probability of being selected. This is the important criterion.

In the actual process of selecting the sampling units from a finite population, a table of random numbers is used and sampling is *without replacement*. Apart from this, the sampling units are drawn independently.

Notation and definitions. Because we are now dealing mainly with finite populations, some new notation and definitions are required. Notation and definitions will not be found to be completely consistent in sampling literature. We shall attempt to use capital letters for population quantities and lower-case letters for sample quantities; σ^2 is also used.

To begin, let Y_i be the ith observation in the population. We also use Y_i as the ith sample observation when there is not likely to be confusion.

Population size: N
Sample size: n
Population mean:

$$\bar{Y} = \frac{\sum_i Y_i}{N} = \frac{Y}{N}, \quad \text{continuous variable}$$

$$P = \frac{A}{N}, \text{ proportion}$$

For a proportion, $Y_i = 0$ or 1; $\Sigma Y_i/N$ is the proportion of individuals possessing a specified characteristic, so can also serve as a definition of the population mean. It is more common to replace $\sum_i Y_i$ by A. For a percentage, $100P$ is the appropriate mean.

Sample mean:

$$\hat{\bar{Y}} = \bar{y} = \frac{\sum_i Y_i}{n} = \frac{y}{n}, \quad \text{continuous variable}$$

$$\hat{P} = p = \frac{a}{n}, \text{ proportion}$$

For a proportion, a replaces $\sum_i Y_i$. Since population totals and their estimates are often of interest, the quantities Y, A, y, a are fairly common. Notice that y is no longer a deviation but is, rather, a total.

Population variance:

$$\sigma^2 = \frac{\sum_i (Y_i - \bar{Y})^2}{N}$$

$$S^2 = \frac{\sum_i (Y_i - \bar{Y})^2}{N - 1} \tag{22.1}$$

We shall use Eq. (22.1) to define the population variance, because our definition of s^2, Eq. (22.4), gives an unbiased estimate of S^2.

Population variance of a mean:

$$S_{\bar{y}}^2 = \frac{S^2}{n}\left(\frac{N-n}{N}\right) \tag{22.2}$$

$$S_p^2 = \frac{PQ}{n}\left(\frac{N-n}{N-1}\right) \quad \text{where } Q = 1 - P \tag{22.3}$$

Sample variance:

$$s^2 = \frac{\sum_i (Y_i - \bar{y})^2}{n - 1}, \quad \text{an unbiased estimate of } S^2 \tag{22.4}$$

The numerator will be computed as $\Sigma Y_i^2 - (\Sigma Y_i)^2/n$.

Sample variance of a mean:

$$s_{\bar{y}}^2 = \frac{s^2}{n}\left(\frac{N-n}{N}\right) \tag{22.5}$$

$$s_p^2 = \frac{pq}{n-1} \frac{N-n}{N} \quad \text{where } q = 1 - p \tag{22.6}$$

Equation (22.6) gives an unbiased estimate of S_p^2 but is not generally used when computing confidence intervals. The more familiar form is implied in Eq. (22.8).

The quantity $(N - n)/N$ is known as the *finite population correction* or *fpc*. It may also be written as $1 - n/N$ and n/N is called the *sampling fraction*. If the sampling fraction is small, say less than 5%, it may be neglected. It is of interest to note that $pq/(n - 1)$ is an unbiased estimate of the population variance regardless of whether or not the population is finite. In other words, we use a biased estimate of the population variance in Chaps. 18 to 20 when we use $\hat{p}(1 - \hat{p})/n$. (Recall that in Chaps. 18 to 20 p is used as the parameter, $\hat{p}$ as the estimate.)

The confidence interval for a mean is given by Eq. (22.7). Note that it makes use of the *fpc*.

$$\text{CI} = \bar{y} \pm t\left[\frac{s}{\sqrt{n}} \sqrt{\frac{N-n}{N}}\right] \tag{22.7}$$

Obviously, we are assuming that $\bar{y}$ is normally distributed, knowing that the population of Y's is not normal since it is a finite population. Moreover, it is sampled without replacement.

The confidence interval for a proportion requires use of the hypergeometric distribution, Sec. 20.2, if it is to be completely valid. The interested investigator is referred to charts by Chung and DeLury (22.3). A common approximation is given by Eq. (22.8).

$$\text{CI} = p \pm t\sqrt{\frac{pq}{n}\frac{N-n}{N-1}} \qquad (22.8)$$

Notice that the estimated standard deviation is not that given by Eq. (22.6) but is comparable to the population quantity given as Eq. (22.3). The estimate used in Eq. (22.8) is the more common. Table 18.1 may be used to judge the appropriateness of Eq. (22.8).

Simple random sampling is used when the population is known to be not highly variable or when the true proportion lies between 20 and 80%. When there is considerable variation, the sampling units should be grouped into strata in such a way that variation within strata can be expected to be less than variation among strata. This leads to stratified sampling. Much the same idea leads to an among-groups and within-groups analysis of variance.

Exercise 22.4.1 Consider the finite population consisting of the numbers $1, 2, \ldots, 6$. Compute the mean and variance. Consider all possible samples of two observations when this population is sampled without replacement. Make a table of sample means and variances and the frequency with which each value occurs. Show that the sample mean and variance of Eq. (22.5) are unbiased estimates of the population mean and variance of Eq. (22.1). If the exercise had said "sample with replacement," what changes would be necessary in the computations?

Exercise 22.4.2 A buying population consisting of 6,000 furniture buyers is to be sampled, by mailed questionnaire, concerning an appliance preference. A random sample of 250 individuals is drawn and the questionnaire is mailed. Since the preference involves an inexpensive attachment, provision is made for Yes and No responses only. All questionnaires are returned and 187 Yeses counted. Estimate the true proportion of Yeses in the population by means of a 95% confidence interval.

Exercise 22.4.3 For the same population as in Exercise 22.4.2, a lengthy questionnaire was sent to a random sample of 750 buyers. Only 469 questionnaires were returned. Estimate the true proportion of *respondents* in the population by means of a 90% confidence interval.

22.5 Stratified sampling. The estimated variance of a population mean is given by Eq. (22.5) and, for a proportion, by Eq. (22.6) or the more frequent alternative implied by Eq. (22.8). To decrease the length of the confidence interval which estimates the population mean, we may increase n or decrease the population variance. Obviously, both possibilities must be considered.

The obvious way to decrease a population variance is to construct *strata* from the sampling units, the total variation being partitioned in such a way that as much as possible is assigned to differences among strata. In this way, variation within strata is kept small. Variation among the strata means in the population does not contribute to the sampling error of the estimate, Eq. (22.15), of the population mean.

Reduction in the variation of the estimate of the population mean is a very important reason for stratification. However, many surveys involve several variables and good stratification for one variable may not be so for another.

Thus we find that strata are often constructed on a purely geographical basis. This generally works out well and we find that townships, counties, and land resource areas are often used as strata. This kind of stratification is often convenient for administrative reasons since it may be possible to obtain the cooperation of town, county, or other conveniently located agencies.

In addition to increasing the precision with which means are measured, stratification permits an efficient job of allocating resources since we may use any method for deciding how many sampling units are to be taken from each stratum. It is assumed that each stratum will be sampled. Estimates of strata means are often desired and, in such cases, stratification is essential.

Notation and definitions. Notation and definitions for stratified random sampling are fairly obvious relatives and extensions of what was given in Sec. 22.4 under the same heading.

Let Y_{ki} be the ith observation in the kth stratum, $k = 1, \ldots, s$. Strata sizes, means, and variances will be designated by N_k, $\bar{Y}_k$, or P_k, and S_k^2 with corresponding sample values of n_k, $\bar{y}_k$ or p_k, and s_k^2.

Stratum mean and variance:

$$\bar{Y}_k = \frac{\sum_{i=1}^{N_k} Y_{ki}}{N_k} = \frac{Y_k}{N_k}$$

$$P_k = \frac{A_k}{N_k}$$

$$S_k^2 = \frac{\sum_{i=1}^{N_k} (Y_{ki} - \bar{Y}_k)^2}{N_k - 1}$$

from Eq. (22.1).

Sample mean and variance for kth stratum:

$$\hat{\bar{Y}}_k = \bar{y}_k = \frac{\sum_{i=1}^{n_k} Y_{ki}}{n_k} = \frac{y_k}{n_k}$$

$$\hat{P}_k = p_k = \frac{a_k}{n_k}$$

$$s_k^2 = \frac{\sum_{i=1}^{n_k} (Y_{ki} - \bar{y}_k)^2}{n_k - 1}$$

from Eq. (22.4).

Parameters and statistics for the complete population are also required. Let $N = \sum_k N_k$ and $n = \sum_k n_k$. The ratio N_k/N occurs frequently enough that we give it the symbol W_k, that is, $W_k = N_k/N$ where W stands for weight.

Population mean (st means stratified):

$$\bar{Y}_{st} = \frac{\sum_k N_k \bar{Y}_k}{N} = \sum_k W_k \bar{Y}_k \qquad (22.9)$$

$$P_{st} = \frac{\sum_k N_k P_k}{N} = \sum_k W_k P_k \qquad (22.10)$$

Estimate of population mean:

$$\hat{\bar{Y}}_{st} = \bar{y}_{st} = \frac{\sum_k N_k \bar{y}_k}{N} = \sum_k W_k \bar{y}_k \qquad (22.11)$$

$$\hat{P}_{st} = p_{st} = \frac{\sum_k N_k p_k}{N} = \sum_k W_k p_k \qquad (22.12)$$

(Sample means are $\bar{y} = \Sigma n_k \bar{y}_k / n$ and $p = \sum_k n_k p_k / n$.)

Variance of the estimate of the population mean:

$$\begin{aligned}\sigma^2(\bar{y}_{st}) &= \sum_k \left(\frac{N_k}{N}\right)^2 \frac{S_k^2}{n_k} \frac{N_k - n_k}{N_k} \\ &= \frac{1}{N^2} \Sigma N_k (N_k - n_k) \frac{S_k^2}{n_k}\end{aligned} \qquad (22.13)$$

Compare Eq. (22.5) with the first expression for $\sigma^2(\bar{y}_{st})$.

$$\sigma^2(p_{st}) = \sum_k \frac{N_k^2}{N^2} \frac{P_k Q_k}{n_k} \frac{N_k - n_k}{N_k - 1} \qquad (22.14)$$

Compare the variance given in Eq. (22.3) with $\sigma^2(p_{st})$.
Sample variance of the estimate of the population mean:

$$\begin{aligned}s^2(\bar{y}_{st}) &= \frac{1}{N^2} \sum_k N_k (N_k - n_k) \frac{s_k^2}{n_k} \\ &= \sum_k W_k^2 \frac{s_k^2}{n_k} - \frac{1}{N} \sum_k W_k s_k^2\end{aligned} \qquad (22.15)$$

The first form is obtained from Eq. (22.13) by substituting estimates for parameters. It is an unbiased estimate of $\sigma^2(\bar{y}_{st})$. The second form may be used for computing.

$$s^2(p_{st}) = \sum_k W_k^2 \frac{p_k q_k}{n_k} \frac{N_k - n_k}{N_k - 1} \qquad (22.16)$$

This equation is obtained from Eq. (22.14). It gives a biased estimate of $\sigma^2(p_{st})$ but is commonly used.

In all formulas in which the finite population correction or *fpc* appears, it is ignored if small when confidence intervals are being computed.

In estimating the population mean, weights are used. For this reason, the estimate of the population mean and the sample mean, $\bar{y}_{st}$ and $\bar{y}$, respectively, need not be the same. However, when $n_1/N_1 = \cdots = n_s/N_s = n/N$, then $\bar{y}_{st} = \bar{y}$. This is called *proportional allocation* and the sample is said to be *self-weighting*.

When proportional allocation is used and the *within strata* variances are homogeneous, the results relative to variances may be summarized in an analysis of variance table with sources of variation for total, among strata, and within strata. The value of the particular stratification can be estimated by comparing the standard deviation of $\bar{y}_{st}$, as computed from the within-strata mean square, with that of $\bar{y}$, as computed from the total mean square. Sound procedures are given by Cochran (22.2) and by Hansen et al. (22.4).

Exercise 22.5.1 The problem of *nonrespondents* plagues users of mailed questionnaires and even interviewers. Suppose 900 buyers are chosen randomly from a population of 6,000. Replies are received from 250 and, of these, 195 favor one suggestion. Estimate the proportion of "favorables" in the population consisting of respondents; do this by means of a 95% confidence interval.

From the known nonrespondents, a random sample of 50 is drawn and interviewed. These show 30 "favorables" in the sample from the population of nonrespondents. Estimate, by means of a 95% confidence interval, the proportion of "favorables" in this population.

Response and nonresponse to mailed questionnaires are sometimes used as a criterion for stratification. Suppose the above results are considered to be from such strata. Estimate the population proportion of "favorables" and the standard deviation of the estimate. Give one practical and two theoretical criticisms of your procedure.

How might you use the sample results to test whether or not the two strata differed in response to the suggestion?

22.6 Optimum allocation. Stratification generally results in decreasing the variance of the estimate of the population mean. However, proportional allocation is not always *optimum allocation* and, for this, a *variable sampling fraction* may be necessary.

Fixed cost. In some sampling experiments, the cost of obtaining an observation from a sampling unit does not vary to any extent from one stratum to another and can be ignored when determining sampling fractions for the various strata. The problem is to minimize $\sigma^2(\bar{y}_{st})$ as given by Eq. (22.13) or $\sigma^2(p_{st})$ as given by Eq. (22.14). Sampling fractions are determined by the size of the strata and their variability and it is fairly clear that a larger stratum will call for a larger number of observations, as will a stratum with a high variability. It has been shown that the optimum allocation is obtained when the number of observations made in any stratum is determined by Eq. (22.17).

$$n_k = n \frac{N_k S_k}{\sum_k N_k S_k} \tag{22.17}$$

Note that the denominator is the sum over all strata.

While application of this formula calls for the parameters S_k, $k = 1, \ldots, s$, it will often be necessary to use estimates. Thus, if sample information is desired for the years between censuses, standard deviations from the nearest preceding census can be used. In other cases, it may be necessary to make use of information from related sample surveys. When no estimates of the S_k's are available, proportional allocation is recommended.

Occasionally, Eq. (22.17) will give one or more n_k values which are greater than their corresponding N_k's. In such cases, 100% sampling is done in these strata and the remaining sampling fractions are adjusted so that the total sample is of the size originally planned. For example, if $n_s > N_s$ on application of Eq. (22.17), then we set $n_s = N_s$ for sampling purposes and recompute the remaining n_k's by the equation

$$n_k = \frac{(n - N_s) N_k S_k}{\sum_{k=1}^{s-1} N_k S_k}, \quad k = 1, \ldots, s - 1$$

When stratified sampling is for *proportions*, the n_k may be determined by Eq. (22.18).

$$n_k = n \frac{N_k \sqrt{P_k Q_k}}{\sum_k N_k \sqrt{P_k Q_k}} \quad (22.18)$$

This equation is an approximation; it is similar to Eq. (22.17) although the former is not an approximation.

Gains in precision as a result of using optimum rather than proportional allocation are not likely to be as great for estimating proportions as for estimating means of continuous variables. Since proportional allocation gives the convenience of self-weighting samples, it is usually recommended when proportions are to be estimated.

Variable cost. When the cost of obtaining an observation varies from stratum to stratum, some *cost function* is needed to give the total cost. A simple cost function is given by Eq. (22.19).

$$\text{Cost} = C = a + \Sigma c_k n_k \quad (22.19)$$

where a is a fixed cost, regardless of the allocation of the sampling to the strata, and c_k represents the cost per observation in the kth stratum. For this cost function, the minimum $\sigma^2(\bar{y}_{st})$ is obtained if we take a large sample in a large stratum, a large sample when the stratum variance is high, and a small sample when the stratum cost is high. In other words, the sample size for any stratum is proportional to $N_k S_k / \sqrt{c_k}$.

In the actual survey, we may have to work with a fixed budget or may be required to estimate the variance of the population mean with a specified degree of accuracy. The latter requirement determines the sample size and, in turn, the budget.

For a *fixed budget*, Eq. (22.20) gives the optimum sample size for each stratum.

$$n_k = \frac{N_k S_k / \sqrt{c_k} (C - a)}{\sum_k N_k S_k \sqrt{c_k}} \quad (22.20)$$

For a *fixed variance*, Eq. (22.21) gives the optimum sample size for each stratum. Here, we minimize the cost for a fixed or predetermined variance.

$$n_k = \frac{N_k S_k}{\sqrt{c_k}} \frac{\sum_k N_k S_k \sqrt{c_k}}{N^2 \sigma^2(\bar{y}_{st}) + \sum_k N_k S_k^2}$$

$$= \frac{W_k S_k}{\sqrt{c_k}} \frac{\sum_k W_k S_k \sqrt{c_k}}{\sigma^2(\bar{y}_{st}) + \sum_k W_k S_k^2/N}$$

(22.21)

where $W_k = N_k/N$.

Rough estimates of the S_k^2's and c_k's are usually quite adequate for estimating optimum sample sizes for the strata. When sampling is to estimate *proportions*, S_k may be replaced by $\sqrt{P_k Q_k}$ in Eqs. (22.20) and (22.21) to give approximately optimum n_k's.

Exercise 22.6.1 Show that Eqs. (22.17) and (22.20) give the same results as proportional allocation when the strata variances are homogeneous and the cost c_k does not vary.

Exercise 22.6.2 The following data are from R. J. Jessen, "Statistical investigation of a sample survey for obtaining farm facts," *Iowa Agr. Expt. Sta. Research Bull. No.* 304, 1942, Ames, Iowa. The strata are type of farming areas, N_k is number of rural farms, $S_k^2(1)$ is a variance for number of swine, and $S_k^2(2)$ is a variance for number of sheep. The N_k's are 1939 census data but the S_k^2's are only estimates obtained from a 1939 sample.

For each set of data, compute sample sizes using proportional and optimum allocation for a total sample size of 800. Compare the results.

Stratum	1	2	3	4	5	State
N_k	39,574	38,412	44,017	36,935	41,832	200,770
$S_k^2(1)$	1,926	2,352	2,767	1,967	2,235	2,303
$S_k^2(2)$	764	20	618	209	87	235

22.7 Multistage or cluster sampling.

In some sampling schemes, the sampling units are in groups of equal or unequal sizes and the groups, rather than the units, are randomly sampled. Such groups are called *primary sampling units*, or *psu*'s. Observations may be obtained on all elementary units or these, in turn, may be sampled. For example, we may be interested in individuals, the elementary units, but may obtain them by drawing a random sample of families, the primary sampling unit, and observing all units within. This would be a *simple cluster sampling* plan or a *single-stage* sampling plan. In sampling the soil in a field to be used for an experiment, we might divide the field into experimental plots, place a grid over each plot to define the sampling units, then obtain several observations from each plot. This would be *two-stage* sampling or *subsampling* where the first stage was essentially a census.

Many types of cluster sampling can obviously be devised. Most will have some obvious advantages related to cost or practicability, since the cost of

SAMPLING FINITE POPULATIONS

getting from one *psu* to another is likely to be larger than that of getting from one subunit to another and since identifying the *psu* may be simpler than identifying the subunit. When a cluster is defined by association with an area, we have *area sampling*. For example, we might sample quarter sections of land, the cluster or *psu*, and enumerate all farms in the *psu*.

Suppose a population consists of N *psu*'s from which we draw a random sample of size n; each *psu* consists of M subunits from which we draw a sample of size m for each of the n *psu*'s. (The letters M and N, and m and n are often interchanged in sampling literature.) Then there are MN *elements* in the population and mn in the sample. The plan is two-stage sampling or subsampling.

Computations are usually carried out on a per-element basis just as we are accustomed to in the analysis of variance. An observation is denoted by Y_{ij}, where j refers to the element and i to the *psu*. The mean of all elements in a *psu* is designated by $\bar{Y}_{i.}$ or simply $\bar{Y}_i$, and the population mean by $\bar{\bar{Y}}_{..}$ or simply $\bar{\bar{Y}}$. Corresponding sample means are given the symbols $\bar{y}_{i.}$ or $\bar{y}_i$ and $\bar{y}_{..}$ or $\bar{y}$.

Let us now assume that N and M are infinite and define an element by Eq. (22.22).

$$Y_{ij} = \bar{\bar{Y}} + \delta_i + \epsilon_{ij} \tag{22.22}$$

This is a linear model such as was discussed in Secs. 7.12 and 7.13 with slightly different notation. If we denote the variance of the δ's by S_a^2 and that of the ϵ's by S_w^2, then sample mean squares defined as in Table 22.1 are estimates of

TABLE 22.1
ANALYSIS OF VARIANCE AND AVERAGE VALUES IN TWO-STAGE SAMPLING

Source of variation	df	Mean square	Average value of mean square
Among *psu*'s	$n - 1$	$s_a^2 = \dfrac{m \sum_i (\bar{y}_i - \bar{y})^2}{n - 1}$	$S_w^2 + mS_a^2$
Within *psu*'s	$n(m - 1)$	$s_w^2 = \dfrac{\sum_i \sum_j (Y_{ij} - \bar{y}_i)^2}{n(m - 1)}$	S_w^2
Total	$nm - 1$	$\dfrac{\sum_{i,j} (Y_{ij} - \bar{y})^2}{nm - 1}$	

the quantities given in the average value column. Notice that s_a^2 is not intended to be an estimate of S_a^2.

The sums of squares are usually computed from the following computation formulas rather than from the definition formulas of Table 22.1.

Among *psu*'s: $\quad (n - 1)s_a^2 = \dfrac{\sum_i Y_{i.}^2}{m} - \dfrac{Y_{..}^2}{nm}$

Within *psu*'s: $n(m-1)s_w^2 = \sum_i \left(\sum_{j=1}^m Y_{ij}^2 - \frac{Y_{i.}^2}{m} \right)$

or $\qquad\qquad\qquad\qquad =$ total SS $-$ among *psu*'s SS

$$\text{Total SS} = \sum_{i,j} Y_{ij}^2 - \frac{Y_{..}^2}{nm}$$

If we relate the analysis of variance given in Table 22.1 to Secs. 7.11 and 7.13, we see that the mean square within *psu*'s may be called the *sampling error* and the mean square among *psu*'s may be called the *experimental error*.

Experimental error rather than sampling error is appropriate in terms of estimating $\bar{\bar{Y}}$ by means of a confidence interval. Experimental error is based on the unit chosen at random at the first stage of sampling and corresponds to the plot to which a treatment is applied at random in a field or laboratory experiment. In the usual course of events, we might expect sampling error to be smaller than experimental error because we expect more homogeneity within *psu*'s than among *psu*'s. Hence, sampling error would not be appropriate for computing a confidence interval for $\bar{\bar{Y}}$.

The variance of the sampling mean, $s^2(y)$, is estimated by s_a^2/nm, an unbiased estimate of the true variance. We can now estimate both S_w^2 and S_a^2 and construct estimates of the variances of treatment means for different allocations of our efforts. Thus, for the present scheme,

$$\sigma^2(\bar{y}) = \frac{S_w^2}{nm} + \frac{S_a^2}{n}$$

We do not decrease S_a^2/n by taking more subsamples, yet S_a^2 is likely to be the larger contributor to $\sigma^2(\bar{y})$. If we were to increase n, we would decrease both contributions. Hence, in theory, the best allocation of our effort is to take as many *psu*'s as possible and very few elements within each *psu* if they require an appreciable effort; of course, we require two such elements from each *psu* if we have to estimate either S_w^2 or S_a^2 and retain computational ease.

Theory is also available for finite populations and when the *psu*'s differ in the number of elements they contain. The interested reader is referred to Cochran (22.2) and to Hansen et al. (22.4).

In sampling for proportions with N clusters and M elements per cluster, suppose we draw n clusters and completely enumerate them. Then an observed proportion is a true proportion P_i for the ith cluster and is not subject to sampling variation. We estimate the population proportion by

$$\hat{P} = p_{nM} = \frac{\sum_i P_i}{n}$$

and its variance by

$$s^2(p_{nM}) = \frac{N-n}{N} \frac{1}{n} \frac{\sum_i (P_i - p_{nM})^2}{n-1}$$

SAMPLING FINITE POPULATIONS

When the sampling scheme involves taking only m of the M elements in a cluster, then the P_i are only estimated and a term for sampling variation within clusters must be introduced into the variance of the estimate of the population proportion P. We now have

$$\hat{P} = p_{nm} = \bar{p} = \frac{\sum_i p_i}{n}$$

and $$s^2(\bar{p}) = \frac{M-m}{M-1} \frac{m}{m-1} \frac{1}{Nnm} \sum_i p_i q_i + \frac{N-n}{N} \frac{1}{n} \frac{\sum_i (p_i - \bar{p})^2}{n-1}$$

Ladell (22.5) and Cochran (22.1) describe an interesting sampling experiment where local control, in the form of a restriction on the subsampling, was imposed. The particular example involved superimposing a Latin-square design, initially without treatments, on an experimental area and then obtaining six soil samples from each plot. Wireworm counts were made on the soil samples. It was so arranged that three samples were obtained from each of the north and south halves of the plot. Consequently, nonrandom differences in numbers of wireworms between halves do not influence treatment comparisons or experimental error. The results are shown in Table 22.2;

TABLE 22.2
ANALYSIS OF VARIANCE OF WIREWORM DATA

Source	df	Sum of squares	Mean square
Rows	4	515.44	128.86
Columns	4	523.44	130.86
Experimental error	16	712.16	44.51
Between half plots	25	2,269.00	90.76
Sampling error	100	3,844.00	38.44
Totals	149	7,864.04	

the usual "experimental error" and "treatments" are pooled here because there were no true treatments. It is readily apparent that "local control" greatly increased the precision of the experiment.

Exercise 22.7.1 Suppose that we wish to have a preharvest estimate of the wheat yield for a wheat-growing state. The area in wheat is divided, for sampling purposes, into 1-acre plots. A random sample of 250 plots is drawn and two subsamples are obtained from each of the 250 plots. Each subsample is 2 ft square, approximately 1/10,000 acre, so that finite sampling theory need not be applied.

The analysis of variance gives a sampling error of 20 (within psu's) and an experimental error of 70 (among psu's). (Yields were converted to bushels per acre.) Write out the analysis of variance. Estimate components of variance. Compute the variance of a treatment mean. Estimate the variance of a treatment mean assuming that the rate of subsampling is doubled (four subsamples instead of two). Does this give an appreciable gain in precision? (Express the estimated variance as a percentage of that observed.)

Exercise 22.7.2 Use the data of Table 22.2 to compute the experimental error as if no design had been superimposed on the plots. (Use a weighted average of row, column, and experimental error mean squares.) What would sampling error have been if no local control had been used? (Use a weighted average of between half plots and sampling error mean squares.) What would experimental error have been without the design and without local control? (Add the values just computed for experimental and sampling errors. This includes sampling error without local control twice so that sampling error with local control must now be subtracted.) Compute standard deviations from each of the three variances you have just computed.

Exercise 22.7.3 What is the minimum number of subsamples per half plot with local control if it is required to estimate sampling error and, at the same time, not destroy computational ease?

References

22.1 Cochran, W. G.: "The information supplied by the sampling results," *Ann. Appl. Biol.*, **25:** 383–389 (1938).

22.2 Cochran, W. G.: *Sampling Techniques*, John Wiley & Sons, Inc., New York, 1953.

22.3 Chung, J. H., and D. B. DeLury: *Confidence Limits for the Hypergeometric Distribution*, University of Toronto Press, Toronto, Ontario, 1950.

22.4 Hansen, M. H., W. N. Hurwitz, and W. G. Madow: *Sample Survey Methods and Theory*, 2 vols., John Wiley & Sons, Inc., New York, 1953.

22.5 Ladell, W. R. S.: "Field experiments on the control of wireworms," *Ann. Appl. Biol.*, **25:** 341–382 (1938).

APPENDIX

List of Tables

A.1.	Ten thousand random digits	428
A.2.	Values of the ratio, range divided by the standard deviation σ, for sample sizes from 20 to 1,000	432
A.3.	Values of t	433
A.4.	Probability of a random value of $z = (X - \mu)/\sigma$ being greater than the values tabulated in the margins	434
A.5.	Values of χ^2	435
A.6.	Values of F	436
A.7.	Significant studentized ranges for 5% and 1% level new multiple-range test	442
A.8.	Upper percentage points of the studentized range, $q_\alpha = (\bar{x}_{max} - \bar{x}_{min})/s_{\bar{x}}$	444
A.9.	Table of t for one-sided and two-sided comparisons between p treatment means and a control for a joint confidence coefficient of $P = .95$ and $P = .99$	446
A.10.	The arcsin $\sqrt{\text{percentage}}$ transformation	448
A.11.	Confidence belts for the correlation coefficient ρ: $\rho = .95$ and $\rho = .99$	450
A.12.	Transformation of r to z	452
A.13.	Significant values of r and R	453
A.14.	Binomial confidence limits	454
A.15.	Confidence belts for proportions: confidence coefficients of .95 and .99	458
A.16.	Mosteller-Tukey binomial probability paper	460
A.17.	Sample size and the probability of making wrong decisions for a limited set of alternatives	461
A.18.	Wilcoxon's signed rank test	469
A.19.	Critical points of rank sums	470
A.20.	Working significance levels for magnitudes of quadrant sums	471

Greek Alphabet

(Letter and Name)

A	α	Alpha	H	η	Eta	N	ν	Nu	T	τ	Tau
B	β	Beta	Θ	θ	Theta	Ξ	ξ	Xi	Υ	υ	Upsilon
Γ	γ	Gamma	I	ι	Iota	O	o	Omicron	Φ	ϕ	Phi
Δ	δ	Delta	K	κ	Kappa	Π	π	Pi	X	χ	Chi
E	ϵ	Epsilon	Λ	λ	Lambda	P	ρ	Rho	Ψ	ψ	Psi
Z	ζ	Zeta	M	μ	Mu	Σ	σ	Sigma	Ω	ω	Omega

Table A.1
Ten Thousand Random Digits

	00–04	05–09	10–14	15–19	20–24	25–29	30–34	35–39	40–44	45–49
00	88758	66605	33843	43623	62774	25517	09560	41880	85126	60755
01	35661	42832	16240	77410	20686	26656	59698	86241	13152	49187
02	26335	03771	46115	88133	40721	06787	95962	60841	91788	86386
03	60826	74718	56527	29508	91975	13695	25215	72237	06337	73439
04	95044	99896	13763	31764	93970	60987	14692	71039	34165	21297
05	83746	47694	06143	42741	38338	97694	69300	99864	19641	15083
06	27998	42562	63402	10056	81668	48744	08400	83124	19896	18805
07	82685	32323	74625	14510	85927	28017	80588	14756	54937	76379
08	18386	13862	10988	04197	13770	72757	71418	81133	69503	44037
09	21717	13141	22707	68165	58440	19187	08421	23872	03036	34208
10	18446	83052	31842	08634	11887	86070	08464	20565	74390	36541
11	66027	75177	47398	66423	70160	16232	67343	36205	50036	59411
12	51420	96779	54309	87456	78967	79638	68869	49062	02196	55109
13	27045	62626	73159	91149	96509	44204	92237	29969	49315	11804
14	13094	17725	14103	00067	68843	63565	93578	24756	10814	15185
15	92382	62518	17752	53163	63852	44840	02592	88572	03107	90169
16	16215	50809	49326	77232	90155	69955	93892	70445	00906	57002
17	09342	14528	64727	71403	84156	34083	35613	35670	10549	07468
18	38148	79001	03509	79424	39625	73315	18811	86230	99682	82896
19	23689	19997	72382	15247	80205	58090	43804	94548	82693	22799
20	25407	37726	73099	51057	68733	75768	77991	72641	95386	70138
21	25349	69456	19693	85568	93876	18661	69018	10332	83137	88257
22	02322	77491	56095	03055	37738	18216	81781	32245	84081	18436
23	15072	33261	99219	43307	39239	79712	94753	41450	30944	53912
24	27002	31036	85278	74547	84809	36252	09373	69471	15606	77209
25	66181	83316	40386	54316	29505	86032	34563	93204	72973	90760
26	09779	01822	45537	13128	51128	82703	75350	25179	86104	40638
27	10791	07706	87481	26107	24857	27805	42710	63471	08804	23455
28	74833	55767	31312	76611	67389	04691	39687	13596	88730	86850
29	17583	24038	83701	28570	63561	00098	60784	76098	84217	34997
30	45601	46977	39325	09286	41133	34031	94867	11849	75171	57682
31	60683	33112	65995	64203	18070	65437	13624	90896	80945	71987
32	29956	81169	18877	15296	94368	16317	34239	03643	66081	12242
33	91713	84235	75296	69875	82414	05197	66596	13083	46278	73498
34	85704	86588	82837	67822	95963	83021	90732	32661	64751	83903
35	17921	26111	35373	86494	48266	01888	65735	05315	79328	13367
36	13929	71341	80488	89827	48277	07229	71953	16128	65074	28782
37	03248	18880	21667	01311	61806	80201	47889	83052	31029	06023
38	50583	17972	12690	00452	93766	16414	01212	27964	02766	28786
39	10636	46975	09449	45986	34672	46916	63881	83117	53947	95218
40	43896	41278	42205	10425	66560	59967	90139	73563	29875	79033
41	76714	80963	74907	16890	15492	27489	06067	22287	19760	13056
42	22393	46719	02083	62428	45177	57562	49243	31748	64278	05731
43	70942	92042	22776	47761	13503	16037	30875	80754	47491	96012
44	92011	60326	86346	26738	01983	04186	41388	03848	78354	14964
45	66456	00126	45685	67607	70796	04889	98128	13599	93710	23974
46	96292	44348	20898	02227	76512	53185	03057	61375	10760	26889
47	19680	07146	53951	10935	23333	76233	13706	20502	60405	09745
48	67347	51442	24536	60151	05498	64678	87569	65066	17790	55413
49	95888	59255	06898	99137	50871	81265	42223	83303	48694	81953

Table A.1 (Continued)
Ten Thousand Random Digits

	50–54	55–59	60–64	65–69	70–74	75–79	80–84	85–89	90–94	95–99
00	70896	44520	64720	49898	78088	76740	47460	83150	78905	59870
01	56809	42909	25853	47624	29486	14196	75841	00393	42390	24847
02	66109	84775	07515	49949	61482	91836	48126	80778	21302	24975
03	18071	36263	14053	52526	44347	04923	68100	57805	19521	15345
04	98732	15120	91754	12657	74675	78500	01247	49719	47635	55514
05	36075	83967	22268	77971	31169	68584	21336	72541	66959	39708
06	04110	45061	78062	18911	27855	09419	56459	00695	70323	04538
07	75658	58509	24479	10202	13150	95946	55087	38398	18718	95561
08	87403	19142	27208	35149	34889	27003	14181	44813	17784	41036
09	00005	52142	65021	64438	69610	12154	98422	65320	79996	01935
10	43674	47103	48614	70823	78252	82403	93424	05236	54588	27757
11	68597	68874	35567	98463	99671	05634	81533	47406	17228	44455
12	91874	70208	06308	40719	02772	69589	79936	07514	44950	35190
13	73854	19470	53014	29375	62256	77488	74388	53949	49607	19816
14	65926	34117	55344	68155	38099	56009	03513	05926	35584	42328
15	40005	35246	49440	40295	44390	83043	26090	80201	02934	49260
16	46686	29890	14821	69783	34733	11803	64845	32065	14527	38702
17	02717	61518	39583	72863	50707	96115	07416	05041	36756	61065
18	17048	22281	35573	28944	96889	51823	57268	03866	27658	91950
19	75304	53248	42151	93928	17343	88322	28683	11252	10355	65175
20	97844	62947	62230	30500	92816	85232	27222	91701	11057	83257
21	07611	71163	82212	20653	21499	51496	40715	78952	33029	64207
22	47744	04603	44522	62783	39347	72310	41460	31052	40814	94297
23	54293	43576	88116	67416	34908	15238	40561	73940	56850	31078
24	67556	93979	73363	00300	11217	74405	18937	79000	68834	48307
25	86581	73041	95809	73986	49408	53316	90841	73808	53421	82315
26	28020	86282	83365	76600	11261	74354	20968	60770	12141	09539
27	42578	32471	37840	30872	75074	79027	57813	62831	54715	26693
28	47290	15997	86163	10571	81911	92124	92971	80860	41012	58666
29	24856	63911	13221	77028	06573	33667	30732	47280	12926	67276
30	16352	24836	60799	76281	83402	44709	78930	82969	84468	36910
31	89060	79852	97854	28324	39638	86936	06702	74304	39873	19496
32	07637	30412	04921	26471	09605	07355	20466	49793	40539	21077
33	37711	47786	37468	31963	16908	50283	80884	08252	72655	58926
34	82994	53232	58202	73318	62471	49650	15888	73370	98748	69181
35	31722	67288	12110	04776	15168	68862	92347	90789	66961	04162
36	93819	78050	19364	38037	25706	90879	05215	00260	14426	88207
37	65557	24496	04713	23688	26623	41356	47049	60676	72236	01214
38	88001	91382	05129	36041	10257	55558	89979	58061	28957	10701
39	96648	70303	18191	62404	26558	92804	15415	02865	52449	78509
40	04118	51573	59356	02426	35010	37104	98316	44602	96478	08433
41	19317	27753	39431	26996	04465	69695	61374	06317	42225	62025
42	37182	91221	17307	68507	85725	81898	22588	22241	80337	89033
43	82990	03607	29560	60413	59743	75000	03806	13741	79671	25416
44	97294	21991	11217	98087	79124	52275	31088	32085	23089	21498
45	86771	69504	13345	42544	59616	07867	78717	82840	74669	21515
46	26046	55559	12200	95106	56496	76662	44880	89457	84209	01332
47	39689	05999	92290	79024	70271	93352	90272	94495	26842	54477
48	83265	89573	01437	43786	52986	49041	17952	35035	88985	84671
49	15128	35791	11296	45319	06330	82027	90808	54351	43091	30387

TABLE A.1 (Continued)

TEN THOUSAND RANDOM DIGITS

	00-04	05-09	10-14	15-19	20-24	25-29	30-34	35-39	40-44	45-49
50	54441	64681	93190	00993	62130	44484	46293	60717	50239	76319
51	08573	52937	84274	95106	89117	65849	41356	65549	78787	50442
52	81067	68052	14270	19718	88499	63303	13533	91882	51136	60828
53	39737	58891	75278	98046	52284	40164	72442	77824	72900	14886
54	34958	76090	08827	61623	31114	86952	83645	91786	29633	78294
55	61417	72424	92626	71952	69709	81259	58472	43409	84454	88648
56	99187	14149	57474	32268	85424	90378	34682	47606	89295	02420
57	13130	13064	36485	48133	35319	05720	76317	70953	50823	06793
58	65563	11831	82402	46929	91446	72037	17205	89600	59084	55718
59	28737	49502	06060	52100	43704	50839	22538	56768	83467	19313
60	50353	74022	59767	49927	45882	74099	18758	57510	58560	07050
61	65208	96466	29917	22862	69972	35178	32911	08172	06277	62795
62	21323	38148	26696	81741	25131	20087	67452	19670	35898	50636
63	67875	29831	59330	46570	69768	36671	01031	95995	68417	68665
64	82631	26260	86554	31881	70512	37899	38851	40568	54284	24056
65	91989	39633	59039	12526	37730	68848	71399	28513	69018	10289
66	12950	31418	93425	69756	34036	55097	97241	92480	49745	42461
67	00328	27427	95474	97217	05034	26676	49629	13594	50525	13485
68	63986	16698	82804	04524	39919	32381	67488	05223	89537	59490
69	55775	75005	57912	20977	35722	51931	89565	77579	93085	06467
70	24761	56877	56357	78809	40748	69727	56652	12462	40528	75269
71	43820	80926	26795	57553	28319	25376	51795	26123	51102	89853
72	66669	02880	02987	33615	54206	20013	75872	88678	17726	60640
73	49944	66725	19779	50416	42800	71733	82052	28504	15593	51799
74	71003	87598	61296	95019	21568	86134	66096	65403	47166	78638
75	52715	04593	69484	93411	38046	13000	04293	60830	03914	75357
76	21998	31729	89963	11573	49442	69467	40265	56066	36024	25705
77	58970	96827	18377	31564	23555	86338	79250	43168	96929	97732
78	67592	59149	42554	42719	13553	48560	81167	10747	92552	19867
79	18298	18429	09357	96436	11237	88039	81020	00428	75731	37779
80	88420	28841	42628	84647	59024	52032	31251	72017	43875	48320
81	07627	88424	23381	29680	14027	75905	27037	22113	77873	78711
82	37917	93581	04979	21041	95252	62450	05937	81670	44894	47262
83	14783	95119	68464	08726	74818	91700	05961	23554	74649	50540
84	05378	32640	64562	15303	13168	23189	88198	63617	58566	56047
85	19640	96709	22047	07825	40583	99500	39989	96593	32254	37158
86	20514	11081	51131	56469	33947	77703	35679	45774	06776	67062
87	96763	56249	81243	62416	84451	14696	38195	70435	45948	67690
88	49439	61075	31558	59740	52759	55323	95226	01385	20158	54054
89	16294	50548	71317	32168	86071	47314	65393	56367	46910	51269
90	31381	94301	79273	32843	05862	36211	93960	00671	67631	23952
91	98032	87203	03227	66021	99666	98368	39222	36056	81992	20121
92	40700	31826	94774	11366	81391	33602	69608	84119	93204	26825
93	68692	66849	29366	77540	14978	06508	10824	65416	23629	63029
94	19047	10784	19607	20296	31804	72984	60060	50353	23260	58909
95	82867	69266	50733	62630	00956	61500	89913	30049	82321	62367
96	26528	28928	52600	72997	80943	04084	86662	90025	14360	64867
97	51166	00607	49962	30724	81707	14548	25844	47336	57492	02207
98	97245	15440	55182	15368	85136	98869	33712	95152	50973	98658
99	54998	88830	95639	45104	72676	28220	82576	57381	34438	24565

SOURCE: Prepared by Fred Gruenberger, Numerical Analysis Laboratory, University of Wisconsin, Madison, Wis., 1952.

Table A.1 (Continued)
Ten Thousand Random Digits

	50–54	55–59	60–64	65–69	70–74	75–79	80–84	85–89	90–94	95–99
50	58649	85086	16502	97541	76611	94229	34987	86718	87208	05426
51	97306	52449	55596	66739	36525	97563	29469	31235	79276	10831
52	09942	79344	78160	11015	55777	22047	57615	15717	86239	36578
53	83842	28631	74893	47911	92170	38181	30416	54860	44120	73031
54	73778	30395	20163	76111	13712	33449	99224	18206	51418	70006
55	88381	56550	47467	59663	61117	39716	32927	06168	06217	45477
56	31044	21404	15968	21357	30772	81482	38807	67231	84283	63552
57	00909	63837	91328	81106	11740	50193	86806	21931	18054	49601
58	69882	37028	41732	37425	80832	03320	20690	32653	90145	03029
59	26059	78324	22501	73825	16927	31545	15695	74216	98372	28547
60	38573	98078	38982	33078	93524	45606	53463	20391	81637	37269
61	70624	00063	81455	16924	12848	23801	55481	78978	26795	10553
62	49806	23976	05640	29804	38988	25024	76951	02341	63219	75864
63	05461	67523	48316	14613	08541	35231	38312	14969	67279	50502
64	76582	62153	53801	51219	30424	32599	49099	83959	68408	20147
65	16660	80470	75062	75588	24384	27874	20018	11428	32265	07692
66	60166	42424	97470	88451	81270	80070	72959	26220	59939	31127
67	28953	03272	31460	41691	57736	72052	22762	96323	27616	53123
68	47536	86439	95210	96386	38704	15484	07426	70675	06888	81203
69	73457	26657	36983	72410	30244	97711	25652	09373	66218	64077
70	11190	66193	66287	09116	48140	37669	02932	50799	17255	06181
71	57062	78964	44455	14036	36098	40773	11688	33150	07459	36127
72	99624	67254	67302	18991	97687	54099	94884	42283	63258	50651
73	97521	83669	85968	16135	30133	51312	17831	75016	80278	68953
74	40273	04838	13661	64757	17461	78085	60094	27010	80945	66439
75	57260	06176	49963	29760	69546	61336	39429	41985	18572	98128
76	03451	47098	63495	71227	79304	29753	99131	18419	71791	81515
77	62331	20492	15393	84270	24396	32962	21632	92965	38670	44923
78	32290	51079	06512	38806	93327	80086	19088	59887	98416	24918
79	28014	80428	92853	31333	32648	16734	43418	90124	15086	48444
80	18950	16091	29543	65817	07002	73115	94115	20271	50250	25061
81	17403	69503	01866	13049	07263	13039	83844	80143	39048	62654
82	27999	50489	66613	21843	71746	65868	16208	46781	93402	12323
83	87076	53174	12165	84495	47947	60706	64034	31635	65169	93070
84	89044	45974	14524	46906	26052	51851	84197	61694	57429	63395
85	98048	64400	24705	75711	36232	57624	41424	77366	52790	84705
86	09345	12956	49770	80311	32319	48238	16952	92088	51222	82865
87	07086	77628	76195	47584	62411	40397	71857	54823	26536	56792
88	93128	25657	46872	11206	06831	87944	97914	64670	45760	34353
89	85137	70964	29947	27795	25547	37682	96105	26848	09389	64326
90	32798	39024	13814	98546	46585	84108	74603	94812	73968	68766
91	62496	26371	89880	52078	47781	95260	83464	65942	91761	53727
92	62707	81825	40987	97656	89714	52177	23778	07482	91678	40128
93	05500	28982	86124	19554	80818	94935	61924	31828	79369	23507
94	79476	31445	59498	85132	24582	26024	24002	63718	79164	43556
95	10653	29954	97568	91541	33139	84525	72271	02546	64818	14381
96	30524	06495	00886	40666	68574	49574	19705	16429	90981	08103
97	69050	22019	74066	14500	14506	06423	38332	34191	82663	85323
98	27908	78802	63446	07674	98871	63831	72449	42705	26513	19883
99	64520	16618	47409	19574	78136	46047	01277	79146	95759	36781

TABLE A.2

VALUES OF THE RATIO, RANGE DIVIDED BY THE STANDARD DEVIATION σ, FOR SAMPLE SIZES FROM 20 TO 1,000

Number in sample	Range/σ	Number in sample	Range/σ
20	3.7	200	5.5
30	4.1	300	5.8
50	4.5	400	5.9
70	4.8	500	6.1
100	5.0	700	6.3
150	5.3	1,000	6.5

SOURCE: Abridged, with permission of the *Biometrika* trustees and the editors, from E. S. Pearson and H. O. Hartley, *Biometrika Tables for Statisticians*, vol. 1, Cambridge University Press, 1954. Original table by L. H. C. Tippett, "On the extreme individuals and the range of samples taken from a normal population," *Biometrika*, **17**: pp. 364–387, (1925).

TABLE A.3

VALUES OF t

df	Probability of a larger value of t, sign ignored								
	0.5	0.4	0.3	0.2	0.1	0.05	0.02	0.01	0.001
1	1.000	1.376	1.963	3.078	6.314	12.706	31.821	63.657	636.619
2	.816	1.061	1.386	1.886	2.920	4.303	6.965	9.925	31.598
3	.765	.978	1.250	1.638	2.353	3.182	4.541	5.841	12.941
4	.741	.941	1.190	1.533	2.132	2.776	3.747	4.604	8.610
5	.727	.920	1.156	1.476	2.015	2.571	3.365	4.032	6.859
6	.718	.906	1.134	1.440	1.943	2.447	3.143	3.707	5.959
7	.711	.896	1.119	1.415	1.895	2.365	2.998	3.499	5.405
8	.706	.889	1.108	1.397	1.860	2.306	2.896	3.355	5.041
9	.703	.883	1.100	1.383	1.833	2.262	2.821	3.250	4.781
10	.700	.879	1.093	1.372	1.812	2.228	2.764	3.169	4.587
11	.697	.876	1.088	1.363	1.796	2.201	2.718	3.106	4.437
12	.695	.873	1.083	1.356	1.782	2.179	2.681	3.055	4.318
13	.694	.870	1.079	1.350	1.771	2.160	2.650	3.012	4.221
14	.692	.868	1.076	1.345	1.761	2.145	2.624	2.977	4.140
15	.691	.866	1.074	1.341	1.753	2.131	2.602	2.947	4.073
16	.690	.865	1.071	1.337	1.746	2.120	2.583	2.921	4.015
17	.689	.863	1.069	1.333	1.740	2.110	2.567	2.898	3.965
18	.688	.862	1.067	1.330	1.734	2.101	2.552	2.878	3.922
19	.688	.861	1.066	1.328	1.729	2.093	2.539	2.861	3.883
20	.687	.860	1.064	1.325	1.725	2.086	2.528	2.845	3.850
21	.686	.859	1.063	1.323	1.721	2.080	2.518	2.831	3.819
22	.686	.858	1.061	1.321	1.717	2.074	2.508	2.819	3.792
23	.685	.858	1.060	1.319	1.714	2.069	2.500	2.807	3.767
24	.685	.857	1.059	1.318	1.711	2.064	2.492	2.797	3.745
25	.684	.856	1.058	1.316	1.708	2.060	2.485	2.787	3.725
26	.684	.856	1.058	1.315	1.706	2.056	2.479	2.779	3.707
27	.684	.855	1.057	1.314	1.703	2.052	2.473	2.771	3.690
28	.683	.855	1.056	1.313	1.701	2.048	2.467	2.763	3.674
29	.683	.854	1.055	1.311	1.699	2.045	2.462	2.756	3.659
30	.683	.854	1.055	1.310	1.697	2.042	2.457	2.750	3.646
40	.681	.851	1.050	1.303	1.684	2.021	2.423	2.704	3.551
60	.679	.848	1.046	1.296	1.671	2.000	2.390	2.660	3.460
120	.677	.845	1.041	1.289	1.658	1.980	2.358	2.617	3.373
∞	.674	.842	1.036	1.282	1.645	1.960	2.326	2.576	3.291
df	0.25	0.2	0.15	0.1	0.05	0.025	0.01	0.005	0.0005
	Probability of a larger value of t, sign considered								

SOURCE: This table is abridged from Table III of Fisher and Yates, *Statistical Tables for Biological, Agricultural, and Medical Research*, published by Oliver and Boyd Ltd., Edinburgh, 1949, by permission of the authors and publishers.

Table A.4

Probability of a Random Value of $z = (X - \mu)/\sigma$ Being Greater Than the Values Tabulated in the Margins

z	.00	.01	.02	.03	.04	.05	.06	.07	.08	.09
.0	.5000	.4960	.4920	.4880	.4840	.4801	.4761	.4721	.4681	.4641
.1	.4602	.4562	.4522	.4483	.4443	.4404	.4364	.4325	.4286	.4247
.2	.4207	.4168	.4129	.4090	.4052	.4013	.3974	.3936	.3897	.3859
.3	.3821	.3783	.3745	.3707	.3669	.3632	.3594	.3557	.3520	.3483
.4	.3446	.3409	.3372	.3336	.3300	.3264	.3228	.3192	.3156	.3121
.5	.3085	.3050	.3015	.2981	.2946	.2912	.2877	.2843	.2810	.2776
.6	.2743	.2709	.2676	.2643	.2611	.2578	.2546	.2514	.2483	.2451
.7	.2420	.2389	.2358	.2327	.2296	.2266	.2236	.2206	.2177	.2148
.8	.2119	.2090	.2061	.2033	.2005	.1977	.1949	.1922	.1894	.1867
.9	.1841	.1814	.1788	.1762	.1736	.1711	.1685	.1660	.1635	.1611
1.0	.1587	.1562	.1539	.1515	.1492	.1469	.1446	.1423	.1401	.1379
1.1	.1357	.1335	.1314	.1292	.1271	.1251	.1230	.1210	.1190	.1170
1.2	.1151	.1131	.1112	.1093	.1075	.1056	.1038	.1020	.1003	.0985
1.3	.0968	.0951	.0934	.0918	.0901	.0885	.0869	.0853	.0838	.0823
1.4	.0808	.0793	.0778	.0764	.0749	.0735	.0721	.0708	.0694	.0681
1.5	.0668	.0655	.0643	.0630	.0618	.0606	.0594	.0582	.0571	.0559
1.6	.0548	.0537	.0526	.0516	.0505	.0495	.0485	.0475	.0465	.0455
1.7	.0446	.0436	.0427	.0418	.0409	.0401	.0392	.0384	.0375	.0367
1.8	.0359	.0351	.0344	.0336	.0329	.0322	.0314	.0307	.0301	.0294
1.9	.0287	.0281	.0274	.0268	.0262	.0256	.0250	.0244	.0239	.0233
2.0	.0228	.0222	.0217	.0212	.0207	.0202	.0197	.0192	.0188	.0183
2.1	.0179	.0174	.0170	.0166	.0162	.0158	.0154	.0150	.0146	.0143
2.2	.0139	.0136	.0132	.0129	.0125	.0122	.0119	.0116	.0113	.0110
2.3	.0107	.0104	.0102	.0099	.0096	.0094	.0091	.0089	.0087	.0084
2.4	.0082	.0080	.0078	.0075	.0073	.0071	.0069	.0068	.0066	.0064
2.5	.0062	.0060	.0059	.0057	.0055	.0054	.0052	.0051	.0049	.0048
2.6	.0047	.0045	.0044	.0043	.0041	.0040	.0039	.0038	.0037	.0036
2.7	.0035	.0034	.0033	.0032	.0031	.0030	.0029	.0028	.0027	.0026
2.8	.0026	.0025	.0024	.0023	.0023	.0022	.0021	.0021	.0020	.0019
2.9	.0019	.0018	.0018	.0017	.0016	.0016	.0015	.0015	.0014	.0014
3.0	.0013	.0013	.0013	.0012	.0012	.0011	.0011	.0011	.0010	.0010
3.1	.0010	.0009	.0009	.0009	.0008	.0008	.0008	.0008	.0007	.0007
3.2	.0007	.0007	.0006	.0006	.0006	.0006	.0006	.0005	.0005	.0005
3.3	.0005	.0005	.0005	.0004	.0004	.0004	.0004	.0004	.0004	.0003
3.4	.0003	.0003	.0003	.0003	.0003	.0003	.0003	.0003	.0003	.0002
3.6	.0002	.0002	.0001	.0001	.0001	.0001	.0001	.0001	.0001	.0001
3.9	.0000									

Table A.5
Values of χ^2

df	.995	.990	.975	.950	.900	.750	.500	.250	.100	.050	.025	.010	.005
1	.0⁴393	.0³157	.0³982	.0²393	.0158	.102	.455	1.32	2.71	3.84	5.02	6.63	7.88
2	.0100	.0201	.0506	.103	.211	.575	1.39	2.77	4.61	5.99	7.38	9.21	10.6
3	.0717	.115	.216	.352	.584	1.21	2.37	4.11	6.25	7.81	9.35	11.3	12.8
4	.207	.297	.484	.711	1.06	1.92	3.36	5.39	7.78	9.49	11.1	13.3	14.9
5	.412	.554	.831	1.15	1.61	2.67	4.35	6.63	9.24	11.1	12.8	15.1	16.7
6	.676	.872	1.24	1.64	2.20	3.45	5.35	7.84	10.6	12.6	14.4	16.8	18.5
7	.989	1.24	1.69	2.17	2.83	4.25	6.35	9.04	12.0	14.1	16.0	18.5	20.3
8	1.34	1.65	2.18	2.73	3.49	5.07	7.34	10.2	13.4	15.5	17.5	20.1	22.0
9	1.73	2.09	2.70	3.33	4.17	5.90	8.34	11.4	14.7	16.9	19.0	21.7	23.6
10	2.16	2.56	3.25	3.94	4.87	6.74	9.34	12.5	16.0	18.3	20.5	23.2	25.2
11	2.60	3.05	3.82	4.57	5.58	7.58	10.3	13.7	17.3	19.7	21.9	24.7	26.8
12	3.07	3.57	4.40	5.23	6.30	8.44	11.3	14.8	18.5	21.0	23.3	26.2	28.3
13	3.57	4.11	5.01	5.89	7.04	9.30	12.3	16.0	19.8	22.4	24.7	27.7	29.8
14	4.07	4.66	5.63	6.57	7.79	10.2	13.3	17.1	21.1	23.7	26.1	29.1	31.3
15	4.60	5.23	6.26	7.26	8.55	11.0	14.3	18.2	22.3	25.0	27.5	30.6	32.8
16	5.14	5.81	6.91	7.96	9.31	11.9	15.3	19.4	23.5	26.3	28.8	32.0	34.3
17	5.70	6.41	7.56	8.67	10.1	12.8	16.3	20.5	24.8	27.6	30.2	33.4	35.7
18	6.26	7.01	8.23	9.39	10.9	13.7	17.3	21.6	26.0	28.9	31.5	34.8	37.2
19	6.84	7.63	8.91	10.1	11.7	14.6	18.3	22.7	27.2	30.1	32.9	36.2	38.6
20	7.43	8.26	9.59	10.9	12.4	15.5	19.3	23.8	28.4	31.4	34.2	37.6	40.0
21	8.03	8.90	10.3	11.6	13.2	16.3	20.3	24.9	29.6	32.7	35.5	38.9	41.4
22	8.64	9.54	10.9	12.3	14.0	17.2	21.3	26.0	30.8	33.9	36.8	40.3	42.8
23	9.26	10.2	11.7	13.1	14.8	18.1	22.3	27.1	32.0	35.2	38.1	41.6	44.2
24	9.89	10.9	12.4	13.8	15.7	19.0	23.3	28.2	33.2	36.4	39.4	43.0	45.6
25	10.5	11.5	13.1	14.6	16.5	19.9	24.3	29.3	34.4	37.7	40.6	44.3	46.9
26	11.2	12.2	13.8	15.4	17.3	20.8	25.3	30.4	35.6	38.9	41.9	45.6	48.3
27	11.8	12.9	14.6	16.2	18.1	21.7	26.3	31.5	36.7	40.1	43.2	47.0	49.6
28	12.5	13.6	15.3	16.9	18.9	22.7	27.3	32.6	37.9	41.3	44.5	48.3	51.0
29	13.1	14.3	16.0	17.7	19.8	23.6	28.3	33.7	39.1	42.6	45.7	49.6	52.3
30	13.8	15.0	16.8	18.5	20.6	24.5	29.3	34.8	40.3	43.8	47.0	50.9	53.7
40	20.7	22.2	24.4	26.5	29.1	33.7	39.3	45.6	51.8	55.8	59.3	63.7	66.8
50	28.0	29.7	32.4	34.8	37.7	42.9	49.3	56.3	63.2	67.5	71.4	76.2	79.5
60	35.5	37.5	40.5	43.2	46.5	52.3	59.3	67.0	74.4	79.1	83.3	88.4	92.0

Probability of a larger value of χ^2

SOURCE: This table is abridged from "Table of percentage points of the χ^2 distribution," *Biometrika*, **32:** 188–189 (1941), by Catherine M. Thompson. It is published here with kind permission of the author and the editor of *Biometrika*.

Table A.6
Values of F

Denominator df	Probability of a larger F	Numerator df								
		1	2	3	4	5	6	7	8	9
1	.100	39.86	49.50	53.59	55.83	57.24	58.20	58.91	59.44	59.86
	.050	161.4	199.5	215.7	224.6	230.2	234.0	236.8	238.9	240.5
	.025	647.8	799.5	864.2	899.6	921.8	937.1	948.2	956.7	963.3
	.010	4052	4999.5	5403	5625	5764	5859	5928	5982	6022
	.005	16211	20000	21615	22500	23056	23437	23715	23925	24091
2	.100	8.53	9.00	9.16	9.24	9.29	9.33	9.35	9.37	9.38
	.050	18.51	19.00	19.16	19.25	19.30	19.33	19.35	19.37	19.38
	.025	38.51	39.00	39.17	39.25	39.30	39.33	39.36	39.37	39.39
	.010	98.50	99.00	99.17	99.25	99.30	99.33	99.36	99.37	99.39
	.005	198.5	199.0	199.2	199.2	199.3	199.3	199.4	199.4	199.4
3	.100	5.54	5.46	5.39	5.34	5.31	5.28	5.27	5.25	5.24
	.050	10.13	9.55	9.28	9.12	9.01	8.94	8.89	8.85	8.81
	.025	17.44	16.04	15.44	15.10	14.88	14.73	14.62	14.54	14.47
	.010	34.12	30.82	29.46	28.71	28.24	27.91	27.67	27.49	27.35
	.005	55.55	49.80	47.47	46.19	45.39	44.84	44.43	44.13	43.88
4	.100	4.54	4.32	4.19	4.11	4.05	4.01	3.98	3.95	3.94
	.050	7.71	6.94	6.59	6.39	6.26	6.16	6.09	6.04	6.00
	.025	12.22	10.65	9.98	9.60	9.36	9.20	9.07	8.98	8.90
	.010	21.20	18.00	16.69	15.98	15.52	15.21	14.98	14.80	14.66
	.005	31.33	26.28	24.26	23.15	22.46	21.97	21.62	21.35	21.14
5	.100	4.06	3.78	3.62	3.52	3.45	3.40	3.37	3.34	3.32
	.050	6.61	5.79	5.41	5.19	5.05	4.95	4.88	4.82	4.77
	.025	10.01	8.43	7.76	7.39	7.15	6.98	6.85	6.76	6.68
	.010	16.26	13.27	12.06	11.39	10.97	10.67	10.46	10.29	10.16
	.005	22.78	18.31	16.53	15.56	14.94	14.51	14.20	13.96	13.77
6	.100	3.78	3.46	3.29	3.18	3.11	3.05	3.01	2.98	2.96
	.050	5.99	5.14	4.76	4.53	4.39	4.28	4.21	4.15	4.10
	.025	8.81	7.26	6.60	6.23	5.99	5.82	5.70	5.60	5.52
	.010	13.75	10.92	9.78	9.15	8.75	8.47	8.26	8.10	7.98
	.005	18.63	14.54	12.92	12.03	11.46	11.07	10.79	10.57	10.39
7	.100	3.59	3.26	3.07	2.96	2.88	2.83	2.78	2.75	2.72
	.050	5.59	4.74	4.35	4.12	3.97	3.87	3.79	3.73	3.68
	.025	8.07	6.54	5.89	5.52	5.29	5.12	4.99	4.90	4.82
	.010	12.25	9.55	8.45	7.85	7.46	7.19	6.99	6.84	6.72
	.005	16.24	12.40	10.88	10.05	9.52	9.16	8.89	8.68	8.51
8	.100	3.46	3.11	2.92	2.81	2.73	2.67	2.62	2.59	2.56
	.050	5.32	4.46	4.07	3.84	3.69	3.58	3.50	3.44	3.39
	.025	7.57	6.06	5.42	5.05	4.82	4.65	4.53	4.43	4.36
	.010	11.26	8.65	7.59	7.01	6.63	6.37	6.18	6.03	5.91
	.005	14.69	11.04	9.60	8.81	8.30	7.95	7.69	7.50	7.34
9	.100	3.36	3.01	2.81	2.69	2.61	2.55	2.51	2.47	2.44
	.050	5.12	4.26	3.86	3.63	3.48	3.37	3.29	3.23	3.18
	.025	7.21	5.71	5.08	4.72	4.48	4.32	4.20	4.10	4.03
	.010	10.56	8.02	6.99	6.42	6.06	5.80	5.61	5.47	5.35
	.005	13.61	10.11	8.72	7.96	7.47	7.13	6.88	6.69	6.54
10	.100	3.29	2.92	2.73	2.61	2.52	2.46	2.41	2.38	2.35
	.050	4.96	4.10	3.71	3.48	3.33	3.22	3.14	3.07	3.02
	.025	6.94	5.46	4.83	4.47	4.24	4.07	3.95	3.85	3.78
	.010	10.04	7.56	6.55	5.99	5.64	5.39	5.20	5.06	4.94
	.005	12.83	9.43	8.08	7.34	6.87	6.54	6.30	6.12	5.97
11	.100	3.23	2.86	2.66	2.54	2.45	2.39	2.34	2.30	2.27
	.050	4.84	3.98	3.59	3.36	3.20	3.09	3.01	2.95	2.90
	.025	6.72	5.26	4.63	4.28	4.04	3.88	3.76	3.66	3.59
	.010	9.65	7.21	6.22	5.67	5.32	5.07	4.89	4.74	4.63
	.005	12.23	8.91	7.60	6.88	6.42	6.10	5.86	5.68	5.54
12	.100	3.18	2.81	2.61	2.48	2.39	2.33	2.28	2.24	2.21
	.050	4.75	3.89	3.49	3.26	3.11	3.00	2.91	2.85	2.80
	.025	6.55	5.10	4.47	4.12	3.89	3.73	3.61	3.51	3.44
	.010	9.33	6.93	5.95	5.41	5.06	4.82	4.64	4.50	4.39
	.005	11.75	8.51	7.23	6.52	6.07	5.76	5.52	5.35	5.20
13	.100	3.14	2.76	2.56	2.43	2.35	2.28	2.23	2.20	2.16
	.050	4.67	3.81	3.41	3.18	3.03	2.92	2.83	2.77	2.71
	.025	6.41	4.97	4.35	4.00	3.77	3.60	3.48	3.39	3.31
	.010	9.07	6.70	5.74	5.21	4.86	4.62	4.44	4.30	4.19
	.005	11.37	8.19	6.93	6.23	5.79	5.48	5.25	5.08	4.94
14	.100	3.10	2.73	2.52	2.39	2.31	2.24	2.19	2.15	2.12
	.050	4.60	3.74	3.34	3.11	2.96	2.85	2.76	2.70	2.65
	.025	6.30	4.86	4.24	3.89	3.66	3.50	3.38	3.29	3.21
	.010	8.86	6.51	5.56	5.04	4.69	4.46	4.28	4.14	4.03
	.005	11.06	7.92	6.68	6.00	5.56	5.26	5.03	4.86	4.72

TABLE A.6 (Continued)
VALUES OF F

Numerator df											
10	12	15	20	24	30	40	60	120	∞	P	df
60.19	60.71	61.22	61.74	62.00	62.26	62.53	62.79	63.06	63.33	.100	1
241.9	243.9	245.9	248.0	249.1	250.1	251.1	252.2	253.3	254.3	.050	
968.6	976.7	984.9	993.1	997.2	1001	1006	1010	1014	1018	.025	
6056	6106	6157	6209	6235	6261	6287	6313	6339	6366	.010	
24224	24426	24630	24836	24940	25044	25148	25253	25359	25465	.005	
9.39	9.41	9.42	9.44	9.45	9.46	9.47	9.47	9.48	9.49	.100	2
19.40	19.41	19.43	19.45	19.45	19.46	19.47	19.48	19.49	19.50	.050	
39.40	39.41	39.43	39.45	39.46	39.46	39.47	39.48	39.49	39.50	.025	
99.40	99.42	99.43	99.45	99.46	99.47	99.47	99.48	99.49	99.50	.010	
199.4	199.4	199.4	199.4	199.5	199.5	199.5	199.5	199.5	199.5	.005	
5.23	5.22	5.20	5.18	5.18	5.17	5.16	5.15	5.14	5.13	.100	3
8.79	8.74	8.70	8.66	8.64	8.62	8.59	8.57	8.55	8.53	.050	
14.42	14.34	14.25	14.17	14.12	14.08	14.04	13.99	13.95	13.90	.025	
27.23	27.05	26.87	26.69	26.60	26.50	26.41	26.32	26.22	26.13	.010	
43.69	43.39	43.08	42.78	42.62	42.47	42.31	42.15	41.99	41.83	.005	
3.92	3.90	3.87	3.84	3.83	3.82	3.80	3.79	3.78	3.76	.100	4
5.96	5.91	5.86	5.80	5.77	5.75	5.72	5.69	5.66	5.63	.050	
8.84	8.75	8.66	8.56	8.51	8.46	8.41	8.36	8.31	8.26	.025	
14.55	14.37	14.20	14.02	13.93	13.84	13.75	13.65	13.56	13.46	.010	
20.97	20.70	20.44	20.17	20.03	19.89	19.75	19.61	19.47	19.32	.005	
3.30	3.27	3.24	3.21	3.19	3.17	3.16	3.14	3.12	3.10	.100	5
4.74	4.68	4.62	4.56	4.53	4.50	4.46	4.43	4.40	4.36	.050	
6.62	6.52	6.43	6.33	6.28	6.23	6.18	6.12	6.07	6.02	.025	
10.05	9.89	9.72	9.55	9.47	9.38	9.29	9.20	9.11	9.02	.010	
13.62	13.38	13.15	12.90	12.78	12.66	12.53	12.40	12.27	12.14	.005	
2.94	2.90	2.87	2.84	2.82	2.80	2.78	2.76	2.74	2.72	.100	6
4.06	4.00	3.94	3.87	3.84	3.81	3.77	3.74	3.70	3.67	.050	
5.46	5.37	5.27	5.17	5.12	5.07	5.01	4.96	4.90	4.85	.025	
7.87	7.72	7.56	7.40	7.31	7.23	7.14	7.06	6.97	6.88	.010	
10.25	10.03	9.81	9.59	9.47	9.36	9.24	9.12	9.00	8.88	.005	
2.70	2.67	2.63	2.59	2.58	2.56	2.54	2.51	2.49	2.47	.100	7
3.64	3.57	3.51	3.44	3.41	3.38	3.34	3.30	3.27	3.23	.050	
4.76	4.67	4.57	4.47	4.42	4.36	4.31	4.25	4.20	4.14	.025	
6.62	6.47	6.31	6.16	6.07	5.99	5.91	5.82	5.74	5.65	.010	
8.38	8.18	7.97	7.75	7.65	7.53	7.42	7.31	7.19	7.08	.005	
2.54	2.50	2.46	2.42	2.40	2.38	2.36	2.34	2.32	2.29	.100	8
3.35	3.28	3.22	3.15	3.12	3.08	3.04	3.01	2.97	2.93	.050	
4.30	4.20	4.10	4.00	3.95	3.89	3.84	3.78	3.73	3.67	.025	
5.81	5.67	5.52	5.36	5.28	5.20	5.12	5.03	4.95	4.86	.010	
7.21	7.01	6.81	6.61	6.50	6.40	6.29	6.18	6.06	5.95	.005	
2.42	2.38	2.34	2.30	2.28	2.25	2.23	2.21	2.18	2.16	.100	9
3.14	3.07	3.01	2.94	2.90	2.86	2.83	2.79	2.75	2.71	.050	
3.96	3.87	3.77	3.67	3.61	3.56	3.51	3.45	3.39	3.33	.025	
5.26	5.11	4.96	4.81	4.73	4.65	4.57	4.48	4.40	4.31	.010	
6.42	6.23	6.03	5.83	5.73	5.62	5.52	5.41	5.30	5.19	.005	
2.32	2.28	2.24	2.20	2.18	2.16	2.13	2.11	2.08	2.06	.100	10
2.98	2.91	2.85	2.77	2.74	2.70	2.66	2.62	2.58	2.54	.050	
3.72	3.62	3.52	3.42	3.37	3.31	3.26	3.20	3.14	3.08	.025	
4.85	4.71	4.56	4.41	4.33	4.25	4.17	4.08	4.00	3.91	.010	
5.85	5.66	5.47	5.27	5.17	5.07	4.97	4.86	4.75	4.64	.005	
2.25	2.21	2.17	2.12	2.10	2.08	2.05	2.03	2.00	1.97	.100	11
2.85	2.79	2.72	2.65	2.61	2.57	2.53	2.49	2.45	2.40	.050	
3.53	3.43	3.33	3.23	3.17	3.12	3.06	3.00	2.94	2.88	.025	
4.54	4.40	4.25	4.10	4.02	3.94	3.86	3.78	3.69	3.60	.010	
5.42	5.24	5.05	4.86	4.76	4.65	4.55	4.44	4.34	4.23	.005	
2.19	2.15	2.10	2.06	2.04	2.01	1.99	1.96	1.93	1.90	.100	12
2.75	2.69	2.62	2.54	2.51	2.47	2.43	2.38	2.34	2.30	.050	
3.37	3.28	3.18	3.07	3.02	2.96	2.91	2.85	2.79	2.72	.025	
4.30	4.16	4.01	3.86	3.78	3.70	3.62	3.54	3.45	3.36	.010	
5.09	4.91	4.72	4.53	4.43	4.33	4.23	4.12	4.01	3.90	.005	
2.14	2.10	2.05	2.01	1.98	1.96	1.93	1.90	1.88	1.85	.100	13
2.67	2.60	2.53	2.46	2.42	2.38	2.34	2.30	2.25	2.21	.050	
3.25	3.15	3.05	2.95	2.89	2.84	2.78	2.72	2.66	2.60	.025	
4.10	3.96	3.82	3.66	3.59	3.51	3.43	3.34	3.25	3.17	.010	
4.82	4.64	4.46	4.27	4.17	4.07	3.97	3.87	3.76	3.65	.005	
2.10	2.05	2.01	1.96	1.94	1.91	1.89	1.86	1.83	1.80	.100	14
2.60	2.53	2.46	2.39	2.35	2.31	2.27	2.22	2.18	2.13	.050	
3.15	3.05	2.95	2.84	2.79	2.73	2.67	2.61	2.55	2.49	.025	
3.94	3.80	3.66	3.51	3.43	3.35	3.27	3.18	3.09	3.00	.010	
4.60	4.43	4.25	4.06	3.96	3.86	3.76	3.66	3.55	3.44	.005	

Table A.6 (Continued)
Values of F

Denominator df	Probability of a larger F	\multicolumn{9}{c}{Numerator df}								
		1	2	3	4	5	6	7	8	9
15	.100	3.07	2.70	2.49	2.36	2.27	2.21	2.16	2.12	2.09
	.050	4.54	3.68	3.29	3.06	2.90	2.79	2.71	2.64	2.59
	.025	6.20	4.77	4.15	3.80	3.58	3.41	3.29	3.20	3.12
	.010	8.68	6.36	5.42	4.89	4.56	4.32	4.14	4.00	3.89
	.005	10.80	7.70	6.48	5.80	5.37	5.07	4.85	4.67	4.54
16	.100	3.05	2.67	2.46	2.33	2.24	2.18	2.13	2.09	2.06
	.050	4.49	3.63	3.24	3.01	2.85	2.74	2.66	2.59	2.54
	.025	6.12	4.69	4.08	3.73	3.50	3.34	3.22	3.12	3.05
	.010	8.53	6.23	5.29	4.77	4.44	4.20	4.03	3.89	3.78
	.005	10.58	7.51	6.30	5.64	5.21	4.91	4.69	4.52	4.38
17	.100	3.03	2.64	2.44	2.31	2.22	2.15	2.10	2.06	2.03
	.050	4.45	3.59	3.20	2.96	2.81	2.70	2.61	2.55	2.49
	.025	6.04	4.62	4.01	3.66	3.44	3.28	3.16	3.06	2.98
	.010	8.40	6.11	5.18	4.67	4.34	4.10	3.93	3.79	3.68
	.005	10.38	7.35	6.16	5.50	5.07	4.78	4.56	4.39	4.25
18	.100	3.01	2.62	2.42	2.29	2.20	2.13	2.08	2.04	2.00
	.050	4.41	3.55	3.16	2.93	2.77	2.66	2.58	2.51	2.46
	.025	5.98	4.56	3.95	3.61	3.38	3.22	3.10	3.01	2.93
	.010	8.29	6.01	5.09	4.58	4.25	4.01	3.84	3.71	3.60
	.005	10.22	7.21	6.03	5.37	4.96	4.66	4.44	4.28	4.14
19	.100	2.99	2.61	2.40	2.27	2.18	2.11	2.06	2.02	1.98
	.050	4.38	3.52	3.13	2.90	2.74	2.63	2.54	2.48	2.42
	.025	5.92	4.51	3.90	3.56	3.33	3.17	3.05	2.96	2.88
	.010	8.18	5.93	5.01	4.50	4.17	3.94	3.77	3.63	3.52
	.005	10.07	7.09	5.92	5.27	4.85	4.56	4.34	4.18	4.04
20	.100	2.97	2.59	2.38	2.25	2.16	2.09	2.04	2.00	1.96
	.050	4.35	3.49	3.10	2.87	2.71	2.60	2.51	2.45	2.39
	.025	5.87	4.46	3.86	3.51	3.29	3.13	3.01	2.91	2.84
	.010	8.10	5.85	4.94	4.43	4.10	3.87	3.70	3.56	3.46
	.005	9.94	6.99	5.82	5.17	4.76	4.47	4.26	4.09	3.96
21	.100	2.96	2.57	2.36	2.23	2.14	2.08	2.02	1.98	1.95
	.050	4.32	3.47	3.07	2.84	2.68	2.57	2.49	2.42	2.37
	.025	5.83	4.42	3.82	3.48	3.25	3.09	2.97	2.87	2.80
	.010	8.02	5.78	4.87	4.37	4.04	3.81	3.64	3.51	3.40
	.005	9.83	6.89	5.73	5.09	4.68	4.39	4.18	4.01	3.88
22	.100	2.95	2.56	2.35	2.22	2.13	2.06	2.01	1.97	1.93
	.050	4.30	3.44	3.05	2.82	2.66	2.55	2.46	2.40	2.34
	.025	5.79	4.38	3.78	3.44	3.22	3.05	2.93	2.84	2.76
	.010	7.95	5.72	4.82	4.31	3.99	3.76	3.59	3.45	3.35
	.005	9.73	6.81	5.65	5.02	4.61	4.32	4.11	3.94	3.81
23	.100	2.94	2.55	2.34	2.21	2.11	2.05	1.99	1.95	1.92
	.050	4.28	3.42	3.03	2.80	2.64	2.53	2.44	2.37	2.32
	.025	5.75	4.35	3.75	3.41	3.18	3.02	2.90	2.81	2.73
	.010	7.88	5.66	4.76	4.26	3.94	3.71	3.54	3.41	3.30
	.005	9.63	6.73	5.58	4.95	4.54	4.26	4.05	3.88	3.75
24	.100	2.93	2.54	2.33	2.19	2.10	2.04	1.98	1.94	1.91
	.050	4.26	3.40	3.01	2.78	2.62	2.51	2.42	2.36	2.30
	.025	5.72	4.32	3.72	3.38	3.15	2.99	2.87	2.78	2.70
	.010	7.82	5.61	4.72	4.22	3.90	3.67	3.50	3.36	3.26
	.005	9.55	6.66	5.52	4.89	4.49	4.20	3.99	3.83	3.69
25	.100	2.92	2.53	2.32	2.18	2.09	2.02	1.97	1.93	1.89
	.050	4.24	3.39	2.99	2.76	2.60	2.49	2.40	2.34	2.28
	.025	5.69	4.29	3.69	3.35	3.13	2.97	2.85	2.75	2.68
	.010	7.77	5.57	4.68	4.18	3.85	3.63	3.46	3.32	3.22
	.005	9.48	6.60	5.46	4.84	4.43	4.15	3.94	3.78	3.64
26	.100	2.91	2.52	2.31	2.17	2.08	2.01	1.96	1.92	1.88
	.050	4.23	3.37	2.98	2.74	2.59	2.47	2.39	2.32	2.27
	.025	5.66	4.27	3.67	3.33	3.10	2.94	2.82	2.73	2.65
	.010	7.72	5.53	4.64	4.14	3.82	3.59	3.42	3.29	3.18
	.005	9.41	6.54	5.41	4.79	4.38	4.10	3.89	3.73	3.60
27	.100	2.90	2.51	2.30	2.17	2.07	2.00	1.95	1.91	1.87
	.050	4.21	3.35	2.96	2.73	2.57	2.46	2.37	2.31	2.25
	.025	5.63	4.24	3.65	3.31	3.08	2.92	2.80	2.71	2.63
	.010	7.68	5.49	4.60	4.11	3.78	3.56	3.39	3.26	3.15
	.005	9.34	6.49	5.36	4.74	4.34	4.06	3.85	3.69	3.56
28	.100	2.89	2.50	2.29	2.16	2.06	2.00	1.94	1.90	1.87
	.050	4.20	3.34	2.95	2.71	2.56	2.45	2.36	2.29	2.24
	.025	5.61	4.22	3.63	3.29	3.06	2.90	2.78	2.69	2.61
	.010	7.64	5.45	4.57	4.07	3.75	3.53	3.36	3.23	3.12
	.005	9.28	6.44	5.32	4.70	4.30	4.02	3.81	3.65	3.52

Table A.6 (Continued)
Values of F

			Numerator df								
10	12	15	20	24	30	40	60	120	∞	P	df
2.06	2.02	1.97	1.92	1.90	1.87	1.85	1.82	1.79	1.76	.100	15
2.54	2.48	2.40	2.33	2.29	2.25	2.20	2.16	2.11	2.07	.050	
3.06	2.96	2.86	2.76	2.70	2.64	2.59	2.52	2.46	2.40	.025	
3.80	3.67	3.52	3.37	3.29	3.21	3.13	3.05	2.96	2.87	.010	
4.42	4.25	4.07	3.88	3.79	3.69	3.58	3.48	3.37	3.26	.005	
2.03	1.99	1.94	1.89	1.87	1.84	1.81	1.78	1.75	1.72	.100	16
2.49	2.42	2.35	2.28	2.24	2.19	2.15	2.11	2.06	2.01	.050	
2.99	2.89	2.79	2.68	2.63	2.57	2.51	2.45	2.38	2.32	.025	
3.69	3.55	3.41	3.26	3.18	3.10	3.02	2.93	2.84	2.75	.010	
4.27	4.10	3.92	3.73	3.64	3.54	3.44	3.33	3.22	3.11	.005	
2.00	1.96	1.91	1.86	1.84	1.81	1.78	1.75	1.72	1.69	.100	17
2.45	2.38	2.31	2.23	2.19	2.15	2.10	2.06	2.01	1.96	.050	
2.92	2.82	2.72	2.62	2.56	2.50	2.44	2.38	2.32	2.25	.025	
3.59	3.46	3.31	3.16	3.08	3.00	2.92	2.83	2.75	2.65	.010	
4.14	3.97	3.79	3.61	3.51	3.41	3.31	3.21	3.10	2.98	.005	
1.98	1.93	1.89	1.84	1.81	1.78	1.75	1.72	1.69	1.66	.100	18
2.41	2.34	2.27	2.19	2.15	2.11	2.06	2.02	1.97	1.92	.050	
2.87	2.77	2.67	2.56	2.50	2.44	2.38	2.32	2.26	2.19	.025	
3.51	3.37	3.23	3.08	3.00	2.92	2.84	2.75	2.66	2.57	.010	
4.03	3.86	3.68	3.50	3.40	3.30	3.20	3.10	2.99	2.87	.005	
1.96	1.91	1.86	1.81	1.79	1.76	1.73	1.70	1.67	1.63	.100	19
2.38	2.31	2.23	2.16	2.11	2.07	2.03	1.98	1.93	1.88	.050	
2.82	2.72	2.62	2.51	2.45	2.39	2.33	2.27	2.20	2.13	.025	
3.43	3.30	3.15	3.00	2.92	2.84	2.76	2.67	2.58	2.49	.010	
3.93	3.76	3.59	3.40	3.31	3.21	3.11	3.00	2.89	2.78	.005	
1.94	1.89	1.84	1.79	1.77	1.74	1.71	1.68	1.64	1.61	.100	20
2.35	2.28	2.20	2.12	2.08	2.04	1.99	1.95	1.90	1.84	.050	
2.77	2.68	2.57	2.46	2.41	2.35	2.29	2.22	2.16	2.09	.025	
3.37	3.23	3.09	2.94	2.86	2.78	2.69	2.61	2.52	2.42	.010	
3.85	3.68	3.50	3.32	3.22	3.12	3.02	2.92	2.81	2.69	.005	
1.92	1.87	1.83	1.78	1.75	1.72	1.69	1.66	1.62	1.59	.100	21
2.32	2.25	2.18	2.10	2.05	2.01	1.96	1.92	1.87	1.81	.050	
2.73	2.64	2.53	2.42	2.37	2.31	2.25	2.18	2.11	2.04	.025	
3.31	3.17	3.03	2.88	2.80	2.72	2.64	2.55	2.46	2.36	.010	
3.77	3.60	3.43	3.24	3.15	3.05	2.95	2.84	2.73	2.61	.005	
1.90	1.86	1.81	1.76	1.73	1.70	1.67	1.64	1.60	1.57	.100	22
2.30	2.23	2.15	2.07	2.03	1.98	1.94	1.89	1.84	1.78	.050	
2.70	2.60	2.50	2.39	2.33	2.27	2.21	2.14	2.08	2.00	.025	
3.26	3.12	2.98	2.83	2.75	2.67	2.58	2.50	2.40	2.31	.010	
3.70	3.54	3.36	3.18	3.08	2.98	2.88	2.77	2.66	2.55	.005	
1.89	1.84	1.80	1.74	1.72	1.69	1.66	1.62	1.59	1.55	.100	23
2.27	2.20	2.13	2.05	2.01	1.96	1.91	1.86	1.81	1.76	.050	
2.67	2.57	2.47	2.36	2.30	2.24	2.18	2.11	2.04	1.97	.025	
3.21	3.07	2.93	2.78	2.70	2.62	2.54	2.45	2.35	2.26	.010	
3.64	3.47	3.30	3.12	3.02	2.92	2.82	2.71	2.60	2.48	.005	
1.88	1.83	1.78	1.73	1.70	1.67	1.64	1.61	1.57	1.53	.100	24
2.25	2.18	2.11	2.03	1.98	1.94	1.89	1.84	1.79	1.73	.050	
2.64	2.54	2.44	2.33	2.27	2.21	2.15	2.08	2.01	1.94	.025	
3.17	3.03	2.89	2.74	2.66	2.58	2.49	2.40	2.31	2.21	.010	
3.59	3.42	3.25	3.06	2.97	2.87	2.77	2.66	2.55	2.43	.005	
1.87	1.82	1.77	1.72	1.69	1.66	1.63	1.59	1.56	1.52	.100	25
2.24	2.16	2.09	2.01	1.96	1.92	1.87	1.82	1.77	1.71	.050	
2.61	2.51	2.41	2.30	2.24	2.18	2.12	2.05	1.98	1.91	.025	
3.13	2.99	2.85	2.70	2.62	2.54	2.45	2.36	2.27	2.17	.010	
3.54	3.37	3.20	3.01	2.92	2.82	2.72	2.61	2.50	2.38	.005	
1.86	1.81	1.76	1.71	1.68	1.65	1.61	1.58	1.54	1.50	.100	26
2.22	2.15	2.07	1.99	1.95	1.90	1.85	1.80	1.75	1.69	.050	
2.59	2.49	2.39	2.28	2.22	2.16	2.09	2.03	1.95	1.88	.025	
3.09	2.96	2.81	2.66	2.58	2.50	2.42	2.33	2.23	2.13	.010	
3.49	3.33	3.15	2.97	2.87	2.77	2.67	2.56	2.45	2.33	.005	
1.85	1.80	1.75	1.70	1.67	1.64	1.60	1.57	1.53	1.49	.100	27
2.20	2.13	2.06	1.97	1.93	1.88	1.84	1.79	1.73	1.67	.050	
2.57	2.47	2.36	2.25	2.19	2.13	2.07	2.00	1.93	1.85	.025	
3.06	2.93	2.78	2.63	2.55	2.47	2.38	2.29	2.20	2.10	.010	
3.45	3.28	3.11	2.93	2.83	2.73	2.63	2.52	2.41	2.29	.005	
1.84	1.79	1.74	1.69	1.66	1.63	1.59	1.56	1.52	1.48	.100	28
2.19	2.12	2.04	1.96	1.91	1.87	1.82	1.77	1.71	1.65	.050	
2.55	2.45	2.34	2.23	2.17	2.11	2.05	1.98	1.91	1.83	.025	
3.03	2.90	2.75	2.60	2.52	2.44	2.35	2.26	2.17	2.06	.010	
3.41	3.25	3.07	2.89	2.79	2.69	2.59	2.48	2.37	2.25	.005	

TABLE A.6 (Continued)

VALUES OF F

Denominator df	Probability of a larger F	Numerator df								
		1	2	3	4	5	6	7	8	9
29	.100	2.89	2.50	2.28	2.15	2.06	1.99	1.93	1.89	1.86
	.050	4.18	3.33	2.93	2.70	2.55	2.43	2.35	2.28	2.22
	.025	5.59	4.20	3.61	3.27	3.04	2.88	2.76	2.67	2.59
	.010	7.60	5.42	4.54	4.04	3.73	3.50	3.33	3.20	3.09
	.005	9.23	6.40	5.28	4.66	4.26	3.98	3.77	3.61	3.48
30	.100	2.88	2.49	2.28	2.14	2.05	1.98	1.93	1.88	1.85
	.050	4.17	3.32	2.92	2.69	2.53	2.42	2.33	2.27	2.21
	.025	5.57	4.18	3.59	3.25	3.03	2.87	2.75	2.65	2.57
	.010	7.56	5.39	4.51	4.02	3.70	3.47	3.30	3.17	3.07
	.005	9.18	6.35	5.24	4.62	4.23	3.95	3.74	3.58	3.45
40	.100	2.84	2.44	2.23	2.09	2.00	1.93	1.87	1.83	1.79
	.050	4.08	3.23	2.84	2.61	2.45	2.34	2.25	2.18	2.12
	.025	5.42	4.05	3.46	3.13	2.90	2.74	2.62	2.53	2.45
	.010	7.31	5.18	4.31	3.83	3.51	3.29	3.12	2.99	2.89
	.005	8.83	6.07	4.98	4.37	3.99	3.71	3.51	3.35	3.22
60	.100	2.79	2.39	2.18	2.04	1.95	1.87	1.82	1.77	1.74
	.050	4.00	3.15	2.76	2.53	2.37	2.25	2.17	2.10	2.04
	.025	5.29	3.93	3.34	3.01	2.79	2.63	2.51	2.41	2.33
	.010	7.08	4.98	4.13	3.65	3.34	3.12	2.95	2.82	2.72
	.005	8.49	5.79	4.73	4.14	3.76	3.49	3.29	3.13	3.01
120	.100	2.75	2.35	2.13	1.99	1.90	1.82	1.77	1.72	1.68
	.050	3.92	3.07	2.68	2.45	2.29	2.17	2.09	2.02	1.96
	.025	5.15	3.80	3.23	2.89	2.67	2.52	2.39	2.30	2.22
	.010	6.85	4.79	3.95	3.48	3.17	2.96	2.79	2.66	2.56
	.005	8.18	5.54	4.50	3.92	3.55	3.28	3.09	2.93	2.81
∞	.100	2.71	2.30	2.08	1.94	1.85	1.77	1.72	1.67	1.63
	.050	3.84	3.00	2.60	2.37	2.21	2.10	2.01	1.94	1.88
	.025	5.02	3.69	3.12	2.79	2.57	2.41	2.29	2.19	2.11
	.010	6.63	4.61	3.78	3.32	3.02	2.80	2.64	2.51	2.41
	.005	7.88	5.30	4.28	3.72	3.35	3.09	2.90	2.74	2.62

SOURCE: A portion of "Tables of percentage points of the inverted beta (F) distribution," *Biometrika*, vol. 33 (1943) by M. Merrington and C. M. Thompson and from Table 18 of *Biometrika Tables for Statisticians*, vol. 1, Cambridge University Press, 1954, edited by E. S. Pearson and H. O. Hartley. Reproduced with permission of the authors, editors, and *Biometrika* trustees.

TABLE A.6 (Continued)

VALUES OF F

				Numerator df							
10	12	15	20	24	30	40	60	120	∞	P	df
1.83	1.78	1.73	1.68	1.65	1.62	1.58	1.55	1.51	1.47	.100	29
2.18	2.10	2.03	1.94	1.90	1.85	1.81	1.75	1.70	1.64	.050	
2.53	2.43	2.32	2.21	2.15	2.09	2.03	1.96	1.89	1.81	.025	
3.00	2.87	2.73	2.57	2.49	2.41	2.33	2.23	2.14	2.03	.010	
3.38	3.21	3.04	2.86	2.76	2.66	2.56	2.45	2.33	2.21	.005	
1.82	1.77	1.72	1.67	1.64	1.61	1.57	1.54	1.50	1.46	.100	30
2.16	2.09	2.01	1.93	1.89	1.84	1.79	1.74	1.68	1.62	.050	
2.51	2.41	2.31	2.20	2.14	2.07	2.01	1.94	1.87	1.79	.025	
2.98	2.84	2.70	2.55	2.47	2.39	2.30	2.21	2.11	2.01	.010	
3.34	3.18	3.01	2.82	2.73	2.63	2.52	2.42	2.30	2.18	.005	
1.76	1.71	1.66	1.61	1.57	1.54	1.51	1.47	1.42	1.38	.100	40
2.08	2.00	1.92	1.84	1.79	1.74	1.69	1.64	1.58	1.51	.050	
2.39	2.29	2.18	2.07	2.01	1.94	1.88	1.80	1.72	1.64	.025	
2.80	2.66	2.52	2.37	2.29	2.20	2.11	2.02	1.92	1.80	.010	
3.12	2.95	2.78	2.60	2.50	2.40	2.30	2.18	2.06	1.93	.005	
1.71	1.66	1.60	1.54	1.51	1.48	1.44	1.40	1.35	1.29	.100	60
1.99	1.92	1.84	1.75	1.70	1.65	1.59	1.53	1.47	1.39	.050	
2.27	2.17	2.06	1.94	1.88	1.82	1.74	1.67	1.58	1.48	.025	
2.63	2.50	2.35	2.20	2.12	2.03	1.94	1.84	1.73	1.60	.010	
2.90	2.74	2.57	2.39	2.29	2.19	2.08	1.96	1.83	1.69	.005	
1.65	1.60	1.55	1.48	1.45	1.41	1.37	1.32	1.26	1.19	.100	120
1.91	1.83	1.75	1.66	1.61	1.55	1.50	1.43	1.35	1.25	.050	
2.16	2.05	1.94	1.82	1.76	1.69	1.61	1.53	1.43	1.31	.025	
2.47	2.34	2.19	2.03	1.95	1.86	1.76	1.66	1.53	1.38	.010	
2.71	2.54	2.37	2.19	2.09	1.98	1.87	1.75	1.61	1.43	.005	
1.60	1.55	1.49	1.42	1.38	1.34	1.30	1.24	1.17	1.00	.100	∞
1.83	1.75	1.67	1.57	1.52	1.46	1.39	1.32	1.22	1.00	.050	
2.05	1.94	1.83	1.71	1.64	1.57	1.48	1.39	1.27	1.00	.025	
2.32	2.18	2.04	1.88	1.79	1.70	1.59	1.47	1.32	1.00	.010	
2.52	2.36	2.19	2.00	1.90	1.79	1.67	1.53	1.36	1.00	.005	

DUNCAN'S

TABLE A.7

SIGNIFICANT STUDENTIZED RANGES FOR 5% AND 1% LEVEL NEW MULTIPLE-RANGE TEST

p = number of means for range being tested

Error df	Protection level	2	3	4	5	6	7	8	9	10	12	14	16	18	20
1	.05	18.0	18.0	18.0	18.0	18.0	18.0	18.0	18.0	18.0	18.0	18.0	18.0	18.0	18.0
	.01	90.0	90.0	90.0	90.0	90.0	90.0	90.0	90.0	90.0	90.0	90.0	90.0	90.0	90.0
2	.05	6.09	6.09	6.09	6.09	6.09	6.09	6.09	6.09	6.09	6.09	6.09	6.09	6.09	6.09
	.01	14.0	14.0	14.0	14.0	14.0	14.0	14.0	14.0	14.0	14.0	14.0	14.0	14.0	14.0
3	.05	4.50	4.50	4.50	4.50	4.50	4.50	4.50	4.50	4.50	4.50	4.50	4.50	4.50	4.50
	.01	8.26	8.5	8.6	8.7	8.8	8.9	8.9	9.0	9.0	9.0	9.1	9.2	9.3	9.3
4	.05	3.93	4.01	4.02	4.02	4.02	4.02	4.02	4.02	4.02	4.02	4.02	4.02	4.02	4.02
	.01	6.51	6.8	6.9	7.0	7.1	7.1	7.2	7.2	7.3	7.3	7.4	7.4	7.5	7.5
5	.05	3.64	3.74	3.79	3.83	3.83	3.83	3.83	3.83	3.83	3.83	3.83	3.83	3.83	3.83
	.01	5.70	5.96	6.11	6.18	6.26	6.33	6.40	6.44	6.5	6.6	6.6	6.7	6.7	6.8
6	.05	3.46	3.58	3.64	3.68	3.68	3.68	3.68	3.68	3.68	3.68	3.68	3.68	3.68	3.68
	.01	5.24	5.51	5.65	5.73	5.81	5.88	5.95	6.00	6.0	6.1	6.2	6.2	6.3	6.3
7	.05	3.35	3.47	3.54	3.58	3.60	3.61	3.61	3.61	3.61	3.61	3.61	3.61	3.61	3.61
	.01	4.95	5.22	5.37	5.45	5.53	5.61	5.69	5.73	5.8	5.8	5.9	5.9	6.0	6.0
8	.05	3.26	3.39	3.47	3.52	3.55	3.56	3.56	3.56	3.56	3.56	3.56	3.56	3.56	3.56
	.01	4.74	5.00	5.14	5.23	5.32	5.40	5.47	5.51	5.5	5.6	5.7	5.7	5.8	5.8
9	.05	3.20	3.34	3.41	3.47	3.50	3.52	3.52	3.52	3.52	3.52	3.52	3.52	3.52	3.52
	.01	4.60	4.86	4.99	5.08	5.17	5.25	5.32	5.36	5.4	5.5	5.5	5.6	5.7	5.7
10	.05	3.15	3.30	3.37	3.43	3.46	3.47	3.47	3.47	3.47	3.47	3.47	3.47	3.47	3.48
	.01	4.48	4.73	4.88	4.96	5.06	5.13	5.20	5.24	5.28	5.36	5.42	5.48	5.54	5.55
11	.05	3.11	3.27	3.35	3.39	3.43	3.44	3.45	3.46	3.46	3.46	3.46	3.46	3.47	3.48
	.01	4.39	4.63	4.77	4.86	4.94	5.01	5.06	5.12	5.15	5.24	5.28	5.34	5.38	5.39
12	.05	3.08	3.23	3.33	3.36	3.40	3.42	3.44	3.44	3.46	3.46	3.46	3.46	3.47	3.48
	.01	4.32	4.55	4.68	4.76	4.84	4.92	4.96	5.02	5.07	5.13	5.17	5.22	5.24	5.26
13	.05	3.06	3.21	3.30	3.35	3.38	3.41	3.42	3.44	3.45	3.45	3.46	3.46	3.47	3.47
	.01	4.26	4.48	4.62	4.69	4.74	4.84	4.88	4.94	4.98	5.04	5.08	5.13	5.14	5.15
14	.05	3.03	3.18	3.27	3.33	3.37	3.39	3.41	3.42	3.44	3.45	3.46	3.46	3.47	3.47
	.01	4.21	4.42	4.55	4.63	4.70	4.78	4.83	4.87	4.91	4.96	5.00	5.04	5.06	5.07
15	.05	3.01	3.16	3.25	3.31	3.36	3.38	3.40	3.42	3.43	3.44	3.45	3.46	3.47	3.47
	.01	4.17	4.37	4.50	4.58	4.64	4.72	4.77	4.81	4.84	4.90	4.94	4.97	4.99	5.00

Table A.7 (Continued)
Significant Studentized Ranges for 5% and 1% Level New Multiple-range Test

p = number of means for range being tested

Error df	Protection level	2	3	4	5	6	7	8	9	10	12	14	16	18	20
16	.05	3.00	3.15	3.23	3.30	3.34	3.37	3.39	3.41	3.43	3.44	3.45	3.46	3.47	3.47
	.01	4.13	4.34	4.45	4.54	4.60	4.67	4.72	4.76	4.79	4.84	4.88	4.91	4.93	4.94
17	.05	2.98	3.13	3.22	3.28	3.33	3.36	3.38	3.40	3.42	3.44	3.45	3.46	3.47	3.47
	.01	4.10	4.30	4.41	4.50	4.56	4.63	4.68	4.72	4.75	4.80	4.83	4.86	4.88	4.89
18	.05	2.97	3.12	3.21	3.27	3.32	3.35	3.37	3.39	3.41	3.43	3.45	3.46	3.47	3.47
	.01	4.07	4.27	4.38	4.46	4.53	4.59	4.64	4.68	4.71	4.76	4.79	4.82	4.84	4.85
19	.05	2.96	3.11	3.19	3.26	3.31	3.35	3.37	3.39	3.41	3.43	3.44	3.46	3.47	3.47
	.01	4.05	4.24	4.35	4.43	4.50	4.56	4.61	4.64	4.67	4.72	4.76	4.79	4.81	4.82
20	.05	2.95	3.10	3.18	3.25	3.30	3.34	3.36	3.38	3.40	3.43	3.44	3.46	3.46	3.47
	.01	4.02	4.22	4.33	4.40	4.47	4.53	4.58	4.61	4.65	4.69	4.73	4.76	4.78	4.79
22	.05	2.93	3.08	3.17	3.24	3.29	3.32	3.35	3.37	3.39	3.42	3.44	3.45	3.46	3.47
	.01	3.99	4.17	4.28	4.36	4.42	4.48	4.53	4.57	4.60	4.65	4.68	4.71	4.74	4.75
24	.05	2.92	3.07	3.15	3.22	3.28	3.31	3.34	3.37	3.38	3.41	3.44	3.45	3.46	3.47
	.01	3.96	4.14	4.24	4.33	4.39	4.44	4.49	4.53	4.57	4.62	4.64	4.67	4.70	4.72
26	.05	2.91	3.06	3.14	3.21	3.27	3.30	3.34	3.36	3.38	3.41	3.43	3.45	3.46	3.47
	.01	3.93	4.11	4.21	4.30	4.36	4.41	4.46	4.50	4.53	4.58	4.62	4.65	4.67	4.69
28	.05	2.90	3.04	3.13	3.20	3.26	3.30	3.33	3.35	3.37	3.40	3.43	3.45	3.46	3.47
	.01	3.91	4.08	4.18	4.28	4.34	4.39	4.43	4.47	4.51	4.56	4.60	4.62	4.65	4.67
30	.05	2.89	3.04	3.12	3.20	3.25	3.29	3.32	3.35	3.37	3.40	3.43	3.44	3.46	3.47
	.01	3.89	4.06	4.16	4.22	4.32	4.36	4.41	4.45	4.48	4.54	4.58	4.61	4.63	4.65
40	.05	2.86	3.01	3.10	3.17	3.22	3.27	3.30	3.33	3.35	3.39	3.42	3.44	3.46	3.47
	.01	3.82	3.99	4.10	4.17	4.24	4.30	4.34	4.37	4.41	4.46	4.51	4.54	4.57	4.59
60	.05	2.83	2.98	3.08	3.14	3.20	3.24	3.28	3.31	3.33	3.37	3.40	3.43	3.45	3.47
	.01	3.76	3.92	4.03	4.12	4.17	4.23	4.27	4.31	4.34	4.39	4.44	4.47	4.50	4.53
100	.05	2.80	2.95	3.05	3.12	3.18	3.22	3.26	3.29	3.32	3.36	3.40	3.42	3.45	3.47
	.01	3.71	3.86	3.98	4.06	4.11	4.17	4.21	4.25	4.29	4.35	4.38	4.42	4.45	4.48
∞	.05	2.77	2.92	3.02	3.09	3.15	3.19	3.23	3.26	3.29	3.34	3.38	3.41	3.44	3.47
	.01	3.64	3.80	3.90	3.98	4.04	4.09	4.14	4.17	4.20	4.26	4.31	4.34	4.38	4.41

Source: Abridged from D. B. Duncan, "Multiple range and multiple F tests," *Biometrics*, **11**: 1–42 (1955), with the permission of the editor and the author.

TABLE A.8

UPPER PERCENTAGE POINTS OF THE STUDENTIZED RANGE, $q_\alpha = \dfrac{\bar{x}_{max} - \bar{x}_{min}}{s_{\bar{x}}}$

Error df	α	\multicolumn{10}{c}{p = number of}
		2	3	4	5	6	7	8	9	10	11
5	.05	3.64	4.60	5.22	5.67	6.03	6.33	6.58	6.80	6.99	7.17
	.01	5.70	6.97	7.80	8.42	8.91	9.32	9.67	9.97	10.24	10.48
6	.05	3.46	4.34	4.90	5.31	5.63	5.89	6.12	6.32	6.49	6.65
	.01	5.24	6.33	7.03	7.56	7.97	8.32	8.61	8.87	9.10	9.30
7	.05	3.34	4.16	4.68	5.06	5.36	5.61	5.82	6.00	6.16	6.30
	.01	4.95	5.92	6.54	7.01	7.37	7.68	7.94	8.17	8.37	8.55
8	.05	3.26	4.04	4.53	4.89	5.17	5.40	5.60	5.77	5.92	6.05
	.01	4.74	5.63	6.20	6.63	6.96	7.24	7.47	7.68	7.87	8.03
9	.05	3.20	3.95	4.42	4.76	5.02	5.24	5.43	5.60	5.74	5.87
	.01	4.60	5.43	5.96	6.35	6.66	6.91	7.13	7.32	7.49	7.65
10	.05	3.15	3.88	4.33	4.65	4.91	5.12	5.30	5.46	5.60	5.72
	.01	4.48	5.27	5.77	6.14	6.43	6.67	6.87	7.05	7.21	7.36
11	.05	3.11	3.82	4.26	4.57	4.82	5.03	5.20	5.35	5.49	5.61
	.01	4.39	5.14	5.62	5.97	6.25	6.48	6.67	6.84	6.99	7.13
12	.05	3.08	3.77	4.20	4.51	4.75	4.95	5.12	5.27	5.40	5.51
	.01	4.32	5.04	5.50	5.84	6.10	6.32	6.51	6.67	6.81	6.94
13	.05	3.06	3.73	4.15	4.45	4.69	4.88	5.05	5.19	5.32	5.43
	.01	4.26	4.96	5.40	5.73	5.98	6.19	6.37	6.53	6.67	6.79
14	.05	3.03	3.70	4.11	4.41	4.64	4.83	4.99	5.13	5.25	5.36
	.01	4.21	4.89	5.32	5.63	5.88	6.08	6.26	6.41	6.54	6.66
15	.05	3.01	3.67	4.08	4.37	4.60	4.78	4.94	5.08	5.20	5.31
	.01	4.17	4.83	5.25	5.56	5.80	5.99	6.16	6.31	6.44	6.55
16	.05	3.00	3.65	4.05	4.33	4.56	4.74	4.90	5.03	5.15	5.26
	.01	4.13	4.78	5.19	5.49	5.72	5.92	6.08	6.22	6.35	6.46
17	.05	2.98	3.63	4.02	4.30	4.52	4.71	4.86	4.99	5.11	5.21
	.01	4.10	4.74	5.14	5.43	5.66	5.85	6.01	6.15	6.27	6.38
18	.05	2.97	3.61	4.00	4.28	4.49	4.67	4.82	4.96	5.07	5.17
	.01	4.07	4.70	5.09	5.38	5.60	5.79	5.94	6.08	6.20	6.31
19	.05	2.96	3.59	3.98	4.25	4.47	4.65	4.79	4.92	5.04	5.14
	.01	4.05	4.67	5.05	5.33	5.55	5.73	5.89	6.02	6.14	6.25
20	.05	2.95	3.58	3.96	4.23	4.45	4.62	4.77	4.90	5.01	5.11
	.01	4.02	4.64	5.02	5.29	5.51	5.69	5.84	5.97	6.09	6.19
24	.05	2.92	3.53	3.90	4.17	4.37	4.54	4.68	4.81	4.92	5.01
	.01	3.96	4.54	4.91	5.17	5.37	5.54	5.69	5.81	5.92	6.02
30	.05	2.89	3.49	3.84	4.10	4.30	4.46	4.60	4.72	4.83	4.92
	.01	3.89	4.45	4.80	5.05	5.24	5.40	5.54	5.65	5.76	5.85
40	.05	2.86	3.44	3.79	4.04	4.23	4.39	4.52	4.63	4.74	4.82
	.01	3.82	4.37	4.70	4.93	5.11	5.27	5.39	5.50	5.60	5.69
60	.05	2.83	3.40	3.74	3.98	4.16	4.31	4.44	4.55	4.65	4.73
	.01	3.76	4.28	4.60	4.82	4.99	5.13	5.25	5.36	5.45	5.53
120	.05	2.80	3.36	3.69	3.92	4.10	4.24	4.36	4.48	4.56	4.64
	.01	3.70	4.20	4.50	4.71	4.87	5.01	5.12	5.21	5.30	5.38
∞	.05	2.77	3.31	3.63	3.86	4.03	4.17	4.29	4.39	4.47	4.55
	.01	3.64	4.12	4.40	4.60	4.76	4.88	4.99	5.08	5.16	5.23

SOURCE: This table is abridged from Table 29, *Biometrika Tables for Statisticians*, vol. 1, Cambridge University Press, 1954. It is reproduced with permission of the *Biometrika* trustees and the editors, E. S. Pearson and H. O. Hartley. The original work appeared in a paper by J. M. May, "Extended and corrected tables of the upper percentage points of the 'Studentized' range," *Biometrika*, **39**: 192–193 (1952).

Table A.8 (Continued)

Upper Percentage Points of the Studentized Range, $q_\alpha = \dfrac{\bar{x}_{max} - \bar{x}_{min}}{s_{\bar{x}}}$

treatment means								α	Error df
12	13	14	15	16	17	18	19	20		
7.32	7.47	7.60	7.72	7.83	7.93	8.03	8.12	8.21	.05	5
10.70	10.89	11.08	11.24	11.40	11.55	11.68	11.81	11.93	.01	
6.79	6.92	7.03	7.14	7.24	7.34	7.43	7.51	7.59	.05	6
9.49	9.65	9.81	9.95	10.08	10.21	10.32	10.43	10.54	.01	
6.43	6.55	6.66	6.76	6.85	6.94	7.02	7.09	7.17	.05	7
8.71	8.86	9.00	9.12	9.24	9.35	9.46	9.55	9.65	.01	
6.18	6.29	6.39	6.48	6.57	6.65	6.73	6.80	6.87	.05	8
8.18	8.31	8.44	8.55	8.66	8.76	8.85	8.94	9.03	.01	
5.98	6.09	6.19	6.28	6.36	6.44	6.51	6.58	6.64	.05	9
7.78	7.91	8.03	8.13	8.23	8.32	8.41	8.49	8.57	.01	
5.83	5.93	6.03	6.11	6.20	6.27	6.34	6.40	6.47	.05	10
7.48	7.60	7.71	7.81	7.91	7.99	8.07	8.15	8.22	.01	
5.71	5.81	5.90	5.99	6.06	6.14	6.20	6.26	6.33	.05	11
7.25	7.36	7.46	7.56	7.65	7.73	7.81	7.88	7.95	.01	
5.62	5.71	5.80	5.88	5.95	6.03	6.09	6.15	6.21	.05	12
7.06	7.17	7.26	7.36	7.44	7.52	7.59	7.66	7.73	.01	
5.53	5.63	5.71	5.79	5.86	5.93	6.00	6.05	6.11	.05	13
6.90	7.01	7.10	7.19	7.27	7.34	7.42	7.48	7.55	.01	
5.46	5.55	5.64	5.72	5.79	5.85	5.92	5.97	6.03	.05	14
6.77	6.87	6.96	7.05	7.12	7.20	7.27	7.33	7.39	.01	
5.40	5.49	5.58	5.65	5.72	5.79	5.85	5.90	5.96	.05	15
6.66	6.76	6.84	6.93	7.00	7.07	7.14	7.20	7.26	.01	
5.35	5.44	5.52	5.59	5.66	5.72	5.79	5.84	5.90	.05	16
6.56	6.66	6.74	6.82	6.90	6.97	7.03	7.09	7.15	.01	
5.31	5.39	5.47	5.55	5.61	5.68	5.74	5.79	5.84	.05	17
6.48	6.57	6.66	6.73	6.80	6.87	6.94	7.00	7.05	.01	
5.27	5.35	5.43	5.50	5.57	5.63	5.69	5.74	5.79	.05	18
6.41	6.50	6.58	6.65	6.72	6.79	6.85	6.91	6.96	.01	
5.23	5.32	5.39	5.46	5.53	5.59	5.65	5.70	5.75	.05	19
6.34	6.43	6.51	6.58	6.65	6.72	6.78	6.84	6.89	.01	
5.20	5.28	5.36	5.43	5.49	5.55	5.61	5.66	5.71	.05	20
6.29	6.37	6.45	6.52	6.59	6.65	6.71	6.76	6.82	.01	
5.10	5.18	5.25	5.32	5.38	5.44	5.50	5.54	5.59	.05	24
6.11	6.19	6.26	6.33	6.39	6.45	6.51	6.56	6.61	.01	
5.00	5.08	5.15	5.21	5.27	5.33	5.38	5.43	5.48	.05	30
5.93	6.01	6.08	6.14	6.20	6.26	6.31	6.36	6.41	.01	
4.91	4.98	5.05	5.11	5.16	5.22	5.27	5.31	5.36	.05	40
5.77	5.84	5.90	5.96	6.02	6.07	6.12	6.17	6.21	.01	
4.81	4.88	4.94	5.00	5.06	5.11	5.16	5.20	5.24	.05	60
5.60	5.67	5.73	5.79	5.84	5.89	5.93	5.98	6.02	.01	
4.72	4.78	4.84	4.90	4.95	5.00	5.05	5.09	5.13	.05	120
5.44	5.51	5.56	5.61	5.66	5.71	5.75	5.79	5.83	.01	
4.62	4.68	4.74	4.80	4.85	4.89	4.93	4.97	5.01	.05	∞
5.29	5.35	5.40	5.45	5.49	5.54	5.57	5.61	5.65	.01	

TABLE A.9A

TABLE OF t FOR ONE-SIDED COMPARISONS BETWEEN p TREATMENT MEANS AND A CONTROL FOR A JOINT CONFIDENCE COEFFICIENT OF $P = .95$ AND $P = .99$

Error df	P	p = number of treatment means, excluding control								
		1	2	3	4	5	6	7	8	9
5	.95	2.02	2.44	2.68	2.85	2.98	3.08	3.16	3.24	3.30
	.99	3.37	3.90	4.21	4.43	4.60	4.73	4.85	4.94	5.03
6	.95	1.94	2.34	2.56	2.71	2.83	2.92	3.00	3.07	3.12
	.99	3.14	3.61	3.88	4.07	4.21	4.33	4.43	4.51	4.59
7	.95	1.89	2.27	2.48	2.62	2.73	2.82	2.89	2.95	3.01
	.99	3.00	3.42	3.66	3.83	3.96	4.07	4.15	4.23	4.30
8	.95	1.86	2.22	2.42	2.55	2.66	2.74	2.81	2.87	2.92
	.99	2.90	3.29	3.51	3.67	3.79	3.88	3.96	4.03	4.09
9	.95	1.83	2.18	2.37	2.50	2.60	2.68	2.75	2.81	2.86
	.99	2.82	3.19	3.40	3.55	3.66	3.75	3.82	3.89	3.94
10	.95	1.81	2.15	2.34	2.47	2.56	2.64	2.70	2.76	2.81
	.99	2.76	3.11	3.31	3.45	3.56	3.64	3.71	3.78	3.83
11	.95	1.80	2.13	2.31	2.44	2.53	2.60	2.67	2.72	2.77
	.99	2.72	3.06	3.25	3.38	3.48	3.56	3.63	3.69	3.74
12	.95	1.78	2.11	2.29	2.41	2.50	2.58	2.64	2.69	2.74
	.99	2.68	3.01	3.19	3.32	3.42	3.50	3.56	3.62	3.67
13	.95	1.77	2.09	2.27	2.39	2.48	2.55	2.61	2.66	2.71
	.99	2.65	2.97	3.15	3.27	3.37	3.44	3.51	3.56	3.61
14	.95	1.76	2.08	2.25	2.37	2.46	2.53	2.59	2.64	2.69
	.99	2.62	2.94	3.11	3.23	3.32	3.40	3.46	3.51	3.56
15	.95	1.75	2.07	2.24	2.36	2.44	2.51	2.57	2.62	2.67
	.99	2.60	2.91	3.08	3.20	3.29	3.36	3.42	3.47	3.52
16	.95	1.75	2.06	2.23	2.34	2.43	2.50	2.56	2.61	2.65
	.99	2.58	2.88	3.05	3.17	3.26	3.33	3.39	3.44	3.48
17	.95	1.74	2.05	2.22	2.33	2.42	2.49	2.54	2.59	2.64
	.99	2.57	2.86	3.03	3.14	3.23	3.30	3.36	3.41	3.45
18	.95	1.73	2.04	2.21	2.32	2.41	2.48	2.53	2.58	2.62
	.99	2.55	2.84	3.01	3.12	3.21	3.27	3.33	3.38	3.42
19	.95	1.73	2.03	2.20	2.31	2.40	2.47	2.52	2.57	2.61
	.99	2.54	2.83	2.99	3.10	3.18	3.25	3.31	3.36	3.40
20	.95	1.72	2.03	2.19	2.30	2.39	2.46	2.51	2.56	2.60
	.99	2.53	2.81	2.97	3.08	3.17	3.23	3.29	3.34	3.38
24	.95	1.71	2.01	2.17	2.28	2.36	2.43	2.48	2.53	2.57
	.99	2.49	2.77	2.92	3.03	3.11	3.17	3.22	3.27	3.31
30	.95	1.70	1.99	2.15	2.25	2.33	2.40	2.45	2.50	2.54
	.99	2.46	2.72	2.87	2.97	3.05	3.11	3.16	3.21	3.24
40	.95	1.68	1.97	2.13	2.23	2.31	2.37	2.42	2.47	2.51
	.99	2.42	2.68	2.82	2.92	2.99	3.05	3.10	3.14	3.18
60	.95	1.67	1.95	2.10	2.21	2.28	2.35	2.39	2.44	2.48
	.99	2.39	2.64	2.78	2.87	2.94	3.00	3.04	3.08	3.12
120	.95	1.66	1.93	2.08	2.18	2.26	2.32	2.37	2.41	2.45
	.99	2.36	2.60	2.73	2.82	2.89	2.94	2.99	3.03	3.06
∞	.95	1.64	1.92	2.06	2.16	2.23	2.29	2.34	2.38	2.42
	.99	2.33	2.56	2.68	2.77	2.84	2.89	2.93	2.97	3.00

SOURCE: This table is reproduced from "A multiple comparison procedure for comparing several treatments with a control," *J. Am. Stat. Assn.*, **50**: 1096–1121 (1955), with permission of the author, C. W. Dunnett, and the editor.

Table A.9B

Table of t for Two-sided Comparisons between p Treatment Means and a Control for a Joint Confidence Coefficient of $P = .95$ and $P = .99$

Error df	P	\multicolumn{9}{c}{p = number of treatment means, excluding control}								
		1	2	3	4	5	6	7	8	9
5	.95	2.57	3.03	3.39	3.66	3.88	4.06	4.22	4.36	4.49
	.99	4.03	4.63	5.09	5.44	5.73	5.97	6.18	6.36	6.53
6	.95	2.45	2.86	3.18	3.41	3.60	3.75	3.88	4.00	4.11
	.99	3.71	4.22	4.60	4.88	5.11	5.30	5.47	5.61	5.74
7	.95	2.36	2.75	3.04	3.24	3.41	3.54	3.66	3.76	3.86
	.99	3.50	3.95	4.28	4.52	4.71	4.87	5.01	5.13	5.24
8	.95	2.31	2.67	2.94	3.13	3.28	3.40	3.51	3.60	3.68
	.99	3.36	3.77	4.06	4.27	4.44	4.58	4.70	4.81	4.90
9	.95	2.26	2.61	2.86	3.04	3.18	3.29	3.39	3.48	3.55
	.99	3.25	3.63	3.90	4.09	4.24	4.37	4.48	4.57	4.65
10	.95	2.23	2.57	2.81	2.97	3.11	3.21	3.31	3.39	3.46
	.99	3.17	3.53	3.78	3.95	4.10	4.21	4.31	4.40	4.47
11	.95	2.20	2.53	2.76	2.92	3.05	3.15	3.24	3.31	3.38
	.99	3.11	3.45	3.68	3.85	3.98	4.09	4.18	4.26	4.33
12	.95	2.18	2.50	2.72	2.88	3.00	3.10	3.18	3.25	3.32
	.99	3.05	3.39	3.61	3.76	3.89	3.99	4.08	4.15	4.22
13	.95	2.16	2.48	2.69	2.84	2.96	3.06	3.14	3.21	3.27
	.99	3.01	3.33	3.54	3.69	3.81	3.91	3.99	4.06	4.13
14	.95	2.14	2.46	2.67	2.81	2.93	3.02	3.10	3.17	3.23
	.99	2.98	3.29	3.49	3.64	3.75	3.84	3.92	3.99	4.05
15	.95	2.13	2.44	2.64	2.79	2.90	2.99	3.07	3.13	3.19
	.99	2.95	3.25	3.45	3.59	3.70	3.79	3.86	3.93	3.99
16	.95	2.12	2.42	2.63	2.77	2.88	2.96	3.04	3.10	3.16
	.99	2.92	3.22	3.41	3.55	3.65	3.74	3.82	3.88	3.93
17	.95	2.11	2.41	2.61	2.75	2.85	2.94	3.01	3.08	3.13
	.99	2.90	3.19	3.38	3.51	3.62	3.70	3.77	3.83	3.89
18	.95	2.10	2.40	2.59	2.73	2.84	2.92	2.99	3.05	3.11
	.99	2.88	3.17	3.35	3.48	3.58	3.67	3.74	3.80	3.85
19	.95	2.09	2.39	2.58	2.72	2.82	2.90	2.97	3.04	3.09
	.99	2.86	3.15	3.33	3.46	3.55	3.64	3.70	3.76	3.81
20	.95	2.09	2.38	2.57	2.70	2.81	2.89	2.96	3.02	3.07
	.99	2.85	3.13	3.31	3.43	3.53	3.61	3.67	3.73	3.78
24	.95	2.06	2.35	2.53	2.66	2.76	2.84	2.91	2.96	3.01
	.99	2.80	3.07	3.24	3.36	3.45	3.52	3.58	3.64	3.69
30	.95	2.04	2.32	2.50	2.62	2.72	2.79	2.86	2.91	2.96
	.99	2.75	3.01	3.17	3.28	3.37	3.44	3.50	3.55	3.59
40	.95	2.02	2.29	2.47	2.58	2.67	2.75	2.81	2.86	2.90
	.99	2.70	2.95	3.10	3.21	3.29	3.36	3.41	3.46	3.50
60	.95	2.00	2.27	2.43	2.55	2.63	2.70	2.76	2.81	2.85
	.99	2.66	2.90	3.04	3.14	3.22	3.28	3.33	3.38	3.42
120	.95	1.98	2.24	2.40	2.51	2.59	2.66	2.71	2.76	2.80
	.99	2.62	2.84	2.98	3.08	3.15	3.21	3.25	3.30	3.33
∞	.95	1.96	2.21	2.37	2.47	2.55	2.62	2.67	2.71	2.75
	.99	2.58	2.79	2.92	3.01	3.08	3.14	3.18	3.22	3.25

Source: This table is reproduced from "A multiple comparison procedure for comparing several treatments with a control," *J. Am. Stat. Assn.*, **50:** 1096–1121 (1955), with permission of the author, C. W. Dunnett, and the editor.

Table A.10

The Arcsin $\sqrt{\text{Percentage}}$ Transformation

Transformation of binomial percentages, in the margins, to angles of equal information in degrees. The + or − signs following angles ending in 5 are for guidance in rounding to one decimal.

%	0	1	2	3	4	5	6	7	8	9
0.0	0	0.57	0.81	0.99	1.15−	1.28	1.40	1.52	1.62	1.72
0.1	1.81	1.90	1.99	2.07	2.14	2.22	2.29	2.36	2.43	2.50
0.2	2.56	2.63	2.69	2.75−	2.81	2.87	2.92	2.98	3.03	3.09
0.3	3.14	3.19	3.24	3.29	3.34	3.39	3.44	3.49	3.53	3.58
0.4	3.63	3.67	3.72	3.76	3.80	3.85−	3.89	3.93	3.97	4.01
0.5	4.05+	4.09	4.13	4.17	4.21	4.25+	4.29	4.33	4.37	4.40
0.6	4.44	4.48	4.52	4.55+	4.59	4.62	4.66	4.69	4.73	4.76
0.7	4.80	4.83	4.87	4.90	4.93	4.97	5.00	5.03	5.07	5.10
0.8	5.13	5.16	5.20	5.23	5.26	5.29	5.32	5.35+	5.38	5.41
0.9	5.44	5.47	5.50	5.53	5.56	5.59	5.62	5.65+	5.68	5.71
1	5.74	6.02	6.29	6.55−	6.80	7.04	7.27	7.49	7.71	7.92
2	8.13	8.33	8.53	8.72	8.91	9.10	9.28	9.46	9.63	9.81
3	9.98	10.14	10.31	10.47	10.63	10.78	10.94	11.09	11.24	11.39
4	11.54	11.68	11.83	11.97	12.11	12.25−	12.39	12.52	12.66	12.79
5	12.92	13.05+	13.18	13.31	13.44	13.56	13.69	13.81	13.94	14.06
6	14.18	14.30	14.42	14.54	14.65+	14.77	14.89	15.00	15.12	15.23
7	15.34	15.45+	15.56	15.68	15.79	15.89	16.00	16.11	16.22	16.32
8	16.43	16.54	16.64	16.74	16.85−	16.95+	17.05+	17.16	17.26	17.36
9	17.46	17.56	17.66	17.76	17.85+	17.95+	18.05−	18.15−	18.24	18.34
10	18.44	18.53	18.63	18.72	18.81	18.91	19.00	19.09	19.19	19.28
11	19.37	19.46	19.55+	19.64	19.73	19.82	19.91	20.00	20.09	20.18
12	20.27	20.36	20.44	20 53	20.62	20.70	20.79	20.88	20.96	21.05−
13	21.13	21.22	21.30	21.39	21.47	21.56	21.64	21.72	21.81	21.89
14	21.97	22.06	22.14	22.22	22.30	22.38	22.46	22.55−	22.63	22.71
15	22.79	22.87	22.95−	23.03	23.11	23.19	23.26	23.34	23.42	23.50
16	23.58	23.66	23.73	23.81	23.89	23.97	24.04	24.12	24.20	24.27
17	24.35+	24.43	24.50	24.58	24.65+	24.73	24.80	24.88	24.95+	25.03
18	25.10	25.18	25.25+	25.33	25.40	25.48	25.55−	25.62	25.70	25.77
19	25.84	25.92	25.99	26.06	26.13	26.21	26.28	26.35−	26.42	26.49
20	26.56	26.64	26.71	26.78	26.85+	26.92	26.99	27.06	27.13	27.20
21	27.28	27.35−	27.42	27.49	27.56	27.63	27.69	27.76	27.83	27.90
22	27.97	28.04	28.11	28.18	28.25−	28.32	28.38	28.45+	28.52	28.59
23	28.66	28.73	28.79	28.86	28.93	29.00	29.06	29.13	29.20	29.27
24	29.33	29.40	29.47	29.53	29.60	29.67	29.73	29.80	29.87	29.93
25	30.00	30.07	30.13	30.20	30.26	30.33	30.40	30.46	30.53	30.59
26	30.66	30.72	30.79	30.85+	30.92	30.98	31.05−	31.11	31.18	31.24
27	31.31	31.37	31.44	31.50	31.56	31.63	31.69	31.76	31.82	31.88
28	31.95−	32.01	32.08	32.14	32.20	32.27	32.33	32.39	32.46	32.52
29	32.58	32.65−	32.71	32.77	32.83	32.90	32.96	33.02	33.09	33.15−
30	33.21	33.27	33.34	33.40	33.46	33.52	33.58	33.65−	33.71	33.77
31	33.83	33.89	33.96	34.02	34.08	34.14	34.20	34.27	34.33	34.39
32	34.45−	34.51	34.57	34.63	34.70	34.76	34.82	34.88	34.94	35.00
33	35.06	35.12	35.18	35.24	35.30	35.37	35.43	35.49	35.55−	35.61
34	35.67	35.73	35.79	35.85−	35.91	35.97	36.03	36.09	36.15+	36.21
35	36.27	36.33	36.39	36.45+	36.51	36.57	36.63	36.69	36.75+	36.81
36	36.87	36.93	36.99	37.05−	37.11	37.17	37.23	37.29	37.35−	37.41
37	37.47	37.52	37.58	37.64	37.70	37.76	37.82	37.88	37.94	38.00
38	38.06	38.12	38.17	38.23	38.29	38.35+	38.41	38.47	38.53	38.59
39	38.65−	38.70	38.76	38.82	38.88	38.94	39.00	39.06	39.11	39.17
40	39.23	39.29	39.35−	39.41	39.47	39.52	39.58	39.64	39.70	39.76
41	39.82	39.87	39.93	39.99	40.05−	40.11	40.16	40.22	40.28	40.34
42	40.40	40.46	40.51	40.57	40.63	40.69	40.74	40.80	40.86	40.92
43	40.98	41.03	41.09	41.15−	41.21	41.27	41.32	41.38	41.44	41.50
44	41.55+	41.61	41.67	41.73	41.78	41.84	41.90	41.96	42.02	42.07
45	42.13	42.19	42.25−	42.30	42.36	42.42	42.48	42.53	42.59	42.65−
46	42.71	42.76	42.82	42.88	42.94	42.99	43.05−	43.11	43.17	43.22
47	43.28	43.34	43.39	43.45−	43.51	43.57	43.62	43.68	43.74	43.80
48	43.85+	43.91	43.97	44.03	44.08	44.14	44.20	44.25+	44.31	44.37
49	44.43	44.48	44.54	44.60	44.66	44.71	44.77	44.83	44.89	44.94

SOURCE: This table appeared in *Plant Protection* (Leningrad), **12:** 67 (1937), and is reproduced with permission of the author, C. I. Bliss.

Table A.10 (Continued)
The Arcsin √Percentage Transformation

%	0	1	2	3	4	5	6	7	8	9
50	45.00	45.06	45.11	45.17	45.23	45.29	45.34	45.40	45.46	45.52
51	45.57	45.63	45.69	45.75−	45.80	45.86	45.92	45.97	46.03	46.09
52	46.15−	46.20	46.26	46.32	46.38	46.43	46.49	46.55−	46.61	46.66
53	46.72	46.78	46.83	46.89	46.95+	47.01	47.06	47.12	47.18	47.24
54	47.29	47.35+	47.41	47.47	47.52	47.58	47.64	47.70	47.75+	47.81
55	47.87	47.93	47.98	48.04	48.10	48.16	48.22	48.27	48.33	48.39
56	48.45−	48.50	48.56	48.62	48.68	48.73	48.79	48.85+	48.91	48.97
57	49.02	49.08	49.14	49.20	49.26	49.31	49.37	49.43	49.49	49.54
58	49.60	49.66	49.72	49.78	49.84	49.89	49.95+	50.01	50.07	50.13
59	50.18	50.24	50.30	50.36	50.42	50.48	50.53	50.59	50.65+	50.71
60	50.77	50.83	50.89	50.94	51.00	51.06	51.12	51.18	51.24	51.30
61	51.35+	51.41	51.47	51.53	51.59	51.65−	51.71	51.77	51.83	51.88
62	51.94	52.00	52.06	52.12	52.18	52.24	52.30	52.36	52.42	52.48
63	52.53	52.59	52.65+	52.71	52.77	52.83	52.89	52.95+	53.01	53.07
64	53.13	53.19	53.25−	53.31	53.37	53.43	53.49	53.55−	53.61	53.67
65	53.73	53.79	53.85−	53.91	53.97	54.03	54.09	54.15+	54.21	54.27
66	54.33	54.39	54.45+	54.51	54.57	54.63	54.70	54.76	54.82	54.88
67	54.94	55.00	55.06	55.12	55.18	55.24	55.30	55.37	55.43	55.49
68	55.55+	55.61	55.67	55.73	55.80	55.86	55.92	55.98	56.04	56.11
69	56.17	56.23	56.29	56.35+	56.42	56.48	56.54	56.60	56.66	56.73
70	56.79	56.85+	56.91	56.98	57.04	57.10	57.17	57.23	57.29	57.35+
71	57.42	57.48	57.54	57.61	57.67	57.73	57.80	57.86	57.92	57.99
72	58.05+	58.12	58.18	58.24	58.31	58.37	58.44	58.50	58.56	58.63
73	58.69	58.76	58.82	58.89	58.95+	59.02	59.08	59.15−	59.21	59.28
74	59.34	59.41	59.47	59.54	59.60	59.67	59.74	59.80	59.87	59.93
75	60.00	60.07	60.13	60.20	60.27	60.33	60.40	60.47	60.53	60.60
76	60.67	60.73	60.80	60.87	60.94	61.00	61.07	61.14	61.21	61.27
77	61.34	61.41	61.48	61.55−	61.62	61.68	61.75+	61.82	61.89	61.96
78	62.03	62.10	62.17	62.24	62.31	62.37	62.44	62.51	62.58	62.65+
79	62.72	62.80	62.87	62.94	63.01	63.08	63.15−	63.22	63.29	63.36
80	63.44	63.51	63.58	63.65+	63.72	63.79	63.87	63.94	64.01	64.08
81	64.16	64.23	64.30	64.38	64.45+	64.52	64.60	64.67	64.75−	64.82
82	64.90	64.97	65.05−	65.12	65.20	65.27	65.35−	65.42	65.50	65.57
83	65.65+	65.73	65.80	65.88	65.96	66.03	66.11	66.19	66.27	66.34·
84	66.42	66.50	66.58	66.66	66.74	66.81	66.89	66.97	67.05+	67.13
85	67.21	67.29	67.37	67.45+	67.54	67.62	67.70	67.78	67.86	67.94
86	68.03	68.11	68.19	68.28	68.36	68.44	68.53	68.61	68.70	68.78
87	68.87	68.95+	69.04	69.12	69.21	69.30	69.38	69.47	69.56	69.64
88	69.73	69.82	69.91	70.00	70.09	70.18	70.27	70.36	70.45−	70.54
89	70.63	70.72	70.81	70.91	71.00	71.09	71.19	71.28	71.37	71.47
90	71.56	71.66	71.76	71.85+	71.95+	72.05−	72.15−	72.24	72.34	72.44
91	72.54	72.64	72.74	72.84	72.95−	73.05−	73.15+	73.26	73.36	73.46
92	73.57	73.68	73.78	73.89	74.00	74.11	74.21	74.32	74.44	74.55−
93	74.66	74.77	74.88	75.00	75.11	75.23	75.35−	75.46	75.58	75.70
94	75.82	75.94	76.06	76.19	76.31	76.44	76.56	76.69	76.82	76.95−
95	77.08	77.21	77.34	77.48	77.61	77.75+	77.89	78.03	78.17	78.32
96	78.46	78.61	78.76	78.91	79.06	79.22	79.37	79.53	79.69	79.86
97	80.02	80.19	80.37	80.54	80.72	80.90	81.09	81.28	81.47	81.67
98	81.87	82.08	82.29	82.51	82.73	82.96	83.20	83.45+	83.71	83.98
99.0	84.26	84.29	84.32	84.35−	84.38	84.41	84.44	84.47	84.50	84.53
99.1	84.56	84.59	84.62	84.65−	74.68	84.71	84.74	84.77	84.80	84.84
99.2	84.87	84.90	84.93	84.97	85.00	85.03	85.07	85.10	85.13	85.17
99.3	85.20	85.24	85.27	85.31	85.34	85.38	85.41	85.45−	85.48	85.52
99.4	85.56	85.60	85.63	85.67	85.71	85.75−	85.79	85.83	85.87	85.91
99.5	85.95−	85.99	86.03	86.07	86.11	86.15−	86.20	86.24	86.28	86.33
99.6	86.37	86.42	86.47	86.51	86.56	86.61	86.66	86.71	86.76	86.81
99.7	86.86	86.91	86.97	87.02	87.08	87.13	87.19	87.25+	87.31	87.37
99.8	87.44	87.50	87.57	87.64	87.71	87.78	87.86	87.93	88.01	88.10
99.9	88.19	88.28	88.38	88.48	88.60	88.72	88.85+	89.01	89.19	89.43
100.0	90.00									

Table A.11A
Confidence Belts for the Correlation Coefficient ρ: $P = .95$

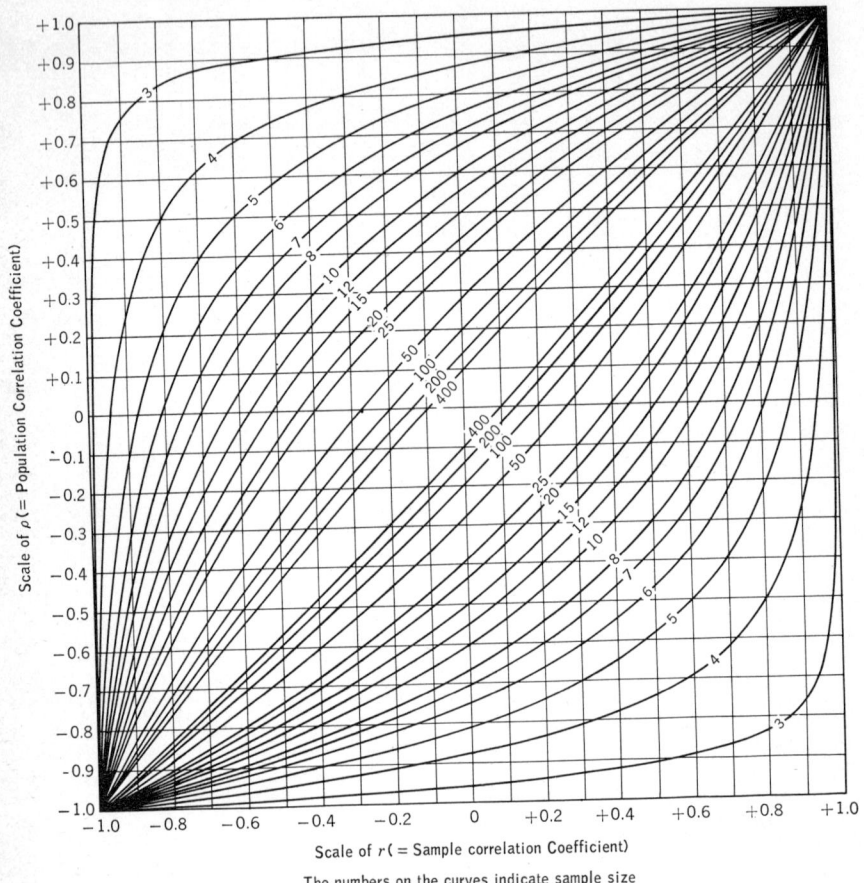

The numbers on the curves indicate sample size

SOURCE: This table is reproduced with the permission of E. S. Pearson, from F. N. David, *Tables of the Ordinates and Probability Integral of the Distribution of the Correlation Coefficient in Small Samples*, Cambridge University Press for the *Biometrika* trustees, 1938.

TABLE A.11B
CONFIDENCE BELTS FOR THE CORRELATION COEFFICIENT ρ: $P = .99$

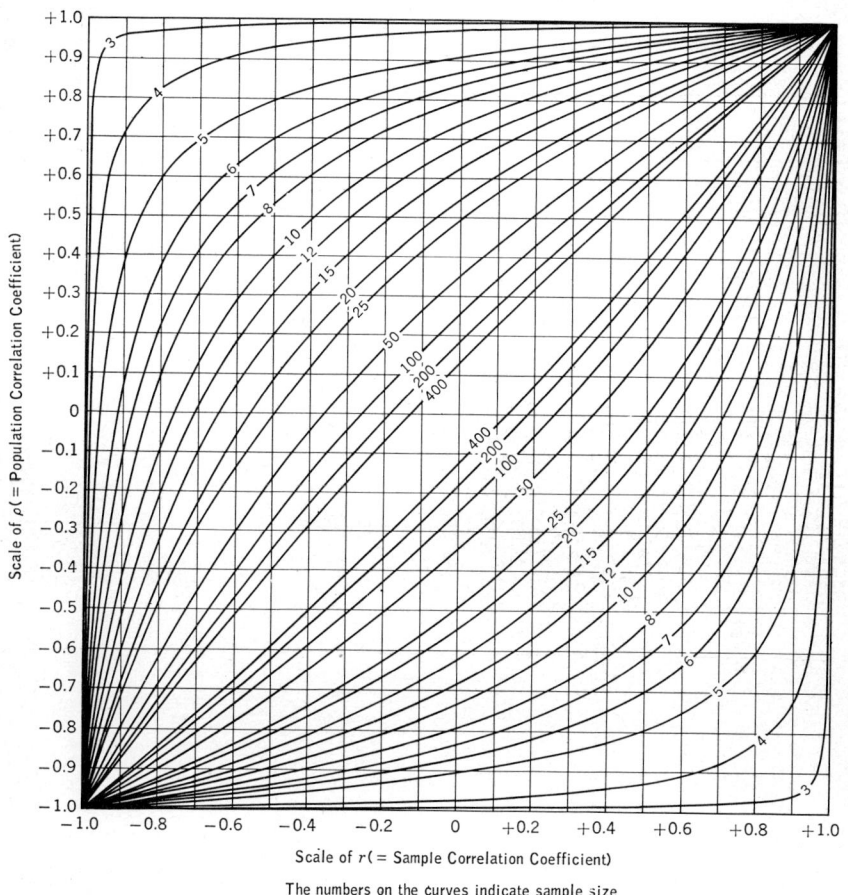

The numbers on the curves indicate sample size

SOURCE: This table is reproduced with the permission of E. S. Pearson, from F. N. David, *Tables of the Ordinates and Probability Integral of the Distribution of the Correlation Coefficient in Small Samples*, Cambridge University Press for the *Biometrika* trustees, 1938.

TABLE A.12
TRANSFORMATION OF r TO z

Values of $z = .5 \ln (1 + r)/(1 - r) = \tanh^{-1} r$ appear in the body of the table for corresponding values of r, the correlation coefficient, in the margins.

r	.00	.01	.02	.03	.04	.05	.06	.07	.08	.09
.0	.00000	.01000	.02000	.03001	.04002	.05004	.06007	.07012	.08017	.09024
.1	.10034	.11045	.12058	.13074	.14093	.15114	.16139	.17167	.18198	.19234
.2	.20273	.21317	.22366	.23419	.24477	.25541	.26611	.27686	.28768	.29857
.3	.30952	.32055	.33165	.34283	.35409	.36544	.37689	.38842	.40006	.41180
.4	.42365	.43561	.44769	.45990	.47223	.48470	.49731	.51007	.52298	.53606
.5	.54931	.56273	.57634	.59014	.60415	.61838	.63283	.64752	.66246	.67767
.6	.69315	.70892	.72500	.74142	.75817	.77530	.79281	.81074	.82911	.84795
.7	.86730	.88718	.90764	.92873	.95048	.97295	.99621	1.02033	1.04537	1.07143
.8	1.09861	1.12703	1.15682	1.18813	1.22117	1.25615	1.29334	1.33308	1.37577	1.42192
.9	1.47222	1.52752	1.58902	1.65839	1.73805	1.83178	1.94591	2.09229	2.29756	2.64665

SOURCE: This table is abridged from Table XII of *Standard Four-figure Mathematical Tables*, 1931, by L. M. Milne-Thomson and L. J. Comrie, with permission of the authors and the publishers, Macmillan and Company, London.

TABLE A.13
SIGNIFICANT VALUES OF r AND R

Error df	P	Independent variables				Error df	P	Independent variables			
		1	2	3	4			1	2	3	4
1	.05	.997	.999	.999	.999	24	.05	.388	.470	.523	.562
	.01	1.000	1.000	1.000	1.000		.01	.496	.565	.609	.642
2	.05	.950	.975	.983	.987	25	.05	.381	.462	.514	.553
	.01	.990	.995	.997	.998		.01	.487	.555	.600	.633
3	.05	.878	.930	.950	.961	26	.05	.374	.454	.506	.545
	.01	.959	.976	.983	.987		.01	.478	.546	.590	.624
4	.05	.811	.881	.912	.930	27	.05	.367	.446	.498	.536
	.01	.917	.949	.962	.970		.01	.470	.538	.582	.615
5	.05	.754	.836	.874	.898	28	.05	.361	.439	.490	.529
	.01	.874	.917	.937	.949		.01	.463	.530	.573	.606
6	.05	.707	.795	.839	.867	29	.05	.355	.432	.482	.521
	.01	.834	.886	.911	.927		.01	.456	.522	.565	.598
7	.05	.666	.758	.807	.838	30	.05	.349	.426	.476	.514
	.01	.798	.855	.885	.904		.01	.449	.514	.558	.591
8	.05	.632	.726	.777	.811	35	.05	.325	.397	.445	.482
	.01	.765	.827	.860	.882		.01	.418	.481	.523	.556
9	.05	.602	.697	.750	.786	40	.05	.304	.373	.419	.455
	.01	.735	.800	.836	.861		.01	.393	.454	.494	.526
10	.05	.576	.671	.726	.763	45	.05	.288	.353	.397	.432
	.01	.708	.776	.814	.840		.01	.372	.430	.470	.501
11	.05	.553	.648	.703	.741	50	.05	.273	.336	.379	.412
	.01	.684	.753	.793	.821		.01	.354	.410	.449	.479
12	.05	.532	.627	.683	.722	60	.05	.250	.308	.348	.380
	.01	.661	.732	.773	.802		.01	.325	.377	.414	.442
13	.05	.514	.608	.664	.703	70	.05	.232	.286	.324	.354
	.01	.641	.712	.755	.785		.01	.302	.351	.386	.413
14	.05	.497	.590	.646	.686	80	.05	.217	.269	.304	.332
	.01	.623	.694	.737	.768		.01	.283	.330	.362	.389
15	.05	.482	.574	.630	.670	90	.05	.205	.254	.288	.315
	.01	.606	.677	.721	.752		.01	.267	.312	.343	.368
16	.05	.468	.559	.615	.655	100	.05	.195	.241	.274	.300
	.01	.590	.662	.706	.738		.01	.254	.297	.327	.351
17	.05	.456	.545	.601	.641	125	.05	.174	.216	.246	.269
	.01	.575	.647	.691	.724		.01	.228	.266	.294	.316
18	.05	.444	.532	.587	.628	150	.05	.159	.198	.225	.247
	.01	.561	.633	.678	.710		.01	.208	.244	.270	.290
19	.05	.433	.520	.575	.615	200	.05	.138	.172	.196	.215
	.01	.549	.620	.665	.698		.01	.181	.212	.234	.253
20	.05	.423	.509	.563	.604	300	.05	.113	.141	.160	.176
	.01	.537	.608	.652	.685		.01	.148	.174	.192	.208
21	.05	.413	.498	.522	.592	400	.05	.098	.122	.139	.153
	.01	.526	.596	.641	.674		.01	.128	.151	.167	.180
22	.05	.404	.488	.542	.582	500	.05	.088	.109	.124	.137
	.01	.515	.585	.630	.663		.01	.115	.135	.150	.162
23	.05	.396	.479	.532	.572	1,000	.05	.062	.077	.088	.097
	.01	.505	.574	.619	.652		.01	.081	.096	.106	.115

SOURCE: Reproduced from G. W. Snedecor, *Statistical Methods*, 4th ed, The Iowa State College Press, Ames, Iowa, 1946, with permission of the author and publisher.

Table A.14A
Binomial Confidence Limits†

Number with characteristic	P	Sample size 10	Sample size 15	Sample size 20	Sample size 25
0	.95	.0000–.3085	.0000–.2180	.0000–.1685	.0000–.1372
	.99	.0000–.4113	.0000–.2976	.0000–.2327	.0000–.1910
1	.95	.0025–.4450	.0017–.3200	.0013–.2485	.0010–.2036
	.99	.0005–.5440	.0003–.4027	.0002–.3170	.0002–.2624
2	.95	.0252–.5560	.0166–.4049	.0124–.3170	.0098–.2605
	.99	.0108–.6480	.0071–.4871	.0053–.3870	.0042–.3208
3	.95	.0667–.6520	.0433–.4807	.0321–.3793	.0255–.3124
	.99	.0370–.7350	.0239–.5607	.0177–.4505	.0140–.3748
4	.95	.1220–.7380	.0780–.5514	.0575–.4365	.0455–.3610
	.99	.0768–.8091	.0488–.6278	.0358–.5065	.0283–.4241
5	.95	.1870–.8130	.1185–.6162	.0868–.4913	.0684–.4072
	.99	.1280–.8720	.0803–.6889	.0585–.5605	.0460–.4700
6	.95		.1633–.6774	.1190–.5430	.0935–.4514
	.99		.1167–.7440	.0845–.6095	.0662–.5138
7	.95		.2129–.7338	.1538–.5920	.1206–.4938
	.99		.1587–.7954	.1140–.6570	.0890–.5556
8	.95			.1910–.6395	.1496–.5350
	.99			.1460–.7010	.1136–.5954
9	.95			.2305–.6848	.1797–.5748
	.99			.1808–.7430	.1401–.6336
10	.95			.2720–.7280	.2112–.6132
	.99			.2175–.7825	.1680–.6704
11	.95				.2441–.6506
	.99				.1975–.7055
12	.95				.2781–.6869
	.99				.2284–.7393
13	.95				
	.99				
14	.95				
	.99				

† Confidence intervals found from these tables are such that probabilities of approximately .025 and .005 are associated with unusual events at each extreme. Hence, the confidence probabilities are at least .95 and .99 and are .975 and .995 when the number possessing the characteristic is zero.

To interpolate in this table, use the formula $\text{CI}(T)n(T)/n$, where $\text{CI}(T)$ and $n(T)$ are the

TABLE A.14A (Continued)
BINOMIAL CONFIDENCE LIMITS

Sample size					P	Number with characteristic
30	50	100	500	1,000		
.0000–.1157 .0000–.1619	.0000–.0711 .0000–.1005	.0000–.0362 .0000–.0516	.0000–.0074 .0000–.0105	.0000–.0037 .0000–.0053	.95 .99	0
.0009–.1779 .0002–.2233	.0005–.1066 .0001–.1398	.0002–.0545 .0000–.0721	.0001–.0111 .0000–.0148	.0000–.0056 .0000–.0074	.95 .99	1
.0082–.2209 .0035–.2735	.0049–.1372 .0021–.1721	.0024–.0704 .0010–.0894	.0005–.0144 .0002–.0184	.0002–.0072 .0001–.0092	.95 .99	2
.0211–.2653 .0116–.3203	.0126–.1657 .0069–.2032	.0062–.0853 .0034–.1057	.0012–.0174 .0007–.0218	.0006–.0087 .0003–.0109	.95 .99	3
.0377–.3074 .0234–.3639	.0223–.1925 .0138–.2313	.0110–.0993 .0068–.1208	.0022–.0204 .0013–.0250	.0011–.0102 .0007–.0125	.95 .99	4
.0564–.3474 .0379–.4044	.0332–.2182 .0222–.2580	.0164–.1129 .0110–.1353	.0032–.0232 .0022–.0281	.0016–.0116 .0011–.0141	.95 .99	5
.0770–.3856 .0543–.4426	.0454–.2431 .0318–.2842	.0224–.1260 .0156–.1493	.0044–.0259 .0031–.0310	.0022–.0130 .0015–.0156	.95 .99	6
.0992–.4229 .0729–.4801	.0582–.2675 .0425–.3092	.0286–.1390 .0208–.1628	.0056–.0286 .0041–.0339	.0028–.0144 .0020–.0170	.95 .99	7
.1229–.4589 .0930–.5158	.0717–.2912 .0540–.3336	.0351–.1516 .0263–.1761	.0069–.0313 .0052–.0368	.0035–.0157 .0026–.0185	.95 .99	8
.1473–.4940 .1143–.5500	.0858–.3144 .0660–.3573	.0420–.1640 .0321–.1892	.0083–.0339 .0063–.0396	.0041–.0170 .0031–.0199	.95 .99	9
.1729–.5280 .1369–.5835	.1004–.3372 .0786–.3804	.0490–.1762 .0382–.2020	.0096–.0365 .0075–.0423	.0048–.0183 .0037–.0213	.95 .99	10
.1993–.5613 .1606–.6157	.1154–.3595 .0920–.4032	.0562–.1883 .0445–.2145	.0110–.0390 .0087–.0450	.0050–.0196 .0043–.0226	.95 .99	11
.2266–.5939 .1850–.6469	.1307–.3817 .1056–.4256	.0636–.2002 .0510–.2269	.0125–.0416 .0099–.0477	.0062–.0209 .0050–.0240	.95 .99	12
.2546–.6256 .2107–.6772	.1463–.4034 .1198–.4473	.0711–.2120 .0577–.2392	.0139–.0441 .0112–.0504	.0069–.0221 .0056–.0253	.95 .99	13
.2835–.6566 .2373–.7066	.1623–.4248 .1342–.4688	.0787–.2237 .0646–.2513	.0154–.0465 .0126–.0530	.0077–.0234 .0063–.0267	.95 .99	14

tabled confidence interval and sample size next below that for the observed sample size n. For example, if three individuals out of 40 possess the characteristic, then the lower value of the 99% confidence interval is calculated as .0116(30)/40 = .0087.

SOURCE: Abridged from *Statistical Tables for Use with Binomial Samples*, published by D. Mainland, L. Herrera, and M. I. Sutcliffe, New York, 1956, with permission of the authors.

Table A.14B
Binomial Confidence Limits

Observed fraction	P	Sample size					
		50	75	150	300	500	1,000
.01	.95				.0021–.0289	.0032–.0232	.0048–.0183
	.99				.0011–.0361	.0022–.0280	.0037–.0213
.02	.95				.0086–.0420	.0106–.0356	.0129–.0301
	.99				.0067–.0500	.0087–.0412	.0113–.0336
.03	.95				.0152–.0550	.0179–.0481	.0211–.0419
	.99				.0122–.0640	.0152–.0544	.0188–.0459
.04	.95				.0217–.0681	.0253–.0605	.0292–.0536
	.99				.0177–.0779	.0217–.0675	.0264–.0582
.05	.95			.0211–.0981	.0283–.0811	.0326–.0729	.0373–.0654
	.99			.0156–.1150	.0232–.0918	.0283–.0807	.0339–.0705
.06	.95			.0283–.1104	.0363–.0928	.0411–.0843	.0463–.0764
	.99			.0219–.1279	.0307–.1040	.0363–.0924	.0425–.0818
.07	.95			.0355–.1227	.0444–.1045	.0496–.0956	.0552–.0873
	.99			.0282–.1408	.0381–.1162	.0443–.1042	.0512–.0931
.08	.95			.0427–.1350	.0524–.1162	.0581–.1070	.0642–.0983
	.99			.0345–.1537	.0455–.1284	.0523–.1160	.0598–.1043
.09	.95			.0499–.1473	.0605–.1280	.0666–.1183	.0732–.1093
	.99			.0408–.1666	.0529–.1406	.0604–.1277	.0684–.1156
.10	.95			.0571–.1595	.0685–.1397	.0751–.1297	.0821–.1203
	.99			.0471–.1796	.0604–.1528	.0684–.1395	.0770–.1269
.11	.95			.0651–.1711	.0771–.1508	.0841–.1406	.0914–.1310
	.99			.0544–.1915	.0685–.1643	.0770–.1507	.0860–.1378
.12	.95			.0730–.1827	.0857–.1620	.0930–.1516	.1006–.1416
	.99			.0617–.2035	.0767–.1758	.0856–.1619	.0951–.1486
.13	.95			.0810–.1942	.0943–.1732	.1020–.1625	.1099–.1523
	.99			.0690–.2155	.0848–.1873	.0942–.1731	.1041–.1595
.14	.95			.0890–.2058	.1030–.1843	.1109–.1734	.1192–.1630
	.99			.0764–.2274	.0930–.1988	.1028–.1843	.1131–.1704
.15	.95		.0780–.2512	.0970–.2174	.1116–.1955	.1198–.1844	.1284–.1737
	.99		.0628–.2844	.0837–.2394	.1012–.2103	.1114–.1955	.1221–.1813
.16	.95		.0857–.2626	.1054–.2285	.1205–.2064	.1290–.1950	.1379–.1842
	.99		.0690–.2961	.0916–.2508	.1097–.2214	.1203–.2063	.1314–.1919
.17	.95		.0934–.2741	.1139–.2396	.1294–.2172	.1382–.2057	.1473–.1947
	.99		.0759–.3078	.0995–.2622	.1183–.2325	.1292–.2172	.1407–.2025
.18	.95		.1011–.2855	.1223–.2508	.1384–.2281	.1474–.2164	.1567–.2052
	.99		.0829–.3195	.1074–.2736	.1269–.2436	.1381–.2281	.1499–.2132
.19	.95		.1088–.2969	.1307–.2619	.1473–.2390	.1566–.2271	.1662–.2157
	.99		.0898–.3312	.1153–.2850	.1355–.2547	.1471–.2390	.1592–.2238
.20	.95		.1165–.3084	.1392–.2731	.1562–.2498	.1658–.2378	.1756–.2262
	.99		.0967–.3429	.1232–.2964	.1440–.2657	.1560–.2499	.1684–.2345
.21	.95		.1246–.3194	.1479–.2839	.1654–.2604	.1752–.2483	.1852–.2365
	.99		.1042–.3541	.1316–.3075	.1529–.2765	.1651–.2605	.1778–.2450
.22	.95		.1327–.3304	.1567–.2947	.1745–.2711	.1845–.2588	.1947–.2469
	.99		.1116–.3652	.1399–.3185	.1618–.2874	.1743–.2712	.1872–.2555
.23	.95		.1409–.3414	.1654–.3055	.1837–.2817	.1939–.2693	.2043–.2573
	.99		.1191–.3764	.1482–.3295	.1706–.2982	.1834–.2818	.1967–.2659
.24	.95		.1490–.3524	.1742–.3164	.1929–.2924	.2033–.2799	.2139–.2677
	.99		.1265–.3876	.1565–.3405	.1795–.3090	.1926–.2925	.2061–.2764
.25	.95	.1384–.3927	.1572–.3634	.1830–.3272	.2020–.3030	.2126–.2904	.2234–.2781
	.99	.1125–.4365	.1340–.3988	.1648–.3516	.1884–.3198	.2017–.3031	.2155–.2869

Table A.14B (Continued)
Binomial Confidence Limits

Observed fraction	P	Sample size					
		50	75	150	300	500	1,000
.26	.95	.1465–.4034	.1656–.3741	.1920–.3378	.2113–.3135	.2221–.3008	.2331–.2883
	.99	.1198–.4472	.1419–.4095	.1734–.3623	.1975–.3303	.2110–.3136	.2250–.2973
.27	.95	.1545–.4140	.1741–.3848	.2010–.3484	.2207–.3239	.2316–.3111	.2427–.2986
	.99	.1271–.4579	.1498–.4203	.1821–.3729	.2066–.3409	.2204–.3241	.2346–.3076
.28	.95	.1626–.4247	.1826–.3955	.2100–.3590	.2300–.3344	.2411–.3215	.2524–.3089
	.99	.1344–.4686	.1577–.4310	.1907–.3836	.2157–.3515	.2297–.3346	.2441–.3180
.29	.95	.1706–.4354	.1911–.4061	.2190–.3695	.2393–.3449	.2506–.3319	.2621–.3192
	.99	.1408–.4792	.1656–.4418	.1994–.3943	.2247–.3621	.2390–.3451	.2537–.3284
.30	.95	.1787–.4461	.1996–.4168	.2280–.3801	.2487–.3553	.2601–.3423	.2717–.3295
	.99	.1491–.4899	.1735–.4525	.2080–.4050	.2338–.3726	.2483–.3555	.2632–.3387
.31	.95	.1871–.4565	.2083–.4272	.2372–.3905	.2582–.3656	.2697–.3525	.2815–.3397
	.99	.1568–.5002	.1818–.4629	.2169–.4155	.2431–.3830	.2578–.3659	.2729–.3490
.32	.95	.1955–.4668	.2171–.4376	.2464–.4009	.2676–.3760	.2793–.3628	.2912–.3499
	.99	.1646–.5105	.1901–.4733	.2258–.4259	.2524–.3934	.2673–.3762	.2825–.3592
.33	.95	.2038–.4772	.2259–.4481	.2556–.4113	.2771–.3863	.2890–.3731	.3009–.3601
	.99	.1723–.5208	.1984–.4838	.2347–.4364	.2617–.4038	.2768–.3865	.2922–.3695
.34	.95	.2122–.4876	.2346–.4585	.2648–.4217	.2866–.3966	.2986–.3833	.3107–.3703
	.99	.1801–.5311	.2067–.4942	.2436–.4468	.2710–.4142	.2862–.3969	.3018–.3797
.35	.95	.2206–.4980	.2434–.4689	.2740–.4320	.2961–.4069	.3082–.3936	.3204–.3805
	.99	.1878–.5414	.2150–.5046	.2525–.4572	.2803–.4246	.2957–.4072	.3114–.3900
.36	.95	.2293–.5080	.2524–.4790	.2834–.4422	.3057–.4171	.3179–.4038	.3302–.3906
	.99	.1960–.5513	.2236–.5147	.2617–.4674	.2897–.4348	.3053–.4174	.3212–.4002
.37	.95	.2380–.5181	.2615–.4892	.2928–.4524	.3153–.4273	.3276–.4139	.3400–.4007
	.99	.2042–.5612	.2322–.5247	.2708–.4776	.2992–.4450	.3149–.4276	.3309–.4103
.38	.95	.2467–.5281	.2705–.4993	.3022–.4626	.3249–.4375	.3373–.4241	.3498–.4109
	.99	.2123–.5710	.2409–.5348	.2800–.4879	.3087–.4552	.3245–.4378	.3407–.4205
.39	.95	.2554–.5382	.2795–.5095	.3116–.4728	.3345–.4477	.3470–.4343	.3597–.4210
	.99	.2205–.5809	.2495–.5449	.2891–.4981	.3181–.4655	.3342–.4480	.3504–.4306
.40	.95	.2641–.5482	.2885–.5196	.3210–.4830	.3441–.4579	.3568–.4444	.3695–.4311
	.99	.2287–.5908	.2581–.5549	.2983–.5083	.3276–.4757	.3438–.4582	.3602–.4408
.41	.95	.2731–.5580	.2978–.5296	.3305–.4931	.3539–.4679	.3666–.4545	.3793–.4412
	.99	.2372–.6004	.2670–.5647	.3076–.5182	.3372–.4857	.3535–.4683	.3700–.4509
.42	.95	.2821–.5678	.3070–.5395	.3401–.5031	.3636–.4780	.3764–.4646	.3892–.4512
	.99	.2457–.6099	.2759–.5745	.3170–.5282	.3468–.4958	.3632–.4783	.3798–.4610
.43	.95	.2910–.5776	.3163–.5494	.3496–.5132	.3733–.4881	.3862–.4746	.3991–.4613
	.99	.2542–.6195	.2849–.5843	.3264–.5382	.3564–.5059	.3729–.4884	.3896–.4710
.44	.95	.3000–.5874	.3256–.5593	.3592–.5232	.3830–.4981	.3960–.4847	.4090–.4714
	.99	.2627–.6290	.2938–.5941	.3357–.5482	.3660–.5159	.3827–.4985	.3995–.4811
.45	.95	.3090–.5971	.3348–.5693	.3687–.5333	.3928–.5082	.4058–.4948	.4189–.4814
	.99	.2712–.6386	.3027–.6038	.3451–.5582	.3756–.5260	.3924–.5086	.4093–.4912
.46	.95	.3183–.6067	.3443–.5790	.3785–.5431	.4026–.5182	.4157–.5048	.4288–.4914
	.99	.2800–.6478	.3119–.6133	.3547–.5680	.3854–.5359	.4022–.5185	.4192–.5012
.47	.95	.3275–.6162	.3538–.5886	.3882–.5530	.4125–.5281	.4256–.5148	.4387–.5014
	.99	.2889–.6569	.3211–.6228	.3642–.5777	.3952–.5458	.4121–.5285	.4291–.5112
.48	.95	.3368–.6257	.3633–.5983	.3979–.5625	.4223–.5381	.4355–.5247	.4487–.5114
	.99	.2978–.6661	.3304–.6323	.3738–.5875	.4049–.5557	.4219–.5385	.4390–.5212
.49	.95	.3461–.6352	.3728–.6080	.4076–.5728	.4321–.5481	.4454–.5347	.4586–.5214
	.99	.3067–.6753	.3396–.6417	.3834–.5973	.4147–.5656	.4318–.5484	.4489–.5312
.50	.95	.3553–.6447	.3823–.6177	.4173–.5827	.4420–.5580	.4553–.5447	.4685–.5315
	.99	.3155–.6845	.3488–.6512	.3929–.6071	.4245–.5755	.4416–.5584	.4589–.5411

Source: Abridged from *Statistical Tables for Use with Binomial Samples*, published by D. Mainland, L. Herrera, and M. I. Sutcliffe, New York, 1956, with permission of the authors.

TABLE A.15A

CONFIDENCE BELTS FOR PROPORTIONS: CONFIDENCE COEFFICIENT OF .95

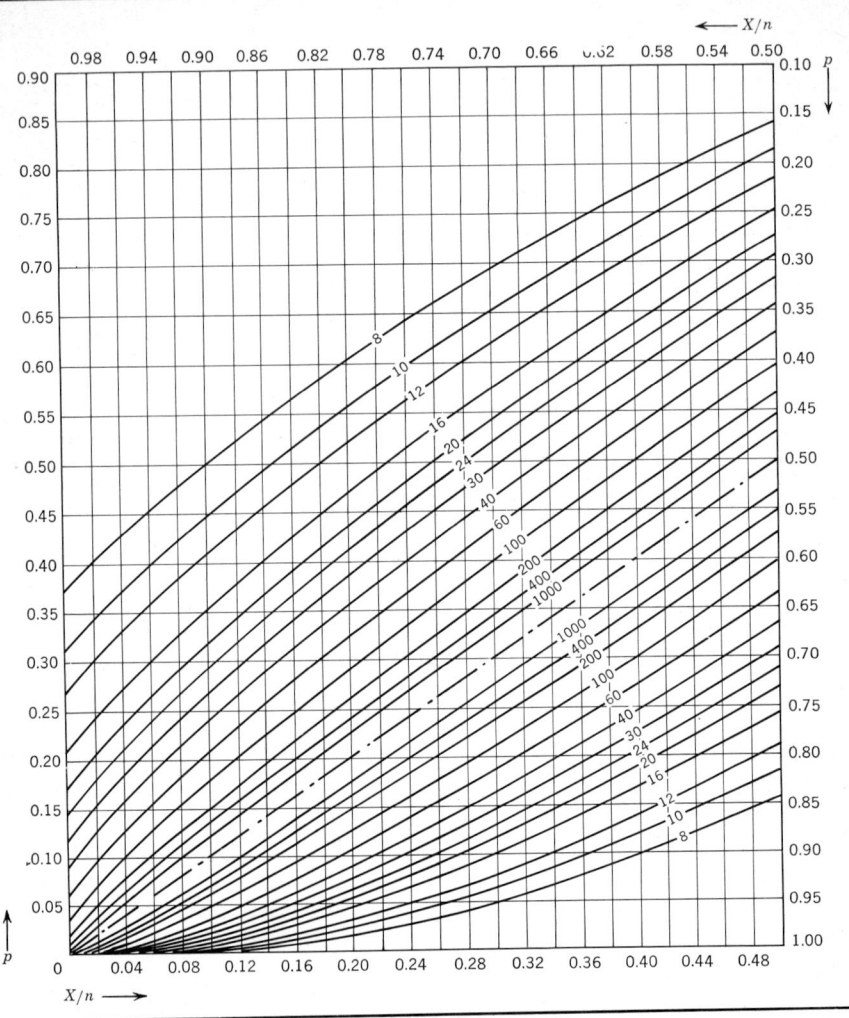

The numbers printed along the curves indicate the sample size n. If for a given value of the abscissa X/n, p_A and p_B are the ordinates read from (or interpolated between) the appropriate lower and upper curves, and $\Pr(p_A \leq p \leq p_B) \leq 1 - 2\alpha$.

SOURCE: Reproduced with permission of the *Biometrika* trustees and the editors, from E. S. Pearson and H. O. Hartley, *Biometrika Tables for Statisticians*, vol. 1, Cambridge University Press, 1954.

Table A.15B
Confidence Belts for Proportions: Confidence Coefficient of .99

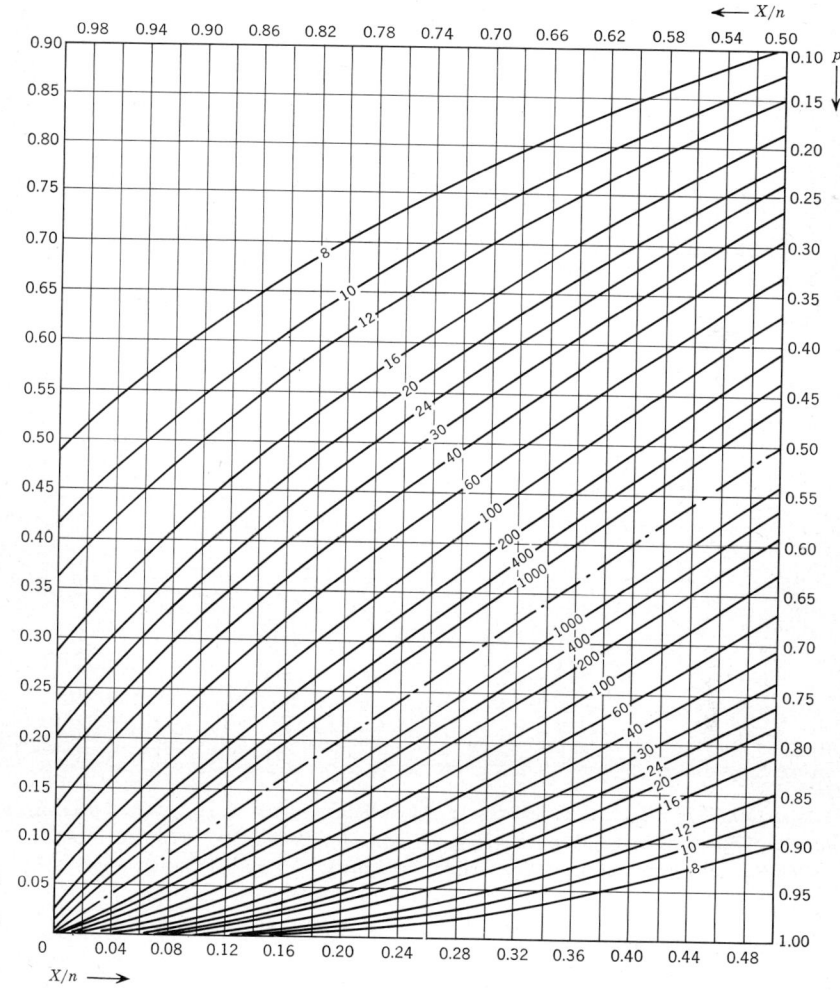

The numbers printed along the curves indicate the sample size n. Note: The process of reading from the curve can be simplified with the help of the right-angled corner of a loose sheet of paper or thin card, along the edges of which are marked off the scales shown in the top left-hand corner of the chart.

Source: Reproduced with permission of the *Biometrika* trustees and the editors, from E. S. Pearson and H. O. Hartley, *Biometrika Tables for Statisticians* vol. 1, Cambridge University Press, 1954.

Table A.16
Mosteller-Tukey Binomial Probability Paper

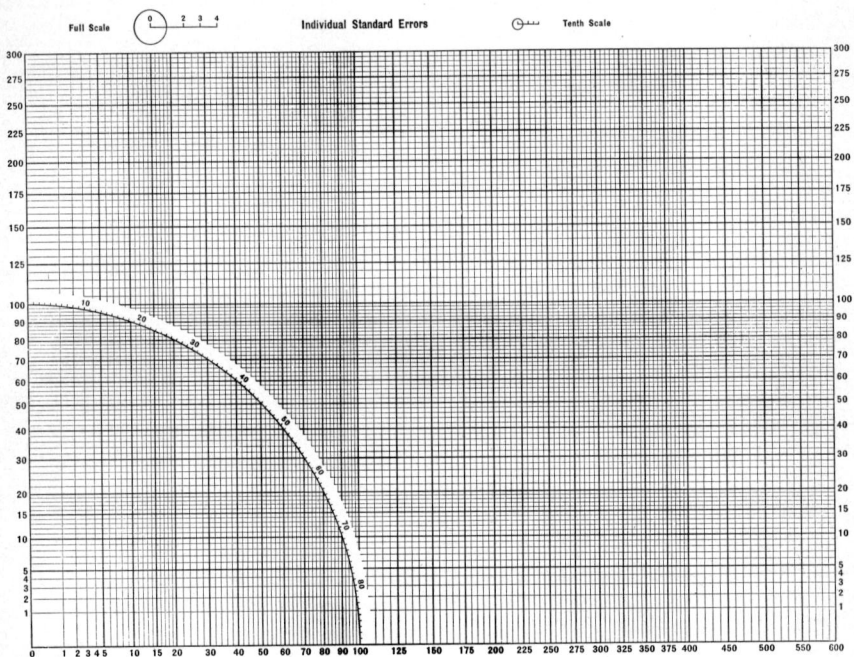

SOURCE: This chart first appeared in F. Mosteller and J. W. Tukey, "The uses and usefulness of binomial probability paper," *J. Am. Stat. Assn.*, **44**: 174–212 (1949). It is reproduced with permission of the authors, editor, and the Codex Book Company, Inc.

TABLE A.17A

SAMPLE SIZE AND THE PROBABILITY OF MAKING A WRONG DECISION BETWEEN THE TEST CROSS RATIOS 1:1 AND 3:1

Sample size n	Ratio accepted	Classes in regions of acceptance	Probability of making a wrong decision
20	1:1	0–12	.1316
	3:1	13–20	.1018
30	1:1	0–18	.1002
	3:1	19–30	.0507
40	1:1	0–25	.0403
	3:1	26–40	.0544
44	1:1	0–27	.0481
	3:1	28–44	.0318
50	1:1	0–31	.0325
	3:1	32–50	.0287
60	1:1	0–37	.0259
	3:1	38–60	.0154
70	1:1	0–44	.0112
	3:1	45–70	.0163
80	1:1	0–50	.0092
	3:1	51–80	.0089

$\left. \begin{array}{l} 1:1 = .5: \ .5 \\ 3:1 = .75: \ .25 \end{array} \right\}$ The dividing line $R/n \doteq .63091$

SOURCE: Reprinted from Prasert NaNagara, *Testing Mendelian Ratios*, M.S. Thesis, 1953, Cornell University, Ithaca, N.Y.

Table A.17B
Sample Size and the Probability of Making a Wrong Decision between the Test Cross Ratios 3:1 and 7:1†

Sample size n	Ratio accepted	Classes in regions of acceptance	Probability of making a wrong decision
20	3:1	0–16	.2252
	7:1	17–20	.2347
30	3:1	0–24	.2026
	7:1	25–30	.1644
40	3:1	0–32	.1820
	7:1	33–40	.1190
50	3:1	0–40	.1637
	7:1	41–50	.0879
60	3:1	0–49	.0859
	7:1	50–60	.1231
70	3:1	0–57	.08
	7:1	58–70	.09
80	3:1	0–65	.08
	7:1	66–80	.06
90	3:1	0–73	.07
	7:1	74–90	.05
100	3:1	0–82	.04
	7:1	83–100	.07
110	3:1	0–90	.04
	7:1	91–110	.05
200	3:1	0–164	.01
	7:1	165–200	.01+
210	3:1	0–171	.011
	7:1	172–210	.006

$3:1 = .75: .25$
$7:1 = .875: .125$ } The dividing line $R/n \doteq .81786$

† See Table A.17A.

Table A.17C

SAMPLE SIZE AND THE PROBABILITY OF MAKING A WRONG DECISION AMONG THE TEST CROSS RATIOS 1:1, 3:1, AND 7:1†

Sample size n	Ratio accepted	Classes in regions of acceptance	Probability of making a wrong decision
20	1:1 3:1 7:1	0–11 12–16 17–20	.2517 .2661* .2347
30	1:1 3:1 7:1	0–17 18–24 25–30	.1808 .2242* .1644
40	1:1 3:1 7:1	0–23 24–32 33–40	.1341 .1936* .1190
50	1:1 3:1 7:1	0–31 32–41 42–50	.0325 .1203 .1660*
60	1:1 3:1 7:1	0–37 38–49 50–60	.0259 .1013 .1231*
80	1:1 3:1 7:1	0–50 51–65 66–80	.01 .08* .06
100	1:1 3:1 7:1	0–63 64–82 83–100	.00 .04 .07*
110	1:1 3:1 7:1	0–69 70–90 91–110	.00 .04 .05*
200	1:1 3:1 7:1	0–126 127–164 165–200	.00 .01 .01+*
210	1:1 3:1 7:1	0–132 133–171 172–210	.000 .011* .006

$1:1 = .5: .5$
$3:1 = .75: .25$ } The dividing line $R/n \doteq .63091$
$7:1 = .875: .125$ } The dividing line $R^1/n \doteq .81786$

† See Table A.17A.

TABLE A.17D

SAMPLE SIZE AND THE PROBABILITY OF MAKING A WRONG DECISION AMONG THE F_2 RATIOS 9:7, 13:3, AND 15:1†

Sample size n	Ratio accepted	Classes in regions of acceptance	Probability of making a wrong decision
20	9:7	0–12	.29
	13:3	13–17	.34*
	15:1	18–20	.15
30	9:7	0–19	.17
	13:3	20–26	.18*
	15:1	27–30	.12
50	9:7	0–34	.03
	13:3	35–44	.09*
	15:1	45–50	.09
75	9:7	0–51	.020
	13:3	52–66	.048*
	15:1	67–75	.044
150	9:7	0–104	.000
	13:3	105–132	.010*
	15:1	133–150	.006

9:7 = .5625 : .4375
13:3 = .8125 : .1875
15:1 = .9375 : .0625

The dividing line $R/n \doteq .69736$
The dividing line $R^1/n \doteq .88478$

† See Table A.17A.

TABLE A.17E

SAMPLE SIZE AND THE PROBABILITY OF MAKING A WRONG DECISION AMONG THE F_2 RATIOS 27:37, 55:9, AND 63:1†

Sample size n	Ratio accepted	Classes in regions of acceptance	Probability of making a wrong decision
20	27:37 55:9 63:1	0–12 13–18 19–20	.03 .21* .04
30	27:37 55:9 63:1	0–19 20–28 29–30	.01 .06 .08*
40	27:37 55:9 63:1	0–26 27–37 38–40	.00 .07* .03
50	27:37 55:9 63:1	0–33 34–47 48–50	.00 .02 .04*
75	27:37 55:9 63:1	0–49 50–70 71–75	.000 .014* .007
90	27:37 55:9 63:1	0–59 60–84 85–90	.000 .008* .003
95	27:37 55:9 63:1	0–63 64–89 90–95	.000 .005* .004
100	27:37 55:9 63:1	0–66 67–94 95–100	.000 .003 .005*

27:37 = .421875 : .578125
55:9 = .859375 : .140625   The dividing line $R/n \doteq .665214$
63:1 = .984375 : .015625   The dividing line $R^1/n \doteq .94178$

† See Table A.17A.

Table A.17F

SAMPLE SIZE AND THE PROBABILITY OF MAKING A WRONG DECISION AMONG THE F_2 RATIOS 27:37, 9:7, 3:1, 13:3, 55:9, 15:1, AND 63:1†

Sample size n	Ratio accepted	Classes in regions of acceptance	Probability of making a wrong decision
50	27:37	0–22	.342
	9:7	23–29	.405
	3:1	30–37	.517
	13:3	38–41	.521*
	55:9	42–45	.415
	15:1	46–48	.372
	63:1	49–50	.177
75	27:37	0–33	.330
	9:7	34–44	.319
	3:1	45–57	.379
	13:3	58–63	.383
	55:9	64–69	.400*
	15:1	70–73	.375
	63:1	74–75	.328
100	27:37	0–49	.059
	9:7	50–66	.096
	3:1	67–77	.314
	13:3	78–84	.372*
	55:9	85–92	.352
	15:1	93–97	.334
	63:1	98–100	.206
500	27:37	0–246	.00
	9:7	247–330	.00
	3:1	331–387	.10
	13:3	388–418	.10*
	55:9	419–451	.09
	15:1	452–483	.01
	63:1	484–500	.00
800	27:37	0–393	.00
	9:7	394–528	.00
	3:1	529–625	.02
	13:3	626–670	.04+*
	55:9	671–722	.04
	15:1	723–772	.00
	63:1	773–800	.00
1375	27:37	0–676	.00
	9:7	677–908	.00
	3:1	909–1075	.00
	13:3	1076–1151	.01
	55:9	1152–1241	.01
	15:1	1242–1328	.00
	63:1	1329–1375	.00

27:37 } The dividing line $R/n \doteq .4921$
9:7 } The dividing line $R/n \doteq .6605$
3:1 } The dividing line $R/n \doteq .7823$
13:3 } The dividing line $R/n \doteq .8368$
55:9 } The dividing line $R/n \doteq .9031$
15:1 } The dividing line $R/n \doteq .9660$
63:1 }

† See Table A.17A.

Table A.17G

SAMPLE SIZE AND THE PROBABILITY OF MAKING A WRONG DECISION AMONG THE TEST CROSS RATIOS 2:1:1, 1:2:1, AND 1:1:2†

Sample size, n	Probability of making a wrong decision
20	.1890
40	.0645
45	.0501
50	.0394
70	.0147
75	.0113
80	.0099

Accept 2:1:1 when the first group is larger than the other two; accept 1:2:1 when the second group is larger than the other two; and accept 1:1:2 when the third group is the largest group.

† See Table A.17A.

Table A.17H

SAMPLE SIZE AND THE PROBABILITY OF MAKING A WRONG DECISION AMONG THE TEST CROSS RATIOS 1:1:2, 1:1:4, AND 1:1:6†

Sample size n	Ratio accepted	Classes in regions of acceptance	Probability of making a wrong decision
50	1:1:2	0–27	.240
	1:1:4	28–35	.305*
	1:1:6	36–50	.252
100	1:1:2	0–54	.18
	1:1:4	55–71	.16
	1:1:6	72–100	.21*
200	1:1:2	0–117	.01
	1:1:4	118–142	.10
	1:1:6	143–200	.11*
330	1:1:2	0–193	.00
	1:1:4	194–234	.05*
	1:1:6	235–330	.04
646	1:1:2	0–377	.000
	1:1:4	378–458	.008
	1:1:6	459–646	.010*

Accept 1:1:2 when the number of individuals in the third group z is less than $.5850n$; accept 1:1:4 when z is between $.5850n$ and $.7095n$; accept 1:1:6 when z is greater than $.7095n$.

† See Table A.17A.

TABLE A.17*I*

SAMPLE SIZE AND THE PROBABILITY OF MAKING A WRONG DECISION AMONG THE F_2 RATIOS 9:6:1, 9:3:4, AND 12:3:1†

Sample size n	Ratio accepted	Classes in regions of acceptance	Probability of making a wrong decision
20	9:6:1	$y \geq 6$ and $x < 14$	.213
	9:3:4	$y < 6$ and $x < 14$	.312*
	12:3:1	$x \geq 14$	.214
50	9:6:1	$y \geq 14$ and $x < 34$	.084
	9:3:4	$y < 14$ and $x < 34$	.134*
	12:3:1	$x \geq 34$	.098
75	9:6:1	$y \geq 21$ and $x < 50$	.054
	9:3:4	$y < 21$ and $x < 50$	.075*
	12:3:1	$x \geq 50$	.039
90	9:6:1	$y \geq 25$ and $x < 60$	.035
	9:3:4	$y < 25$ and $x < 60$	.051*
	12:3:1	$x \geq 60$	.028
150	9:6:1	$y \geq 42$ and $x < 100$	.008
	9:3:4	$y < 42$ and $x < 100$	.009*
	12:3:1	$x \geq 100$	.008

x and y are the numbers of individuals in the first and the second groups of the sample, respectively.

Accept 9:6:1 when $y \geq .27457n$ and $x < .6605n$
Accept 9:3:4 when $y < .27457n$ and $x < .6605n$
Accept 12:3:1 when $x \geq .6605n$
† See Table A.17*A*.

Table A.17J
Sample Size and the Probability of Making a Wrong Decision between the F_2 Ratios 27:9:28 and 81:27:148†

Sample size n	Ratio accepted	Classes of z in regions of acceptance	Probability of making a wrong decision
20	27:9:28 81:27:148	0–10 11–20	.2144 .3125
40	27:9:28 81:27:148	0–20 21–40	.1694 .1998
60	27:9:28 81:27:148	0–30 31–60	.1345 .1370
100	27:9:28 81:27:148	0–50 51–100	.09 .07
135	27:9:28 81:27:148	0–68 69–135	.05+ .05−
200	27:9:28 81:27:148	0–101 102–200	.02 .02
269	27:9:28 81:27:148	0–136 137–269	.010 .009

z is the number of individuals in the third group of the sample. The dividing line $R/n \doteq .50788$.
† See Table A.17A.

Table A.18
Wilcoxon's Signed Rank Test

Tabulated values of T are such that smaller values, regardless of sign, occur by chance with stated probability †

Pairs n	Probability			Pairs n	Probability		
	.05	.02	.01		.05	.02	.01
6	0	—	—	16	30	24	20
7	2	0	—	17	35	28	23
8	4	2	0	18	40	33	28
9	6	3	2	19	46	38	32
10	8	5	3	20	52	43	38
11	11	7	5	21	59	49	43
12	14	10	7	22	66	56	49
13	17	13	10	23	73	62	55
14	21	16	13	24	81	69	61
15	25	20	16	25	89	77	68

† Probabilities are for two-tailed tests. For one-tailed tests, the above probabilities become .025, .01, and .005.

SOURCE: Reproduced from F. Wilcoxon, *Some Rapid Approximate Statistical Procedures*, American Cyanamid Company, Stamford, Conn., 1949, with permission of the author and the American Cyanamid Company. The values in this table were obtained by rounding off values given by Tukey in Memorandum Rept. 17, "The simplest signed rank tests," Stat. Research Group, Princeton Univ., 1949.

TABLE A.19
Critical Points of Rank Sums
(Two-tailed alternatives)

n_2 = larger n	P	\multicolumn{14}{c}{n_1 = smaller n}

n_2 = larger n	P	2	3	4	5	6	7	8	9	10	11	12	13	14	15
4	.05			10											
	.01			—											
5	.05		6	11	17										
	.01		—	—	15										
6	.05		7	12	18	26									
	.01		—	10	16	23									
7	.05		7	13	20	27	36								
	.01		—	10	17	24	32								
8	.05	3	8	14	21	29	38	49							
	.01	—	—	11	17	25	34	43							
9	.05	3	8	15	22	31	40	51	63						
	.01	—	6	11	18	26	35	45	56						
10	.05	3	9	15	23	32	42	53	65	78					
	.01	—	6	12	19	27	37	47	58	71					
11	.05	4	9	16	24	34	44	55	68	81	96				
	.01	—	6	12	20	28	38	49	61	74	87				
12	.05	4	10	17	26	35	46	58	71	85	99	115			
	.01	—	7	13	21	30	40	51	63	76	90	106			
13	.05	4	10	18	27	37	48	60	73	88	103	119	137		
	.01	—	7	14	22	31	41	53	65	79	93	109	125		
14	.05	4	11	19	28	38	50	63	76	91	106	123	141	160	
	.01	—	7	14	22	32	43	54	67	81	96	112	129	147	
15	.05	4	11	20	29	40	52	65	79	94	110	127	145	164	185
	.01	—	8	15	23	33	44	56	70	84	99	115	133	151	171
16	.05	4	12	21	31	42	54	67	82	97	114	131	150	169	
	.01	—	8	15	24	34	46	58	72	86	102	119	137	155	
17	.05	5	12	21	32	43	56	70	84	100	117	135	154		
	.01	—	8	16	25	36	47	60	74	89	105	122	140		
18	.05	5	13	22	33	45	58	72	87	103	121	139			
	.01	—	8	16	26	37	49	62	76	92	108	125			
19	.05	5	13	23	34	46	60	74	90	107	124				
	.01	3	9	17	27	38	50	64	78	94	111				
20	.05	5	14	24	35	48	62	77	93	110					
	.01	3	9	18	28	39	52	66	81	97					
21	.05	6	14	25	37	50	64	79	95						
	.01	3	9	18	29	40	53	68	83						
22	.05	6	15	26	38	51	66	82							
	.01	3	10	19	29	42	55	70							
23	.05	6	15	27	39	53	68								
	.01	3	10	19	30	43	57								
24	.05	6	16	28	40	55									
	.01	3	10	20	31	44									
25	.05	6	16	28	42										
	.01	3	11	20	32										
26	.05	7	17	29											
	.01	3	11	21											
27	.05	7	17												
	.01	4	11												
28	.05	7													
	.01	4													

SOURCE: Reprinted from Colin White, "The use of ranks in a test of significance for comparing two treatments," *Biometrics*, **8**: 33–41 (1950), with permission of the editor and the author.

Table A.20
Working Significance Levels for Magnitudes of Quadrant Sums

Significance level	Magnitude of quadrant sum†
.10	9
.05	11
.02	13
.01	14–15
.005	15–17
.002	17–19
.001	18–21

† The smaller magnitude applies for large sample size, the larger magnitude for small sample size. Magnitude equal to or greater than twice the sample size less 6 should not be used.

SOURCE: Reprinted from P. S. Olmstead and J. W. Tukey, "A corner test for association," *Annals Math. Stat.*, **18:** 495–513 (1947), with permission of the authors, editor, and courtesy of Bell Telephone Laboratories, Inc.

INDEX

Abbreviated Doolittle method, 281, 289-296
Absolute value, 19
Adams, J. L., 276
Addition of variable, 299-301
Additivity, 80, 129
Adjusted values (*see* Analysis of covariance; Tests of hypotheses)
Alternative hypothesis, 65, 71, 117, 119
Analysis, of covariance, assumptions, 308-310
   to estimate missing data, 308, 324, 325
   factorial experiment, 319-323
   linear model, 309
   multiple regression, 325-330
   polynomials, 341
   randomized complete-block design, 311-315
   testing adjusted treatment means, 310, 311, 315, 323
   uses, 305
  of variance, assumptions in model, 128-131
   completely random design, equal replication, 101-106
    two treatments, 74, 75
    unequal replication, 112-115
   disproportionate subclass numbers, fitting constants, 258-265
    $r \times 2$ tables, 269-272
    2×2 tables, 272-276
    weighted squares of means, 265-269
   equally spaced treatments, 224-226
   factorial experiment, 2×2, 200-203
    $3^2 \times 2$, 205-209
   Latin square, 147-149
   linear regression, 172, 287-289, 296
   randomized complete-block design, 134-137
    subsamples, 142-145
   single degree of freedom comparisons, 203, 209, 218
   split-plot design, 236-239, 244, 248
   subsamples, equal numbers, 118-124
    unequal numbers, 125-127
   two-stage sampling, 423

Anderson, R. L., 127, 131, 222, 230, 239, 251, 289, 303, 341, 345
Angular or inverse sine transformation, 158
Arithmetic mean or average (*see* Mean)
Arny, D. C., 236, 237
Attoe, O. J., 282
Aughtry, J. D., 159
Average deviation, 19

Babcock, S. M., 146, 159
Baker, G. A., 65, 66
Baker, R. E., 65, 66
Bancroft, T. A., 127, 131, 289, 303
Bar charts, 10, 11
Bartlett, M. S., 129, 131, 158, 159, 324, 331, 349, 351
Bartlett's test for homogeneity of variances, 347-349
Beeson, K. C., 331
Bias in standard deviation, 55, 56, 61
Bing, A., 137, 159
Binomial data, use, of exact distribution, 353, 358, 379, 380
  of normal approximation, 353, 355, 357
Binomial distribution or population, 353, 389, 390
  fitting of, 390-393
  transformation for, 394
Binomial probability paper, 354-356, 358, 363, 364, 372
Birch, H. F., 283, 303
Bivariate distribution, model II, 177-179
Blischke, W. R., 395, 398
Bliss, C. I., 177, 182
Blocks, 132, 133
  certain misconceptions, 136, 137
  effect, 94
  (*See also* Randomized complete-block design)
Bowers, J. L., 311
Box, G. E. P., 221, 230
Brockington, S. F., 337, 345
Brown, G. W., 403

Brownlee, K. A., 98
Buch, N. C., 258, 276
Buss, I. O., 193

Carroll, F. D., 265, 276
Carson, R. B., 134, 160
Casida, L. E., 272, 275, 276
Central tendency measures, 13-15
Chakravarti, I. M., 397, 399
Chapman, A. B., 276
Charts, bar, 10, 11
  pie, 10, 11
Chebyshev's inequality, 407, 408
Chi-square, additivity, 375-378
  combination of probabilities, 350, 351
  confidence interval for σ^2, 346, 347
  distribution, 41-43
  homogeneity, 376
  test criterion, 352, 357
Chi-square tests, goodness of fit, binomial distribution, 392, 393
  continuous distribution, 348-350
  one-way tables, 364, 365
  Poisson distribution, 397
  two-cell tables, 357, 358
  homogeneity, of correlation coefficients, 191
  of $r \times c$ tables, 366-369
  of variances, 347-349
  of independence, $r \times 2$ tables, 370, 371
  2×2 tables, 371, 372
  for interaction, 371
  linear regression, 381-383
  n-way classification, 384-386
  rank test for two-way classifications, 403, 404
  sign test, 401, 402
Chung, J. H., 417, 426
Class interval, 11, 26-28, 124
Clausen, R. T., 87, 113
Clopper, C. J., 354, 365
Clopper and Pearson charts, 354
Cluster sampling, 422-425
Cochran, F. D., 160, 331
Cochran, W. G., 65, 66, 81, 87, 93-95, 98, 107, 129, 131, 142, 147, 149, 154, 155, 158, 159, 212, 230, 235, 241, 249, 251, 299, 303, 316, 323, 324, 331, 350, 351, 353, 365, 381, 386, 420, 424-426
Coding use, 25, 26, 225
  in frequency distribution, 28
Coefficient, of alienation, 187
  of contingency, 369
  of correlation (*see* Correlation coefficient)
  of determination, 179, 187, 188, 287
  of nondetermination, 187
  of regression (*see* Regression coefficient)
  of variability, 20, 239

Comparisons among treatments (*see* Treatment comparisons)
Completely random design, advantages and disadvantages, 100
  analysis of variance, equal replication, 101-106
  unequal replication, 112-115
  linear model, 116-119
  nonparametric tests, 404-407
  randomization, 99, 100
Components of variance (*see* Variance components)
Comstock, R. E., 160
Concomitant variable, 172
  in error control, 95, 161, 172, 173
Confidence belts for p, 354
Confidence inference, 22, 23
Confidence interval (limits), for correlation coefficient, 188, 189
  for mean differences, 74, 76, 79, 82, 84
  for means, 22, 23, 25, 45-47, 58
    from finite populations, 416
  for partial regression coefficients, 298
  for predicted values, 47, 175
  for proportion or percentage, 353-355, 417
  for regression coefficient, 171
  for regression line, 170, 171
  for variance, 346, 347
Constants, method of fitting, 257-265
Contingency tables (*see* Chi-square tests)
Continuous variable, 8
Corner test of association, 410
Correction, for continuity, 357, 371
  for disproportion, 273, 274
  for finite population, 416
Correction factor or term, 18
Correlation, interclass, 192
  intraclass, 191-193
  between mean and variance, 130, 156-158
  multiple, 277, 285-287, 297
  partial, 277, 285-287, 301-303
  and regression, 187, 188
  total or simple, 277
Correlation coefficient, 183-188
  confidence interval, 188, 189
  sampling distribution, 188
  tests of hypotheses, 189-191
Cost functions in sampling experiments, 421, 422
Covariance, 78, 95
  analysis (*see* Analysis)
  error control, 306
  partition, 305, 317, 318
Cowden, D. J., 303
Cox, G. M., 81, 87, 93-95, 98, 107, 131, 142, 147, 149, 154, 155, 159, 235, 241, 249, 251, 316, 323, 324, 331
Cramer, C. Y., 293, 304
Critical region, 69

INDEX

Crump, S. L., 212, 230
Curvilinear regression (*see* Regression, nonlinear)

Darwin, Charles, 3
David, F. N., 188, 190, 193
Deciles, 15
Degrees of freedom, 18, 44, 56
  single-degree-of-freedom comparisons, 203, 208, 213-220
  for split-plot design, 234
  (*See also* Experimental design)
Deletion of variable, 299-301
DeLury, D. B., 152, 159, 305, 331, 417, 426
Dependent variable, 162
Deviate, 14
Dickson, A. D., 24
Differences, distribution, 60
  among means, 106-112, 114
  between means (*see* Confidence interval; Tests of hypotheses)
  sample size to detect, 84-86, 154-156
  sampling, 58-63
  standard deviation, 61
  variance, 61
Digits, significant, 29
Di Raimondo, F., 370, 371, 377, 386
Discrete variable, 8
Discontinuous variable, 8
Dispersion measures, 16-19
Disproportionate subclass numbers (*see* Subclass numbers)
Distribution, 8
  of arithmetic mean, 40, 41, 53, 54
  binomial (*see* Binomial distribution)
  of chi-square, 41-43
  of correlation coefficient, 188
  derived, 40
  of differences, 60
  frequency, 8, 12, 26-28
  hypergeometric, 388, 389, 417
  normal (*see* Normal distribution)
  parent, 40
  Poisson, 395-397
  of standard deviations, 54, 55, 61
  Students' t, 43, 57, 62, 63
  of variances, 54, 55, 61
  of z, 36, 40
Distribution-free (nonparametric) statistics, 400
Dixon, W. J., 55, 66
Doolittle method, 281
  abbreviated, 281, 289-296
Dorin, H. C., 345
Dot notation, 77, 101, 120, 134, 143
Drapala, W. J., 27, 30, 277, 304, 335
Duncan, D. B., 107, 131

Duncan's new multiple-range test, 107-109, 114
Dunnett, C. W., 111, 131
Dunnett's procedure, 111, 112, 114, 239
Durand, D., 299, 304
Dwyer, P. S., 289, 304

Effects, main, 197
  simple, 196, 197, 202, 203, 209
Efficiency, of covariance, 316, 317
  of Latin square, 152, 153
  of randomized complete blocks, 142
  of small experiments, 93
  of subsamples, 128
Eisenhart, C., 129, 131, 158, 159, 179, 182
Eliassen, Rolf, 20, 30
El Khishen, A. A., 148
Enumeration data, 352
Erdman, L. W., 101, 131
Error, 46
  choice, factorial experiments, 211, 212
  components, orthogonal comparisons, 219
  control, 94-96
    by concomitant variable, 95, 161, 172, 173
  experimental, 119, 123, 124, 126, 127, 143
  heterogeneity, 130, 250
  independence, 130
  rate, experimentwise, 109, 112
    per-comparison, 107
  sampling, 119, 123, 126, 127, 143
  term, nature, 137-139
  Type I (first kind), 70, 71, 84-86
  Type II (second kind), 70, 71, 84-86
Estimators, 44-46
Expected values, 138
  (*See also* Mean square)
Experiment, definition, 88
Experimental design, 88-98
  completely random, 99-105
  error control, 94-96
  factorial experiment (*see* Factorial experiment)
  Latin-square, 146-154
  randomized complete-block, 132-146
  split-plot (*see* Split-plot design)
Experimental error, 90, 91, 102, 119, 123, 124, 126, 127, 143, 167, 424
  heterogeneity, 82, 83, 130, 131, 156, 250
Experimental unit, 90
  size and shape, 95
Experiments, factorial (*see* Factorial experiment)
  objectives, 89
  planning, 127, 128
  series of similar, 249, 250
  size (*see* Size of experiment or sample)
  small, relative efficiency, 93
Exponential curves, 333-336

F (variance ratio), among means, 72, 73, 104, 117, 123, 135, 213
  one-tailed, 105
  relation to t, 72, 75, 289
  test of homogeneity of variance, 82, 83
  two-tailed, 83, 105
Factor, 194
Factorial experiment, 96
  analysis, of covariance, 319-323
    of variance, 200-203, 205, 209, 210, 214, 218, 224-226
  average value of mean square, rules for, 211, 212
  error term, choice, 211, 212
  linear models, 211-213
  n-way classification, 220-222
  notation and definitions, 195-199
  response surfaces, 220-222
  2×2, 199-203
  $3^2 \times 2$, 204-209
Federer, W. T., 75, 94, 95, 98, 178, 189, 190, 227, 230, 235, 249, 251, 307, 331, 366, 386
Fiducial inference, 23
Finite populations, 154
  notations and definitions, 415, 416, 418, 419
  sampling methods (*see* Sampling)
Finney, D. J., 46-48, 179, 182, 316, 331
Fisher, R. A., 1, 5, 93, 97-99, 147, 160, 189, 193, 222, 230, 341, 345, 350, 351, 395, 397, 399
Fixed model (*see* Model I)
Foote, R. J., 304
Forester, H. C., 307, 331
Freeman, L. C., 5
Frequency polygon, 11, 12
Frequency table, 12, 13, 26-28
  calculations from, 26-30
Friedman, J., 304
Friedman, M., 403, 411

Galton, F., 385
Gauss multipliers, 301
Genetic ratios, limited set of alternatives, 358-362
  sample size, 362-364
  test of goodness of fit, 357, 364, 365
Geometric mean, 15
Goodness of fit (*see* Chi-square tests)
Gosset, W. S., 3, 43
Goulden, C. H., 27, 30
Green, J. M., 369
Greenberg, B. G., 98
Gregory, P. W., 276

Hale, R. W., 152, 160
Hansen, M. H., 420, 424, 426

Harmonic mean, 15, 274
Harris, M. D., 154, 160
Hart, J. S., 79, 87
Harter, H. L., 109, 131, 154, 160, 230
Hartley, H. O., 110, 131, 341, 345, 380, 387
Hasler, A. D., 303, 304
Healy, W. C., Jr., 156, 160
Henderson, C. R., 213, 231
Herrera, L., 365, 386
Heterogeneity, of error, 82, 83, 130, 131, 156, 250
  of interactions, 250
  (*See also* Tests of hypotheses, homogeneity)
Hierarchal classification, 120, 213, 214
Histogram, 11, 12
Homogeneity tests (*see* Tests of hypotheses)
Honestly significant difference, 109, 110, 114
Hoppe, P. E., 217
Horner, T. W., 337, 345
Horvitz, D. G., 160
Hotelling, Harold, 5
Houseman, E. E., 222, 230, 341, 345
Howerton, H. K., 345
Hunter, J. S., 230
Hurwitz, W. N., 426
Hypergeometric distribution, 388, 389, 417
Hypothesis, alternative, 65, 71, 117, 119
  null, 65, 68, 71, 117, 119, 123, 124, 135
  test (*see* Tests of hypotheses)

Improvement factor, 187
Incomplete-block designs, error control, 95
Independence, test for (*see* Chi-square tests)
  of $\bar{x}$ and s, 43, 128-131
Independent variable, 162
Individual degrees of freedom (*see* Single degree of freedom, comparisons)
Inference, statistical, about populations, 9, 22, 23, 44-47, 58, 64, 65, 67-71, 98
  (*See also* Confidence intervals; Tests of hypotheses)
Interaction, 145, 197-199, 206-209
  heterogeneity, 250
  orthogonal components, 227-228
Interclass correlation, 192
Intraclass correlation, 191-193
Irwin, J. O., 177, 182, 377, 386

Jafar, S. M., 265, 276
Jessen, R. J., 422
Johnson, W. E., 303, 304

Kabat, C., 187, 193
Kempthorne, O., 94, 98, 211, 231
Kermack, W. O., 384, 386
Khargonkar, S. A., 241, 249

INDEX 477

Kimball, A. W., 377, 386
King, S. C., 161, 164
Knodt, C. B., 75, 87
Kramer, C. Y., 114, 131
Kruskal, W. H., 406, 411
Kuesel, D., 169

Ladell, W. R. S., 425, 426
Lambert, J. W., 223, 224, 240
Lancaster, H. O., 377, 386
Latin-square design, 146-154
  analysis of variance, 147-149
  efficiency, 152, 153
  linear model, 153-154
  missing data, 150-152
  randomization, 148
Least-significant difference, 106, 107, 114
Least-squares method, 256
Lemke, C., 269
Level, 194
Levels of significance, 46, 68, 69
Limited set of alternatives, tests of hypotheses, 358-362
Limits (*see* Confidence interval)
Linear correlation (*see* Correlation coefficient)
Linear equation in more than two dimensions, 278, 279
Linear model, completely random design, 115-119
  covariance, 309
  disproportionate subclass numbers, 255, 263
  factorial experiments, 211-213
  Latin-square design, 153, 154
  linear regression, 164, 165
  paired comparisons, 80-81
  proportional subclass numbers, 253-254
  randomized complete-block design, 145, 146
  single sample, 20-22
  split-plot design, 245-247
  subsamples, 123-125
  unpaired samples, 76-78
Linear regression (*see* Regression)
Logarithmic curves, 333-336
Love, H. H., 307, 331
Lum, M. D., 230
Lyell, C., 2

McIntyre, C. B., 158
McKendrick, A. G., 384, 386
McMullen, Launce, 5
Madow, W. G., 426
Mahalanobis, P. C., 5
Main effects, 197
Mainland, D., 5, 358, 365, 380, 386
Mann, H. B., 405, 411

Massey, F. J., Jr., 55, 66
Mather, K., 5
Matrices, 290
Matrone, G., 311, 331
Mean, arithmetic, 13, 14
  distribution, 40, 41, 53, 54
  of finite population, 415, 418, 419
  unweighted, 254-257, 267
  weighted, 253-257, 271, 274
  correction for, 18
  geometric, 15
  harmonic, 15, 274
Mean square, 17, 24
  average value, factorial experiments, 210, 214
  Latin square, 153
  mixed model, 144
  model I, fixed effects, 116, 144, 153
  model II, random effects, 116, 144, 153
  randomized complete-block design, 144
  rules for, 211, 212
  split-plot design, 246, 247
  two-stage sampling, 423
Median, 15
Median test, completely random design, 407
  for two populations, 405, 406
  for two-way classifications, 403, 404
Mendel, G., 374, 375, 386
Menger, Karl, 5
Meyer, R. K., 193
Missing data, estimation by covariance, 308, 324, 325
  Latin-square design, 150-152
  randomized complete-block design, 139-141
  split-plot design, 239-242
Mode, 15
Model, linear (*see* Linear model)
  mixed, 144-146, 210-214, 246
Model I (fixed), 115-119, 144, 145, 165, 210-214, 246
  predicted X from Y, 177
Model II (random), 115-119, 144, 145, 165, 210-214, 246
  bivariate distribution, 177-179
Mood, A. M., 160, 403, 405, 411
More, D. M., 5
Mosteller, F., 354, 365, 386
Multiple correlation, 277, 285-287, 297
Multiple-range tests, Duncan's, 107-109, 114
  Student-Newman-Keuls', 110, 111, 114
Multistage sampling, 420-425
Myers, W. M., 347, 351

n-way classification, binomial data, 384-386
  factorial experiments, 220-222
NaNagara, P., 359, 365
Newman, D., 110, 131

Neyman, J., 3
Nielsen, E. L., 373, 387
Nonparametric statistics, advantages, 400
Nonparametric tests, for completely random design, 404-407
    corner, of association, 410
    median, 403-406
    for randomized complete blocks, 402-404
    rank, 402-407
    sign, 401, 402
Nested classification, 120, 213, 214
Nonadditivity, single degree of freedom, 229-230
Nonlinear regression (see Regression)
Nonrespondents, 420
Normal approximation, applied to binomial data, 353, 355-357, 371, 372
    for correlation coefficient, 190, 191
Normal distribution, 2, 35, 49
    cumulative, 35
    fitting, 348-350
    probability calculations, 36-40
    random sample from, 51-53
    sampling from, 49-66
Normal equations, 281, 283, 290
Null hypothesis (see Hypothesis)

Odds, 31
Olmstead, P. S., 410, 411
One-tailed test, 70, 82-86
One-way classification (see Completely random design)
Optimum allocation, 420-425
Orthogonal comparisons, 106
Orthogonal components, 224-227
Orthogonal polynomials, 341-343
    equally spaced treatments, 222-227
    table, 223
Orthogonality, single degree of freedom, 216
Outhwaite, A. D., 307, 331

Parameter, 13
Partial correlation, 277, 285-287, 301-303
Partial regression, 280, 294-299
Pascal's triangle, 392
Patton, R. F., 382
Paulson, E., 387
Pearson, E. S., 6, 354, 365, 380, 387
Pearson, Karl, 2, 3
Percentage data (angular transformation) 158
    proportion or, confidence limit, 353-355
Percentile, 15
Permenter, L. P., 311
Perrotta, C., 373
Peterson, H. L., 80, 87
Peterson, W. J., 149, 160, 331

Pie charts, 10, 11
Poisson distribution, 395-397
    fitting, 395, 396
Polynomial curves, 333-334
Polynomials, analysis of covariance, 341
    higher-degree, 340
    orthogonal (see Orthogonal polynomials)
    second-degree, 338-340
Population, definition, 9
Power of test, 71, 85
Precision, relation to replicates, 92
    (See also Efficiency)
Predicted value, confidence interval, 175
    variance, 175
    X from Y, 177
Primary sampling units, 422
Probability, 2
    definition, 32
Probability distribution, continuous variable, 32, 33
    discrete variable, 33, 34
Probability sampling, 414, 415
Proportion or percentage, confidence limit, 353-355
Proportional allocation, 420
Proportional subclass numbers, 252-255

Quality inspection, 389
Quartiles, 15

$r \times c$ table, disproportionate subclass numbers, methods for, 257-268
$r \times 2$ table, disproportionate subclass numbers, methods for, 269-272
Rabson, R., 118
Rao, C. R., 397, 399
Random model (see Model II)
Random number table, 9, 10
Random sample, 9
    method of drawing, 9, 10
Random sampling, 413
    simple, 415-417
Randomization, function, 97, 98
    methods, completely random design, 99, 100
    Latin-square design, 148
    randomized complete-block design, 133
Randomized complete-block design, 132-146
    analysis, of covariance, 311-314
    of variance, 134-137
    efficiency, 142
    error control, 94
    linear model, 145, 146
    missing data, 139-141
    nature of error term, 137-139
    nonparametric tests, 402-404
    randomization, 133
    subsamples, 142-145

INDEX

Range, relation to standard deviation, 18
Rank correlation, 409
Rank tests, completely random design, 404-407
  for two-way classifications, 403-404
  Wilcoxon's, 402, 405
Rasmussen, D. C., 358, 365
Region, of acceptance, 69
  of rejection, 69
Regressed values, 167-169
Regression, and correlation, 187, 188
  linear, 161-182
    adjusted values, 167-169
    analysis of variance, 172, 287-289, 296
    assumptions, 165, 166
    enumeration data, $r \times 2$ tables, 381-383
    equation, 163
    models, 165-166
    multiple, 279, 280, 289-296
    through origin, 179
    partial, 279, 280, 289-296
    sources of variation, 166, 167
    weighted analysis, 180, 181
  nonlinear, 332
    logarithmic or exponential, 333-336
    polynomial, higher-degree, 340
      second-degree, 338-340
Regression coefficient, 163
  difference between two, 173, 174
  equally spaced treatments, 225
  partial, 280, 294-299
  standard partial, 284, 285, 299
  test of homogeneity, 320
Regression equation, multiple, 280-283, 296
Reid, R. D., 6
Relative efficiency (see Efficiency)
Replication, functions, 90-93
  number, estimating (see Size of experiment or sample)
  factors affecting, 92, 93
Response surfaces, factorial experiment, 220-222
Rinke, E. H., 369, 387
Robertson, D. W., 364, 365
Robson, D. S., 341, 345
Rollins, W. C., 276
Ross, R. H., 75, 87
Rowles, W., 81, 87
Rutherford, A., 307, 331

Sackston, W. E., 134, 160
Sample, definition, 9
  prediction of results, 47, 48
  random, 9, 51-53
  size (see Size of experiment or sample)
Sample survey, organization, 413, 414

Sampling, finite populations, authoritative, 412
  fraction, 416
  multistage or cluster, 420-425
  probability, 414, 415
  random, 413, 415-417
  stratified, 417-420
  systematic, 412
Sampling error, 119, 120, 123, 126, 127, 143, 424
Sampling units, 119
Satterthwaite, F. E., 212, 231
Scheffé, H., 211, 231
Schlottfeldt, C. S., 307, 331
Schultz, E. F., Jr., 212, 231
Scientific method, 3, 4
Self, H. L., 173
Shuel, R. W., 78, 87
Significance level, 46, 68, 69
Significance tests (see Tests of hypotheses)
Significant digits, 29, 30
Significant studentized ranges, 108
Sign test, 401, 402
Simple effects, 196, 197, 202, 203, 209
Single degree of freedom, comparisons, 203, 208, 213-220, 222-227
  for nonadditivity, 229, 230
Size of experiment or sample, for binomial data, 362-364
  Chebyshev's inequality, 408
  for detection of differences, 84-86, 154-156
  for finite populations, 420-422
  Stein's two-stage, 86-87, 156
  2×2 tables, 383, 384
Skewed distribution, 15, 55
Skory, J., 368, 387
Smith, D. C., 373, 387
Smith, F. H., 331
Smith, H. B., 373, 387
Smith, H. H., 187
Snedecor, G. W., 127, 131, 227, 231
Spearman's coefficient, 409
Split-plot design, analysis of variance, 236-239
  error control, 95
  linear models, 245-247
  missing data, 240-242
  partition of degrees of freedom, 234
  standard errors, 236, 239, 242, 244, 245
  in time, 242-245
    and space, 247-249
  uses, 233
Square-root transformation, 157
Standard deviate, 40
Standard deviation, 16-19, 24, 26, 28
  bias, 55, 56, 61
  distribution, 55
  for proportions, 408
  relation to range, 27

480 INDEX

Standard error, of adjusted mean, 316
  of difference, 61, 73, 75, 78, 81, 104, 135, 139, 141, 149, 151, 152, 267, 271
    adjusted means, 316, 329, 330
    missing data, 141, 151, 152, 241, 242
    split-plot design, 236, 241, 242, 244, 245
  of estimate, 169
  of mean, 19, 20, 25, 53, 56, 104, 149
  of partial regression coefficient, 297-299
  of standard regression coefficient, 289
Standard normal deviate, 40
Standard partial regression coefficients, 284, 285, 299
Standard variable, 185
Statistical inference (see Inference)
Statistics, 13
  definition, 1
  history, 2
  studying, 4, 5
Steel, R. G. D., 245, 249, 251, 325, 331, 398, 399
Stein, C., 86, 87
Stein two-stage sample procedure, 86, 87
Stratification, 417
Stratified sampling, 417-420
Student, 3, 43, 395, 399
Student's t (see t)
Student-Newman-Keuls' test, 110, 111, 114
Subclass numbers, disproportionate, general, 252-257
    linear model, 255, 263
    method of fitting constants, 257-265
    $r \times 2$ tables, 269-272
    2×2 tables, 272-276
    weighted squares of means, 265-269
  proportional, 252-255
Subsamples, 119
  analysis of variance, 119-127, 143
  factorial experiment, average value of mean square, 214
  linear model, 123-125
  randomized complete-block design, 142-145
  variance components in planning experiments, 127, 128
Subunit or subplot, 232
Sum, of products, 163
  of squares, 17
    adjusted, 263
    error, 102, 113, 121, 126
      adjusted by covariance, 310, 313, 322, 328
    components, 102
    regression, 59
    weighted, 252-255
Sutcliffe, M. I., 365, 386
Systematic sampling, 412

t, distribution, 43, 57, 62-63
  relation to F, 72, 75, 289
  uses (see Confidence interval; Tests of hypotheses)
  variances, different, 81, 235, 239
Tag-recapture problems, 389
Tang, P. C., 154, 160
Technique refinement, 96, 97
Test criterion, 68
  (See also Chi-square; F; t; z distribution)
Tests of hypotheses, 64-72
  among adjusted means, 310, 311, 315, 323, 328, 329
  correlation coefficient, homogeneity, 190, 191
    significance, 190
  goodness of fit, 349, 350, 352, 357, 358, 364, 365
    binomial distribution, 392, 393
    Poisson distribution, 395-397
  homogeneity, of two-cell samples, 373, 374
    of variances, 82, 83, 347-349
  independence and association, 366-372
  limited set of alternatives, 358-362
  regression, linear, partial and multiple, 287-289, 296-299, 302
  regression coefficients, difference between two, 173, 174
    homogeneity, 320
    significance, 171
  two-cell tables, 355-358
  two means, paired observations, 78-81
    unpaired observations, and equal variance, 73-76
    and unequal variance, 81, 82
    when σ^2 is known, 75
  two or more means, 72
  among unadjusted means, 313
  (See also Chi-square tests)
Tintner, G., 6
Tippett, L. H. C., 27, 30
Transformation, 129
Transformations, angular or inverse sine, 158
  for binomial distribution, 394
  general considerations, 158
  logarithmic, 157
  nonlinear regression, 335
  r to z, 189-191
  square-root, 157
Treatment, definition, 90
  means, adjusted, 315
  sum of squares and mean square, 102, 103
Treatment comparisons, by Duncan's new multiple-range test, 107-109, 114
  by Dunnett's procedure, 111, 112, 114, 239
  for equally spaced treatments, 222-227
  by least-significant difference, 106, 107, 114

INDEX

481

Treatment comparisons, orthogonal set, 216-218
  single degree of freedom 213-220
  by Student-Newman-Keuls' test, 110, 111, 114
  by Tukey's w procedure, 109, 110, 114
  unequal variance, 81, 82, 235, 239
  (*See also* Tests of hypotheses)
Treatment effects, additivity, 129
  adjusted, 263, 306, 307
    by covariance, 315, 316, 328
Treatments, choice of, 96
Tucker, H. P., 137, 160, 324, 331
Tukey, J. W., 110, 131, 154, 155, 160, 227, 229, 231, 245, 249, 251, 354, 365, 386, 410, 411
Tukey's w procedure, 109, 110, 114
Two-tailed test, 69
Tyler, W. J., 276
Type I error, 70, 71, 84-86, 155
Type II error, 70, 71, 84-86, 156

Unbiased estimate, 49
Unbiasedness, 44
  s^2, 56
Unequal subclass numbers (*see* Subclass numbers)
Uniformity trial, 95, 307
Universe, 9
Unweighted mean, 254-257, 267
Urey, F. R., 105

van Tienhoven, A., 344
Variable, 7, 8
  continuous, 8
    probability distribution, 33, 34
  discontinuous, 8
  discrete, 8
    probability distribution, 32, 33
  independent, deletion and addition, 299-301
  standard, 40, 185, 188, 284
Variance, 16-19
  analysis (*see* Analysis)
  finite population, 416, 418, 419, 423-425
  homogeneity, 82, 83
    test, 82, 83, 347-349
  minimum, 44
  ratio (*see* F)
  unbiasedness, 56
  weighted average, 73

Variance components, coefficients, unequal subsamples, 126
  completely random design, 115-119
  in experiment planning, subsamples, 127, 128
  factorial experiments, 210, 214
  Latin-square, 153
  randomized complete-block design, 144
  split-plot design, 246, 247
  subsamples, equal size, 121, 123, 124
    unequal size, 126

Wagner, R. E., 205, 231
Wakeley, R. E., 160, 330
Wald, A., 3
Walker, H. M., 6
Wallis, W. A., 351, 387, 406, 411
Watson, C. J., 74, 87
Weber, C. R., 330, 337, 345
Wedin, W. F., 326
Weighted mean, 253-257, 271, 274
Weighted regression, linear, 180, 181
Weighted squares of means, 265-269
Weighted sum of squares, 252-255
Weldon, V. B., 331
Wexelsen, H., 115, 131
White, C., 405, 411
Whitney, D. R., 405, 411
Whole unit or whole plot, 232
Wilcoxon, F., 402, 405, 411
Wilcoxon's signed rank test, 402
Wilcoxon's two-sample test, 405
Wilk, M. B., 211, 231
Wilkinson, W. S., 199, 231
Winsor, C. P., 182
Wishart, J., 316, 319, 331
Wojta, A., 203
Wolfowitz, J., 6
Woodhouse, W. W., Jr., 331
Woodward, R. W., 358, 365

Y intercept, 164
Yates, F., 6, 139, 141, 147, 151, 152, 160, 195, 222, 230, 231, 265, 276, 354, 365, 395, 397
Youden, W. J., 6, 152, 160

z distribution, 36, 40
z transformation for r, 189-191
Zelle, M. R., 180
Zick, W., 225, 231